The Chinese Coal Industry: An Economic History

The coal industry has been and continues to be of critical importance to China's economic modernization. With its huge labour force, countrywide infrastructure, and vital strategic importance to the economy, the industry presents special problems for reformers. It also epitomizes the problems of reform in the state industrial sector as a whole.

This book examines the changes in the structure and operation of the Chinese coal industry from the mid-nineteenth century to the present, concentrating on the years of reform. Although the focus is on the economics of the industry, *The Chinese Coal Industry: An Economic History* also provides many insights into China's sociopolitical development.

Elspeth Thomson received her PhD from the School of Oriental and African Studies, University of London. Her main interests are the energy sector and development of transportation infrastructure in China. She has taught courses on the economies of China, Hong Kong, Taiwan and other Asian countries at Simon Fraser University, Vancouver and Lingnan University, Hong Kong. She is currently a Visiting Fellow at the East Asian Institute, National University of Singapore.

RoutledgeCurzon Studies on the Chinese Economy 2
Series Editors Peter Nolan, *University of Cambridge* and
Dong Fureng, *Chinese Academy of Social Sciences*

The aim of this series is to publish original, high-quality, research-level work by both new and established scholars in the West and the East, on all aspects of the Chinese economy, including studies of business and economic history.

The Chinese Coal Industry:
An Economic History

Elspeth Thomson

LONDON AND NEW YORK

First published 2003 by RoutledgeCurzon
This edition published 2013 by Routledge, an imprint of Taylor & Francis

2 Park Square, Milton Park, Abingdon, Oxfordshire OX14 4RN
711 Third Avenue, New York, NY 10017
Routledge is an imprint of the Taylor & Francis Group, an informa business

First issued in paperback 2018

Typeset in Times New Roman by
Newgen Imaging Systems (P) Ltd, Chennai, India

British Library Cataloguing in Publication Data
A catalogue record for this book is available from the British Library

Library of Congress Cataloging in Publication Data
Thomson, Elspeth, 1960–
     The Chinese coal industry: an economic history/Elspeth Thomson.
       p. cm. – (RoutledgeCurzon studies on the Chinese economy; 2)
     Includes bibliographical references and index.
       1. Coal trade – China – History. 2. Coal mines and mining – China – History.
   I. Title. II. Series.
HD9556.C52 T495 2003
338.2′724′0951–dc21
                                                                        2002068283

ISBN: 978-0-7007-1727-9 (hbk)
ISBN: 978-0-367-02687-5 (pbk)

Contents

List of illustrations

Figures

Plates

Tables

Some explanatory notes

Statistics

Pitfalls and shortcomings

- figures deliberately inflated or deflated by low level bureaucrats trying to get some message across to superiors. For example, in order to get bonuses they would report that the targets were reached when they weren't, or they would under-report to try to get more funding next year, etc.
- figures for public consumption at home or abroad, deliberately inflated or deflated by authorities, for example, downplaying of accidents.
- discrepancies in figures supplied by Chinese official sources, for example, many typos, and variations found in the statistical yearbooks.
- much fuller information given in Chinese sources, for example, English and Chinese versions of the Coal Industry Yearbook.
- discrepancies in figures from the international organizations, for example, World Bank, International Energy Agency, etc.
- all figures from the non-state mines are estimates, for example, great variation in the approximate number of mines, and their output.
- international comparisons complicated by geopolitical changes, for example, collapse of Soviet Union and creation of CIS.

Gaps

- no central administration of statistics before 1949. All that exists are some regional, local, and individual mine records, some data compiled by the Geological Survey of China, and by individual researchers often hired by commercial enterprises who used the usual quantitative techniques of extrapolation for periods when there was absolutely no data at all.
- governmental statistics collection, analysis, and publication had been carried out in the early years of the PRC, but ended with the GLF and was not resumed by the State Statistical Bureau until 1981.
- only four editions of the *Zhongguo nengyuan tongji nianjian* (China Energy Statistics Yearbook) were published: 1986, 1989, 1991, and 1991–96.
- the hiatus between collection of figures and publishing of them.

Problems in translation

−   in material translated in China, precise meanings and/or nuance were often not clear.

Measurements

−   'tons' means metric tons.

Romanizations

−   the pinyin system of romanization has been used throughout, with the exception of some place and company names in Chapter 1.

Acknowledgements

Every effort has been made to contact copyright holders for their permission to reprint material in this book. The publishers would be grateful to hear from any copyright holder who is not here acknowledged and will undertake to rectify any errors or omissions in future editions of this book.

Abbreviations

Abbreviations used in the Bibliography, Notes, Tables and Figures

BR	*Beijing Review* Known as *Peking Review* until 5 January 1979. Note that page numbering differs between North American and Asian editions. Both editions were used.
CBR	*China Business Review*
CD	*China Daily*, Beijing
CDSP	*Current Digest of the Soviet Press*
CIYB	*China Coal Industry Yearbook*, Hong Kong: Economic Information and Agency, 1983–2000. See *MTGYNJ* for Chinese version.
CM	*China Market*
CMJJ	*Cai mao jingji* (Finance and Trade Economics)
CMS	*China Monthly Statistics*
CQ	*China Quarterly*
CSE	Liu Suinian and Wu Qungan (eds), *China's Socialist Economy: An Outline History 1949–1984* (Beijing: Beijing Review, 1986)
CWI	*Coal Week International*
CZ	*Caizheng* (Finance)
DL	*Dili* (Geography)
DLKX	*Dili kexue* (Scientia Geographica Sinica)
DLNJ	Zhongguo dianli nianjian bianji bubian (China Electric Power Yearbook Compiling Dept) *Zhongguo dianli nianjian 1996–1997* (China Electric Power Yearbook 1996–1997), Beijing: China Electric Power Publishing House, 1997.
DLXB	*Dili xuebao* (Acta Geographica Sinica)
DLYJ	*Dili yanjiu* (Geographical Research)
ER	*Economic Reporter*
FBIS	China Daily Report, *Foreign Broadcast Information Service*, Washington, DC: US National Technical Information Service, 1981–95.

FT	*Financial Times* (of London)
GDTZZL	Guojia tongji ju guding zichan touzi tongji ju (State Statistical Bureau, National Fixed Investments Section), *Zhongguo guding zichan touzi tongji ziliao 1950–1985* (China Fixed Asset Investment Statistical Data 1950–1985), Beijing: Zhongguo tongji chubanshe, 1987.
GRRB	*Gongren ribao* (Workers' Daily)
GYJJ	*Zhongguo gongye jingji* (China Industrial Economics)
GYFZBG	*Zhongguo gongye fazhan baogao* (China Industrial Development Report)
GYJJGLCK	*Gongye jingji guanli congkan* (Industrial Economics Management Collection)
GYJJTJNJ	Guojia tongji ju, gongye jiaotong tongji si (Industrial Transportation and Communications Statistics Company of the State Statistical Bureau) (ed.), *Zhongguo gongye jingji tongji nianjian 1998* (China Industrial Economics Statistics 1998), Beijing: Zhongguo tongji chubanshe, 1998.
GYJJTJZL	Guojia tongji ju, gongye jiaotong wuzi tongji sibian, *Gongye jingji tongji ziliao 1949–1984* (Industrial Economics Statistics 1949–1984), Beijing: Zhongguo tongji chubanshe, 1985.
ICR	*International Coal Report*
ICW	*International Coal Weekly*
IEA	International Energy Agency
JDYJ	*Jidi yanjiu* (Base Research). Refers to Shanxi Energy Base.
JEC	Joint Economic Committee Report, Congress of the United States.
JFRB	*Jiefang ribao* (Liberation Daily)
JGLLSJ	*Jiage lilun yu shijian* (Pricing Theory and Practice)
JHJJ	*Jihua jingji* (Planned Economies)
JJCKB	*Jingji cankao bao* (Economic Information Daily)
JJDL	*Jingji dili* (Economic Geography)
JJGL	*Jingji guanli* (Economic Management)
JJJH	*Jingji jihua* (Economic Planning)
JJDB	*Zhongguo jingji daobao* (China Economic Herald)
JJNJ	*Zhongguo jingji nianjian* (Almanac of China's Economy), Beijing: Jingji guanli chubanshe, 1981–2000.
JJRB	*Jingji ribao* (Economic Daily)
JJSB	*Zhongguo jingji shibao* (China Economic Times)
JJWT	*Jingji wenti* (Economic Problems)
JJYJ	*Jingji yanjiu* (Economics Research)

JPRS China Report	Joint Publications Research Service, Congress of the United States
JTNJ	*Zhongguo jiaotong nianjian* (China Transportation and Communications Yearbook)
LDGZZL	Guojia tongji ju shehui tongji ju bian (State Statistical Bureau, Social Statistics Department), *Zhongguo laudong gongzi tongji ziliao, 1949–1985* (China Labour and Wages Statistical Data 1949–1985), Beijing: Zhongguo tongji chubanshe, 1987.
LW	*Liaowang* (Outlook)
MTB	*Zhongguo meitan bao* (China Coal News)
MTGY	*Meitan gongye* (Coal Industry)
MTGYJMSC	Fan Weitang et al. (eds), *Zhongguo meitan gongye jianming shouce* (Concise Handbook of China Coal Industry), Beijing: Meitan gongye chubanshe, 1995.
MTGYNJ	*Zhongguo meitan gongye bu bian* (China Coal Industry Ministry)*, Zhongguo meitan gongye nianjian* (China Coal Industry Yearbook), Beijing: Meitan gongye chubanshe, 1982–2000. See CIYB for English version.
MTJJYJ	*Meitan jingji yanjiu* (Coal Economics Research)
MTKXJS	*Meitan kexue jishu* (Coal Science and Technology)
MTQYGL	*Meitan qiye guanli* (Coal Enterprise Management)
NY	*Nengyuan* (Energy). *Zhongguo nengyuan* (China Energy) before 1989, abbreviated to *ZGNY*.
NYFZBG	Yan Changyue, *Zhongguo nengyuan fazhan baogao 1997* (China Energy Development Report 1997), Beijing: Jingji guanli chubanshe, 1997.
NYTJNJ	Guojia tongji ju gongye jiaotong tongji sibian (State Statistical Bureau, Industrial Communications Statistics), *Zhongguo nengyuan tongji nianjian* (China Energy Statistics Yearbook), Beijing: Zhongguo tongji chubanshe. (Only four yearbooks published, 1986, 1989, 1991, and 1991–96.)
PR	*Peking Review* Known as *Beijing Review* as of 5 January 1979.
RMRB	*Renmin ribao* (People's Daily), Beijing
RMRBHWB	*Renmin ribao haiwai ban* (People's Daily, Overseas Edition), Beijing
SCMP	*South China Morning Post*, Hong Kong
SHKX	*Zhongguo shehui kexue* (China Social Sciences)
SJMYJS	*Shijie meitan jishu* (World Coal Technology)
ST	*The Straits Times*, Singapore
SWB	*BBC Summary of World Broadcasts*
SXJDYJ	Shanxi sheng shehui kexue yuan, Taiyuan shi jishu jingji yanjiu zhongxin (Shanxi Academy of Social

	Sciences, Taiyuan City Technological Economics Research Centre), *Shanxi nengyuan zhonghua gong jidi zonghe kaifa yanjiu* (Research on the Development of the Shanxi Energy and Chemicals Base), Taiyuan: Shanxi renmin chubanshe, 1984.
SXJJNJ	*Shanxi jingji nianjian* (Shanxi Province Economic Yearbook), Taiyuan: Shanxi renmin chubanshe, 1985–2000.
SXRB	*Shanxi ribao* (Shanxi Daily)
TDYSJJ	*Tiedao yunshu yu jingji* (Railway Transport and Economy)
TJNJ	Zhongguo tongji jubian (State Statistical Bureau), *Zhongguo tongji nianjian* (China Statistical Yearbook), Beijing: Zhongguo tongji chubanshe, 1981–2001. English title varies slightly: sometimes appears on spine and in forepaging as Statistical Yearbook of China.
TJZY	Zhongguo tongji jubian (State Statistical Bureau), *Zhongguo tongji zhaiyao 1999* (Summary of China Statistics 1999), Beijing: Zhongguo tongji chubanshe, 1999.
UPIA	United Press International Agency
WZB	*Zhongguo wuzi bao* (China Materials Newspaper)
WZGL	*Wuzi guanli* (Materials Management)
XNA	*Xinhua News Agency*
XZQYNJ	*Zhongguo xiangzhen qiye nianjian* (China Village and Township Enterprise Yearbook)
ZGJJSB	*Zhongguo jingji shibao* (China Economic Times)
ZGNY	*Zhongguo nengyuan* (China Energy). *Nengyuan* (Energy) beginning 1989, abbreviated to *NY*
ZGWJ	*Zhongguo wujia* (China Prices)

Abbreviations used in the text

CCT	Clean Coal Technology
CMAs	Central Mining Administrations – large mines under the jurisdiction of the central goverment
CNCIEC	China National Coal Import and Export Corporation (used to be China National Coal Industry Import and Export Corporation CNCIIEC)
CNCC	China National Coal Corporation
CPC	Communist Party of China
GLF	Great Leap Forward
GVIO	Gross Value of Industrial Output
ICCC	Island Creek Coal Company Ltd.

ICOR	Incremental Capital Output Ratio
ILOR	Incremental Labour Output Ratio
kcal	Kilocalorie
kV	Kilovolt
LNS mines	Local non-state mines: under the jurisdiction of townships, villages, collectives, individual households, others
LS mines	Local state mines: under the jurisdiction of provinces, prefectures, counties
MCI	Ministry of Coal Industry
NPC	National People's Congress
NEPA	National Environmental Protection Agency (became SEPA in 1998)
OMS	Output per manshift
PSF	Pingshuo First Coal Company Ltd.
SCE	Standard coal equivalent
SCIB	State Coal Industry Bureau
SEPA	State Environmental Protection Administration (formerly NEPA)
SPC	State Planning Commission
SPMAs	State Priority Mining Administration (formerly CMAs)
WHO	World Health Organization
WTO	World Trade Organization

Introduction

One of the main constraining factors in China's quest to raise living standards, modernize, and become a major world power has been a persistent shortage of energy. More than twenty years after Deng Xiaoping launched the reform and 'opening up' programme in 1978, per capita energy consumption is still less than half of the average for what the World Bank categorizes as 'lower middle-income economies', and about an eighth of what it is for 'high-income economies'. As the main source of energy in China has always been coal, whether for electric power generation, railway transport, an input to a vast array of industries, or as the principal heating fuel in the residential and commercial sectors, the energy shortage problem until the mid-1990s was essentially a coal industry problem, difficulties or bottlenecks having occurred at all stages – production, transportation, or conversion to other energy forms.

Energy shortages have greatly hindered the industrial, agricultural, and social development of China. They have caused tremendous financial losses in foregone potential production and foreign investment. Capital spent on machinery that has been damaged by power failures, or which was forced to operate at only a fraction of its capacity, has been wasted. The lack of an alternative to coal has perpetuated the gathering of burnable vegetation in the rural areas, resulting in the permanent loss of millions of acres of once fertile land through erosion, and thereby greatly contributing to recurring and devastating floods.

Since 1989 China has been by far the world's largest producer and consumer of coal. In approximate terms, China's 1,001.9 billion tons of proven coal reserves represent 50 per cent of the world total. However, only 114.5 billion tons are classified as proven *recoverable* reserves, representing about 11 per cent of the world total, third largest after the Commonwealth of Independent States and the United States, which each have about 23 per cent. About 74 per cent of the reserves are bituminous, 13 per cent anthracite, and 13 per cent lignites. Some 83 per cent of the total can be used as steam coals, the remainder as coking and gas coals.

While coal use in the developed world peaked in the first decades of the 1900s, and after the Second World War was being replaced with higher efficiency energy forms – oil, gas, and later hydro and nuclear electric power – China's use of coal reached its zenith only in the late 1990s. There was no foreign exchange with which to import oil and gas, and domestic supplies of these were no where near sufficient. In 1949, being a poor, war-exhausted communist country, in fear of foreign domination and exploitation, and under a United Nations embargo, it had no option but to depend on coal as the main source of energy.

The government of the People's Republic did make efforts to replace coal, this relatively low-grade and inconvenient fuel. For a brief period in the 1960s China's oil production ranked among the highest in the world and there was great optimism that new major finds would be discovered. Some consumption equipment was actually converted to the use of oil during those years, but as no new large and easily exploitable fields were found China was forced to revert to coal and to export what oil there was to earn desperately needed foreign exchange. Though there was abundant hydropower potential, materials and technology were sufficient to build only small- and medium-sized plants. Similarly, uranium deposits were adequate, but lack of capital, expertise, and long construction lead-times precluded the nuclear power option.

When the reform and opening up programme began in the late 1970s, the Chinese leadership reviewed its energy options, and concluded a switch to the more modern and efficient fuels would still have to be deferred. While political and security considerations were now less significant factors, costs continued to be a formidable deterrent. Switching from one fuel to one or more other fuels would require huge investments in new equipment and technology. The fear attached to the displacement of a very large work force was probably another factor. It was not until the mid-1990s that there were clear indications that the government was ready to begin a gradual phasing out of the use of coal.

The evolution of the coal industry in China constitutes an important case study because it mirrors the country's economic development as a whole, first with the initial industrialization by foreign capitalists, then a brief period of Soviet-style central planning, followed by several decades of Mao's 'socialism with Chinese characteristics', including two sociopolitical campaigns which led to widespread loss of life and economic breakdown. Then Deng Xiaoping initiated his reform and opening up programme for which there was no blueprint, indeed not even complete agreement among the leadership. Pervasive corruption in government and great disparities in living standards developed very quickly, contributing to mass social unrest and ultimately many deaths near Tiananmen Square in 1989. At the time of writing, the coal industry, indeed the economy as a whole, is in a highly precarious state with overwhelming rates of unemployment combined with ever increasing awareness of Western living standards.

Most analysts would agree that the non-grain agricultural and consumer goods sectors have been virtually fully marketized, and quite successfully so, but that the economic reform of the state industrial sector has lagged far behind. Raising the output, productivity, and profit levels of the state enterprises, where the labour force is very large and many layers of government are involved, is proving extremely difficult. In the cases of sectors it considers strategic, such as energy, transportation, and telecommunications, the government has been particularly cautious about restructuring and opening up to foreign investment.

China's enormous size, population, and diversity make the country's economic problems unique, and the Chinese government had to devise its own plan for development. It could learn from the successes and failures of other countries, particularly from the other formerly centrally planned economies, but could not integrate into its own strategy the entire administrative and regulatory framework of any one country. It was obliged to develop incrementally and suffer some inevitable setbacks resulting from economic experimentation.

China embarked on a radical, though gradual economic transformation with the introduction of the reform programme in late 1978, by which time it had become all too apparent that Mao's adaptation of Soviet central planning and his revolutionary motivational schemes were not working. Deng Xiaoping called for separation of the responsibilities of government and enterprise, incentives for efficiency, and accountability. Opening up of China to the world in order that the Chinese could interact with and learn from other countries and, most importantly, attract foreign investment and technological aid was a top priority. This necessitated developing internationally acceptable standards in the financial and legal sectors, a process which is still far from complete. Of paramount importance, however, was that the reforms in no way lead to an exploitation of the Chinese people and/or their resources by foreign countries, nor to a widening of the difference in income levels between rural and urban areas. This last condition has proved impossible to meet.

After more than twenty years of reforms, the government's principal tenet is still public ownership. One of the main messages of the 15th National Congress, held in September 1997, was that the policy of building socialism with Chinese characteristics must be continued. Socialism was re-confirmed as the primary stage of the ultimate economic goal, communism. However, it was recognized that the concept of public ownership had to be modified somehow in order to accommodate market principles. The dilemma persists.

The most urgent task of the mid-1990s was restructuring the industrial state enterprise sector, especially the textiles, paper products, armaments, machinery, gas, and coal industries, and the process is still ongoing. These dinosaurs of central planning were losing billions of yuan each year. For decades their inputs, outputs, distribution, and product pricing had been determined not by dynamic market forces but by government decree. Local managers and workers had no input into decisions, there were few incentives and little accountability at any level, and the government routinely extended loans to keep them going. After not being able to pay workers' wages and pensions for months, many declared bankruptcy, resulting in, by 1997, a frighteningly high level of unemployment – 14 million in the cities and 130 million in the rural areas – creating not only an economic problem, but nationwide social instability and unrest. Indeed, after the announcement of the massive sale of state owned enterprises (SOEs), labour staged many demonstrations, and these have been accelerating.

Considerable research has been done in the 1980s and 1990s on the economic reforms as a whole, or those pertaining to a given sector, but there have been few case studies of individual industries.[1] A case study of the coal industry is of

enormous importance to a better understanding not only of China's modernization and economic growth, but of its political economy. The industry has deep
historical roots, is widely dispersed over the country, involves a multiplicity of
levels of government, interfaces with many other industries, is classified as
'strategic', and has one of the largest labour forces. Not least of all, China has
been cast in the international spotlight as a result of the link between its incomparable consumption of coal and the potentially devastating effects of global
warming on the planet as a whole.

This book is an economic and sociopolitical history of China's coal industry,
tracing its development since the mid-nineteenth century to the present, with
greatest emphasis on the reform era. The concurrent though far from synchronous
development of the railways and thermal electric plants is also considered in some
detail because they provide the critical links between coal producers and energy
consumers. The study will be of interest not only to China specialists, but to students of comparative and development economics, energy economics, energy
geography, reformation of centrally planned economies, the locational dynamics
of industry, and the history of the coal industry generally. Production, financing,
transportation, and consumption – the four components of coal industry economics – provide the framework for discussion. International comparisons have
been given wherever pertinent and possible. Based in large part on state publications, the Chinese press, reports, and studies by Chinese scholars, and the author's
own field work, the presentation is longitudinal with topical subdivision.

Chapter 1 is a summary, based almost entirely on the very thorough research
of Wright and Ikonnikov, of the first century of development of the modern coal
industry in China, beginning around 1840, and highlights the major involvement
of foreign governments and commercial ventures in mine and railway development.[2] It establishes the level of progress at the time of the proclamation of the
People's Republic of China (PRC) on 1 October 1949.

The first part of Chapter 2 explains why Soviet central planning was chosen,
how the material balances system worked, and the specifics of the plans, formulated with Soviet assistance, for China's coal industry. The material for this part
came mainly from a wide variety of Chinese language journals and collection of
articles. The second part gives the reasons why Mao Zedong believed it was necessary to modify Soviet central planning, and describes how the coal industry was
transformed completely and disastrously by his Great Leap Forward campaign in
the late 1950s and his second and equally damaging sociopolitical experiment,
the Cultural Revolution, beginning in 1966 and lasting for ten years. Throughout
his active tenure a debate raged between him and his supporters on the one
hand, who favoured decentralized, small-scale industrial development, and the
'pragmatists', including Liu Shaoqi, Zhou Enlai, and Deng Xiaoping on the other,
who argued at this time for centralized, large-scale industrial development. This
debate, and others stemming from the radical differences in approaches to economic
and social reform, were clearly played out in the government's policies for the
coal industry. The main sources of information here were the *Zhongguo meitan
gongye nianjian* (China Coal Industry Yearbook),[3] *Zhonghua renmin gongheguo*

jingji da shiji 1949–1980 (Major Economic Events in Communist China 1949–1980),[4] and *China's Socialist Economy – An Outline History (1949–1984).*[5]

Chapter 3 is an assessment of production, financing, transportation, and consumption in 1978, after Mao's death two years earlier, and on the eve of Deng Xiaoping's launching of the programme for reform and opening up. A variety of standard performance indicators is employed, and the weaknesses of central planning – both inherent but especially as applied to the Chinese coal industry – are discussed, as are the pivotal problems of inter-industry linkages.

The total growth in production that was achieved between 1949 and 1978 despite the collapse of the industry after the Great Leap Forward and major disruptions during the Cultural Revolution is astounding, especially considering the out-of-date, ill-adapted, and worn machinery used at the large state mines and the very little, if any, machinery available at the local mines. The cost to the country was astronomical in terms of the thousands of lives lost and the pollution caused by the burning of unparalleled amounts of mostly untreated coal.

In the early 1980s an avalanche of new newspapers, journals, and yearbooks began publication in China, or resumed after a hiatus of some fifteen to twenty years.[6] The government wanted the outside world to know as much as possible about China's economy in order to promote and facilitate foreign trade and investment. Chinese academics were encouraged to write and publish their ideas about how the reforms should be carried out. The volume and detail of the information, some of which is available in English, but mostly only in Chinese, and the exposure of the hitherto hidden sharp dissent and debate over government policy presented foreign China scholars the unprecedented problem of vast amounts of information to digest and assess. Compared to the pre-reform years, the amount of material becoming available in the 1980s and 1990s was not only ever-increasing, but came from a wide variety of often disagreeing sources, making the writing of Chapters 4 to 7 more challenging.

The energy industry was a core component of the Sixth Five Year Plan (1981–85), the first under the leadership of Deng Xiaoping. Chapter 4 recounts how the coal industry was affected by Deng's attacking the problems of too much centralized control and the pervasive lack of motivation and responsibility by separating government from enterprise management, and introducing financial incentives and market forces. His determination to open up China was characterized by persistent pursuit of foreign capital and technology through such instruments as the China National Coal Import and Export Corporation created in 1982, to be responsible for trade, importing foreign technology, and the establishment of joint ventures.

Chapter 5 relates how the widening and deepening of the reforms over the next fifteen years – the Seventh, Eighth and Ninth Five Year Plans – combined to transform the coal industry. The 1989 Tiananmen Incident caused a relatively brief period of retrenchment and consolidation, which was followed in 1992 by Deng's trip to South China where he was so impressed with the rapid modernization that he called upon the whole country to pursue reform and opening up more vigorously, and had the phrase 'socialist market economy' entrenched in the constitution to replace 'central planning'. The coal industry would undergo major organizational changes, with individual administrations and enterprises having greater decision-making authority and responsibility for profit and loss. Coal enterprises would participate in labour contracts, profit retention schemes, price reforms, revenue-sharing systems, new measures to attract investment, the sale of shares, and listing on stock exchanges.

However, at the very time many of these reforms were being introduced the demand for coal began to decline, and in the late 1990s the government, for the first time ever, actually sought to curtail production, and to do so sharply. They also forced many enterprises, including coal, to declare bankruptcy, an unheard of measure hitherto.

The reduction in coal consumption was the result of a combination of a number of more or less equally important factors: the bankruptcy of thousands of state owned industrial enterprises meant far less coal was required; urban residents were switching to gas for cooking and heating as soon as it became available; the demand for light industry products was surpassing that for heavy; many industries by this time, in the interests of efficiency, had installed modern coal-burning equipment or replaced coal entirely with oil or electricity in order to compete internationally; the alarm being expressed worldwide about the local and global effects of pollution caused by the burning of so much untreated coal could no longer be ignored; and two decades of buoyant trade and the securing of global export market niches, particularly in electric equipment and accessories and textile materials and products, assured continuing sizable reserves of foreign exchange with which to import large quantities of oil products.

In the late 1990s, economists were no longer surmising 'if', but 'when' coal consumption in China would fall from accounting for over 70 per cent of total energy consumption to the 20–30 per cent average in the rest of the world. At the time of writing they expected that coal would still be required for at least 50 per cent of national energy requirements until 2020.

Chapter 6 is an assessment of the coal industry after twenty years of reform, using the same performance indicators employed in Chapter 3, and some additional ones. Over the 1980s and 1990s the structure of production changed dramatically, with the proportion of total production coming from the state mines decreasing from nearly 85 per cent in 1978 to a low of about 54 in 1995, while that from the small, local non-state mines increased from about 15 per cent to 46. Great strides had been made at the former in raising average technical standards including safety levels, as well as the quality of output. However, the high ratio of coal from the local non-state mines kept the average safety and quality statistics for the country as a whole still alarmingly low, internationally speaking. From approximately 1996 the demand for coal began to fall quickly and the government's two-pronged strategy since then has been to close thousands of the small mines that had been opened in the 1980s and to begin to consolidate total production at a greatly reduced number of large, modern operations. In 2000, the state: non-state ratio of production stood at about 70:30 per cent. Attainment country-wide of international standards for production methods, quality and safety would finally be a possibility.

The first section of Chapter 7 summarizes the development of China's coal industry over some 150 years, while the second is a conjectural look towards the early decades of the new millennium. The issues to be resolved in the industry at the turn of the century, the expected development of other forms of energy, and some energy consumption structure scenarios up to 2050 put forward by several Chinese energy analysts are discussed.

This much is certain: China has abandoned its policy of energy self-sufficiency. At the beginning of the reform period, the leadership reassessed the role of coal in the hopes of decreasing, if not replacing, its use, and concluded it was not feasible at that time. About sixteen years later, they began to compete eagerly for a share of international liquid fossil fuel supplies and to build the necessary infrastructure for importing these. This testifies in some measure to both the success of the reform programme and the depth of integration into the global economy.

1 The industry before 1949

Development by foreign interests

Historical sources are more or less in agreement that coal was first used in various parts of the world during the Bronze Age, 3,000–4,000 years ago, that the Chinese began to use coal for heating and smelting in the Warring States Period (475–221 BCE), and that they are credited with organizing production and consumption to the extent that by the year 1,000 this activity could be called an industry. China remained the world's largest producer and consumer of coal until the perfecting and patenting of the steam engine by James Watt in 1769.

Europe's coal industry began in England and Belgium in the 1200s. Initially, coal was used almost exclusively for smelting and forging of metals. Wood and charcoal made from wood remained the main forms of fuel for residential heating as late as the 1600s when a switch to coal for most energy needs was necessitated by the exhaustion of accessible forests.

The Industrial Revolution, which began in Britain in the 1700s and later spread to Europe, North America, and Japan, was based on the availability of coal, and human and animal powered industries were transformed in scale and scope with the invention of the steam engine. International trade expanded exponentially when these coal-fed engines were built for the railways and ships, but this technology did not reach China until the mid- to late-1800s.

The demand for Chinese coal greatly increased when foreign steamships began to arrive following the opening of several treaty ports in the 1840s. Initially, coal for the return voyage was brought from the US or Britain, or wherever the home or last refuelling port happened to be, but this was, of course, impractical. At that time, China's coal was very expensive due to the high transportation costs in combination with heavy government taxation – the mineral production tax, mining area tax, land tax, provincial transit dues, coast trade and maritime customs duties, and others. The poor quality of the coal from the native pits was also a drawback. Coal bought from Japan was a cheaper alternative, but by the mid-1800s the demand for coal by the Chinese navy, international shipping companies, and economic growth generally, was such that both Chinese and foreigners sought larger-scale development of China's coal industry.[1]

The first foreign proposal for development was made by a Briton, John William Howell, in 1864, and construction was started on a mine at Jilong

(Taiwan) in 1874 which was to supply the Fuzhou shipyard and the Chinese navy.[2] In the late 1800s British, German, Belgian, and Russian interests built, invested capital and/or operated mines in Hebei, Shanxi, Shandong, Liaoning, and Inner Mongolia.[3]

The Treaty of Shimonoseki in 1895, the Sino-British Treaty of 1902, and the Sino-American Treaty of 1903, *inter alia*, clarified the regulations regarding foreign involvement in the mining sector and investors were quick to respond. Between 1895 and 1912 thirty-three coal mining agreements were signed with foreigners and many Chinese mines were taken over by foreign interests.[4] In 1898 mining rights ten miles on either side of the Jiaozhou-Jinan Railway in Shandong were given to the Germans, and similar rights were awarded to the Russians along the Chinese Eastern Railway in Jilin and Heilongjiang. In the same year, the Peking Syndicate, an Anglo-Italian Corporation, was granted mining rights in Shanxi and Henan. Concessions were also given to Great Britain at Tungkuanshan (Anhui), and to the Japanese at Fushun, Yantai, and Benxi (all in Liaoning).

However, the Chinese soon began to understand fully both the economic and strategic value of their coal industry and in 1906 the government sought to redeem some of these concessions. The regulations that were officially promulgated in 1908 stipulated that foreign capital in any enterprise was limited to 50 per cent of the total. The same policy was perpetuated in the mining regulations of 1914, and it was further specified that the general manager of any Sino-foreign company had to be Chinese. It was later ruled that foreigners could not invest in small mines unless they introduced machinery.[5] Generally, the fifty : fifty arrangement functioned well enough. However, many foreigners believed that enterprises could never be profitable if managed predominantly by Chinese and found ways to disregard the regulation.

When the Qing Dynasty collapsed in 1911, the development and modernization of the coal industry were disrupted. However, the effects of the civil wars that took place in the 1920s were far worse. Only a few railway lines outside of Manchuria continued to operate at full capacity, if at all. During this period no new agreements with foreigners were signed, and most of those who had already been operating mines, apart from the Japanese who by now had established ascendancy, lost their rights.

Production

Quantity, geographical distribution, and quality

The Japanese involvement in the industry gained momentum after their defeat of the Russians in the Russo-Japanese War of 1904–05 and soon thereafter they were the most active foreign participants in China's coal industry. The Fushun mine (Liaoning), which had been developed by the Russians before the turn of the century, was now taken over completely by the Japanese South Manchurian Railway Company. It had some of the world's thickest seams, and of all the mines involving foreign interests was by far the largest. In fact it was one of only two entirely foreign-controlled coal enterprises, the other being Yantai (Sino-Russian from 1899–1905, Japanese after the War), south of Mukden City (now Shenyang). The Fushun mine exported to southern China, Formosa (now Taiwan), Japan, and to the East Indies.[6]

The mines operated by the Japanese, by 1926, accounted for almost all of the output from the foreign-connected mines, and of the output from the Japanese mines alone, never less than 65 per cent came from Fushun. Production was evidently highly concentrated. Indeed, the Kailuan Mining Administration accounted for never less than 70 per cent of production from the British-connected mines.[7] Many of the foreign companies enjoyed several advantages, for not only did they control some of the best coal seams and had the most advantageous locations in China in terms of transportation and markets, they had diplomatic protection against extra taxes and imposts levied by zealous local authorities and warlords demanding lump sums for military funds.[8]

The Japanese occupied Manchuria in 1931 and set up a puppet state called Manchoukuo. The second Sino-Japanese War began in 1937. By 1942 the Japanese controlled all the major coal centres in the country, not only in Manchuria, but in the north, northeast, as well as the east, and central south, all together accounting for 90 per cent of all the coal produced. The British enterprises accounted for 7–9 per cent, the German 0.8–1.6 per cent, and the Russian 0.2–0.3 per cent.[9]

Table 1.1 gives the percentages of total output from Chinese enterprises, joint Chinese-foreign enterprises, and foreign enterprises from 1913 to 1942.[10] Chinese enterprises accounted for over 50 per cent of the total output in 1919, while joint Chinese-foreign and foreign together made up the remainder. Of the joint Chinese-foreign enterprises, those with English interests had the largest outputs. The 'other' category included Russian, German, French, and Belgium interests. By 1936 the Chinese share had fallen to 43 per cent and in 1942, almost 90 per cent of the production was in the hands of the Japanese.

The mines owned and operated solely by the Chinese were generally small pits. Most were worked only during the winter as harvesting diverted the labour force over the summer months, and because there was less flooding and better ventilation in the colder season. They lacked mechanized equipment and productivity was very low and accidents frequent. In all they contributed about one-sixth to one-quarter of the total output.[11] In some cases, small and large mines worked together in harmony, the large mines buying coal from the small and organizing its sale. Often up to 10 per cent of the output sold by the large mines actually came from the surrounding pits.[12] In other situations, small mines competed with the larger ones, putting coal on the local market at comparable or lower prices, and although all mines were supposed to have a licence from the government, many of the smaller operations were mining illegally on territory granted to large companies.[13]

Table 1.2 gives the annual coal production in China between 1896 and 1949. In 1936 it accounted for 2.7 per cent of the world's bituminous output, 5.6 per cent of its anthracite, and less than one-fifth of one per cent of its lignite.[14] The pre-1949 coal production peak was 65.7 million tons in 1942, over one-third of this having been produced in areas occupied by the Japanese.[15] For perspective, the

Table 1.1 Output of coal by ownership of enterprise, 1913–42 (%)

	1913	1919	1926	1936	1942
National total	100.00	100.00	100.00	100.00	100.00
Chinese enterprises	44.00	50.26	44.92	42.52	10.06
Private	43.29	42.83	42.25	35.11	8.12
Joint government–private	0.17	4.21	2.02	4.83	0.67
Government	0.54	3.22	0.65	2.58	1.27
Joint Chinese-foreign enterprises	32.37	26.54	23.50	19.80	0.60
Chinese–Japanese	11.76	6.84	7.05	1.71	0.0
Chinese–English	15.66	17.96	15.08	15.95	0.60
Chinese–Other	4.95	1.74	1.37	2.14	0.0
Foreign enterprises	23.64	23.20	31.58	37.68	89.34
Japanese	14.35	20.87	31.11	37.68	89.34
English	3.20	2.33	0.47	0.00	0.00
Other	6.09	0.00	0.00	0.00	0.00

Sources: Alexander B. Ikonnikov, *The Coal Industry of China*, Canberra: Research School of Pacific Studies, Australian National University, 1977, p. 23. Adaptation of Table 1. Ikonnikov used data from The Geological Survey of China, *Chungkuo k'uangyeh chiyao no. 7* (7th General Statement on the Mining Industry of China), Chungking: 1945, pp. 53–72; Taiwan yinhang, chingchi yenchiushih (The Bank of Taiwan, Economic Research Unit), *Jichushihtai Taiwan chingchishih* (The Economic History of Taiwan Under the Japanese Occupation), Taipei: 1958, vol. II, pp. 182–3; and Yen Chung-ping and others, *Chungkuo chintai chingchishih t'ungchi tzuliao hsuanchi* (Modern Economic History of China, Selected Statistical Data), Peking: 1955, pp. 102–3, 126, and 154. Reproduced with permission from the Department of Economics, Australian National University.

Ikonnikov's explanatory notes:
a. The entire pit output of the mainland part of China has been included in the item 'private' of 'Chinese enterprises', with the exception of 1936 and 1942; for these two years, the total output (mines and pits) in the areas under Japanese occupation, has been included in the item 'Japanese' of 'Foreign enterprises'.
b. No statistics of output of Taiwan with a break-down by ownership have been found. In 1929 Japanese capital represented 76.45 per cent of the total capital invested in all industries, while Chinese capital represented 21.89 per cent. [Taiwan yinhang, chingchi yenchiushih (The Bank of Taiwan, Economic Research Unit), *Jichushihtai Taiwan chingchishih* (The Economic History of Taiwan Under the Japanese Occupation), Taipei: 1958, vol. I, p. 76.] In this table, 80 per cent of the total output of coal of Taiwan (100 per cent in 1942), has been included in the item 'Japanese' of 'Foreign enterprises', and 20 per cent (none in 1942), in the 'private' of 'Chinese enterprises'.

largest quantity the coal industry in the UK ever produced in one year was 287.4 million tons in 1913.[16]

The production peak in Manchuria alone was in 1944, at 25.6 million tons.[17] During the war with the Japanese, between 1937 and 1945, despite shortages in the south, which was forced to import, northern coal was shipped to Japan for their war effort. Output in 1946 fell dramatically, but recovered in 1949. Nationalization of the industry began immediately after evacuation of the Japanese forces in 1945. Between 1945 and 1947, government capital in the industry rose from 12 to 29 per cent.[18]

Between August 1945 and May 1946, the Soviets removed whatever industrial equipment they could find in Manchuria. What they could not take back to the

Table 1.2 Production, consumption, imports and exports, 1864–1949 (mt)

Year	Production	Consumption	Imports	Exports
1864			0.1175	
1865			0.1005	0.0001
1866			0.1217	0.0512
1867			0.1152	0.0007
1868			0.1603	0.0034
1869			0.1288	0.0001
1870			0.0797	0.0002
1871			0.0869	0.0000
1872			0.1365	0.0010
1873			0.1170	0.0022
1874			0.1182	0.0005
1875			0.1456	0.0000
1876			0.1296	0.0007
1877			0.1709	0.0001
1878			0.2070	
1879			0.1786	0.0003
1880			0.2179	0.0000
1881			0.2568	0.0001
1882			0.2572	0.0006
1883			0.2457	0.0000
1884			0.2676	0.0004
1885			0.3068	0.0005
1886			0.3160	0.0003
1887			0.3098	0.0018
1888			0.2726	0.0054
1889			0.3765	0.0156
1890			0.3109	0.0129
1891			0.3759	0.0020
1892			0.4046	0.0045
1893			0.4358	0.0027
1894			0.1941	0.0057
1895			0.5816	0.0046
1896	0.489		0.6349	0.0088
1897	0.539		0.5581	0.0272
1898	0.783		0.7423	0.0160
1899	0.842		0.8731	0.0070
1900			0.8780	0.0031
1901	0.521		1.1714	0.0030
1902	0.954		1.1924	0.0268
1903	1.026		1.4301	0.0770
1904	1.274		1.2706	0.0105
1905	1.345		1.2292	0.0117
1906	1.696		1.5831	0.0023
1907	2.189		1.4259	0.0065
1908	2.788		1.5286	0.0283
1909	3.868		1.5409	0.1991
1910	4.237		1.4670	0.3232
1911	5.266		1.5656	0.3318

Table 1.2 (Continued)

Year	Production	Consumption	Imports	Exports
1912	8.988	9.8377	1.5411	0.6914
1913	12.800	12.8261	1.7179	1.6918
1914	14.102	13.6909	1.6266	2.0377
1915	13.417	13.5042	1.4238	1.3366
1916	15.903	16.0118	1.4447	1.3359
1917	16.902	16.7683	1.4672	1.6009
1918	18.340	17.6966	1.0923	1.7355
1919	20.055	19.7455	1.1916	1.5011
1920	21.260	20.5329	1.2746	2.0017
1921	20.459	19.9260	1.3836	1.9166
1922	21.097	19.8513	1.1698	2.4155
1923	24.552	22.7816	1.3880	3.1584
1924	25.781	24.1632	1.6358	3.2536
1925	24.255	24.0062	2.8021	3.0509
1926	23.040	22.8498	2.9439	3.1341
1927	24.172	12.4496	2.3559	4.0783
1928	25.092	23.6101	2.4655	3.9474
1929	25.437	23.5654	2.3177	4.1893
1930	26.037	24.9822	2.5065	3.5613
1931	27.245	25.5379	1.9333	3.6404
1932	26.376	24.2292	1.4544	3.6913
1933	28.359	28.8978	1.9854	4.4656
1934	32.725	29.0258	1.0619	4.7611
1935	38.032	34.0247	0.8292	4.8365
1936	39.342	35.0417	0.6152	4.9155
1937	36.913	31.8812	0.4911	5.5229
1938	31.943	30.9574	1.0918	2.0774
1939	38.542	36.9911	1.4137	2.9646
1940	46.828	43.9984	2.0084	4.8380
1941	58.823	54.0947	1.1318	5.8601
1942	65.686	59.2589	0.2694	6.6965
1943	56.687		0.0819	5.1140
1944	53.782		0.0160	
1945	29.206		0.0003	
1946	18.898		0.0731	0.0560
1947	21.904		0.2154	0.0162
1948	20.106		0.1382	0.0173
1949	32.430			

Sources: All figures from are Zhongguo jindai meikuang shi bianxue zu (The History of Coal Mining in Contemporary China Compiling and Study Group), *Zhongguo jindai meikuang shi* (The History of Coal Mining in Contemporary China), Beijing: Meitan gongye chubanshe, 1990, pp. 268, 269, 271, and 538 except the production figures for 1896 to 1911, which are for large mines only, and are from Tim Wright, *Coal Mining in China's Economy and Society, 1895–1937*, Cambridge Studies in Chinese History, Literature and Institutions, Cambridge: Cambridge University Press, 1984, p. 10. Ikonnikov, and Wu and Ling gave incomplete and different sets of figures. The 1949 production figure is from *CIYB 1982*, p. 24.

Soviet Union they destroyed. It has been estimated that Soviet looting and destruction reduced the plant and equipment capacity of the coal industry by some 80–90 per cent and further losses throughout the rest of the country occurred as a result of the civil war in 1947–48.[19] By 1949 there was little operable machinery in any industry. Yet the mine development that took place before 1949 had lasting significance. Six of the country's largest mines in the 1990s, Fushun and Fuxin (Liaoning), Hegang and Jixi (Heilongjiang), Huainan (Anhui), and Kailuan (Hebei), were built in the early 1900s, largely with foreign capital. Other large mines constructed around that time were Benxi (Liaoning), Jingxing (Hebei), Zhongxing and Luta (both in Shandong), Chiaotso (Henan), and Pingxiang (Jiangxi). For the sake of comparison, while China's total output in 1949 was 32.4 million tons, and ranked ninth in the world, it was 435.9 million tons in the US, 218.6 million tons in the UK, 176.1 in Germany, 74.1 in Poland, 52.9 in France, 37.9 in Japan, 32.0 in India, 27.9 in Belgium, 25.0 in South Africa, and 14.3 in Australia.[20]

Table 1.3 shows the geographical distribution of large mine production between 1912 and 1936. Production was concentrated in the north and northeast regions which together accounted for over 90 per cent of the country's large mine output, and of this, two provinces, Liaoning and Hebei, alone accounted for a minimum of 66 per cent (see Figure 1.1). Kailuan and Fushun mines almost consistently accounted for over half of the total large mine output and actually accounted for 71 per cent in 1927.[21] According to the management, Kailuan's

Table 1.3 Geographical distribution of large mine output, 1912–36 (% of total output)

	1912–16	1917–21	1922–26	1927–31	1932–36
Northeast	33.2	29.5	39.1	50.2	42.1
of which:					
Liaoning	30.8	27.5	37.3	37.3	38.2
North	57.7	61.9	55.4	46.5	50.8
of which:					
Hebei	39.3	39.6	34.4	34.0	27.9
Northwest	—	—	—	—	—
Centraleast	9.1	8.5	5.5	2.9	5.7
West and Southwest	—	—	—	—	0.4
Southeast	—	—	—	0.3	1.0

Source: Tim Wright, *Coal Mining in China's Economy and Society 1895–1937*, Cambridge Studies in Chinese History, Literature and Institutions, Cambridge: Cambridge University Press, 1984, Table 24, p. 78. Copyright © 1984 Cambridge University Press. Reprinted by permission of Cambridge University Press.

Wright's notes:
Northeast: Liaoning, Heilongjiang, Jilin, Rehe.
North: Hebei, Shanxi, Shandong, Henan, Chahar, Suiyuan.
Northwest: Shaanxi, Gansu, Ningxia.
Centraleast: Hunan, Hubei, Anhui, Jiangxi, Jiangsu.
West and Southwest: Sichuan, Yunnan, Guizhou, Guangxi.
Southeast: Zhejiang, Fujian, Guangdong.
The percentages may not all add up to 100 because of rounding errors.

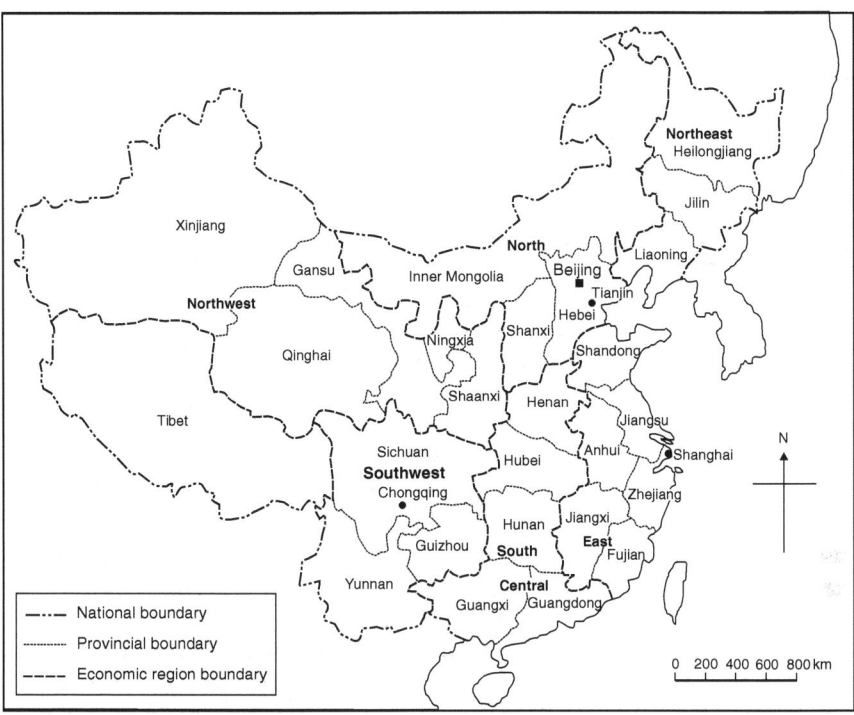

Figure 1.1 Provinces and autonomous regions of China

Zhaogezhuang, with a capacity of some two million tons, was the largest underground colliery outside the US in the early 1920s.[22] The result of this concentration in the north and northeast was that the country's overall industrial development became increasingly imbalanced. Not surprisingly, most of the coal was consumed where the industrial development was greatest, namely, Manchuria, Shanghai, and the Treaty Ports. In 1936, 36.0 per cent was consumed in the north, 26.4 per cent in the northeast, 22.9 per cent in the centre and east, 7.9 in the southeast, 5.6 in the northwest, and 4.2 in the west and southwest.[23]

In the early 1910s and early 1920s Manchuria produced over 30 per cent of China's coal even though it had under 2 per cent of China's coal reserves and only about 7 per cent of its population. In 1925–26, when the civil wars disrupted production in the north, it accounted for over 50 per cent of the total coal production. Output from large mines in Liaoning, Jilin, Heilongjiang, and Rehe (a province in the pre-1949 period) grew at 10.9 per cent per year between 1907 and 1936, while in the rest of China the rate was 5.8 per cent.[24] Compared to growth in China proper, the coal industry's expansion in Manchuria was rapid. The area enjoyed relative political stability, GNP was higher, and per capita consumption of coal was twice as high. Japan's intention was that Manchuria serve not just as a raw materials hinterland, but also as a new industrial and population base.

In the 1930s Shanxi, China's most coal-rich province and one of the largest deposits in the world, was producing less than 4 per cent of total output. Reasons that have been given for this include: shortage of water for steam engines and timber for pit props; shortage of capital, which not only limited growth potential but resulted in horrific working conditions; and the high cost of transportation to coastal markets.[25] Moreover, before 1937 the area was continually plagued with political and military upheavals which disrupted railway traffic. The Peking Syndicate did plan to develop the resources, but was unsuccessful in gaining permission to build a railway to Pukou (Jiangsu), where the coal would have been unloaded onto ships and transported on the Changjiang.[26]

The ratio between production in Manchuria and China proper in 1942 was 63 : 37. In 1947 it was 75 : 25, but by the time of the Communist takeover in 1949 it was down to 60 : 40.[27] In order to reduce the demand on northern coal, the Nationalist government had opened the Huainan mine (Anhui) in 1929 and by 1936 it was supplying much of Shanghai's needs. The southwest region, held by the Nationalists and sometimes referred to as 'Free China', was developed after 1937, when the wars disrupted the supplies in the north.

Washing coal removes various impurities such as shale, clay, pyrites, etc., and may involve either physical separation, which mechanically distinguishes the relative gravities and/or surface properties of the coal, or magnetic or chemical separation. Often after the coal is washed, it is processed, which entails the crushing of the large lumps, and screening them into various standard sizes. Up to two-thirds more raw than clean coal is required to achieve the necessary heat levels, and up to one-third more space is occupied in transit and storage.

A good quality washed coal has about 7,000 kcal per kg. The average caloric value before washing at the best mines was about 6,600 kcal, but from the least good may have been as low as 3,500.[28] Before 1949 little attention was given to improving the quality of the coal. Generally, it was simply used in its raw form without being washed, though some was sized.

Mechanization and use of labour

Apart from opencast mines at the Guchengzi site at Fushun (Liaoning) and at Zhalainuoer and Hegang (Heilongjiang), almost all the mines before the Sino-Japanese War (1937–45) were underground and used the room and pillar system.[29] The longwall system was not implemented until the 1950s.[30] Compared to European mines, which had been built much earlier, the mines in China were relatively shallow. The average depth in the deepest shafts at China's largest mines was 163 m, comparable to that in the US.[31]

The level of technology used for cutting the coal was rudimentary. While the proportion of coal cut by mechanized means was less than 1 per cent at the time the PRC was established, 55 per cent of the coal was cut by mechanized means in the UK in 1936, 97 per cent in Germany's Ruhr Valley, 99 per cent in Belgium, and 88 per cent in France's Pas-de-Calais mines in 1935.[32] The main reason for the lack of mechanized operations in China was the abundant and cheap labour available to the contractors. Other factors were the shallowness of the mines, and the fear on the part of foreign developers that the prevalent political turmoil could at any time disrupt mining or coal transport. It was simply not in the interests of the foreigners to bring the best and most expensive equipment available. As the mines became deeper, however, not only was extraction more dangerous, but the miners needed training. This was not easy to provide to the illiterate gangs of men from the farms who worked only for a few years at the mines, and for only part of the year at that.

The hauling of the coal to the surface was generally the first mining process to be mechanized, and at the Japanese-operated mines in particular, it was relatively sophisticated. While over 50,000 pit ponies were in use in British mines in the late 1930s, the Japanese-operated mines had switched to electrically powered ropes or locomotives.[33]

Most mines had contractors to oversee the coal cutting and hauling. They were charged with supervising crews of men to bring to the surface, by a certain target date, an agreed amount of coal. Often bonuses were given if the amount was exceeded. Under this contract system, the contractor hired the men, mainly peasants from rural areas, and provided them lodging, meals, and pay. For the mine owners, this system was very convenient, but for the gangs of men it was abusive. The living and working conditions were primitive and the pay paltry. The contractors tended to cut corners wherever possible and to request extra fees for the smallest of services to both the workers and the mine owners. Initially, the workers had no rights (apart from a free coffin!) or recourse whatsoever, though as time went on it was necessary to improve the conditions and terms in order to reduce the high turnover of men.[34] A minimum wage was eventually instituted and the responsibility of paying the workers was taken out of the hands of the contractors and assumed by the mine owners.

Most of the labourers were male rural peasants who returned to their fields during the planting and harvest seasons. Though often ignored, there was a law prohibiting the hiring of workers under the age of sixteen. A few women hauled coal to the surface, while many worked at the surface level. Unlike the mines in Japan, where convicts made up a large proportion of the miners, it seems in China convict labour was used only in Sichuan Province.[35] There was some use of slaves and instances of kidnapped workers. Others were tricked into amassing gambling or drinking debts and then forced to sell their labour. Some signed up on the basis of feigned promises.

While unskilled labour was rarely in short supply, an insufficiency of skilled labour characterized the industry until the 1920s. Mining companies were forced to seek out foreign engineers and technical staff. This raised production costs considerably. When the Japanese took over all the Chinese-owned mines in the northeast, they brought their own engineers.

The average output per manshift at the largest mines in the 1930s and 1940s was 0.592 tons.[36] By contrast, in 1937 it was 1.534 tons in the UK, 2.002 in West Germany, 1.435 in the Saar, 1.236 in France, 2.413 in Poland, 1.139 in Belgium, 2.550 in the Netherlands, and 1.837 in Czechoslovakia.[37]

The accident rate was very high. While there were between 5 and 21 fatalities per million tons of output at the American, Indian, and Japanese mines in the late 1800s and early 1900s, there were between 25 and 212 at the main mines in China between 1907 and 1934.[38] The difference would be very much greater if the small Chinese-run mines were included. A gas explosion at Benxi (Liaoning) in 1942 killed 1,662. This was the biggest ever coal mine accident in world history to date. The rate of lung and skin disease was also very high, and the average working life of the miners was but a decade.

Wages, though barely enough to survive on, were not the lowest in the industrial category. The mainly women workers in the cotton spinning, silk reeling, and match industries were paid less. Relatively speaking, though, working conditions in these industries were somewhat better. Unlike in other countries, the Chinese mine workers were not highly organized. There were several altercations – mainly protests against the barbaric treatment accorded them by the contractors – but until the 1920s the workers were insufficiently integrated to form unions. There were a number of reasons for this. The mines were spread over a large geographical area, the men were not at the mines all year round, and due to the wretched working conditions and exploitative contract system the worker turnover rate was high. They were illiterate and completely uninformed about alternative industrial management structures, far less the militancy in other countries' coal mining industries.

Mention must be made, however, of an escalation in organized labour protests in the 1920s, especially among miners at the larger operations. Marxist Study Groups, the Chinese Labour Organization Secretariat, and after its formation in 1921, the Chinese Communist Party had been organizing labour in those areas where it was concentrated. Miners staged strikes at Pingxiang (Jiangxi), Kailuan (Hebei), Tangshan (Hebei), Zhaogezhuang (Hebei), Linxi (Hebei), Majiagou (Shanxi), and several other sites. The issue was usually low or non-payment of wages, maltreatment of workers, or some aspect of working conditions or terms. This heightened agitation of the miners lasted throughout the decade.

This localized and short-lived intensification of labour unrest apart, the figures for mine worker unionization were never large when considered over the whole pre-1949 period. In China proper never more than 40 per cent of the workers were unionized. This was half the rate of unionization at the American mines in the same period. If the work force in the large mines in Manchuria were factored in, the percentage would be reduced to 20 per cent, while inclusion of the workers at the native pits would further reduce it to less than 10.[39] In the 1940s the key strategic industries such as the railroads and communications were nationalized for defence reasons and in 1942 a draft labour policy forbade 'workers in military industries' (all heavy industries including coal) to strike or participate in collective bargaining.[40]

Financing

The big mines either made direct sales to retailers or consumers at the ports or at stops along railway lines, or sold it through distributing companies or agents who earned commissions. It was the responsibility of the mines to organize the transport and handling of the coal, while the agent managed storage and sales. The coal that was not sold directly to consumers often went through several middlemen, who in turn raised the price higher and higher.

Managers of all urban coal distribution centres with capital amounting to over a certain level belonged to 'coal guilds'. The Peiping Coal Guild was registered with the Peiping Chamber of Commerce and its members met monthly.[41] In Manchuria, distribution was managed by the Japan–Manchukuo Trading Corporation, while marketing came under the jurisdiction of the main railway, metallurgical, and other heavy industry companies.

The blockage of railway traffic during the civil wars in the late 1920s nearly forced the closure of some mines in the interior, but mines with foreign connections such as Kailuan and Fushun were spared and continued to make high profits. In the 1930s, the depression affected Manchuria first, and China proper later.

Coal occupied a very important role in the economy. Compared to consumer goods or other mining products its percentage contribution to total industrial net value-added was fairly consistent (see Table 1.4). It accounted for about one-quarter to one-third of the total between 1912 and 1945, dropped sharply to about one-sixth in 1946, but came back up to about one-quarter in 1949.

Table 1.4 China's industrial structure 1912–49 (% of total net value-added in 1933)

	Consumer goods	Coal	Ferrous metals	Other mining products	Electric power
1912	21.9	31.2	1.5	40.0	3.3
1917	29.2	31.0	7.2	25.6	3.7
1921	44.1	29.8	7.7	11.7	5.4
1922	32.7	34.1	6.8	16.3	7.3
1926	40.6	24.2	4.5	15.6	12.0
1930	44.3	23.5	5.3	8.9	13.5
1933	41.6	20.9	5.2	7.4	20.2
1936	30.7	23.7	8.5	8.6	22.1
1937	25.4	26.4	13.0	11.4	16.1
1940	8.7	30.5	13.8	9.5	23.6
1943	5.7	30.4	19.5	3.1	32.3
1945	5.3	26.4	5.0	3.9	50.4
1946	37.5	16.5	0.7	4.7	37.7
1949	29.3	24.6	3.3	3.3	35.2

Source: John K. Chang, *Industrial Development in Pre-Communist China: A Quantitative Analysis*, Edinburgh: University of Edinburgh Press, 1969, p. 76. Adaptation of Table 20.

Chang's note:
Consumer goods: cotton yarn and cotton cloth; ferrous metals: iron ore, pig iron, and steel; other mining products: antimony, copper, gold, mercury, tin, and tungsten. Cement and crude oil, accounting for only negligible fractions of the total, have been omitted; the percentages, therefore, do not add up to 100.

Coal exports and imports between 1864 and 1948 are given in Table 1.2. Before 1915, imports were greater than exports. During the early 1800s coal had been brought from Japan, India, America, Britain, and Australia. Between 1917 and 1943 the situation was reversed. The volume of exports depended largely on the supply situation in Japan, and concomitantly, the relative prices of Chinese and Japanese coal. Exports as a proportion of total national coal output were as high as 14–16 per cent in some years, though as southern China imported some coal from Vietnam (mainly anthracite), India, the Netherlands, East Indies, and Japan, the adjusted net export figures were somewhat less.[42]

Most but not all of the export shipments came from Japanese-operated mines and most – up to one-third of their production – were destined for Japan. Some shipments were also sent to Hong Kong and Korea. In 1936 exports from China proper amounted to 7 per cent of total production and 1.5 per cent of total export trade. Imports amounted to 3 per cent of domestic consumption and 0.7 per cent of total import trade.[43] For perspective, 4.92 million tons were exported from China in 1936, compared to 51.2 million tons from the UK in 1935.[44]

The demand for extra coal supplies from China rose sharply in the years leading up to the Second World War. Ironically, it was Chinese coal which fuelled the Japanese war in Southeast Asia, including China. One of the first acts of the invading Japanese soldiers was to commandeer all the largest mines.[45] Of particular interest was the coking coal, used in the manufacture of steel, at Kailuan. Its output was shipped north to Manchuria and to Japan. In 1944 the Kailuan mines were incapacitated by bombs, forcing China proper to import coal from Indo-China.

Transport

As in the majority of other countries, the most efficient way to move coal from the mines to points of consumption in China was the railways. There was a symbiotic relationship between the coal industry and the railways – neither could survive without the other. The cost of railway transport was as low as less than a fifth of that for the other forms of land transport, and, in any case, humans or pack animals were capable of pulling or carrying this heavy bulky commodity only limited distances.[46] Wright calculates that for cart transport of coal, the approximate economic radius in 1913 was 75 km, and for railways well over 500 km.[47] The development and modernization of railway transport made it feasible for foreign mining companies to operate profitably, and immediately created in the cities an insatiable demand for coal from manufacturing and commercial enterprises. Factories grew in such number and scope that imports of manufactured items dropped sharply. The development of railways also made domestic coal competitive with coal imports, which came to be used only in emergencies.

A proposal for a railway line from Shanghai to Suzhou had been made in 1863 by a group of British businessmen. The Chinese government, however, fearful that this technology would make the Chinese vulnerable to foreign economic and political dominance, refused permission. It was not until 1876 that a ten-mile line

Table 1.5 Railway trackage, 1876–1949 (km)

Year	Annual increment	Total to date
1876	16	16
1877	−16	0
1878	0	0
1879	0	0
1880	0	0
1881	10	10
1882	0	10
1883	0	10
1884	0	10
1885	0	10
1886	32	42
1887	0	42
1888	103	145
1889	0	145
1890	0	145
1891	0	145
1892	45	190
1893	98	288
1894	122	410
1895	0	410
1896	0	410
1897	134	544
1898	30	574
1899	287	861
1900	209	1,070
1901	243	1,313
1902	2,645	3,958
1903	645	4,603
1904	948	5,551
1905	494	6,045
1906	201	6,246
1907	246	6,492
1908	256	6,748
1909	800	7,548
1910	769	8,317
1911	983	9,300
1912	322	9,622
1913	75	9,697
1914	39	9,736
1915	768	10,504
1916	145	10,649
1917	0	10,649
1918	534	11,183
1919	14	11,197
1920	84	11,281
1921	461	11,742
1922	305	12,047
1923	244	12,291
1924	398	12,689
1925	278	12,967

Table 1.5 (Continued)

Year	Annual increment	Total to date
1926	417	13,384
1927	698	14,082
1928	455	14,537
1929	201	14,738
1930	226	14,964
1931	120	15,084
1932	424	15,508
1933	296	15,804
1934	1,255	17,059
1935	1,834	18,893
1936	1,849	20,742
1937	973	21,715
1938		
1939		
1940		
1941		
1942		
1943		
1944		
1945		
1946		
1947		
1948		
1949		21,800

Sources: Figures for 1876–1937 are calculated from Ralph W. Huenemann, *The Dragon and the Iron Horse: The Economics of Railroads in China 1876–1937*, Cambridge, MA: Council on East Asian Studies, Harvard University, 1984, pp. 76–7, Table 3. The 1949 figure is from *TJNJ 1985*, p. 385.

was actually built – by a Briton, Gabriel James Morrison – connecting Shanghai with Wusung.[48] However, the authorities tore it up the following year and shipped it to Taiwan, never to be reassembled. The second line was begun in 1878 by an English engineer, Claude Kinder, from the Kaiping mines to Xugezhuang (Hebei), which is situated on a canal. It was later extended to Tianjin and then Peking, Mukden, now called Shenyang (Liaoning), and Hankou, now part of Wuhan (Hubei) becoming a major trunk line.[49]

France, in 1885, was the first to obtain a railway concession which bordered Indo-China. Russia was given rights the following year in Manchuria, which permitted the construction of the Chinese Eastern Railway and the South Manchurian Railway. The Germans, British, and Japanese were also granted railway concessions for both trunk and feeder lines and 9,300 km had been built by the end of the Qing Dynasty in 1911 (see Table 1.5). Though the differences in gauge were problematic, taken together these lines became a vital network and coal was a main cargo.

Only another 11,442 km were laid before the end of the Republican period, 40 per cent of which were in Manchuria. The First World War, the civil wars, and the Japanese invasion of Manchuria, and the Second World War all severely affected railway construction, the military and/or warlords expropriating all the rolling stock. Whereas India in 1936 had about 72,000 km of railway trackage and 33.5 billion ton-km of freight, China had only 20,742 and 14.6 billion, respectively.[50] The Nationalists constructed several railways in the southwest during the 1940s. By 1949 the per capita national trackage was still very low compared to that in other countries.

In the 1930s coal shipments occupied about half of the country's railway capacity and on average accounted for 17 per cent of the railway revenues. On some lines they accounted for much more of the revenues, for example, about 31 per cent on the Kiaochow-Jinan (Shandong) line, 33 per cent on the Peking-Mukden line, 41 per cent on the Chengting-Taiyuan (Shanxi) line, and 68 per cent on the Taokow-Chinghua line.[51] The trackage and rolling stock were not only inadequate relative to production rates, but there was a lack of coordination between the mining and railway sectors. Railway capacity often determined coal production.[52] Coal was also shipped by water, southwards along the coast, and on the lower Changjiang and Wei River, and on various small rivers throughout the southwest. There were some 40,000 miles of canals at that time including the Grand Canal.[53] Total shipments on the Changjiang and along the coast were about one-quarter those on the railways and were for the most part overseen by the foreign-controlled mines.[54]

Consumption

Total coal consumption figures for 1912 to 1942 are given in Table 1.2. The highest figure here is 59.3 million tons in 1942. For perspective, total coal consumption in the UK in 1925 was 176.0 million tons.[55] Before the introduction of railways in China, coal use was necessarily restricted to a small radius around the mines. The demand for coal to fuel ships rose sharply with the opening of the Treaty Ports, beginning in 1840, which hastened both mining and railway development. Total potential demand was about 300,000 tons per annum in the early 1870s and rose to some 2 million tons in the early 1900s.[56] The demand for bunker coal slowed later due to a downturn in world trade, more efficient use of coal by the steamships, and the beginning of their conversion to the use of fuel oil. About 7 per cent of the coal from the large mines was sold to shipping companies in the mid-1930s.[57]

When fuel oil was replacing bunker coal, the new railways became a major consumer of coal and some railway owners invested directly in mines to guarantee their supplies. Among these were the South Manchurian Railway, Chinese Eastern Railway, and the Jing-Han line, connecting Peking and Hankou (Hubei). Several railway companies actually built their own mines.

Table 1.6 gives a breakdown of coal consumption in China from 1915 to 1933. For contrast, the corresponding breakdown in 1929 in the UK was: gas works 10 per cent, electricity generation 6 per cent, railways 7 per cent, coastal shipping

Table 1.6 Estimates of coal consumption breakdown, 1915, 1923, and 1933 (% sectoral shares)

	1915	*1923*	*1933*
Rural use	43.2	33.3	29.5
Urban commercial and household use	13.2	12.1	13.2
Industrial	14.3	15.7	21.5
of which:			
Electric Utilities	1.2	3.9	6.0
Iron and Steel	5.6	4.0	4.8
Textiles	4.5	4.1	4.3
Railways	6.6	8.8	9.0
Bunker	8.6	9.1	6.5
Exports	7.7	12.9	13.0
Used at mines	6.3	8.1	7.5

Source: Tim Wright, 'Growth of the Modern Chinese Coal Industry: An Analysis of Supply and Demand, 1896–1936', *Modern China*, 7, 3 (July 1981): 333. Copyright © 1981 Sage Publications, Inc. Reprinted by permission of Sage Publications, Inc.

Wright's note:
These figures are only intended to give an approximate indication of the magnitudes involved. The first line in particular is subject to a wide margin of error. That category could alternately be estimated as residual, thus eliminating the discrepancy between the use and supply of coal. This discrepancy could also be the result of changes in stocks, or of incorrect estimation of small mine output by the Geological Survey of China.

1 per cent, iron works (using blast furnaces) 8 per cent, other iron works and steel works 5 per cent, collieries 8 per cent, and general manufactures and all other purposes including domestic use 55 per cent.[58] In China, the transport sector, comprising both the railways and steamships, was the largest consumer of coal up to the 1930s. Purchases of coal by railway companies sustained many mines, and the main business of some railways was shipping coal. From the mid-1910s the proportion of output from the large mines used by the railways was steady at about 10 per cent.[59] The advent of cheaper transportation of coal led to an immediate rise in industrial activity in the urban areas.

The urban commercial and household sector accounted for 12 to 13 per cent of total coal consumption between 1915 and 1933. Per capita consumption was very low, only 0.07 tons in 1936 versus 3.9 tons in Great Britain, 3.5 tons in the US, 0.8 in the USSR, 0.6 in Japan, 0.2 in Spain, and 0.1 in India.[60] Coal was used in the large cities only. Residents in towns and medium-sized cities relied on straw and firewood.[61]

The first industry to require large quantities of coal in the mid- to late-1800s was the manufacture of armaments. Later, coal was used to fuel and provide heat for light industries such as cotton, silk, salt, sugar, tobacco, alcoholic beverages, paper, ceramics, glass, and oil-pressing. Most of these industries were located in

China proper. The demand from heavy industries such as lime, brick-making, sulphur, antimony, cement, and the metallurgical industries was much higher and predominantly from Manchuria.

Use of coal to generate electricity quickly increased from about 1910 onwards. At the time of the establishment of the PRC, about 10 per cent of total coal production was devoted to power generation.[62] In all there were 1.85 gigawatts (GW) of installed electric capacity, of which 1.69 GW (91.4 per cent) was thermal and 0.16 GW (8.6 per cent) was hydro. Total electricity generation amounted to 4.3 billion kilowatt hours (kWh), of which 83.7 per cent came from thermal plants and 16.3 per cent came from hydro. Total electricity production peaked in 1943 at 59.5 billion kWh. For comparison total installed capacity in India in 1947 was about 1.36 GW and generation was 4.1 billion kWh.[63]

The first thermal plant was built in 1882 by an American company and the first hydro plant in 1912. Up until the war with Japan most of the generation and transmission equipment that was installed had been brought from the UK, Germany, France, the US, and Japan. Thermal plants produced almost all of the electricity until the Japanese began to build a few large hydro plants in the northeast. China proper had 460 thermal plants with a generating capacity of 542,000 kilowatts (kW) and an energy output of 1,694 million kWh in 1934. During the Second World War these were either taken over by the Japanese or destroyed, while in Manchuria, at the end of the war, many plants which had been built by the Japanese were stripped of their equipment by the Soviet armies.

The efficiency of the thermal plants was not high in 1949. By international standards the generators were very small and little of the equipment was operating at full capacity. The average efficiency for the country was 18 per cent, and the utilization rate was also low – over a third of all the power produced was consumed by the plants themselves.[64] In 1949 there were a total of 6,475 km of transmission lines over 35 kilovolt (kV), of which 70.1 per cent were 35–66 kV, 5.3 per cent were 110 kV, 12.8 per cent were 154 kV, and 11.8 per cent were 220 kV.[65] Losses on the lines were very high at 25 per cent.[66]

Electricity consumption per capita was a mere 7 kWh in 1947, compared to 291 in the Soviet Union, 2,130 in the US, 888 in Britain, 642 in France, and 418 in Japan.[67] Many industries operated their own power plants. The percentage split between civilian and industrial use in 1949 expressed as a ratio was 35 : 65. Of the latter, the railways accounted for 2.7 per cent, metal processing 3.5 per cent, and chemical processing 5.4 per cent.[68]

The pre-1949 legacy

On the eve of the Communist takeover in late 1949 the coal industry, as virtually all industries and the railways, was semi-paralysed from bombings and lootings. Though much reconstruction and re-equipping would be required, a relatively rapid restoration of what had been a profitable coal industry would be possible however.

Six of the country's largest mines in the 1990s had been built in the early 1900s and at least a dozen other large mines date from the years of foreign development.

As a percentage of total industrial net value-added, the coal industry had accounted for one-quarter to one-third between 1912 and 1945. China ranked ninth in the world as a producer of coal in 1949, and the annual growth rate before 1937 had compared well with that of other countries.

Abundant cheap labour had kept mechanization at a low level, the work force was largely unskilled, and the quality of most of the coal was poor due to a lack of washing and sorting. As in most countries the location of coal deposits very much affected the development of the railways. Several railway lines built before 1911 became main transport arteries.

The most serious problem of the industry on the eve of the Communist takeover was the lack of a national plan. The disparate foreign interests had wanted quick and large returns on their investments and took a short-term view of development, thereby contributing to regional industrial imbalance.

2 Mao's adaptation of Soviet central planning

Effects on the industry

Soviet central planning applied to the Chinese coal industry

The People's Republic of China (PRC) was proclaimed on 1 October 1949. The Communist government inherited an economy in which the industrial sector had been decimated by wars. Inflation was rampant, there was a large budgetary deficit, transportation and communication channels had been disrupted for years, and there was high unemployment – the unemployment rate for males in non-agricultural areas was possibly as high as 32 per cent.[1] The largely rural population – nearly 90 per cent – was illiterate and unskilled.

The government's key priority was immediate restoration of the economy, to be followed by rapid growth. Bitterness over the country's history of exploitation by foreign capitalists spurred on achievement of the goal of making China a world power and capable of militarily defending itself. A United Nations embargo on strategic goods had been placed on China in 1950 and, as a result, there was little option but to seek assistance from the Soviet Union.[2] The Soviets were nearby and willing to offer technical expertise, equipment, and even whole industrial plants. The Chinese adopted, almost in total, the Soviet system of central or command planning. It was believed that given the country's stage of development and the problems left by the Nationalist Government and the ravages of the years of war, the economy could best be brought under control and regional disparities diminished by unification under a highly centralized form of government. It seemed that Soviet central planning was the most appropriate, if not the only available method of simultaneously guaranteeing the population a basic level of subsistence and achieving a high rate of capital formation. The material balances system provided the framework of Soviet central planning.

The material balances system

In the Soviet economic planning system production was organized hierarchically, with targets set by the central authorities. The total plan was in the form of a matrix in which the supply of and demand for raw materials, intermediate goods, and finished goods all had to balance exactly. Included in the matrix were goods flowing through the international trade sector and those held in inventories.

Enterprises did not themselves arrange for their supplies of inputs. Planners calculated how for every state enterprise these were derived, and how and when they were to be distributed. If either the plan or its execution were faulty, cumulative dislocation occurred. For example, if one coal producer failed to reach its output target, or if the coal was delivered late or was of the wrong type, this would in turn affect the designated users in their attempts to attain their output targets.

In the final quarter of each year, the State Planning Commission (SPC) prepared preliminary balances for all commodities and capital goods currently under state control. These goods were referred to as being 'under unified distribution' or simply 'under plan', and plans were based on a combination of recent production figures and forecasts of potential output based on capital productive capacity and labour inputs. Each ministry was given proposed output targets which were communicated to all the various sub-units. These were expected to calculate the quantities of all the inputs they required to fulfil their production targets. This information was then passed back up through the ministerial hierarchy to the SPC, which considered all of these reports and constructed a balance sheet for each of the commodities under plan.

Considerable adjustments had to be made to the output and consumption targets for each commodity before the balances could be closed, that is, before the total allocations of any commodity precisely equalled the total supply of that particular commodity. The result was a matrix which had demand for each commodity along one axis and the supply on the other. When all of the individual commodity balances were completed, they served as the basis for the annual and five-year plans which were presented to the State Council for approval. Compared to the Soviet Union, China had relatively few commodities under unified distribution. In the 1970s there were some 60,000 such commodities in the Soviet Union, whereas in China there were only 28 items under state distribution in 1952, 96 in 1953, 235 in 1956, 417 in 1958, 432 in 1960, 579 in 1965, 217 in 1971, 226 in 1972, and 210 in 1978.[3]

Balance tables were also drawn up to schedule the mode and route of transportation for every commodity under plan. The transportation volumes were determined only for the bulk commodities under plan, that is, coal, charcoal, mineral ores, iron and steel, timber, etc. The thousands of other commodities were not included in the national and regional balances tables. Of all the railway cargoes, the shipment of coal took precedence.

Production

The ministry responsible for the coal industry was variously designated as the Ministry of Fuel Industry (1949–55), Ministry of Coal Industry (1955–70), Ministry of Fuel and Chemical Industry (1970–75), and again as Ministry of Coal Industry from 1975–88 when there was a major ministerial restructuring (see Chapter 5).

China's coal mines were divided into three main categories, the Central Mining Administrations (CMAs), the local state (LS) mines, and the local non-state

(LNS) mines.[4] The CMAs were those which were controlled by the central government and most of their production was allocated under plan. The LS mines were operated either by provincial, prefectural, or county governments, while LNS mines were operated by townships, communes, collectives, or the People's Liberation Army. The output from the LS mines was also generally under unified distribution, while that from the LNS mines was not.

Figure 2.1 depicts the planning hierarchy of the Chinese coal industry and Figures 2.2 and 2.3 show the planning structures of CMAs and LS mines, respectively. In the coal industry long-term plans were set every five years in conjunction with the five-year national development plans and provided only general policy statements and, usually, output target ranges rather than specific targets. It was the highly detailed yearly plans which guided each department in the ministry responsible for coal towards the ultimate fulfilment of the long-term targets. The yearly plan had two parts, production and capital construction. The first of these had four major sections: output, personnel, material supply, and financing.

The production component was the basis upon which all of the other projections were formulated. Depending on the workable coal reserves available, and the amount and type of prepared capacity, the output section specified the rate and amount of production for each type of coal for the year at all mines under the Ministry's control. It was further subdivided into quarterly, monthly, ten-day, day,

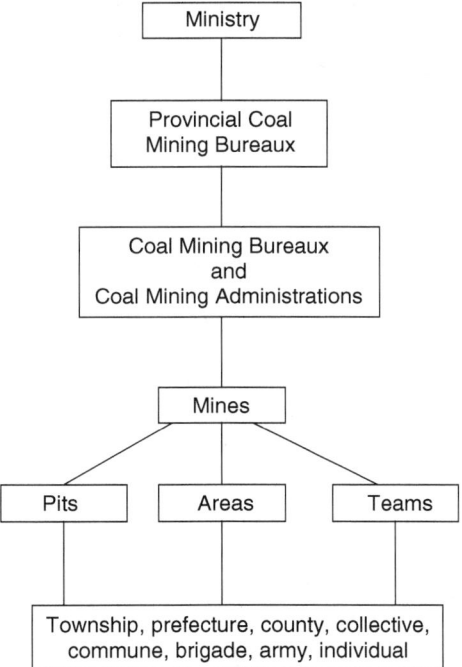

Figure 2.1 Planning hierarchy of the Chinese coal industry

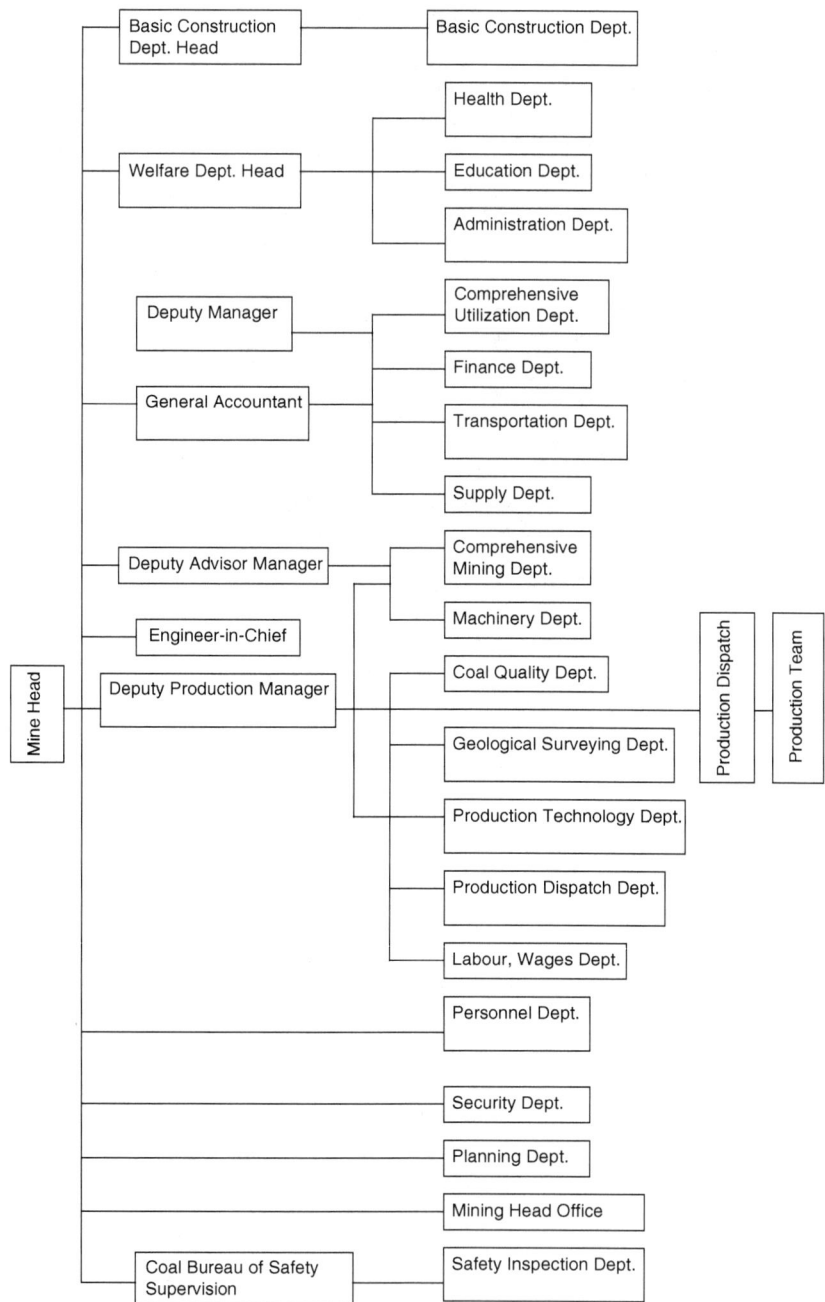

Figure 2.2 Planning structure of the CMAs

Source: Wang Maolin (ed.), *Meitan gongye qiye guanli* (Coal Industry Enterprise Management), Taiyuan: Shanxi renmin chubanshe, 1982, p. 54.

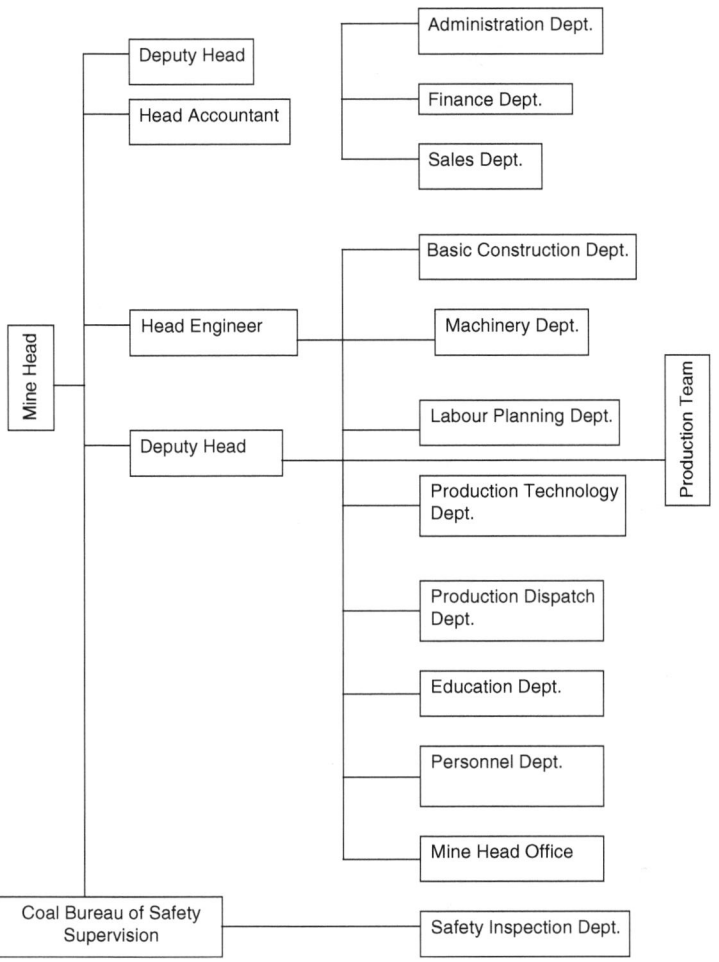

Figure 2.3 Planning structure of the LS mines

Source: Wang Maolin (ed.), *Meitan gongye qiye guanli* (Coal Industry Enterprise Management), Taiyuan: Shanxi renmin chubanshe, 1982, p. 55.

and even shift plans, and also stipulated what machinery was to be used at each site, from where it was to be obtained, and how it was to be maintained. Figure 2.4 is a sample monthly production plan matrix. On the left margin are output, quality, and production norms, and along the top are output targets.

The personnel section gave details of the number of employees hired by the Ministry, their recruitment, training before and during employment, remuneration, and pensions. There were six categories of personnel: workers, apprentices, engineers and technicians, administrative personnel, service personnel, and others. The first four categories were further broken down into drilling and tunnelling

	Total output	Coal face	Tunneling	Opencast	Preparation	Transport (mine)	Washed coal	Machine maintenance	Roadway maintenance
Output month (tons) day (tons)									
Quality: ash (%) silica (%)									
Drivage (metres)									
Transport (tons/km)									
Mechanical repairs platforms: % in good repair									
Roadway maintenance (metres)									
Labour allocation number: labour/day labour/month average productivity									
Pit timber consumption (cubic metres/ton)									
Explosives consumption (kg/ton)									
Detonators consumption (kg/ton)									
Timber substitutes consumption/ton									
Production costs (yuan/ton)									

Figure 2.4 Monthly production plan matrix

Source: Wang Maolin (ed.), *Meitan gongye qiye guanli* (Coal Industry Enterprise Management), Taiyuan: Shanxi renmin chubanshe, 1982, p. 110.

workers, face workers, surface workers, machine repair men, geologists, survey-ors, and designers. For each type of worker there were eight ranks based on expe-rience, responsibility, and technical expertise. The men worked in brigades. For some tasks the men were paid individually, while for others, the brigade was paid on the basis of the work it did collectively. Theoretically, extra wages were given for above-quota production, experience, and the less desirable work, for example, underground work, construction work, and night shifts. Productivity norms, the proportion of individual and collective time-rate and piece-rate wages, bonuses, allowance structures, and welfare benefits were outlined.

The material balances system was used to determine the allocation of all the inputs required by the Ministry. After it had been informed of its quotas of materials – steel, iron, timber, etc. – for the year, it would in turn distribute these to the sub-units of the bureaux according to their planned consumption. The material supply section also gave consumption norms for each item and norms for maintaining reserves of materials.

The remainder of the production component was devoted to training, research into new production and/or management techniques, and to projects handled independently by the mines, such as residential, commercial or cultural buildings, schools, medical facilities, etc.

Initially there was no government department or agency responsible for the LNS mines. When the Communist government came to power most of the small pits were closed or combined to make larger mines. All remaining ones were put under Party control. A special office was established to coordinate coal produc-tion for the Great Leap Forward (GLF), but this was disbanded soon after. In the years that followed, when coal supplies were especially tight, encouragement in the form of grants was given for the development of small mines. However, throughout most of the 1960s and 1970s there was no governing body which monitored the opening, operation, and output of the non-state mines.

Financing

Capital construction investment

The capital construction investment component was divided into productive and non-productive. Productive investment included funds for the prospecting, sur-veying, design, and construction of new mine capacity and expansion of existing capacity; construction of coal treatment plants, railway lines from the mines to stations, and machinery plants; funds for research and the operation of training colleges. Non-productive investment included the construction of buildings for the miners and their families, such as housing, clinics and hospitals, schools, sport, and cultural amenities.

A year-end target for total capacity specified where and how it should be achieved, by new mine construction or renovation or expansion of existing mines. Prospecting, surveying, and designing teams were assigned for each project. The amount of new capacity created each year was directly related to the investment funds allocated to the Ministry from the state budget, that is, 'above-norm' projects.[5] This component also dealt with investment in coal washing plants, machinery plants, and railway branch lines.

The setting of prices

In centrally planned economies prices had three distinct functions.[6] The first was control and evaluation. Prices were used as accounting devices in the tabulation of the annual and five-year plans. The inputs and outputs of each enterprise, as well as targets for sales, profits, taxes, etc., were calculated by the central planners and expressed in value terms, and enterprise managers were evaluated on the basis of their fulfilment of these targets. Second, prices were used for allocation purposes and were integral to the process of drawing up national product, inter-sectoral accounts, and capital–output ratios. Value calculations were also used in deciding the relative merits of investment, modernization, or innovation projects. The third function was income distribution. Wages, transfer payments, and income taxes determined the distribution of income.

Prices could generally be classified as being 'cost-plus non-scarcity prices', meaning they were based on average production costs, and may have included a turnover tax or subsidy and a wholesale and/or retail trade profit margin.[7] Prices in Soviet style command economies were set with the assumption that supply would dictate demand. Very low prices, often lower than the production costs, were attached to basic consumer goods in order to ensure their affordability for all. Due to perpetual shortages and rationing, most industries were not 'price takers' but 'quantity takers', that is, hoarding when supplies were available. Thus, there was no clear and consistent relationship between prices and demand. It was believed that in the interests of economic stability, that is, to help prevent inflation, domestic prices should not be influenced by world market conditions; therefore the monetary 'value' attached to each commodity bore little relation to the international price.

When the Communist government took power in 1949, in the interests of stabilizing the economy, no major changes were made to the price of coal from the recently nationalized state mines. The government's key reason for maintaining the status quo was the fear of inflation.[8] During the war years the country had suffered hyper-inflation, and the new government did not want to take the chance of having prices go out of control again. As coal was a basic raw material for most industry, especially for heavy industry, which was the focus of the country's development programme at that time, the effects of a too rapid price increase could reverberate throughout the economy. Furthermore, a substantial increase could potentially cause many of the large coal-consuming enterprises to default.

Since the early 1950s the Chinese conformed to the Soviet planning doctrine of *shengchan ziliao bu shi chanpin*,[9] that is, materials needed for production were a constituent part of production and were therefore not to be sold with the aim of making profits.[10] According to Marxist theory, raw materials should be priced as low as possible because they themselves are not responsible for accumulation.

The 'producer' or 'supply' price of coal in each of the provinces was strictly a function of production costs plus the cost of taking it to the railway station and loading it onto the cars, but did not include the cost of shipping it to its final destination. The 'retail' prices that the consumers actually paid were therefore higher and varied greatly around the country because they included the transportation costs. The Soviet system for setting shipping prices was very rudimentary. There was essentially a flat ton–km rate and fixed terminal charges which reflected neither differences in handling costs nor the relative differences in terrain.

China's coal prices did not include any depletion or rent component whatsoever, whereas in most market economies the price for any non-renewable fuel includes a scarcity factor, that is, a depletion component. Though there was a price differential between washed and unwashed coal, there was no variation between different qualities of coal, that is, ash, moisture, sulphur content, etc., were not taken into consideration.[11]

What are referred to here as production costs are average annual operating costs per ton incurred by the state underground mines operating at peak levels, and exclude any investment costs sustained in the preliminary stages of mine development, for example, geological surveying and mine planning. In terms of Chinese accounting conventions, the largest component of total production costs in 1957 was wages at 36.8 per cent, followed by materials at 32.8, fuel at 0.7, depreciation at 6.1, electricity at 12.6, and other at 11.0.[12] For the sake of comparison, in the West labour costs in underground mines typically accounted for at least half of total operating costs.

Socialism with Chinese characteristics: Mao's frustration with central planning

Though he retained the basic tenets of socialism, Mao Zedong, Chairman of the Communist Party, shunned Soviet central planning as early as 1956. He sought to create a special brand of socialism, 'socialism with Chinese characteristics', that would, he thought, more closely meet China's particular needs. To this end he carried out experiments which he hoped would lead concurrently to economic growth and the levelling of class barriers. He did this by launching a series of campaigns.

Between 1958 and 1978 the entire Chinese economy, not least of all the coal industry and the railways, was dramatically affected, especially by the 'Great Leap Forward' and the 'Great Proletarian Cultural Revolution'. Industrial policies reflected the power struggle between Mao, who favoured small-scale decentralized development, and the 'pragmatist' leaders who favoured large-scale centralized development. Superimposed on this antagonism were strategic considerations, which, until the early 1970s had a major impact on the coal industry and railway construction.

1949–52: Restoration period

Prior to 1931 there had been considerable industrial development, mainly in the northeast, where large coal mines had been developed first by British, German, and Russian interests, and subsequently by the Japanese. However, much of this valuable infrastructure was destroyed in the wars that occurred between 1937 and 1949. During the restoration period, 1949–52, the major objectives were to gain control of coal supplies, reconstruct the existing mines, and concentrate all new coal mine construction in the northeast.[13]

Yearly plans were drawn up during this period. The *First National Economic Plan for Northeast China* was formulated in 1949, and subsequently control figures were set by the Central Financial Commission for 1950, 1951, and 1952.[14] However, these control figures were only guidelines given to the large coal bureaux; they were not industry-wide targets. By the end of 1952 the state had taken control of 88 per cent of the industry and production had been restored to pre-1949 levels.[15] A target of 36.8 million tons was set for 1950 and 61.34 million tons for 1953.[16]

The government immediately set about repairing the existing railway lines and bridges and commencing the construction of new lines in the northwest, south-central, and southwest regions. At that time there were few lines in the northwest and southwest and there were no lines at all in Qinghai, Ningxia, or Xinjiang. What did exist were not of uniform gauge and the equipment was antiquated. By the end of 1952, the government had centralized the management of the railways and claimed that 22,900 km were functional.

1953–57: the First Five Year Plan

The First Five Year Plan was formulated with the assistance of Soviet advisers. Hundreds of Soviet technicians came to China, and many Chinese were sent to the Soviet Union for training. The latest Soviet mining technology was introduced into China as well as Soviet industrial management practices. In 1949 the Chinese first thought that some 175 shafts would have either to be closed or abandoned by 1955, but in 1952 after consultation with the Soviet advisers, it was estimated that not more than 60 shafts would have to be closed before 1957.[17] As part of their loan package, the Soviets provided almost all the rolling stock needed for industry start-up.[18] Of the 156 projects for which the Soviet Union and the other Eastern Bloc countries were providing assistance during this time, 24 were coal projects and the focus was on large mines and preparation plants.[19] The Poles are also said to have delivered at least five plants, in addition to contracting to supply 'complete equipment for two coal mines, three coal dressing plants and several smaller projects'.[20]

Drafting of the First Five Year Plan began in the spring of 1951, but the Plan was not actually adopted until January 1955 at the Second Plenary Session of the National People's Congress (NPC). (Only the first and second five year plans were actually published, the former in great detail, the latter just in outline.) This highly detailed plan included some very specific guidelines for the development of the industry, with over half of the total investment allocated to the energy sector being for coal. The 1957 target set for coal production was 112,985 million tons, 78 per cent over the 1952 level.

Of the 1957 output target, 71.6 per cent was to come from state mines under central control, 20.0 per cent from state mines under local control, 4.2 per cent from joint state-private mines, 4.1 per cent from private mines, and 0.1 per cent from cooperatively operated mines. The economic utility of the LNS mines was recognized, but it was a goal (at that time) eventually to phase out all private mines either by closure or consolidation. Over the five-year period, construction was to have started on a total of 194 above-norm projects. Most of these were focused in the established northern and coastal areas. It was believed that 'In order to change the irrational distribution of industry [in general, and from both economic and defence points of view], we must build up new industrial bases, but the utilization, reconstruction and extension of existing industrial bases is a prerequisite for the establishment of new industrial bases'.[21] Of the 179 mining construction projects, 57 were old shafts to be restored or reconstructed, while 122 were to be newly built, and 131 were in the northeast, north, and east regions. The Fushun (Liaoning), Fuxin (Liaoning), Kailuan (Hebei), Datong (Shanxi), and Huainan (Anhui) administrations were designated to become enterprises of 'exceptional size' and to be fitted with 'new or even the most up-to-date equipment'.[22]

The Soviet planners regarded the iron and steel industry as the basis upon which all economic development depended and the targets set for them virtually determined the targets set for all the other industrial sectors. Necessarily, planning for the coal and iron and steel industries was carried out simultaneously. With strategic considerations in mind, the Soviets designated three plants – new ones in Wuhan (Hubei) and Baotou (Inner Mongolia), which are in the southcentre and northwest, respectively, and an upgraded Japanese-built plant, Anshan (Liaoning) in the northeast – to serve as the core of the industry. Other plants were built or upgraded at Benxi (Liaoning), Tianjin, Chongqing (Sichuan), Taiyuan (Shanxi), Maanshan (Anhui), and Shijingshan (west of Beijing).

Well before the First Five Year Plan was launched in 1955, it had become very apparent that coal production was lagging far behind demand. In August 1953 there was an urgent call from the Ministry of Fuel Industry and National Coal Mine Trade Union to increase production and intensify mine drivage.[23] However, the shortages were such that in the last half of 1953 and during 1954 all efforts were instead devoted to production, and mine development work was postponed until 1955.[24] Then in 1956, because Chairman Mao was pressing for a 'High Tide' of production, targets for the 1956 Plan were again revised upwards. Thus, the First Five Year Plan was barely launched when Mao began interfering with it. In January of that year the number of above-norm capital construction projects for the First Five Year Plan was increased from 695 to 745, and the number of projects to be completed from 455 to 477. In June these targets were again raised, to 800 and 500, respectively.

Obviously much more coal was required than planned for earlier. The government ordered all government offices, troops, and schools to reduce coal used for heating and cooking by 15–25 per cent and coal miners were asked to work overtime.[25] The Premier himself, Zhou Enlai, wrote several press articles at this time urging people to save coal.[26] Rationing was introduced in 1957 and explicit directives for the assigning of coal supplies were detailed:

> Ensure the minimum, necessary supply to cities lest industrial production and the livelihood of city people should be affected; basically maintain the 1956 supply level in respect to coal used by industries in general, handicrafts and service trades; reduce the coal used by (government) organs and organizations for heating purposes and the coal used for brick and tile kilns.
>
> Appropriately reduce the supply of coal to the countryside ... Economizing in the use of coal and coke should be made an important feature of the production increase and economy campaign. Coal used by industry and service trades must be used economically; coal used by organs, army units, organizations and schools should be placed under strict control ... An extensive propaganda drive should be launched to encourage the people to save coal. Moreover, exchange and popularization of coal-saving experiences should be organized from time to time.[27]

In an effort to correct the imbalances in the supply of raw materials, revenues and expenditures, and bank credits, the 1957 budget for total investment in construction was readjusted to 20 per cent less than in 1956.[28] Other targets were also reduced, but the coal production target remained unchanged.

The details given for railway construction during the First Five Year Plan were also very specific. The main task was to improve and reconstruct the existing railways. Another 3,284 km of new trunk lines were to be built and 800 km of branch lines.[29] Some 692 km were to be rebuilt, spur lines were to be extended by 1,100 km, fourteen stations were to be enlarged, a new double-decker bridge was to be started over the Changjiang at Wuhan, and a considerable amount of new rolling stock was to be acquired.[30]

1958-Spring–1960: the GLF overtakes the Second Five Year Plan

The proposed targets for the Second Five Year Plan, 1958–62, were adopted in September 1956. Among these was a target for 1962 calling for the number of provinces self-sufficient in coal to be raised from seven to seventeen.[31] The target for coal production in the same year was to be 190–210 million tons and for steel 10.5–12 million tons. By the time the Second Five Year Plan began in January 1958 the conservative and realistic plan for 1957, in tandem with a cautious budget, had begun to ameliorate the shortages and imbalances, but just at this time Mao announced his GLF. Mao called for the masses to become involved in a tremendous acceleration of growth in all sectors of the economy nationwide. He had always believed in the psychological efficacy of unstable, wavelike growth. The enormous surge he was calling for now was meant to raise confidence, and to give the people hope that China could break out of its poverty and catch up with the developed world.

Mao was frustrated by what he considered the slow pace of economic development in the preceding years. The Soviet advisers had been introducing relatively modern large-scale plants which inherently had long gestation periods. In the case of mining, some of the large centres they were helping to construct would take at least an entire decade before they reached full-scale production. Mao was not only displeased with the time factor, but also that it was only the urban areas and already developed parts of the country which were benefiting from this approach, while conditions elsewhere had scarcely improved at all.

Mao agreed with the Soviets that 'steel was the key link' *(yi gang wei gang)* as it was imperative for both national defence and industrial development. However, he believed that smaller plants and less modern means of production, which could be used throughout the country, would be more appropriate and would realize faster economic growth. He also believed that there were large numbers of unemployed in the countryside who, if mobilized, could become a powerful productive force, referring to this latent energy as the 'superstructure' of ideology.[32] Mao had a history of distrusting intellectuals and bureaucracy and the GLF was a deliberate affront to them. A key GLF principle was 'walking on two legs' – the simultaneous development of large and small enterprises, the coexistence of modern and traditional methods, and a balancing of heavy industry, light industry, and agricultural sectors. Another of his concerns was that China might become indefinitely reliant on the Soviet Union. Mao called on the country to embrace his vision of socialism, a 'socialism with Chinese characteristics', that is, maintaining central planning, but adapting it to the poor, rural conditions of China, by encouraging local initiative and self-sufficiency.

The structure of coal production in the GLF was completely transformed. In fact there was decentralization of authority in almost every sector of the economy. In January 1958 all enterprises and institutions under the General Administration of Capital Construction and the General Administration of Geological Exploration, and some designing institutes under the General Administration of Mine Designing, were put under the jurisdiction of provincial coal bureaux or CMAs concerned. All the CMAs, enterprises, and institutions were granted greater authority in planning, financing, materials and coal allocation, personnel management, and other matters.[33] In June 1958, apart from the central mines in the provinces of Hebei, Shanxi, Liaoning, Yunnan, and Guizhou, the Ningxia Hui Autonomous Region, and those under the Beijing Mining Administration, all mines attached to units under the direct jurisdiction of the Ministry of Coal Industry (MCI) were transferred to the jurisdiction of people's commissions in the provinces and autonomous regions.

In September 1958 the MCI established a special office to coordinate coal production for the GLF campaign,[34] and in October large- and medium-sized mines were directed to develop satellite small-scale pits.[35] By the mid-1950s all non-state mines had been put under local Party management. Influenced by Soviet central planning with its emphasis on large-scale operations, the government hitherto had offered the operators of these mines little financial or material support and output was consumed locally. However, this had begun to change dramatically in 1957. It had been recognized that output from the non-state mines was growing much faster than from the state mines and increasing output from this sector was now seen to be the best way to solve the immediate coal crisis.[36] Accordingly, explicit instructions were issued:

> Organize small collieries to produce coal and ensure fulfillment and over-fulfillment of the procurement targets for coal and construction supplies companies. Small collieries may be organized to produce coal as long as their operation does not undermine the state resources and as long as coal can be shipped out ... As production and transportation costs are high for coal pro-duced by small collieries, any losses incurred during procurement of coal may be reported to the Ministry of Commerce, which will make good the losses.[37]

In April 1957 the Government allocated 11.6 million yuan for the promotion of forty 'native-style' pits in nineteen provinces.[38] Then at the enlarged conference of the Political Bureau of the Party Central Committee in August 1958 one of the documents promulgated was *The Call on the Whole Party and the Whole People to Strive to Produce 10.7 Million Tons of Steel*.[39] It was proclaimed that if everyone participated in the 'backyard furnace campaign' and the 'native pits campaign', China could surpass Britain's steel output. The MCI had initially set a target of 150.72 million tons for 1958, about 15 per cent over the actual amount produced in 1957, and of which almost one-third was to come from non-state mines.[40] In May of 1958 it was estimated that production would reach 180 million tons, or 38 per cent over the 1957 level.[41] Later the target was raised again, to 220.8 million tons, or 69 per cent over the 1957 level, in order to meet the higher steel output target. The 10.7 million tons steel target was double the 1957 output and the 1959 target was also set very high at 27 to 30 million tons. The capital invest-ment allocated to the coal industry in 1958 was twice what it was in 1957 and in 1959 there was yet another 33 per cent increase.

Slogans were struck to characterize the strategies to be employed. The 'Four Halves' method called on peasants, traders, students, and villagers to work as half-time industrial workers. The 'Four Simultaneous' method called for simulta-neous building and producing, prospecting and mining, mining and smelting, producing and studying. The 'Four Quicks' method urged managers and workers to start quick, build quick, produce quick, sell quick.[42]

By the end of 1958 some 20 million people were working in 110,000 small pits scattered all around the country, not just in the major mining areas. The proportion of total output from the CMAs fell from over 72 per cent in 1957 to 58 per cent in 1958, while that from the LNS mines rose from 5 per cent to 19 per cent. The people used any implements available, including kitchen utensils, to dig the coal, and whereas the Communist government had hitherto forbidden women to work in the mines, during this period this policy was abandoned. Inexperience and inadequate preparation and equipment resulted in the death or injury of many people. The smelting furnaces were small and inefficient and required proportionately more fuel than the large state plants to produce the same heat because the coal from the pits was of such poor quality, much of it being mixed with sand, rocks, and clay. In turn the iron and steel produced were largely unusable.

The phenomenal pace of growth of the LNS mines and the production of inferior quality coal and other basic materials could not continue. The strain on capital and human resources was too great and the mining equipment and structure of the nation's coal seams were actually being damaged. Nor could the diversion of labour from the agricultural sector continue any longer. The economy was beginning to falter. However, the fractious political situation at the time precluded the development of a comprehensive plan, and the steel and coal targets were adjusted and readjusted several times in the coming months.

In November 1958 the target for steel in 1959 was lowered from 27–30 to 18–20 million tons, for pig iron from 40 million tons to 29, for machine tools from 300,000 to 130,000 pieces, and for total investment in capital construction from 50 billon yuan to 36 billion, but the coal target was actually raised from 370 to 380 million tons.[43] At a conference in March 1959, the Minister affirmed that the northeast, north, and east would remain the main development areas.[44] Then in April, while the coal target was unchanged at 380 million tons, the steel target was readjusted downwards from 20 to 18 million tons, total investment in capital construction was reduced from 36 billion yuan to 26–28 billion, and the number of large- and medium-sized projects was reduced from 1,500 to 1,000.[45] Soon after, the steel target was further reduced to 13 million tons, the coal target to 340 million tons, and the targets for copper, aluminium, cement, and timber were also reduced.[46] Finally, in August the steel target was set at 12 million tons, the coal target at 335 million tons, the total capital construction investment target at 24.8 billion yuan, and the number of large- and medium-sized construction projects at 788.[47] During this period, as a result of wildly exaggerated reports from low level managers, and the fact that so many pits were involved, the planners really did not know just how much coal was being produced.

Another great surge was planned for 1960 and the targets for a variety of commodities were set 10–57 per cent above 1959 production levels. The steel target for that year was first put at 18.4 million tons, and then in March raised to 20.4 million tons.[48] Despite the fact that the 1959 coal target had not been met, for 1960 it was set unrealistically high at 462 million tons, then subsequently lowered to 425 million tons, or 33 per cent and 22 per cent, respectively above the 1959 output.[49] The GLF spurt had now passed and coal production was in fact falling, the main cause of the decline being the failure to build new shafts and tunnels. During the GLF the output–capacity ratio had become skewed and mine development work had fallen seriously behind. Despite awareness of this, the Ministry condoned the diversion of capital and labour from construction to production. The miners were now exhausted and cynical towards the government.

In December 1959, in order to regain control of production, the Central Committee and State Council revoked some of the powers it had given to lower jurisdictions the year before. In Heilongjiang, Hunan, Jiangxi, Anhui, Henan,

Shandong, Jilin, and Inner Mongolia, a dual system of jurisdiction was established in which the central government reclaimed the dominant role.[50]

In 1958 and 1959 the increases in the allocations of capital investment to the railways were very high and there were large advances made in new trackage and double-tracking, especially in the northwest and southwest regions. The uncharacteristically high increases in coal production in the GLF years were more than the railways could handle. Some 20 million tons of coal had accumulated at the mines for lack of railway capacity.[51]

Spring 1960–65: departure of the Soviet personnel, a period of readjustment, and the beginning of the Third Front Development Programme

By the spring of 1960 the Soviet advisers had become frustrated and angry with what they considered the reckless policies of the GLF. This, in combination with various political problems, precipitated Krushchev to order the Soviet technical personnel to return home and bring all their blueprints with them.[52] They had been in China for almost a decade. Thus, China abandoned the 'leaning to one side' strategy, that is, participating in a united front with the Soviet Union and other socialist countries against the US and Japan. During the decade now beginning virtually no assistance from either socialist or capitalist countries would be sought or received. Mao pursued a policy of self-reliance, convinced that if Chinese scientists interacted with those from abroad for any length of time, China would become irreversibly dependent on the more developed world, and would never reach the status of a world power.

The termination of the Soviet assistance, especially at this juncture, was a major setback for the coal industry, which was experiencing an assortment of very serious problems. In January 1961, the CPC Central Committee had little choice but to approve a proposal that the daily production quota for the CMAs be reduced from 650,000 to 600,000 tons for the first quarter of the year.[53] Then in May the quota was further reduced to 530,000 tons.[54] In early July, however, only 460,000 tons were actually being mined and at this point the MCI ordered that the daily quota be maintained at 520,000 tons.[55] This was impossible, and at the National Coal Cadres Meeting in September the daily quota for the September–December period was set at 440,000 tons.[56] The *Coal Report to the Central Authorities* (August 1961) stated that at least 122 of the 278 state mines and 788 pits were behind in preparation work.[57] In September the Party Central Committee issued the *Directive on Current Industrial Problems*, which stipulated among other things, that due to the low production, attention should be paid to *both* the quantity and quality of coal.[58] In 1961 and 1962 the government, now dominated by the 'pragmatists', allocated over one-fifth of the total available industrial investment funding to the coal industry and coal investment as a proportion of investment in energy reached a high of almost 57 per cent in 1961. But the *actual* amounts allocated steadily dropped between 1960 and 1963.

Clearly, the GLF had caused planning to collapse. Throughout 1961 and 1962 desperate attempts were made to raise the morale of the miners. In January the miners at the state mines were given increased food rations,[59] and in October they were granted extra supplies of wine, liquor, meat, vegetable oil, cigarettes, sugar, and other items.[60] In April 1962 Mao's rival, Premier Zhou Enlai, called for still further inducements by implementing a voluntary working system in the mines, establishing a 'Miners' Day', providing uniforms, and giving subsidies according to seniority.[61] Due to the lack of capacity, production was still very low in 1962 and in June it was decided to reduce the daily quotas yet again, from 412,200 tons in the third quarter and 420,000 tons in the fourth, to 390,000 and 400,000 tons, respectively.[62]

In January 1962 the government had called for a period of 'readjustment, consolidation, filling out and raising standards' to correct the problems caused by the GLF, and by severe floods and droughts which had occurred in the early 1960s. Due to the failure of the GLF, with its attendant devastation, Mao was forced to step back and allow the 'pragmatists' more decision-making power. In the coal industry the number of new construction projects was reduced, attempts were made to balance the output–capacity ratio, repairs were made to the equipment and roadways, and there was a restructuring of the work organizations and restoring of rules and regulations.[63] At the enlarged Standing Committee of the Political Bureau in February, the original economywide targets for 1962 were scaled down. The coal target was reduced from 251 million tons to 239, and the steel target from 7.5 million tons to 6 million.[64] Also in February the MCI established a Long-Term Planning Committee.[65]

At the end of 1964, in his report to the Third National People's Congress, Zhou Enlai called for 'Four Modernizations' – modernization before the end of the century of agriculture, industry, national defence, and science and technology – to bring China's national economy to the forefront in the world. To accomplish this he advocated the use of modern, foreign technology, thus beginning a reversal of Mao's self-reliance policy. Several efforts were made to re-centralize authority and in August 1963 it was formally recommended that 'production be rationally centralized and new technology be introduced'.[66] In January and February of 1965 various mines were shifted to other jurisdictions or re-named, and the MCI reorganized its ten departments, one bureau, and one office into four offices: Production, Capital Construction, Management, and Planning, and merged the Policy Research Office into the General Office.[67]

By this time almost all of the small pits opened during the GLF had closed and attention was focused on the large state mines which the Soviet advisers had helped to restore or construct during the First Five Year Plan. The output structure of production, that is the proportions of output coming from the various mine types, had almost completely returned to what it had been before the GLF. Over 70 per cent of the total production was now coming from the CMAs, and only 30 per cent from the local mines. Instead of directing the peasants to operate small mines and to use the coal for local industrial and household use, more effort was directed to developing coal-fired thermal plants. No new mines were started.

New railway construction had been limited by the failure of the GLF and the departure of the Soviet engineers. The amount of capital allocated to the railways as a percentage of total investment in transportation and communications had reached an all-time high of almost 72 per cent in 1961, but between 1962 and 1964, as a proportion of total economywide investment, the amount allocated to the railways dropped to only about half of what it had been in the First Five Year Plan period. In 1965, however, with the recovery of the economy in general and the launching of the Third Front Development Programme late the previous year, the amount of investment allocated to the railways rose sharply. Most of this new construction took place in the northwest and southcentral regions, and by the end of 1965, apart from Tibet, all the provinces and autonomous regions were linked by rail.

The purpose of the Third Front Development Programme, which began in late 1964 and ended in late 1971, was to establish an integrated, self-sufficient industrial base in the interior of the country, strategically located and less vulnerable to attack from land or air. At that time, the Chinese leadership was paranoid about potential attacks from both the US and Soviet Union. Atom and hydrogen bombs were tested by the Chinese in 1964 and 1969, respectively. The Guomindang was fully aware of the Mainland's fragile economic situation after the failure of the GLF, and backed by the US, was threatening to base nuclear missiles on Taiwan. Also, by that time, the Americans were fully involved in the Vietnam War. Relations with the Soviet Union since the abrupt break-off in 1960 were worsening. The Chinese Communist leaders believed that the Third Front region, encompassing the provinces of Sichuan, Yunnan, Guizhou, Gansu, Qinghai, and Ningxia, the section of Shaanxi south of the Qinling mountains, and the western mountainous areas of Henan, Hubei, and Hunan would be the most secure against penetration. The coastal region, or First Front, and the Second Front which lay between the First and Third Fronts, were deemed far too vulnerable.

Over the eight-year period, extremely large capital investments were made in remote areas within the Third Front region. There was large-scale transfer of industrial plants and research facilities from the coast and massive construction of new ones. Mao instructed that new railway lines be built in Sichuan, Guizhou, and Yunnan 'at maximum speed, even if this involved tearing up tracks elsewhere to obtain rails'.[68] One of the key projects was a huge iron and steel plant at Panzhihua in a very mountainous area on the border between Sichuan and Yunnan. This installation required new coal mines and new railway lines. Several mines were built in the area and the connecting Chongqing (Sichuan)-Guiyang (Guizhou) and the Kunming-Guiyang railway lines were completed in October 1965 and mid-1966, respectively. Smaller iron and steel plants were also built at Jiuquan (Gansu), Shuicheng (Guizhou), and Wuyang (Henan).

1966–70: the Third Five Year Plan and the beginning of the Cultural Revolution

The draft of the Third Five Year Plan (1966–70) was presented in September 1965, and the target for coal in 1970 was a realistic 280–290 million tons, 21–25 per cent or 48–58 million tons over 1965.[69] Initially, the main goal of this Plan was to continue restoring the economy, laying equal emphasis on agriculture, light industry, and heavy industry in order to provide the population with basic food, clothing, and other consumer goods. However, with the escalation of the Vietnam War there was a return to orienting the economy to heavy industry. Then in 1966, Mao launched the Cultural Revolution.

The Cultural Revolution was a sociopolitical campaign to put power back into the hands of the Marxists and the masses, thus taking it away from those he perceived to be 'counter-revolutionary revisionists' (his 'pragmatic' rivals) and capitalists, feudalists and exploiters of one sort or another. It lasted from 1966 to 76. The first stage, 1966–70, affected the economy more severely than the second and the social and economic consequences, like those of the GLF, were devastating. Thousands of intellectuals, writers, and artists were sent to the countryside to 'learn from the masses', or were tortured or killed. Publication of scientific and literary periodicals was halted, and all religious activities were prohibited. Families were separated, factories were closed, crops were not planted, and the discipline of economics, regarded as the root of the problem, was virtually banned.[70]

Coal industry officials today estimate that the Cultural Revolution set the coal industry back by at least twenty years. The worst years were 1967 and 1968. Production in 1967, compared to that of 1966, dropped 18 per cent and there were repeated calls to use the available supplies sparingly.[71] Average output at the mine faces decreased from 7,342 tons per month in 1966 to 5,800 tons in 1967.[72] The work of the Institute of Geology was seriously interrupted by factionalism within its membership and in January the Minister of Coal Industry, Zhang Linzhi, was killed.[73] Management and administration were thrown into confusion and physical planning norms for the use of materials were abandoned. The military took over several of the key bureaux and, because the political cadres had no industrial expertise, equipment was not maintained. In December the MCI issued strong warnings to individuals trying to procure coal themselves. In digging the coal by whatever means they could, they were destroying the potentially rich coal seams belonging to the state mines. Again, as in the GLF, many people were killed in accidents, and the output was of poor quality. Much of the younger work force was giving more time to political campaigns than to mining. Shortages forced many coal-consuming enterprises to operate well under capacity or halt production altogether.

In May 1968 the CPC Central Committee requested fifteen major mines to hold a conference on increasing production and meeting the basic needs of the economy.[74] It is apparent from the literature that the economists and scientists were aware of the enormous waste in energy consumption and the potential for reducing it.[75] In late 1969 the SPC called a conference to study ways of ensuring a stable output of steel and the MCI was requested to guarantee 4,700 tons of clean coal per day to be used by designated steel plants.[76]

The problems in the coal industry reflected the reality that the Cultural Revolution had upset the entire planning apparatus. The National Conference of Planning, which had been suspended for two years, was replaced by a forum of about thirty people from the military, veteran cadres, and representatives from the mass organizations. This group formulated a draft of *An Outline of the Plan of the 1969 National Economy* and the coal target suggested was 270–280 million tons.[77] In the 1970 plan, which was approved in September of that year by the Second Plenary Session of the Ninth Party Central Committee, the targets were those set in 1965 for the Third Five Year Plan, that is, for coal, 280–290 million tons. The main thrust of the Plan was the promotion of the 'five local industries' – iron and steel, machine-building, chemical fertilizer, coal mining, and cement. The policy for LNS mines was reversed and their development was widely encouraged to make up for the lost production in 1967–68. Attention shifted from concentrating development in the north and northeast to making all regions self-sufficient.[78] Mao was determined to reduce the concentration of industry in the north and eastern regions, or First and Second Fronts, and to shift more industrial strength to the Third Front.[79] Between 1966 and 1970, most of the investment allocated specifically to the coal industry went to the southwest region.[80]

The small pits, which had been built in the late 1950s and had been abandoned, were re-opened. By 1970 much of the centralization of authority reimposed after the GLF was now reversed. Most enterprises and prospecting teams were transferred from the jurisdiction of the Ministry back to local authorities, and other organizations were put under joint jurisdiction.[81] This sharing of authority, even in the largest state mines, proved highly inefficient because the division of responsibilities and accountability were never precisely defined.

As for the railways, in the early years of the Cultural Revolution factionalism resulted in the closure of some lines, while others were permitted only limited traffic. In travelling all over the country 'making revolution', the Red Guards occupied freight capacity and worsened the coal stockpiling problem. They prevented many workers from doing their jobs and there were reports of tracks actually being removed or blown up. The Taiyuan Railway Bureau – Taiyuan was probably the most important coal transfer point – was occupied and used as a bargaining chip. In January 1967 coal shipments on the railways were 3.8 million tons less than planned, and, due to the fighting, in September only 46 per cent of the planned number of cars were loaded per day.[82] There were only minimal increases in new construction and double-tracking between 1966 and 1968, though these were resumed immediately after the worst of the violence was over. Some major lines were built in the Third Front region, the Chengdu-Kunming line and the Jiazuo (Henan)-Zhicheng (Sichuan) line, plus three different routes to Vietnam. The peak that was reached in 1960 of total freight carried on the railways was finally surpassed in 1970, the same year the administration of the railways was decentralized. Instead of having the twenty railway administrations directly under the Ministry of Railways, they were given more autonomy and the Ministry as a whole was merged with the Ministry of Communications.

1971–75: the Fourth Five Year Plan

The Fourth Five Year Plan (1971–75) was discussed in September 1970 at the Second Plenary Session of the Ninth Party Central Committee.[83] A coal target of

400–430 million tons was set for 1975.[84] This modest target (total production in 1970 had been 354 million tons) with a wide range – it would have entailed an annual increment of 9.2 to 15.2 million tons – would seem to indicate that the planners were not wanting to force production. The target for 1971, set soon after, was 360–370 million tons.[85]

In September 1971 Lin Biao was killed and Zhou Enlai, a strong advocate of the methodical Soviet-type planning, was in control.[86] At that time, rapprochement with the US was in process and the military threat was greatly reduced. US President Nixon visited Beijing in February 1972, following China's acquisition of a permanent seat on the United Nations Security Council in October the previous year. This set in motion Japan's reappraisal of its relationship with China and by September that year the newly elected Japanese Prime Minister, Kakuei Tanaka, and Zhou Enlai had signed a joint communique normalizing relations between the two countries, marking the beginning of China's gradual re-establishment of diplomatic ties with the non-Communist world.

On the world energy scene a defining event which later became known as the 'first oil shock' occurred in October 1973 when all members of the Organization of Arab Petroleum Exporting Countries (OAPEC) joined Saudi Arabia in reducing oil production and placing embargoes on exports to pressure the US and its allies into forcing Israel to leave Arab land that had been seized in 1967 and 1973. Oil prices more than quadrupled in less than a year and a lack of fuel supplies led to inflation and recession throughout the capitalist world. Governments scrambled to establish rationing and conservation measures. China, however, relying almost entirely on its own energy supplies, and involved in virtually no trade outside the Communist Bloc, was not affected.

With the Chinese economy suffering from the effects of the Cultural Revolution and over-extension in the backward Third Front region, Zhou immediately set new priorities. The Third Front Programme was ended and attempts were made to complete the largest of the unfinished projects. He also began limited business, training, and cultural exchanges with Western countries. In 1974 some thirty fully mechanized coal mining systems were imported from England and Germany.

In the coal industry, Zhou focused attention on the need to increase mine capacity work which had fallen behind during both the GLF and the early years of the Cultural Revolution. This was the main topic of discussion at various conferences sponsored by the Ministry throughout the year.[87] It was decided to postpone several new projects until the following year and to cancel others. Prospecting projects already underway were to be completed as rapidly as possible. The target for 1973, set in January of that year, was only 390–396 million tons.[88] At the National Conference on Planning in July 1973 some of the industrial targets were adjusted downwards. For example, the steel target was changed from 35–40 to 30 million tons.[89] In September that year an investigation into the country's coal stocks revealed they were sufficient to last only 1.5 months.[90] In October the government again called for economization in coal consumption.[91]

It appeared as though the industry would now enjoy a period of stability when production was again disrupted by political events in September 1974. This time it was the 'Campaign Against Lin Biao and Confucius', which resulted in more factionalism and purges.[92] The 1974 Plan had been drafted in August 1973, but due to the political turmoil it was not possible to convene a National Planning Conference and the Plan was approved by the central authorities without a meeting.[93] The State Council acknowledged in April 1974 that coal shortages were preventing many industries from meeting their targets.[94] The crisis was of such proportions that in June 1975 an emergency meeting was held and the miners were ordered to strive for higher output, despite the hot summer weather, and to make up for the production shortfall experienced in the first half of the year.[95]

In the early 1970s several major extensions were made to the railway network in the Third Front region, including a line connecting Hunan and Guizhou and the Chongqing (Sichuan)-Wuhan (Hubei) line via Ankang (Shaanxi) and Xiangfan (Hubei), but the Campaign Against Lin Biao and Confucius was extremely detrimental to the railways.[96] Few trains were in service and many of the freight handlers in the yards refused to work. The result was stockpiling of coal at the mines and shortages throughout the country. Between January and October 1974 only 57.9 per cent of the freight trains out of the Zhengzhou Railway Bureau ran on schedule.[97] On 5 March 1975, the Party Central Committee issued the *Decision on Strengthening Railroad Work* in which a separate ministry for the railways was re-established and the autonomy given the twenty administrations in 1970 was rescinded. The *Decision* stressed the need for punctual departures and announced severe punishment for persons disrupting service. Several personnel in the Ministry were removed at that time.[98]

The year 1975 marked a turning point for the coal industry and the economy as a whole. The MCI was re-established in January, after being taken over by the Ministry of Fuel and Chemical Industries in November 1970. Xu Jinqiang was appointed Minister. Also, Zhou Enlai that month in his *Report on the Work of the Government*, reaffirmed his two goals given at the First Session of the Fourth National People's Congress 11 years earlier, in December 1964: to build up an independent, comprehensive industrial and national economic system before 1980, and to bring about the Four Modernizations.[99] This was the beginning of a fresh vision for China, an attempt to put the country on a new programme of development.

The coal target set in the National Economic Plan for 1975, and approved at the Party Central Committee in February that year was 430 million tons, 4.7 per cent over 1974, and a target of 480 million tons was set for 1976.[100] The 1975 target for steel was initially 26 million tons, but was soon revised downward to 24 million because problems in the supplies of raw materials continued. Several other targets were readjusted downwards but the 1976 coal target was not changed.[101]

The promotion of LS and LNS mines was accelerated during this period. By 1975 only 58 per cent of the total output came from the CMAs. In 1972 the state began to allot a special fund for technical transformation of LS and LNS mines[102] and in 1973 a conference was convened with the particular purpose of examining production and construction of local mines in nine southern provinces to relieve the pressure on the north–south railway lines.[103]

1976–78: the years immediately prior to the reforms

An earthquake which occurred soon after the beginning of the Fifth Five Year Plan (1976–80) at Tangshan (Hebei) was a serious setback for the industry. Whereas the national production growth rate in 1975 was 16.7 per cent, it was only 0.3 per cent in 1976. Not only was production halted but railway lines in the area had to be rebuilt. It was also about this time it was realized that the major oilfields, Daqing (Heilongjiang) and Shengli (Shandong), had reached their peak production. The coal output target for 1977 was set at only 490–500 million tons.[104]

Also momentous were the deaths of both Zhou Enlai and Mao Zedong in 1976. The country was again in turmoil until the Gang of Four was arrested and Hua Guofeng was finally installed as the new Chairman of the Central Committee of the Chinese Communist Party in 1978.[105] In 1977 the first signs of an outward-looking policy associated with Deng Xiaoping had become apparent.[106] During the 1950s the economy, especially the coal industry, had received considerable assistance from the Soviet Union, but after their personnel left in 1960 there was little collaboration in any field with other countries. Deng saw opening to the rest of the world, especially to the West, as crucial to the country's development and modernization. In the case of the coal industry, he believed that coal exports could potentially be a significant source of foreign exchange, and that in order to increase these it would be necessary to introduce foreign technology and equipment and focus on efficiency.[107] In 1978 alone, 100 sets of comprehensive mechanized mining equipment were purchased from abroad. He also recognized an opportunity for China in the oil crisis that had affected the West and which had caused the international price of coal to more than double from US$18.23 in 1972 to US$46.43 in 1975.[108]

Deng was also very eager to develop the LS and LNS mines, and to this end in 1977 he instructed that tax policies affecting them be relaxed.[109] As coal supplies continued to be inadequate, the MCI held meetings in January and May with the SPC, Beijing Municipal Committee, the Ministry of Communications, and the Ministry of Railways to discuss improving the flow of coal supplies,[110] and in July the State Council decided to raise the coal output target by 10 million tons and to transport an extra 7 million.[111] That month Xiao Han replaced Xu Jinqiang as Minister of the Coal Industry. The question of output-capacity figured prominently at various meetings during the year. In March there was a conference on exploiting the full potential of old mines, and the issue of preparing sufficient mining capacity was raised.[112]

In February 1978 Hua presented *The Outline for the Ten-Year Plan for the Development of the National Economy* and a *Report on the Work of the Government*. Two years of the ten-year period from 1976 to 1985 had, in fact, already passed. In the Report he announced that '...the state plans to build or complete 120 large-scale projects including ten iron and steel complexes, nine non-ferrous metal complexes, eight coal mines, ten oil and gas fields, thirty power stations, ...' and gave notice that 'All provinces, municipalities and autonomous regions must utilize local resources, strive to make a success of medium-scale and small coalfields, small power stations, small mines, small cement factories, small fertilizer factories…'.[113] Energy development was regarded as the key to the success of the Ten-Year Economic Plan as it was an essential input into every industrial sector.

A steel target of 60 million tons was set for 1985, up from 24 million tons in 1977. Steel was still regarded as 'the key link'. Negotiations had begun in 1977 with the Japanese to construct an ultramodern steel complex at Baoshan in Shanghai.[114] However, it was the targets specifically for coal, revealed as part of the Ten-Year Plan by Minister Xiao Han, that were especially ambitious. The stated goal was to double output in ten years, and double it again before the end of the century. It was also planned that every county in China would produce coal, that every area would be self-sufficient in 'about ten years' time'. He suggested five measures to attain these goals: improve management of the industry, fully use old mines, develop more small LNS mines, improve marketing of the coal, and raise the level of mechanization at the mines.[115]

Such high targets were set because the lack of energy supplies continued to affect growth in every sector of the economy, and it was realized that the Four Modernizations could never succeed without reliable and expanding sources.[116] Convening an emergency national conference by telephone in July 1978, Vice Premier Kang Shien urged 'all coal mines to take immediate action to reverse the passive situation of coal production'.[117]

Hua called for recentralization of authority in all ministries. In the coal industry it was decided that efforts to increase mechanization and training were best coordinated by the MCI itself, as opposed to the mine administrators. Several research institutes were put under the dual jurisdiction of the MCI and the provinces, with the former assuming the principal role.[118] Deng advocated that equipment for mechanized mining should be concentrated in a relatively small number of mines in order that it could be used to maximum efficiency and be well maintained.[119]

In August 1978 the Bureau of Local Coal Mines was established by the MCI, and in September Deng affirmed that while in the past the government had advocated 'Two Initiatives' – the central government on the one hand and the provinces and municipalities on the other – more authority should now be given to the 'grass-roots mines and plants'.[120] The government had by then firmly decided to make coal the basis of their long-term energy strategy and this policy was formally proposed in April 1978 at a National Working Conference on Practising Economy.[121] Though the Chinese had at that time really no choice but to use coal, they were not unaware of the potential of nuclear power in helping to solve the country's energy problems. Towards the end of that year negotiations began with the French company Framatone to build a nuclear plant.

It was quickly realized that all the targets set in the Ten-Year Plan (1976–85) were far beyond the capabilities of the economy. The Plan had called for enormous increases in output that were to be brought about with the help of large imports of foreign equipment and technology. But such imports were, in fact, limited because the country had only minimal reserves of foreign exchange. Hua and others evidently believed that the economic problems had been due solely to the political disruptions and that the economic situation would quickly recover with the fall of the Gang of Four.[122] He was either overly optimistic or ignorant of the complexity of the problems, probably a combination of the two. In his work report he stated:

> [The Gang of Four] sabotaged the national economy and disrupted socialist construction in every field ... As a result ... between 1974 and 1976 the nation lost about 100 billion yuan in total value of industrial output, 28 million tons of steel, and 40 billion yuan in state revenues, and the whole economy was on the brink of collapse.[123]

As the results of the first statistical surveys which did not come available until late 1978 and early 1979, Hua's advisers did not have any data upon which to base accurate plans.[124]

There was virtually no new railway construction in 1976 and due to the activities of the Gang of Four, service was greatly disrupted. The situation improved in 1977 and 1978 and whereas the amount of capital construction investment allocated to the railways actually fell sharply in 1976 and 1977, it was substantially increased in 1978. However, as a share of investment in the transportation and communications sector and of total economywide investment, the allocations were far less than what they had been during the First Five Year Plan period.

Mao's revolutionary zeal and manipulative propensity produced periods of economic catastrophe, compounding the deeply rooted systemic problems that began before the Cultural Revolution. By the mid-1970s both agricultural and industrial growth had begun to slow down. The goal of doubling coal output to 1.24 billion tons in ten years would have required an annual growth rate of over 7 per cent. While the average growth rate between 1953 and 1975 was higher at 11.4 per cent, a 7 per cent annual growth rate after 1978 would have been difficult to sustain. Partly due to the unrealistically ambitious goals of the Ten-Year Plan, a power struggle ensued and Deng Xiaoping emerged as paramount leader of the country in late 1978. In April 1979 it was formally decided at a Party Working Conference that the economy should undergo 'readjustment, restructuring, consolidating and improving' for a period of three years. Hua Guofeng's ambitious plans for the economy were abandoned. In the following years a major, if not the most important component of the economic plans dealt with increasing energy supplies.

3 Economic assessment of the industry on the eve of the reform and opening up programme

Production

Quantity, structure, quality, and geographical distribution

In 1978, two years after the death of Mao and on the eve of the reform and opening up programme, China's coal production at nearly 618 million tons ranked third in the world after the US and USSR, up from ninth in 1949. Between 1953, the first year of the First Five Year Plan, and 1978, production in China increased by almost 800 per cent, compared to 124 per cent in the USSR, and 187 per cent in India.

By 1953 most developed countries had begun to replace coal with oil, gas, and electricity, and in the 1970s the USSR had begun to use more gas than coal. Oil went from accounting for 14, 40, and 5 per cent of total energy consumption in Western Europe, US, and Japan, respectively in 1950, to 56, 44, and 69 per cent in 1970.[1] Coal production peaked in France at 60.04 million tons in 1958, in Japan at 55.79 in 1961, in the FRG at 253.15 in 1964, and in Ukraine at 218.16 in 1976.[2] There were long periods of coal over-supply in world markets. Though mining unions fought bitterly to keep their industry alive, the mandates required to continue the huge subsidization were economically and politically impossible, and mines were gradually being closed down beginning in the 1970s.[3]

That China was able to realize such growth, and with minimal capital and skilled labour, is especially remarkable given that almost all the coal came from underground mines. In the 1950s, with Soviet help, the Chinese largely abandoned the room and pillar method of underground mining in favour of the long-wall method (see Chapter 1, endnote no. 30). In 1949 only about 13 per cent of the total output was cut using the longwall method, but by 1957 almost 93 per cent of the total output was so derived.[4] In 1978 opencast mining, which is relatively uncomplicated and therefore less costly in the long run, accounted for only 2.8 per cent of total production in China,[5] whereas it accounted for 61 per cent in the US in 1977, 32 per cent in the USSR in 1975, and 26 per cent in India in 1979.[6]

Total production in 1978 could probably have been much larger had the output–capacity ratio (discussed in more detail below) not been badly upset by the GLF and Cultural Revolution. The effects of these socioeconomic experiments are

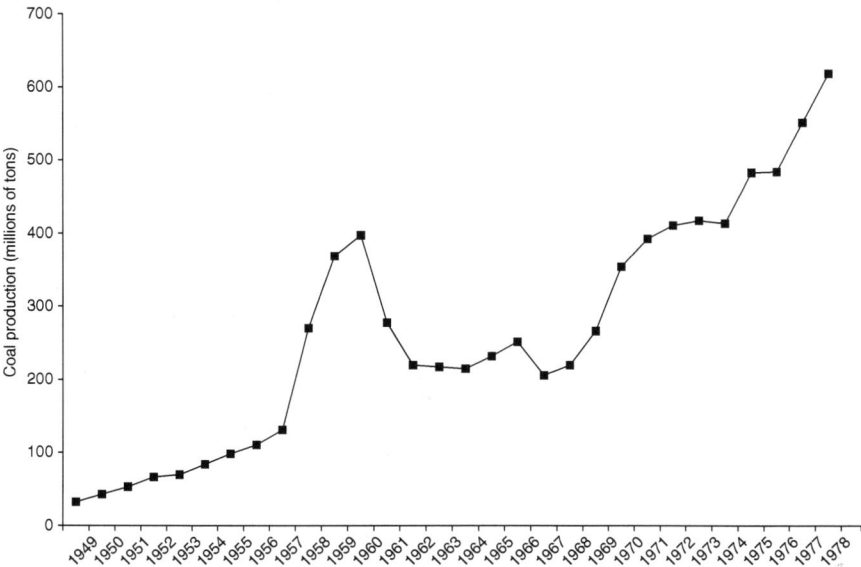

Figure 3.1 Total coal production, 1949–78

clearly apparent in Figure 3.1, which plots total output between 1949 and 1978. The corresponding total output figures and annual increments are given in Table 3.1. The production peak of over 397 million tons, attained in 1960 at the zenith of the GLF, was not surpassed until 1972.

Despite these remarkable outputs and rates of growth, at no point during the almost thirty years was production adequate to meet all needs. By 1952 production had recovered to the pre-1949 peak. However, with the restoration of the economy in the early 1950s demand immediately far outstripped supply, and then Mao's experiments worsened the already critical situation. Planning was in chaos in the late 1960s. Several senior personnel in the Ministries of Coal and Railways had either been killed or sent to the countryside. Moreover, the rigidities of top-down central planning virtually precluded direct communications between the two Ministries. Production was far outpacing transport capacity, but the planners in Beijing seemed oblivious to this.

Table 3.2 gives the output that came from the various types of mines, and Table 3.3 gives this information in percentages of the total. The *1982 Coal Industry Yearbook*, in a retrospective section, gives no figures for output from provincial and county mines in 1949 and 1957. The most likely explanation is that the three distinct mine categories – provincial, prefectural, and county – were not distinguished until the early 1960s. The CMAs accounted for about 72 per cent of total production over the First Five Year Plan, but when peasants nationwide were ordered to mine coal during the GLF their share decreased to 58 per cent. By 1965 the dominant role of the CMAs had almost returned to what it had been in

Table 3.1 Coal production, 1949–78 (mt)

	Production (mt)	Increment (mt)		Production (mt)	Increment (mt)
1949	32.43		1964	214.57	−2.50
1950	42.92	10.49	1965	231.80	17.23
1951	53.08	10.16	1966	251.47	19.67
1952	66.49	13.41	1967	205.70	−45.77
1953	69.68	3.19	1968	219.59	13.89
1954	83.66	13.98	1969	265.95	46.36
1955	98.30	14.64	1970	353.99	88.04
1956	110.36	12.06	1971	392.30	38.31
1957	130.73	20.37	1972	410.47	18.17
1958	270.00	139.27	1973	416.97	6.50
1959	368.79	98.79	1974	413.17	−3.80
1960	397.21	28.42	1975	482.24	69.07
1961	277.62	−119.59	1976	483.45	1.21
1962	219.55	−58.07	1977	550.68	67.23
1963	217.07	−2.48	1978	617.86	67.18

Source: Sun Shangqing and Zhai Ligong, *Zhongguo nengyuan jiegou yanjiu* (Research into China's Energy Structure), Taiyuan: Shanxi renmin chubanshe/Zhongguo shehui kexue chubanshe, 1987, p. 90.

Note: According to the above source 367.21 million tons were produced in 1960. It is assumed that this is a misprint because both *TJNJ 1981*, p. 223 and *GJJJTJZL*, p. 50, give 397.00 million tons (both of these sources supply only rounded figures) and Sinton *et al.* give 397.2 million tons for that year. Sinton, Jonathan E. *et al.* (eds), *China Energy Databook*, Berkeley, CA: Ernest Orlando Lawrence Berkeley National Laboratory Report LBL-32822, Rev. 4, Sept. 1996, Table II.3, pp. II–26, citing *MTGYNJ*, various years; *NYTJNJ*, various years, and *ZGNY*, 2, 1995.

1957, but thereafter the proportion of total output from them steadily declined to only 55 per cent in 1978.

Table 3.4 gives the annual growth rates of total production and production from the CMAs and local mines. Total coal production grew at an average annual rate of 12.8 per cent over the entire period. After the GLF there were four years of negative growth. Due to mass starvation, the main priority of the peasants had to be food production. In 1965 and 1966 the growth of total production averaged over 8 per cent per year, but in 1967, due to the Cultural Revolution, it fell by over 18 per cent. Political disruption affected the industry again in 1973, 1974, and 1976. However, despite the ongoing political instability, remarkable production gains were achieved in 1969, 1970, 1975, 1977, and 1978. The range between lowest and highest annual growth rates was even greater for local production. Most of the growth from this sector occurred at the county, commune, and production brigade mines.

In China the relationship between producers and consumers was made especially difficult by the structure of production. In the USSR the Ministry of Coal was responsible for 99 per cent of total production, and in 1970 there were 842 mines with an average annual output of 514,000 tons per year.[7] However, in China, while all of the output from the CMAs and LS mines was under plan, the

Table 3.2 Coal production by mine type, 1949–78 (mt)

	Total production	State mines			Local mines		
		Total state	CMA	Local state	Total local	Local state	Local non-state
1949	32.43	30.98	23.53	7.45	8.90	7.45	1.45
1950	42.92		30.19		12.74		
1951	53.08		37.14		15.94		
1952	66.49	63.53	48.34	15.19	18.16	15.19	2.96
1953	69.68		52.18		17.50		
1954	83.66		62.28		21.38		
1955	98.30		73.00		25.30		
1956	110.36		81.51		28.85		
1957	130.73	124.24	94.33	29.91	36.40	29.91	6.49
1958	270.00	219.00	157.77	61.23	112.23	61.23	51.00
1959	368.79		216.43		152.36		
1960	397.21		240.36		156.85		
1961	277.62		175.54		102.08		
1962	219.55	211.50	147.55	63.95	72.00	63.95	8.05
1963	217.07		151.29		65.78		
1964	214.57		150.56		64.01		
1965	231.80	222.07	164.28	57.79	67.52	57.79	9.73
1966	251.47		180.72		70.75		
1967	205.70		135.88		69.82		
1968	219.59		147.30		72.29		
1969	265.95		178.57		87.38		
1970	353.99	322.54	226.70	95.84	127.26	95.84	31.42
1971	392.30		246.58	145.72			
1972	410.47		249.22	161.25			
1973	416.97		247.67	169.30			
1974	413.17		242.78	170.39			
1975	482.24	419.17	279.95	139.22	202.28	139.22	63.06
1976	483.45		273.64		209.81		
1977	550.68	471.87	295.27	176.60	255.41	176.60	78.80
1978	617.86	522.54	341.84	180.70	276.02	180.70	95.32

Sources: *CIYB 1982*, p. 25; Dangdai Zhongguo congshu bianzhi bu (Contemporary China Series Editorial Dept.), *Dangdai Zhongguo de meitan gongye* (Contemporary China Coal Industry), Beijing: Zhongguo shehui kexue chubanshe, 1988, p. 128; *Zhongguo jingji nianjian 1985* (Almanac of China's Economy 1985), Beijing: Jingji guanli chubanshe, 1986, pp. v–58; Choh-Ming Li, *The Statistical System of Communist China*, Berkeley, CA: University of California Press, 1962, Table 3, p. 100, citing Chang Lin-chih, 'Strive for a High Growth Rate in the Coal Industry', *RMRB*, 7 Oct. 1959, p. 7.

Notes: Of the 1965 LS output, provincial mines accounted for 25.42 mt, prefectural 13.60, and county 18.77. *CIYB 1982*, p. 25; See text for explanation of categories. Due to rounding some rows do not add up exactly to total shown. 'Total state' and 'total local' do not add up because the local state component is reported twice, once under state mines and once under local mines.

Table 3.3 Proportions of total coal production by mine type, 1949–78 (%)

	State mines			Local mines		
	Total state	CMAs	Local state	Total local	Local state	Local non-state
1949	95.5	72.6	23.0	27.4	23.0	4.5
1950		70.3		29.7		
1951		70.0		30.0		
1952	95.6	72.7	22.8	27.3	22.8	4.5
1953		74.9		25.1		
1954		74.4		25.6		
1955		74.3		25.7		
1956		73.9		26.1		
1957	95.0	72.2	22.9	27.8	22.9	5.0
1958	81.1	58.4	22.7	41.6	22.7	18.9
1959		58.7		41.3		
1960		60.5		39.5		
1961		63.2		36.8		
1962	96.3	67.2	29.1	32.8	29.1	3.7
1963		69.7		30.3		
1964		70.2		29.8		
1965	95.8	70.9	24.9	29.1	24.9	4.2
1966		71.9		28.1		
1967		66.1		33.9		
1968		67.1		32.9		
1969		67.1		32.9		
1970	91.1	64.0	27.1	36.0	27.1	8.9
1971		62.9		37.1		
1972		60.7		39.3		
1973		59.4		40.6		
1974		58.8		41.2		
1975	86.9	58.1	28.9	41.9	28.9	13.1
1976		56.6		43.4		
1977	85.7	53.6	32.1	46.4	32.1	14.3
1978	84.6	55.3	29.2	44.7	29.2	15.4

Sources: Calculated from Table 3.2.

Note: See text for explanation of categories. Due to rounding some rows do not add up exactly to total shown. 'Total state' and 'total local' do not add up because the local state component is reported twice, once under state mines and once under local mines.

output from the commune and collectively run mines was not. It has been estimated that in 1952, 1.4 per cent of the national output came from the latter, but by 1978 this total was 15.4 per cent. In that year there were some 600 state operated mines and 20,000 non-state mines. It was difficult enough for the planners to know exactly how much was being produced by each of the 600 state mines. The production and consumption of the output from the non-state mines was virtually beyond their control prior to the late 1970s.

Table 3.4 Annual growth rates of coal production by
mine type, 1950–78 (%)

	Total	*CMA*	*Total local*
1950	32.35	28.30	43.15
1951	23.67	23.02	25.12
1952	25.26	30.16	13.93
1953	4.80	7.94	−3.63
1954	20.06	19.36	22.17
1955	17.50	17.21	18.33
1956	12.27	11.66	14.03
1957	18.46	15.73	26.17
1958	106.53	67.25	208.32
1959	36.59	37.18	35.76
1960	7.71	11.06	2.95
1961	−30.11	−26.97	−34.92
1962	−20.92	−15.95	−29.47
1963	−1.13	2.53	−8.64
1964	−1.15	−0.48	−2.69
1965	8.03	9.11	5.48
1966	8.49	10.01	4.78
1967	−18.20	−24.81	−1.31
1968	6.75	8.40	3.54
1969	21.11	21.23	20.87
1970	33.10	26.95	45.64
1971	10.82	8.77	14.51
1972	4.63	1.07	10.66
1973	1.58	−0.62	4.99
1974	−0.91	−1.97	0.64
1975	16.72	15.31	18.72
1976	0.25	−2.25	3.72
1977	13.91	7.90	21.73
1978	12.20	15.77	8.07

Sources: Calculated from Table 3.2.

Note: See text for explanation of categories.

The task of coordinating production and consumption was much easier in countries where coal was used by fewer industries. For example, in the US, fifty companies accounted for 65 per cent of the total coal production in 1976 (up from 45 per cent in 1950).[8] Some 154 operations under the jurisdiction of these fifty companies produced over one million tons. Fifty-seven were large mines which sold coal to a variety of consuming industries, while the other mines were controlled by coal-consuming industrial concerns: nineteen oil and gas refineries, nineteen thermal electric, seventeen steel, eleven other metal, eight chemical, and twenty-three other plants.[9]

India's coal industry was nationalized and consolidated in 1973 through the closure or merging of hundreds of small mines. Whereas the public sector accounted for about 11 per cent of total production in 1954, in 1975–76 it accounted for 100 per cent.[10] In 1971 there were nearly 800 collieries in operation with an average annual rate of total production of just over 90,000 tons, but over a third of these were producing less than 6,000 tons per year. After nationalization, through three major companies, the Ministry operated a total of about 300 mines each with an average annual output of approximately 240,000 tons.[11] The national agency directly coordinated the requirements of the main consumers, that is, the steel plants, thermal plants, etc.

Only 17 per cent of China's production in 1978 was washed and processed, mostly using the relatively primitive jiggery method.[12] Of the output from Shanxi, the province with the largest production, only 5 per cent was washed.[13] This proportion was extremely low compared to that of other countries. Since the early 1960s generally at least half of the coal produced in the US, Canada, and Western Europe was washed. Between 1953 and 1957 the proportion of washed coal in China almost doubled, but thereafter the increases were very small. Table 3.5

Table 3.5 Annual coal production capacity increases at the CMAs compared with increases in coal washing capacity, 1953–78 (mt)

	Coal production capacity increases	Washed coal capacity increases
1953	9.69	
1954	10.47	0.90
1955	13.51	3.10
1956	11.28	0.40
1957	18.81	0.40
1958	19.31	4.50
1959	34.80	38.50
1960	31.00	9.61
1961	5.38	0.42
1962	6.27	
1963	4.46	
1964	8.52	0.45
1965	8.57	
1966	12.06	
1967	1.44	
1968	10.38	2.61
1969	7.91	6.95
1970	37.36	9.80
1971	15.40	0.66
1972	11.39	4.10
1973	18.35	5.25
1974	17.02	5.05
1975	19.05	3.00
1976	17.29	0.60
1977	13.91	0.45
1978	11.51	3.00

Source: *NYTJNJ 1991*, p. 56.

Table 3.6 Breakdown of coal types, 1952–79 (%)

	1952	1957	1965	1979
National Total	100.0	100.0	100.0	100.0
Anthracite	11.6	15.4	17.1	19.8
Bituminous	69.4	77.4	79.1	76.2
Consisting of:				
Coal used for coking	24.6	45.3	58.2	51.5
Of which:				
Coking coal		7.7	12.6	13.2
Rich coal		21.1	12.7	8.5
Gas coal		12.7	25.9	20.1
Meagre coal		2.3	5.5	6.5
Lignite	0.1	2.2	3.8	4.0

Source: *MTGYNJ 1982*, p. 25. Note accompanying the table in the year-book: 'In the years 1952 and 1957, the totals do not exactly correspond to the sum of the individual figures due to inadequate definition of coal varieties in some collieries.'

indicates that in some years the increase in washing capacity was less than 5 per cent of the annual increase in production capacity at the CMAs.

Severe air pollution in the form of smoke particulates, carbon dioxide, and sulphur dioxide – all resulting almost entirely from the burning of coal, especially in the winter months – caused respiratory disease in both humans and livestock, and poisoned plant life. China's share of global carbon dioxide emissions went from 1 per cent in 1950 to 8 per cent in 1978. By comparison, India's went from 1 to 2 per cent, Japan's from 2 to 5 per cent, the US' from 42 to 25 per cent, and the USSR's from 11 to 17 per cent.[14]

Table 3.6 gives a breakdown of the coal types from 1952 to 1979. The data for 1952 and 1957 is problematic due to problems with definitions. However, in the late 1960s and 1970s there was a measurable increase in the proportion of anthracite produced while the shares of both rich and gas coals used for coking fell. Whereas in most sectors of the economy the trend was toward modernization, the proportion of coke produced in the inefficient and highly polluting small-scale ovens increased sharply from 1.9 per cent in 1949 to 30.3 per cent in 1978.[15]

China's highest quality coal, that is high in calorific value but low in sulphur and ash levels, is found in the 'three wests' – Shanxi, Shaanxi and *xi Nei Mengu*. (*xi* means west) (see Figure 1.1). Coal from other areas produces far less heat and is considerably more polluting. In 1952 the largest concentration of coal production was in Liaoning (17.7 per cent), followed by Shanxi (16.3 per cent), and Hebei (15.2 per cent). Together the north and northeast regions produced almost 70 per cent of total output, 36 per cent in the north and 33.2 per cent in the northeast. Tables 3.7 and 3.8 give the provincial and regional distribution of production and consumption, respectively. Table 3.9 shows that the north region had the highest surplus (the amount in excess of local demand) followed by the northeast, while the eastern region had the most serious deficit.

Table 3.7 Distribution of coal production, 1952–78 (mt and %)

Region	1952	[Rank]	1965	[Rank]	1970	[Rank]	1978	[Rank]
North	23.93 (36.0)	[1]	75.95 (32.8)	[1]	102.31 (28.9)	[1]	176.88 (28.6)	[1]
Beijing	2.32 (3.5)		5.29 (2.3)		7.79 (2.2)		8.19 (1.3)	
Tianjin	0.00 (0.0)		0.00 (0.0)		0.00 (0.0)		0.00 (0.0)	
Hebei	10.10 (15.2)		24.92 (10.8)		31.86 (9.0)		57.42 (9.3)	
Inner Mongolia	0.64 (1.0)		6.47 (2.8)		8.85 (2.5)		13.02 (2.1)	
Shanxi	10.87 (16.3)		39.27 (16.9)		53.81 (15.2)		98.25 (15.9)	
Northeast	22.12 (33.2)	[2]	55.09 (23.7)	[2]	84.60 (23.9)	[2]	109.48 (17.7)	[2]
Liaoning	11.77 (17.7)		24.33 (10.5)		37.17 (10.5)		48.17 (7.8)	
Jilin	4.20 (6.3)		10.27 (4.4)		15.58 (4.4)		20.74 (3.4)	
Heilongjiang	6.15 (9.2)		20.49 (8.8)		31.85 (9.0)		40.57 (6.6)	
Northwest	2.08 (3.1)	[6]	13.49 (5.9)	[6]	23.01 (6.5)	[6]	49.98 (8.1)	[6]
Shaanxi	1.06 (1.6)		3.94 (1.7)		8.85 (2.5)		16.66 (2.7)	
Gansu	0.55 (0.8)		1.97 (0.9)		3.89 (1.1)		9.80 (1.6)	
Qinghai	0.10 (0.1)		0.63 (0.3)		0.71 (0.2)		2.50 (0.4)	
Xinjiang	0.37 (0.6)		3.87 (1.7)		5.31 (1.5)		10.79 (1.7)	
Ningxia	0.00 (0.0)		3.08 (1.3)		4.25 (1.2)		10.23 (1.6)	
S. Central	5.72 (8.6)	[4]	28.50 (12.3)	[4]	50.98 (14.4)	[4]	105.72 (17.1)	[4]
Henan	3.31 (5.0)		19.03 (8.2)		32.57 (9.2)		58.45 (9.5)	
Hubei	0.42 (0.6)		1.17 (0.5)		2.48 (0.7)		6.44 (1.0)	
Hunan	1.78 (2.7)		5.30 (2.3)		9.91 (2.8)		22.00 (3.6)	
Guangdong	0.12 (0.2)		1.57 (0.7)		3.54 (1.0)		10.54 (1.7)	
Guangxi	0.09 (0.1)		1.43 (0.6)		2.48 (0.7)		8.29 (1.3)	

	[5]		[5]		[5]		[5]	
Southwest	4.05	(6.1)	20.03	(8.6)	34.33	(9.7)	69.47	(11.2)
Sichuan	3.34	(5.0)	11.34	(4.9)	18.76	(5.3)	37.94	(6.1)
Yunnan	0.39	(0.6)	5.36	(2.3)	7.79	(2.2)	14.84	(2.4)
Guizhou	0.32	(0.5)	3.33	(1.4)	7.78	(2.2)	16.69	(2.7)
East [3]	8.62	(12.9)	38.68	(16.7)	58.76	(16.6)	106.24	(17.2)
Shanghai	0.00	(0.0)	0.00	(0.0)	0.00	(0.0)	0.00	(0.0)
Jiangsu	1.20	(1.8)	4.89	(2.1)	7.79	(2.2)	17.94	(2.9)
Zhejiang	negligible	(—)	0.32	(0.1)	0.71	(0.2)	1.59	(0.3)
Anhui	2.59	(3.9)	11.48	(5.0)	18.05	(5.1)	24.53	(4.0)
Jiangxi	1.17	(1.7)	3.95	(1.7)	5.31	(1.5)	15.95	(2.6)
Fujian	negligible	(—)	0.60	(0.3)	1.06	(0.3)	4.23	(0.7)
Shandong	3.66	(5.5)	17.44	(7.5)	25.84	(7.3)	42.00	(6.8)
Total	66.52		231.77		353.99		617.77	

Sources: Calculated from *GYJJTJZL*, p. 166; Cheng Chu-yuan, *The Demand and Supply of Primary Energy in Mainland China*, Mainland China Economic Series, no. 3, Seattle: University of Washington Press, 1985, p. 38, citing the Ministry of Coal.

Note: The total figures given in the *GYJJTJZL* were slightly different from those in Table 3.1.

Table 3.8 Distribution of coal consumption, 1952–78 (mt and %)

Region	1952		[Rank]	1965		[Rank]	1970		[Rank]	1978		[Rank]
North	9.46	(14.2)	[3]	37.32	(16.1)	[3]	55.21	(15.6)	[3]	90.84	(14.7)	[4]
Beijing	2.33	(3.5)		9.50	(4.1)		13.80	(3.9)		22.86	(3.7)	
Tianjin	3.67	(5.5)		11.36	(4.9)		13.45	(3.8)		17.93	(2.9)	
Hebei	2.53	(3.8)		7.42	(3.2)		13.45	(3.8)		26.57	(4.3)	
Inner Mongolia	0.13	(0.2)		2.78	(1.2)		4.60	(1.3)		8.65	(1.4)	
Shanxi	0.80	(1.2)		6.26	(2.7)		9.91	(2.8)		14.83	(2.4)	
Northeast	19.28	(29.0)	[2]	50.30	(21.7)	[2]	71.16	(20.1)	[2]	100.71	(16.3)	[3]
Liaoning	11.77	(17.7)		28.05	(12.1)		37.88	(10.7)		51.28	(8.3)	
Jilin	4.19	(6.3)		10.20	(4.4)		15.58	(4.4)		21.01	(3.4)	
Heilongjiang	3.32	(5.0)		12.05	(5.2)		17.70	(5.0)		28.42	(4.6)	
Northwest	1.60	(2.4)	[6]	11.82	(6.0)	[6]	21.60	(6.1)	[6]	43.24	(7.0)	[6]
Shaanxi	0.53	(0.8)		1.85	(1.7)		7.08	(2.0)		13.59	(2.2)	
Gansu	0.60	(0.9)		2.32	(1.0)		5.31	(1.5)		11.12	(1.8)	
Qinghai	0.07	(0.1)		0.70	(0.3)		0.71	(0.2)		2.47	(0.4)	
Xinjiang	0.40	(0.6)		3.94	(1.7)		5.31	(1.5)		10.50	(1.7)	
Ningxia	0.00	(0)		3.01	(1.3)		3.19	(0.9)		5.56	(0.9)	
S. Central	7.98	(12.0)	[4]	31.3	(13.5)	[4]	53.45	(15.1)	[4]	106.90	(17.3)	[2]
Henan	1.40	(2.1)		6.26	(2.7)		10.97	(3.1)		25.95	(4.2)	
Hubei	1.53	(2.3)		7.42	(3.2)		11.33	(3.2)		21.63	(3.5)	
Hunan	1.86	(2.8)		5.56	(2.4)		11.33	(3.2)		24.10	(3.9)	
Guangdong	2.79	(4.2)		9.97	(4.3)		14.87	(4.2)		24.71	(4.0)	
Guangxi	0.40	(0.6)		2.09	(0.9)		4.95	(1.4)		10.51	(1.7)	

Southwest	4.05	(6.1)	[5]	19.94	(7.7)	[5]	32.57	(9.2)	[5]	60.55	(9.8)	[5]
Sichuan	3.32	(5.0)		11.36	(4.9)		20.18	(5.7)		38.31	(6.2)	
Yunnan	0.40	(0.6)		5.33	(2.3)		8.85	(2.5)		16.06	(2.6)	
Guizhou	0.33	(0.5)		3.25	(0.5)		3.54	(1.0)		6.18	(1.0)	
East	24.12	(36.3)	[1]	81.12	(35.0)	[1]	120.00	(33.9)	[1]	215.62	(34.9)	[1]
Shanghai	12.03	(18.1)		40.80	(17.6)		46.02	(13.0)		72.29	(11.7)	
Jiangsu	5.05	(7.6)		13.91	(6.0)		22.66	(6.4)		39.54	(6.4)	
Zhejiang	1.86	(2.8)		6.03	(2.6)		10.97	(3.1)		19.15	(3.1)	
Anhui	0.66	(1.0)		4.17	(1.8)		9.91	(2.8)		19.77	(3.2)	
Jiangxi	1.26	(1.9)		4.17	(1.8)		8.14	(2.3)		17.30	(2.8)	
Fujian	0.00	(0.0)		2.31	(1.0)		3.54	(1.0)		8.03	(1.3)	
Shandong	3.26	(4.9)		9.73	(4.2)		18.76	(5.3)		39.54	(6.4)	
Total	66.49			231.80			353.99			617.86		

Sources: *Calculated from TJNJ (various years)*.

Note: Percentages are given in parentheses.

Table 3.9 Production deficit/surplus for each region, 1952–78 (%)

	E	N	NE	SC	SW	NW
1952						
Production	13.0	36.0	33.2	8.6	6.1	3.1
Consumption	36.3	14.2	29.0	12.0	6.1	2.4
Difference	−23.3	21.8	4.2	−3.4	0	0.7
Total deficit: −27.7						
1965						
Production	16.7	32.8	23.7	12.3	8.6	5.9
Consumption	35.0	16.1	21.7	13.5	7.7	6.0
Difference	−18.3	16.7	2.0	−1.2	0.9	−0.1
Total deficit: −19.6						
1970						
Production	16.6	28.9	23.9	14.4	9.7	6.5
Consumption	33.9	15.6	20.1	15.1	9.2	6.1
Difference	−17.3	13.3	3.8	−0.7	0.5	0.4
Total deficit: −18.0						
1978						
Production	17.3	28.6	17.8	17.1	11.2	8.0
Consumption	34.9	14.7	16.3	17.3	9.8	7.0
Difference	−17.6	13.9	1.5	−0.2	1.4	1.0
Total deficit: −17.8						

Sources: Tables 3.7 and 3.8.

It must be pointed out that the central authorities divided the country into three zones: heated, transition, and non-heated. Consumption was much higher in all the provinces and autonomous regions north of the Changjiang because they were allocated coal in the winter for heating, whereas those to the south were not. It was a government-decreed line of demarcation which caused people living in the middle transition zone, for example, in Shanghai, to suffer greatly in the winter for lack of the quantities of fuel for heating that they would have desired.

In 1952 Shanghai alone accounted for almost one-fifth of the nation's consumption. The eastern region as a whole accounted for 36.3 per cent, with Jiangsu and Shandong also accounting for sizable proportions. The second largest concentration of consumption was Liaoning. The northeast region as a whole accounted for 29 per cent, the north for 14.2 per cent, the southcentral for 12 per cent, the southwest for 6.1 per cent, and the northwest for 2.4 per cent.

By 1965 Shanxi had become the largest producer (16.9 per cent of total), and Hebei and Liaoning each accounted for about 11 per cent. The north region as a whole still led the country but continued to lose its margin due to efforts to increase production in the northwest, southcentral, and southwest regions. The concentration of consumption in the northeast in 1965 was much less than it had been in 1952. The share of consumption in the east decreased slightly, but the size of the deficit in that region was reduced considerably. The consumption shares in the north, northwest, southcentral, and southwest increased.

Due to the Third Front Development Programme (see Chapter 2), between 1965 and 1970 the shares of total consumption increased in the southcentral, and southwestern regions and stayed about the same or decreased everywhere else. The shares of production in the east and northeast remained about the same, but fell drastically in the north and increased in each of the northwest, southcentral, and southwest regions.

On the eve of the reform era the ten largest mines in terms of total annual output in descending order were Datong (Shanxi), Kailuan (Hebei), Pingdingshan (Henan), Fuxin (Liaoning), Huaibei (Anhui), Jixi (Heilongjiang), Xuzhou (Jiangsu), Hegang (Heilongjiang), Fengfeng (Hebei), and Yangquan (Shanxi).[16] The northwest, northeast, and east regions were all consuming more than in 1970, but the proportions were still highest in Shanghai, Liaoning, Jiangsu, and Shandong. The east region as a whole continued to account for about 35 per cent of total consumption. The concentrations in the north and northeast had begun to give way to the southcentral, northwest, and southwest regions. However, the common slogan *bei mei nan yun*, or transporting north coal to the south, was still apropos.

Mechanization levels

It is difficult to make accurate international comparisons of levels of mechanization because mining conditions around the world are very different. Even within one country, while it is possible for some mines to have almost completely mechanized operations, for geological reasons others must be more labour intensive. However, even taking this into account, the level of mechanization in China's mines was indisputably very low.

In 1957 only 4.1 per cent of the CMAs had some mechanization. As late as 1978 much of the extraction was still done manually using dynamite, picks, and pneumatic drills to dislodge the coal, and shovels to load it into wheeled carts. Some 32.8 per cent of the CMAs had partial mechanization, that is, only extraction and tunnelling were mechanized, and 4.3 per cent had full mechanization. Mechanization was highest in those provinces where pre-1949 foreign involvement had been greatest: 79.62 per cent at Jixi (Heilongjiang), 62.0 per cent at Yangquan (Shanxi), 49.4 at Datong (Shanxi), and 47.0 per cent at Kailuan (Hebei).[17] In the USSR the overall level of mechanization was 94.9 per cent in 1970, though there were a few areas where much of the work was still done manually.[18] In India the overall level of mechanization in 1969 was 16.3 per cent.[19]

Representatives of foreign companies who visited Chinese mines in the 1970s noted that the wide variety of equipment and varying degrees of labour/ mechanization substitution complicated the successful introduction of compatible new equipment. As for opencast technology, the Chinese had not yet employed large excavators, trucks and haulers, and conveyor systems.[20]

More fully mechanized faces would have greatly increased yields. The average face output at fully mechanized large state mines was 33,607 tons per month in

1980, but only 12,186 at the semi-mechanized, and only 9,287 at the non-mechanized.[21] Hydraulic machinery was introduced in China by the Soviets in 1958, though its use never became widespread.[22] In 1980 only 1.9 per cent of all state mines had it and the average face output at these mines was 20,753 tons.

There were very few machinery factories that produced equipment solely for coal mining. Most did not belong to the Ministry, but rather to the Ministry of Machine Building and were old converted military engineering shops. In 1950 there were three coal machinery plants in operation, thirteen by 1965, and thirty-four by 1979, at which time the technical level of the machinery was comparable to that in the US between 1909 and 1912, and in the Soviet Union between 1953 and 1955.[23]

The policy of maintaining low coal prices meant that the industry was incapable of making sufficient profits to upgrade machinery. Low planned rates of depreciation on fixed assets, that is, prolonged use of equipment, also a practice adopted from the Soviet system, exacerbated the situation. Depreciation in China's coal industry ranged annually between 2 and 3 per cent of total costs. By contrast, the rate at which equipment was retired in the US was 3.7 per cent.[24] On the revenue account of British Coal in 1985–86, depreciation accounted for 7.6 per cent of all expenditure.[25]

Prolonged use of capital stock in China's mines meant that it continually had to be repaired, and each repair became more costly and time-consuming. In 1981 only 61 per cent of all the mechanical equipment owned by the Ministry was in operation, 84 per cent of this functional equipment being classified as in good order.[26] This is comparable to the situation in the Soviet Union where it was estimated that one-tenth of the industrial labour force in the late 1970s, and one-third of the machine tool park were absorbed in repair activity.[27] Visitors from the National Coal Board in 1956 to Soviet coal mine workshops observed: 'Judged by the amount of equipment lying about on the shop floor, the work in progress must be exceedingly great, but this does not necessarily mean a high throughput.'[28]

The Chinese mining operations also experienced unreliability of machine and equipment delivery, delivery of incomplete sets, lack of spare parts, incompatibility, and continual breakage due to the use of low-grade steel, all typical problems in the Soviet coal and oil industries.[29] In a large proportion of the mines there was insufficient elevation equipment relative to the rate at which coal was being cut, creating an imbalance in workflow.[30] British visitors to Soviet mines in 1957 noted: 'Far less conveying of coal is done in the USSR than in this country and the development of conveyors is not as advanced as it is here', and 'Of the current stoppages of work at the coalface, some 25 per cent of time lost is said to be attributable to inadequate haulage facilities.'[31]

Another factor reducing the return on capital investment was excessive standardization. In a misguided effort to economize through mass production, the Soviet planners, not appreciating that each mine had its own particular geological conditions, developed a standard set of blueprints for every phase of mine construction and for each piece of equipment. Often the wrong types of machinery were used.

Resource use efficiency

Imbalanced output–capacity ratios, that is, insufficient preparatory work at the mine or oil site relative to the rate of exploitation, was one of the most serious shortcomings of command economies' energy management. The combination of the lack of sufficient investment with the use of gross output targets as the sole success criterion encouraged large production increases in the short run, but did not guarantee steady long-term production, development or full recovery at each site.

The locating and preparing of reserves must be ongoing and, depending on the site, position, and formation of the deposits, the required amount of capital, labour, and preparation time can vary considerably from year to year. Energy planners must balance the projected demand for energy with the amount of these inputs available to bring the resources into productive capacity.

In the early 1950s the Soviet mining engineers devised a carefully considered development programme for China's coal industry. From experience in their own country they understood that about half a ton should be produced for every ton of prepared capacity, that is, an output–capacity ratio of 0.5 : 1 should be maintained. This ratio had not been maintained in the USSR, and a calamity was unfolding at precisely that time in the Soviet coal industry. There was continual discussion of the problem in the Soviet press.[32] The demand for coal had been seriously under-estimated and the miners were being ordered to produce more irrespective of the need to create capacity. The First Deputy Minister of the Soviet Coal Industry wrote in *Pravda* in 1953: 'Above all, a serious lag in carrying out the preparatory work plan must be eliminated. A number of coal combines … have constantly failed to carry out plans for mine excavations … Capital construction is lagging seriously.'[33] The following year it was reported: 'New mines are opened far behind schedule and often with major items unfinished, which delays the development of coal mining. The Ministry of Coal Industry still has not achieved a significant shortening of construction time …'[34] As late as the 1970s almost all of the new mines that were to start operations in the Donets field were 'well behind schedule' and there were also major delays in bringing on new capacity in the Kuznets Basin.[35]

It is significant that the output–capacity ratio for China's state mines was consistently low throughout the First Five Year Plan period (see Table 3.10).[36] The Soviet advisers, aware of the serious situation in their home country, counselled steady production growth based on prudent output–capacity ratios. However, as in the USSR, production in China was not nearly keeping pace with demand, but Mao ignored Soviet advice and called repeatedly for higher production targets at the CMAs. Then, in 1958, as part of the GLF, he launched the backyard furnace campaign in which peasants all over the country were to dig coal in an effort to match Britain's steel production.

The result was an output–capacity ratio in 1958 of 4.3 : 1. The increase in production in 1958 was almost seven times than in 1957, while the increase in capacity was not even two times greater. In 1961 and 1962 due to the exhaustion of the

Table 3.10 Output–capacity ratios, 1954–78

	Capacity increases (mt) *from* Zhongguo tongji nianjian	Total production (mt)	Annual increases in total production (mt)	Output–capacity ratio	Capacity increases (mt) *from* 'Guding'
1954	10.5	83.7	14.0	1.3 : 1	10.5
1955	13.5	98.3	14.6	1.1 : 1	13.5
1956	11.3	110.4	12.1	1.1 : 1	11.3
1957	18.8	130.7	20.3	1.1 : 1	18.8
1958	32.6	270.0	139.3	4.3 : 1	19.3
1959	55.5	368.8	98.8	1.8 : 1	34.8
1960	47.7	397.2	28.4	0.6 : 1	31.0
1961	5.9	277.6	−119.6	−20.3 : 1	5.4
1962	7.5	219.6	−58.0	−7.7 : 1	6.3
1963	5.3	217.1	−2.5	0.5 : 1	4.5
1964	9.0	214.6	−2.5	0.3 : 1	8.5
1965	9.6	231.8	17.2	1.8 : 1	8.6
1966	12.1	251.5	19.7	1.6 : 1	12.1
1967	0.7	205.7	−45.8	−65.4 : 1	1.4
1968	10.4	219.6	13.9	1.3 : 1	10.4
1969	7.6	266.0	46.4	6.1 : 1	7.9
1970	37.4	354.0	88.0	2.4 : 1	37.4
1971	15.4	392.3	38.3	2.5 : 1	15.4
1972	11.4	410.5	18.2	1.6 : 1	11.4
1973	18.4	417.0	6.5	0.4 : 1	18.4
1974	17.0	413.2	−3.8	0.2 : 1	17.0
1975	19.1	482.2	69.0	3.6 : 1	19.1
1976	17.3	483.5	1.3	0.1 : 1	17.3
1977	13.9	550.7	67.2	4.8 : 1	13.9
1978	11.5	617.9	67.2	5.8 : 1	11.5

Sources: Capacity increase figures in column I are from *TJNJ 1983*, p. 348. Production figures are from *TJNJ 1981*, p. 223. Capacity increase figures in column V are from *GDTZZL*, p. 132. The output–capacity ratios were calculated using the *TJNJ* data.

miners, there was negative growth in production and minimal growth in capacity. Between 1963 and 1966 the total increase in production was about 90 per cent of the total increase in capacity. The most harmful year of the Cultural Revolution for the coal industry, 1967, saw negative growth in production and almost no growth in capacity. In 1969, however, the production increase was about six times greater than the capacity increase. Between 1970 and 1976 the total increase in production was 60 per cent greater than the increase in capacity, and in 1977–78, 430 per cent greater. This represents a severe imbalance, and the sizable swings in production and preparation work, and lack of correlation between them, made for less efficiency in the use of labour and capital resources.

The chronic lack of capacity was also related to the fact that mine construction work was frequently prolonged. To build a large mine in the West took about five

Table 3.11 Investment per ton of new capacity, 1953–80 (yuan)

	Mines producing				
	30,000 tons/year	45,000 tons/year	60,000 tons/year	90,000 tons/year	120,000 tons/year
First Five Year Plan (1953–57)	44.0	37.0	52.5	45.0	—
Second Five Year Plan (1958–62)	26.0	28.1	33.4	30.6	28.4
Period of Readjustment (1963–65)	46.3	53.3	46.9	32.8	31.5
Fourth Five Year Plan (1971–75)	40.5	41.9	40.1	44.1	51.8
Fifth Five Year Plan (1976–80)	66.0	70.1	52.0	96.0	36.0

Source: *GDTZZL*, p. 161.

years, but typically at least seven years in China.[37] The main cause for the delays was the allocation of insufficient funds and materials up front – some mine construction projects were given enough funding and materials to provide only partial ventilation and very basic water and electricity supplies – and there was the characteristic lack of incentives to complete on time, and of penalties for time overruns.[38] Table 3.11 indicates that investment per ton of new capacity rose very sharply in the 1970s, by over 25 yuan per ton at mines producing 30,000 and 45,000 tons per year, and by over 50 yuan per ton at mines producing 90,000 tons per year. By 1978 the output–capacity ratio was so out of balance that many workers had to stop cutting and devote all their time to mine preparation.

Use of materials and labour efficiency

Table 3.12 shows that on average the use of timber in the mines between 1952 and 1979 decreased while the consumption of explosives and electric power increased. The consumption of timber would probably have increased had there not been shortages which forced the substitution of cement.

Compared to other countries in 1979, China used a relatively high amount of timber, far more explosives, and large quantities of electricity. The great range in consumption of the materials is mainly a function of the different geological conditions. The generally easier mining conditions in the US were such that very

Table 3.12 International comparison of consumption of materials for every 10,000 tons of coal produced, 1952–79

	Timber (m³)	Explosives (kg)	Electricity (kWh per ton)
China			
1952	244.1	1,333	
1957	248.1	2,147	
1965	167.5	2,925	28.2
1979	100.7	3,386	32.1
USSR			
1979	25.0	—	10.7–76.1
US			
1975 (?)	0.35	10	13.0–21.1
France			
1977	186.0	2,500	40.0
Austria			
1980 (?)	—	—	12.9–15.7

Sources: *CIYB 1982*, p. 27; A. Astakhov and A. Grubler, *Resource Requirements and Economics of the Coal-Mining Process: A Comparative Analysis of Mines in Selected Countries*. Laxenburg, Austria: International Institute for Applied Systems Analysis, 1984, pp. 90, 101, 113–15.

Notes: The French data refers only to mines in the Lorraine Basin, and the Austrian to only the Trimmelkam and Schmitzberg-Hinterschlagen Mines.

little timber, explosives, and electricity were required, unlike in the Lorraine Basin of France, for example, where the great depth of the mines and a high inflow of water necessitated a particularly large consumption of electricity.

The components of the 1978 production costs nationally and for two provinces and one region are given in Table 3.13. It is obvious that the conditions were much more favourable in the northern regions, and hence the materials used cost far less than in the southern and central regions. In all cases, timber, accounting for 31–38 per cent, and wages, accounting for 25–33 per cent, were the largest components of total costs. For comparison, labour costs as a proportion of total costs at underground mines accounted for 50 per cent in the US in the late 1970s, 14 per cent in South Africa, 45 per cent in Australia, and 51 per cent in the UK.[39]

Employment in China's state mines increased by a total of 676 per cent between 1952 and 1978, at an average annual rate of about 13 per cent, though the increase was not steady as can be seen in Table 3.14. There was a 241 per cent increase at the time of the GLF, but as soon as this explosive growth campaign was over many peasants necessarily returned to food production. In 1970, about four years into the Cultural Revolution, there was another large increase because coal shortages were acute.

Table 3.13 Components of 1978 production costs: national average, Shanxi, Inner Mongolia, and Hunan (yuan)

	Total costs	Timber	Wages	Electricity	Renewal	Other
National Average	16.12	5.70	4.71	1.56	2.04	2.11
Shanxi	13.40	5.11	3.30	0.98	2.00	2.01
Inner Mongolia	14.34	4.48	4.57	1.04	2.00	2.24
Hunan	22.70	7.25	7.40	2.59	3.00	2.46

Source: Zhang Liqing, 'Shanxi meitan gongye zai chuanguo de diwei' (The Position of Shanxi's Coal Industry in the Whole Country), in *SXJDYJ*, p. 411.

Notes: The figures for Inner Mongolia are that region's lowest, and those for Hunan are that province's highest. The 'renewal' category is translated from the Chinese *gengxin* meaning renewal or replacement.

Table 3.14 Employment at state mines, 1952–78

	'000 men	% change
1952	488	
1953	537	10.04
1954	542	0.93
1955	560	3.32
1956	646	15.36
1957	740	14.55
1958	2,525	241.22
1959	2,037	−19.33
1960	2,475	21.50
1961	2,281	−7.84
1962	1,730	−24.16
1963	1,558	−9.94
1964	1,463	−6.10
1965	1,535	4.92
1966	1,645	7.17
1967	1,710	3.95
1968	1,850	8.19
1969	2,000	8.11
1970	2,468	23.40
1971	2,746	11.26
1972	2,754	0.29
1973	2,884	4.72
1974	3,040	5.41
1975	3,310	8.88
1976	3,531	6.68
1977	3,759	6.46
1978	3,785	0.69

Source: *LDGZZL*, p. 37.

Note: These figures do not agree with those in the *CIYB*.

International comparisons of productivity are also complicated by geological factors. Notwithstanding, it is safe to say that with the exception of a few of the CMAs, labour productivity was still very low in 1978. The average productivity of face workers increased from 2.341 tons in 1952 to 3.698 in 1978, and average monthly production per face from 4.331 tons to 10,989 tons.[40] However, output per manshift (OMS) in the 1970s was a fraction of what it was in the other large coal-producing countries (see Table 3.15), rising from an average of 0.356 tons in 1949, to 0.661 in 1952, and 0.931 in 1978.

In Shanxi Province, the most developed mining area, the OMS was 43 per cent higher than the national average in 1979, and 73 per cent higher than in the southern provinces.[41] Doyle points out that the OMS in China was calculated according to the number of 'raw coal production workers', whereas when making comparisons with other countries 'personnel in industrial production' would be more appropriate. If this wider definition is used, assuming there are 300 shifts per year, the OMS in China would have been only about half of what the Chinese reported it to be.[42]

The authorities invariably equated more labour with higher production. The Ministry believed that extra workers could substitute for a lack of machinery,

Table 3.15 Average OMS in several coal-producing countries, early 1970s (tons at underground mines)

	1970/71	*1974/75*
West Germany	3.93	4.06
France	2.59	2.74
Belgium	2.59	2.47
UK	3.37	3.41
Total EEC (9 countries)	3.42	3.51
US	—	9.6
Austria	—	4.15
USSR	—	2.44
India	0.73	—
China		0.794

Sources: Neil K. Buxton, *The Economic Development of the British Coal Industry*, London: B.T. Batsford, 1978, p. 258; P. D. Henderson, *India: The Energy Sector*, Delhi: Oxford University Press, 1975, p. 42; A. Astakhov and A. Grubler, *Resource Requirements and Economics of the Coal-Mining Process: A Comparative Analysis of Mines in Selected Countries*, Laxenburg, Austria: International Institute for Applied Systems Analysis, 1984, pp. 90, 97, 115; *GYJJTJNJ*, p. 57.

Notes: The figure for India is for mechanized mines only. The figure for all mines in India is 0.66 tons. The figure for Austria is for 1980 and is not the national average, rather it is the average for two mines. The USSR figure refers to the late 1970s.

albeit machinery that was prone to breaking down. When a group from Britain's National Coal Board visited some Chinese mines in 1976 they reported:

> The detailed figures given at the individual mines clearly indicated what in the west would be regarded as substantial over-manning of the pits. In fact the concepts of productivity, in terms of output per manshift, and of manpower saving, so far play little part in the coal industry's affairs.[43]

In April 1977 it was decided at a national meeting to 'recruit 100,000 miners to the productive front, so that the MCI might cast off the stigma of always making deficits and begin to make profits'.[44] As the overall level of mechanization was increased, it should have been possible to reduce the size of the work force. Rather, capital equipment and labour were simultaneously introduced into the industry with little consideration of how to obtain the highest productivity from each, nor how they should complement one another.

International coal mining employment figures are not always disaggregated into production workers, administration, etc., and the proportions are likely different in each location. However, they can be used to make rough comparisons of labour productivity. In China the total number of employees in the CMAs alone was 1,172,484 in 1979, and their output was 357.77 million tons, meaning that 32.8 men were employed to cut 10,000 tons. For comparison the figures for US coal mines in 1977 were 236,000, 872.4, and 2.7; for the USSR in 1978, 1,035,000, 701.3, and 14.8; and for India in 1978, 660,000, 105.2, and 62.7.[45]

The incremental labour output ratio (ILOR), given in Table 3.16, shows the variation in numbers employed at the state mines per unit increase in output. Over the First Five Year Plan period there was some semblance of balance, but after the GLF this was lost and took several years to be restored. In the early 1970s there were again large increases in labour relative to output. It is very likely that the changes in numbers employed had lagged effects, but it is not possible to make conclusive statements with the limited data available.

According to the general labour productivity index, calculated by dividing the total gross value of industrial output of an industry by the number of its employees, the performance of the coal industry in China declined between 1957 and 1965, reflecting overmanning and low prices though improved somewhat by 1978 (see Table 3.17). For all industries taken together, however, and for the power, petroleum, chemical, machine building, building materials, and textiles industries, the index showed great improvement.

Up to a third of the miners were peasants hired on a temporary basis, and the majority were illiterate and had no training of any kind.[46] The industry did not possess much expertise in 1949 because hitherto production had been almost completely managed by the Japanese, Russians, Germans, and British. The Japanese, for security reasons, actually prohibited the Chinese from assuming any managerial positions with the result that after the Japanese surrender, mining experts from Japan had to be called upon to assist in the mines.

Table 3.16 ICOR and ILOR, 1953–78

	Annual per cent change			ICOR	ILOR
	CMA production	Capital investment	Employees		
1953	7.94		10.04		1.26:1
1954	19.36	24.52	0.93	1.26:1	0.05:1
1955	17.21	29.01	3.32	1.69:1	0.19:1
1956	11.66	35.34	15.36	3.03:1	1.32:1
1957	15.73	−7.86	14.55		0.92:1
1958	67.25	115.14	241.22	1.71:1	3.59:1
1959	37.18	32.63	−19.33	0.88:1	
1960	11.06	9.48	21.50	0.86:1	1.94:1
1961	−26.97	−45.26	−7.84		
1962	−15.95	−45.39	−24.16		
1963	2.53	−1.38	−9.94		
1964	−0.48	16.61	−6.10		
1965	9.11	−14.09	4.92		0.54:1
1966	10.01	4.01	7.17	0.40:1	0.72:1
1967	−24.81	13.42	3.95		
1968	8.40	−0.05	8.19		0.98:1
1969	21.23	12.44	8.11	0.59:1	0.38:1
1970	26.95	0.13	23.40	0.00:1	0.87:1
1971	8.77	31.05	11.26	3.54:1	1.28:1
1972	1.07	23.09	0.29	21.58:1	0.27:1
1973	−0.62	24.55	4.72		
1974	−1.97	14.34	5.41		
1975	15.31	−19.01	8.88		0.58:1
1976	−2.25	−11.60	6.68		
1977	7.90	33.84	6.46	4.28:1	0.82:1
1978	15.77	39.87	0.69	2.53:1	0.04:1

Sources: Calculated from *GDTZZL*, pp. 79–82; Jonathan E. Sinton *et al.* (eds), *China Energy Databook*, Berkeley, CA: Ernest Orlando Lawrence Berkeley National Laboratory Report LBL-32822, Rev. 4, Sept. 1996, Table III.4, p. III.16.

Note: The investment figures are calculated on the basis of totals in constant terms.

In 1957 it was reported that some collieries had no engineers at all and that often the managers and deputy managers had an education level of senior primary school or lower.[47] During the 1950s the Russians offered considerable training, but because conditions were initially so backward, the Chinese were only just beginning to put into practice what they had been taught when the Soviet advisers returned home in 1960. Several projects were left uncompleted, and without the necessary blueprints and expertise, the Chinese were forced to improvise. Some projects were never completed.

The proportion of engineers and technicians in the coal mining workforce was 4.2 per cent in 1957. It fell to 2.7 in 1965 and by 1979 was only 1.3.[48] The number of engineers and technicians in machinery units went from 5.5 per cent in

Table 3.17 Index of overall labour productivity of
state owned independent accounting
industrial enterprises by branch of
industry, 1957–78 (1952 = 100)

	1957	1965	1978
Average for all industries	152.1	214.6	266.0
Coal	150.8	98.9	110.8
Metallurgical	208.2	303.1	233.6
Power	156.3	248.9	386.0
Petroleum	174.9	317.7	624.3
Chemical	231.7	501.2	552.4
Machine building	199.5	287.4	404.0
Building materials	171.7	313.5	328.1
Timber	98.6	95.9	79.7
Food	141.7	162.5	158.2
Textiles	114.5	169.9	208.7
Paper	174.5	209.1	155.4

Source: *TJNJ 1985*, p. 382.

1950 up to 9.8 in 1957, but down to 6.0 in 1965, and 4.5 in 1979.[49] The decline after 1965 was in large part due to the rustication of skilled people during the Cultural Revolution which affected all industries. The lack of trained personnel restricted the industry's efforts to develop, operate, and maintain more sophisticated equipment.

Not only were there not enough skilled workers, but many of the employees with some training were inappropriately deployed. According to one survey, 30,000 people in 1978 were not in posts that used their specializations.[50] This was also a problem in the Soviet Union. It was reported in *Pravda* in 1954: 'many engineers and technicians are spending their time doing work of secondary importance, in offices. In the Kuznets Basin, for example, only about 5 per cent of the engineers in the basin and only 2 per cent of the technicians are working as heads of mining sectors ...'[51]

As in all centrally planned economies remuneration in the production sectors in China was not linked to output. It was assumed that workers would, for the good of their country, offer their skills, energy, and time with little financial recognition. Instead of receiving monetary rewards for special achievements, they were given political honours and promotions.[52] However, as these rewards did not add to physical well-being, the result was that the employees attended their jobs but did as little as possible. In socialist countries life employment was guaranteed; no matter how low their productivity, workers were not dismissed.

Over the thirty-year period coal miners went from among the lowest to the most highly paid industrial workers in China.[53] Besides the lack of correlation between output and wage, there was the fact that a mine's total wage bill was calculated on the basis of the number of miners employed, regardless of the output.[54]

Though a large proportion of the employees were temporary, the core labour force and their children were guaranteed life-long employment irrespective of their productivity.

The difference between the eight ranks of workers was not large. Therefore, there was little incentive for employees to strive to progress from the first to eighth, and progress through the ranks was not on the basis of merit but on years of service. While the coal industries of most Western countries have been distinguished by fierce union militancy, bitter strikes, and lockouts, this was not the case in China. The All China Federation of Trade Unions (ACFTU), including the coal industry unions, served as instruments of the Party and State to promote official goals. After the Cultural Revolution the role and mandate of the unions were unclear.

The rigidly hierarchical nature of the planning system did not facilitate communication between higher and lower levels within the Ministry. Many problems in coal production arose as a result of the fact that most of the planners in Beijing had no first-hand experience in coal mines, apparently the same situation in the USSR. These comments by a director of a Soviet mine in 1954 epitomize the frustrations caused by the difficulties in communications:

> ... the annual plans for mining work are checked and confirmed only in Moscow, by one of the Deputy Ministers ... There are thousands of mines in the country. About ten hours are needed really to get to know the plan of each mine. Consequently, in order to confirm all the plans, the Deputy Minister must have 10,000 hours time, or 1,250 workdays. There's astronomy for you! ... The Ministry plans our output scrupulously and in detail: right down to the last cutting face. Imagine what kind of work is involved in planning for each cutting face from thousands of kilometres away! And we plan production from below, so as to say. And if there is a discrepancy, correspondence goes both ways and we get all tied up; we try to clear it up and reach an agreement, and telegrams fly back and forth. This petty supervision by the ministry hinders us very much.[55]

Safety standards were very low, especially at the commune mines, where there was no mechanical ventilation and open flame torches were used. Most of the accidents were caused by roof-falls or coal and gas explosions. In 1980, fatalities in Shanxi Province, where some of the most technically advanced mines were situated, occurred at the rate of 2.99 per million tons mined, while the rate in West Germany was only 0.87.[56] In 1982 the rate was 0.16 in the US, 0.96 in the USSR, 0.55 in Poland, 0.36 in the UK, and 1.36 in Japan.[57]

Financing

Viability of the industry

International comparisons of investment in energy are problematic because different countries rely on different kinds of energy and therefore face different

Table 3.18 Energy investment in developing countries as a percentage
of total annual public investment, early 1980s

Over 40%	30–40%	20–30%	Below 20%
Argentina	Ecuador	Botswana	Egypt
Brazil	India	China	Ethiopia
Colombia	Pakistan	Costa Rica	Ghana
Korea	Philippines	Liberia	Nigeria
Mexico	Turkey	Nepal	Sudan

Source: Mohan Munasinghe with Peter Meier, *Integrated National Energy Planning and Management: Methodology and Application to Sri Lanka,* World Bank Technical Paper no. 86, Industry and Energy Series, Washington, DC: 1988, Table 1.1, p. 1. Copyright © 1988 International Bank for Reconstruction and Development/The World Bank. Reprinted by permission of the International Bank for Reconstruction and Development/The World Bank.

development problems. Even comparisons limited to the coal industry itself are open to questioning. The amount of capital devoted to mining machinery and washing equipment, for example, varies greatly with the location and geological complexity of the site, type of coal mined, type of mining, method of production, scale of operations, type and degree of technical sophistication of the equipment used in all procedures, and availability of labour.

Before the reform and opening up programme in the late 1970s and 1980s, meaningful conversions of the Chinese currency into a hard currency equivalent were problematic. After suitably weighting various factors, Munasinghe and Meier found that China's investment in the energy sector as a whole, as a percentage of total public investment, was low compared to that of other developing countries (see Table 3.18). Unfortunately, data for an earlier time period was unavailable.

Inadequate funding for energy development has been characteristic of, though by no means unique to, centrally planned economies. Not enough investment went to energy relative to other sectors. Allocations were sporadic and, as many observers of the East Bloc and Soviet Union have noted, were persistently too low. These and the amount of reinvested profits – where there were any – were inadequate to maintain production, let alone to expand at the desired pace.[58] In both the Chinese and Soviet economies there was the lack of an appropriate value for capital. In market economies investment programmes in the public sector are normally evaluated on the basis of potential profits or cost benefit analysis.

Coal was a fundamental requirement for industrial development in many countries, at least until the 1970s, and national governments typically played a major role in financing coal production. Of the developed countries, the coal industry in Britain was completely nationalized in 1946, and though many of the operations in the US, Japan, and Germany were not in the public sector, there was considerable government financial assistance in the form of direct aid, price support, limitations on coal imports, and government-assured long-term agreements between producers and large consumers.[59] In most developing countries, including India, Venezuela, Colombia, Chile, Argentina, Nigeria, Mozambique, Morocco, Turkey,

Greece, and Thailand, the state financed most of the production, while in only a few countries such as Zambia, Zaire, Swaziland, and Brazil, was much of the coal produced in the private sector with both domestic and foreign capital.

Coal industries in very few countries other than Australia, South Africa, and Colombia made large profits. Most governments were subsidizing their coal industries directly or indirectly. In 1982 a government subsidy equivalent to US$17 was provided for every ton produced in Japan, US$19 in Germany, US$26 in Belgium, and US$9 in the UK.[60]

Between 1953 and 1978 China's coal industry received a total of 349.47 billion yuan in state investment compared to 252.49 billion for the oil industry, and 457.73 billion for the electricity industry (see Table 3.19). There were large variations in the yearly allocations. After the Soviet technicians left in 1960, China failed to devise a long-term balanced investment programme for the industry. Investment in coal production as a proportion of total investment in energy decreased from 51.9 per cent in 1953 to 34.2 per cent in 1957, and reached an all-time high of 57.2 per cent in 1961. Thereafter, the average percentage fell from about 40 per cent during the 1963–65 readjustment period to 28 per cent during the Fifth Five Year Plan period (see Table 3.20). Investment in coal as a proportion of total industrial investment was steadier averaging at 10.8 per cent for the entire period (see Table 3.21).

Part of the problem was that beginning in the mid-1960s the government hoped the country could reduce its reliance on coal, and tentatively began conversion to oil and electricity. Investment in the coal industry *vis-à-vis* oil and electricity was lowest in 1976, and coal made the lowest ever contribution, 69.9 per cent, to total energy consumption that year. However, as all easily exploitable sources of oil had been surveyed by then with no major new finds, it was realized there was no alternative but to return to depending on coal as the main fuel.

The Third Front Development Programme (1964–71) had been extremely damaging to the economy, perhaps even more so than the Cultural Revolution.[61] Between 1963 and 1965 it is estimated that 38.2 per cent of the national investment went to the Third Front, 52.7 per cent between 1966 and 1970 (Third Five Year Plan), and 41.1 per cent between 1971 and 1975 (Fourth Five Year Plan).[62] The proportions would be yet higher if only industrial investment were considered. The proportions of national investment allocated to the interior provinces of Sichuan and Guizhou tripled in 1965 and remained high until 1971. Looking at a longer period, Naughton found that between 1953 and 1975 there was more capital construction in Sichuan than in the coastal areas of Shanghai, Tianjin, and Jiangsu combined and the levels in Guizhou and Jiangsu were about on a par.[63] He estimates conservatively: 'if the money had been invested in non-Third Front interior regions (to say nothing of coastal regions) China's industrial output in 1979 would have been 44 billion yuan more than it was. To put it another way, China's annual industrial output is currently 10–15 per cent below what it would have been if the Third Front had never been undertaken ...'[64]

Table 3.22 gives the breakdown by source of investment funds for capital construction. The central government provided over 90 per cent of the funding

Table 3.19 Capital construction investment, 1953–80 (million yuan)

	Total industrial investment	Total energy investment	Coal investment	Oil investment	Electricity investment	of which hydro	of which thermal	Railways investment
1953	28.34	6.90	3.58	0.70	2.62	0.38	1.55	6.48
1954	38.37	9.91	4.57	1.42	3.92	0.66	2.21	9.46
1955	42.95	13.07	5.95	1.77	5.35	1.00	3.78	12.26
1956	68.20	19.55	8.05	4.26	7.24	1.32	4.86	17.56
1957	72.40	22.01	7.53	3.83	10.65	1.87	7.35	13.40
Total First Five Year Plan (1953–57)	250.26	71.44	29.68	11.98	29.78	5.23	19.75	59.16
1958	173.00	40.72	16.24	3.98	20.50	4.86	11.53	20.28
1959	208.85	54.49	21.74	5.20	27.55	7.42	16.64	34.40
1960	229.57	62.86	24.54	8.68	29.64	9.12	15.53	35.74
1961	76.79	27.28	15.61	4.04	7.63	2.26	3.71	10.73
1962	40.09	15.61	8.85	3.20	3.56	1.18	1.20	3.01
Total Second Five Year Plan (1958–62)	728.30	200.96	86.98	25.10	88.88	24.84	48.61	104.16
1963	49.16	16.43	8.22	4.30	3.91	1.47	1.78	4.46
1964	72.06	21.62	9.22	5.92	6.48	2.20	3.09	8.18
1965	88.96	25.61	7.71	6.22	11.68	4.33	6.33	21.31
Total Period of Readjustment (1963–65)	210.18	63.66	25.15	16.44	22.07	8.00	11.20	33.95
1966		27.00	8.00	7.00	12.00			
1967		29.00	9.00	7.00	13.00			
1968		31.00	9.00	8.00	14.00			

Table 3.19 (Continued)

	Total industrial investment	Total energy investment	Coal investment	Oil investment	Electricity investment	of which hydro	of which thermal	Railways investment
1969		32.00	10.00	8.00	14.00			
1970		34.00	10.00	9.00	15.00			
Total Third Five Year Plan (1966–70)	541.51	153.00	46.00	39.00	68.00	23.05	32.65	95.40
1971		44.00	13.00	12.00	19.00			
1972		54.00	16.00	15.00	23.00			
1973		64.00	20.00	18.00	26.00			
1974		75.00	23.00	22.00	30.00			
1975	231.03	70.86	18.65	21.56	30.65	10.67	15.05	32.61
Total Fourth Five Year Plan (1971–75)	977.97	309.13	90.74	89.00	129.39	47.87	61.07	140.47
1976	208.73	69.61	16.54	19.09	33.98	10.09	18.75	24.17
1977	217.36	78.06	22.58	20.76	34.72	11.83	17.54	21.35
1978	273.16	113.83	31.80	31.12	50.91	20.60	23.34	33.38
1979	256.85	109.92	31.86	27.07	50.99	21.62	20.49	31.13
1980	275.61	114.99	33.47	33.38	48.14	17.96	19.13	30.44
Total Fifth Five Year Plan	1,231.71	486.41	136.25	131.42	218.74	82.10	99.25	140.47

Sources: *TJNJ*, various years; *GDTZZL*, pp. 79–87; Jonathan E. Sinton *et al.* (eds), *China Energy Databook*, Berkeley, CA: Ernest Orlando Lawrence Berkeley National Laboratory Report LBL-32822, Rev. 4, Sept. 1996, Table III.4, p. III–16.

Note: These figures are in current terms.

Table 3.20 Breakdown of capital construction investment in energy, 1953–80 (%)

	Investment as % of total energy				
	Coal	*Oil*	*Electricity*	*Hydro*	*Thermal*
1953	51.9	10.1	38.0	5.5	22.5
1954	46.1	14.3	39.6	6.7	22.3
1955	45.5	13.5	41.0	7.7	28.9
1956	41.2	21.8	37.0	6.8	24.9
1957	34.2	17.4	48.4	8.5	33.4
First Five Year Plan (1953–57)	41.5	16.8	41.7	7.3	27.6
1958	39.9	9.8	50.3	11.9	28.3
1959	39.9	9.5	50.6	13.6	30.5
1960	39.0	13.8	47.2	14.5	24.7
1961	57.2	14.8	28.0	8.3	13.6
1962	56.7	20.5	22.8	7.6	7.7
Second Five Year Plan (1958–62)	43.5	12.5	44.2	12.4	24.2
1963	50.0	26.2	23.8	8.9	10.8
1964	42.6	27.4	30.0	10.2	14.3
1965	30.1	24.3	45.6	16.9	24.7
Period of Readjustment (1963–65)	39.5	25.8	34.7	12.6	17.6
1966	29.6	25.9	44.5		
1967	31.0	24.1	44.9		
1968	29.0	25.8	45.2		
1969	31.3	25.0	43.7		
1970	29.4	26.5	44.1		
Third Five Year Plan (1966–70)	30.1	25.5	44.4	15.1	21.3
1971	29.5	27.3	43.2		
1972	29.6	27.8	42.6		
1973	31.3	28.1	40.6		
1974	30.7	29.3	40.0		
1975	26.3	30.4	43.3		
Fourth Five Year Plan (1971–75)	29.4	28.8	41.8	15.5	19.8
1976	23.8	27.4	48.8	14.5	26.9
1977	28.9	26.6	44.5	15.2	22.5
1978	27.9	27.3	44.8	18.1	20.5
1979	29.0	24.6	46.4	19.7	18.6
1980	29.1	29.0	41.9	15.6	16.6
Fifth Five Year Plan (1976–80)	28.0	27.0	45.0	16.9	20.4

Source: Calculated from Table 3.19.

Table 3.21 Breakdown of capital construction investment in industry, 1953–80 (%)

	Investment as % of total industry						
	Total energy	Coal	Oil	Electricity	Thermal	Hydro	Railways
1953	24.3	12.6	2.5	9.2	5.5	1.3	7.2
1954	25.8	11.9	3.7	10.2	5.8	1.7	9.5
1955	30.4	13.9	4.1	12.5	8.8	2.3	12.2
1956	28.7	11.8	6.2	10.6	7.1	1.9	11.3
1957	30.4	10.4	5.3	14.7	10.2	2.6	9.3
First Five Year Plan (1953–57)	28.5	11.9	4.8	11.9	7.9	2.1	10.1
1958	23.5	9.4	2.3	11.8	6.7	2.8	7.5
1959	26.1	10.4	2.5	13.2	8.0	3.6	9.8
1960	27.4	10.7	3.8	12.9	6.8	4.0	9.2
1961	35.5	20.3	5.3	9.9	4.8	2.9	8.4
1962	38.9	22.1	8.0	8.9	3.0	3.0	4.2
Second Five Year Plan (1958–62)	27.6	11.9	3.4	12.2	6.7	3.4	8.6
1963	33.4	16.7	8.7	8.0	3.6	3.0	4.5
1964	30.0	12.8	8.2	9.0	4.3	3.1	5.7
1965	28.8	8.7	7.0	13.1	7.1	4.9	11.9
Period of Readjustment (1963–65)	30.3	12.0	7.8	10.5	5.3	3.8	7.4
1966							
1967							
1968							
1969							
1970							
Third Five Year Plan (1966–70)	28.3	8.6	7.2	12.7	6.0	4.3	9.8
1971							
1972							
1973							
1974							
1975		8.1	9.3	13.3	6.5	4.6	8.0
Fourth Five Year Plan (1971–75)	31.6	9.3	9.1	13.2	6.2	4.9	9.8
1976	33.3	7.9	9.1	16.3	9.0	4.8	6.4
1977	35.9	10.4	9.6	16.0	8.1	5.4	5.6
1978	41.7	11.6	11.4	18.6	8.5	7.5	6.7
1979	42.8	12.4	10.5	19.9	8.0	8.4	5.9
1980	41.7	12.1	12.1	17.5	6.9	6.5	5.4
Fifth Five Year Plan (1976–80)	39.5	11.1	10.7	17.8	8.1	6.7	6.0

Source: Table 3.19.

Table 3.22 Sources of investment for capital construction, 1950–79 (%)

	1950	1952	1957	1965	1975	1979
State investment	100.0	100.0	91.1	97.2	87.5	96.0
Collectives' investment			8.9	2.8	12.5	4.0

Source: Calculated from *MTGYNJ 1982*, Table 9, p. 28.

during the 1950s and 1960s, though in the mid-1970s, the collectives contributed about an eighth of the total.

The incremental capital output ratio (ICOR) is a measure of capital productivity. It compares marginal capital input with marginal output. Table 3.16 gives the ratios for CMAs, that is, the variation in investment (in constant prices) per unit increase in output. The many years when there was a negative change in one or both makes it difficult to draw conclusions. The year 1972 is notable for an increase in investment of over 23 per cent compared to an increase in CMA production of only about one per cent. Without question large investment injections had lagged effects, however, the unsteadiness of the allocations epitomizes the absense of long-term planning after the GLF.

Investment in the coal industry was high in 1977 and 1978 because by then it had been decided that conversion to oil or hydroelectricity was not feasible. Considerable investment was needed immediately to increase both output and capacity and this accounts for the very high ICORS in those years.

The investment allocations were greatly influenced by 'image' factors. The government regarded steel production as a measure of the country's economic stature and was wont to making frequent national and international proclamations about next year's ever higher production. Officials at the various coal bureaux had to provide the necessary coal for coking as best they could. This situation was similar to that which Soviet industry experienced during the Stalin years.

In 1957 half of China's major mines had deficits, and in 1977 the proportion had reached 73 per cent.[65] Table 3.23 shows that far from there being a steady increase in profits per ton, there were large fluctuations. The volatility in profits greatly affected production in that there was never enough capital available on a sustained basis to re-invest, and to increase the levels of mechanization, processing, and storage capacity. The industry never enjoyed a profit of greater than 4.60 yuan per ton, achieved in 1960, prior to the GLF collapse. According to the available data, there were a few successful years in the early 1970s, but the industry again lost ground in the mid-1970s. In 1974, 1976, and 1977 there was a loss on each ton produced.

The table suggests that there was no loss per ton in 1978. However, according to Ye Ruixiang, Deputy Head of the Department of Costing and Pricing of the Institute of Finance and Trade Economics at the Chinese Academy of Social

Table 3.23 Average coal producer prices, production costs, taxes, and profits, 1952–78 (yuan per ton)

	Producer prices	Producer costs		Taxes	Profits
		(a)	(b)		
1952	11.46	9.00	9.78	0.86	1.60
1953					
1954					
1955					
1956			9.64		
1957	12.05	10.90	10.90	0.90	0.25
1958			9.21		
1959			8.76		
1960	14.97	9.17	9.17	1.20	4.60
1961	19.11		14.15		
1962	21.97	17.28	17.28		
1963			16.48		
1964			15.77		
1965	18.00	15.77	16.56	1.44	0.79
1966			15.60		
1967			17.30		
1968			16.71		
1969			15.61		
1970	18.00	13.47	13.47	1.44	3.09
1971	18.00	13.60	13.57	1.44	2.96
1972	18.00	14.08	14.08	1.44	2.43
1973	18.00	14.51	15.41	1.44	2.05
1974	18.00	17.14	17.14	1.44	−0.58
1975	18.00	15.86	15.50	1.44	0.70
1976	18.00	16.70	16.25	1.44	−0.14
1977	18.00	16.61	16.61	1.44	−0.05
1978	18.00	16.12	16.12	1.44	0.44

Sources: The producer prices, taxes, profits, and column (a) of the producer costs are from Xu Yi, Chen Baosen and Liang Wuxia, *Shehui zhuyi jiage wenti* (Socialist Pricing Problems), Beijing: Zhongguo caizheng chubanshe, 1982, Table 9, p. 164, citing *Jingji yanjiu material* (Economic Research Material), 2nd quarter, 11, 1981. The (b) producer cost figures are from *Shanxi shehui kexue yuan, meitan jiage yanjiu keti zu* (Shanxi Academy of Social Sciences, Coal Pricing Research Group), *Meitan jiage gaige yu Shanxi meitan jiage* (Coal Pricing Reform and Shanxi Coal Pricing), unpublished research report prepared by the Academy, Taiyuan, Aug. 1986, p. 49.

Sciences, the cost of producing a ton in 1978 was 16.7 yuan and the producer price was 16.15 yuan.[66] After taxes there was a loss of 1.84 yuan per ton.

Table 3.24 shows that in 1978 the capital profit level of the industrial sector as a whole was 15.5 per cent while for the coal industry it was only 0.3 per cent, but 70.2 per cent for the oil industry.[67] Between 1952 and 1978 the output value of the coal industry grew relatively slowly compared to other industries, expanding from 100 in 1952 (base year index) to only 990.4 in 1978, whereas that of the oil

Table 3.24 Industrial capital profit levels, 1962–78 (%)

	1962	1974	1975	1976	1977	1978
Average for total industry	8.5	13.4	14.1	11.4	12.9	15.5
Coal	−6.0	−2.0	0.1	−1.9	−1.1	0.3
Oil	22.1	75.6	77.5	70.8	74.4	70.2
Electricity	19.8	19.0	17.1	17.4	17.9	18.8
Smelting						9.0
Machinery						12.0
Textiles						37.1
Light industry						24.0

Sources: Lu Qikang, 'Guanyu meitan shengchan chengben yu chu chang jiage wenti de tantao' (Investigation into Coal Production Costs and Producer Prices), *JGLLSJ*, no. 1 (1982): 31; Zhai Ligong, 'Guanyu tiaozheng meitan jiage de tantao' (Investigation into Adjusting Coal Prices), *SXJDYJ*, p. 344.

industry rose to 13,892.4, the chemical industry to 7,625.7, the machinery industry to 4,908.8, and the power industry to 3,493.4.[68]

The low selling price of distributed coal was a major factor in the poor financial performance of the industry. Marxist pricing theory is briefly explained in Chapter 2. Raw materials were not to be sold for profit but at as low a price as possible. The prices did not fully reflect either the different quality characteristics of the coals nor the transport charges. During the first fifteen years of the PRC the average 'producer' or 'supply' price per ton for coal from state mines was raised nominally to cover the differential in production costs across the country. One of the policies during the readjustment years was:

> Strict adherence to the original price levels of eighteen categories of daily necessities accounting for 60 per cent of the spending of industrial and office workers, including grain, edible oils, cotton cloth, cotton textiles and knitwear, fuel coal, basic pharmaceuticals, housing, water and electricity. No fluctuation whatsoever was permitted in these categories.[69]

Early on in the Cultural Revolution, in order to maintain price stability, the Chinese Communist Party Central Committee and the State Council announced the *Regulations on Further Practising Economy in Making Revolution, Restricting the Purchasing Power of Collectives and Strengthening Control Over Funds, Supplies and Prices*. As a result of the enforcement of these regulations there were few changes made to the prices of anything between 1965 and 1977.

From about 1972 the non-state mines were allowed to sell their coal at market prices. However, the prices of output from the state mines were tightly regulated. The single largest increase in average state mine producer prices occurred in 1961, 27.7 per cent over the 1960 level. This increase was an attempt to ameliorate the financial problems of the industry caused by the GLF. Over the same period average production costs per ton ranged from a low of 8.76 yuan in 1959 to a high of 17.30 yuan in 1967, with an annual average of 13.8 yuan. They rose in the early 1970s as existing pits were deepened, more rock was encountered, wages and bonuses were increased, equipment wore out, the prices of other materials increased, and new mines were opened in the difficult terrain of the central and southern regions (see Chapter 2).

The low coal prices in the Soviet Union and East Bloc countries had had the same result as they did in China, that is, the industry's costs were far greater than its income and the government had to subsidize it heavily. In 1967 the Soviet Union's coal industry was in such poor financial shape that the government was forced to raise the price by 80 per cent, which was the largest price increase of all industrial categories, and again in 1982 by another 42 per cent.[70]

A consequence of the low prices and profits in the coal industry, compared to those in the oil and electricity industries, was that the coal industry had less prestige and 'bureaucratic clout'.[71] Had the coal industry made large profits, perhaps it would have been given greater and more consistent attention in terms of planning. Here again is the image problem referred to above.

A major drain on the industry were pensions and other social welfare expenses. Such were the health hazards of working at the coal faces that the men could work only for about twelve years. The average retirement age of 45 years was far lower than for other industries. Also, as most of the mines were situated in remote areas, hospitals, schools, and even universities had to be built and maintained for the miners' families.

It was noted in Chapter 2 that on several occasions the plans were approved and changed after the beginning of the year in which they were meant to be applied. Although control figures were always available, there was an uncertainty about the operations until approved targets were set. Efficiency was affected as a result of decisions being made on the basis of incomplete information. The political convulsions were not the only cause of the frequent revision of targets. The combination of projected rapid growth in all sectors, the impossibility of assessing accurately the requirements of all consumers, and the fact that the planned stockpiles held by producers were small, made coordination a formidable, if not impossible task. Inevitably a miscalculation was made and some consumers were not allocated enough, or were completely overlooked. This meant that several iterations of the balances had to be made and coal targets altered. Late completion and implementation of plans and the continual revision of output targets were also problems of Soviet industry.

Foreign involvement and trade

The Soviet Union had provided technical training to Chinese mine managers both in China and in the USSR, and helped to build mines and install equipment in China. Considerable assistance was also received from Poland, Romania, East Germany, and Czechoslovakia. In the 1970s delegations from various Western countries were permitted to tour the best mines in China, and there were isolated imports of Western machinery, but no large-scale aid or loan contracts were signed.

As for trade, exporting coal was not a priority before 1979. The exports that did occur were mainly part of compensation trade deals with other communist countries, especially North Korea. Between 1953 and 1978 they never amounted to more than 4 million tons per annum, or 2 per cent of total production.[72] Imports, mainly from Australia and North Korea, never exceeded 2.6 million tons.

Transport

By 1936 some 20,740 km of track had been laid in China, mostly in the northeast region, but after the war with Japan and the ongoing civil war from 1945 to 1949, only about 10,000 were serviceable.[73] Thus, one of the first tasks of the Communist government was to repair and consolidate the lines. While some built before 1911 became major trunk lines, the branch lines presented special problems. The foreign business interests had tended to work independently, constructing their own lines for their own specific purposes, resulting in fragmented sub-systems complicated by different gauges. The amount of trackage inherited by the Communist government in 1949 was comparable to what the US had in 1854, and what Russia had in 1878.[74] India and Brazil both had higher railway densities in the early 1950s in terms of kilometres per thousand population and per thousand square kilometres.

As part of their loan package, the Soviets provided almost all rolling stock, which China could not yet produce. After the departure of the Soviet personnel in 1960, railway equipment and expertise were obtained on a barter basis from Poland, Czechoslovakia, and various non-Communist countries. From 1949 to 1979, the main source of capital investment was the central government, and to a lesser extent, the provincial governments.

Due to the 'closed door' policy, and the government's failure to appreciate fully the key economic role of transportation, the total length of the railways in 1978 was relatively small at only 48,618 km (see Table 3.25), and the trackage densities were still relatively low (see Table 3.26). Notwithstanding, apart from the Soviet Union, China had the highest railway freight intensity in the world. Traffic densities of over 25 million ton-miles per route-mile were recorded on the Beijing-Shenyang (Liaoning) and Beijing-Wuhan (Hubei) lines, while in the US in 1972 there was an average of slightly over 3 million tons per route-mile.[75]

Before 1979 China, compared to other countries, invested relatively little in the transportation sector. State capital construction investment allocations were highly erratic (see Tables 3.19 and 3.21), constituting on average 15 per cent of national capital construction investment, which was less than half what it was in many other developing countries.[76]

Investment in railways as a proportion of total capital construction investment averaged 10.8 per cent during the First Five Year Plan (1953–57), 8.8 during the Second (1958–62), 8.4 during the adjustment years of 1963–65, 12.3 during the Third (1966–70), 10.3 during the Fourth (1971–75), and only 6.3 during the Fifth

Table 3.25 Railway trackage and freight carried, 1949–78

	Length 10,000 (km)	Length increment per year (km)	Length increment per year (% change)	Length electrified (km)	Double-tracking (km)	Freight carried (mt)	Freight increment per year (mt)	Freight increment per year (% change)
1949	2.18					55.89		
1950	2.22	400	1.83			99.83	43.94	78.62
1951	2.23	100	0.45			110.83	11.00	11.02
1952	2.29	600	2.69		1,410	132.17	21.34	19.25
1953	2.38	900	3.93			161.31	29.14	22.05
1954	2.45	700	2.94			192.88	31.57	19.57
1955	2.56	1,100	4.49			193.76	0.88	0.46
1956	2.65	900	3.52			246.05	52.29	26.99
1957	2.67	200	0.75		2,203	274.21	28.16	11.44
1958	3.02	3,500	13.11			381.09	106.88	38.98
1959	3.23	2,100	6.95			544.10	163.01	42.77
1960	3.39	1,600	4.95			672.19	128.09	23.54
1961	3.45	600	1.77			449.88	−222.31	−33.07
1962	3.46	100	0.29	100		352.61	−97.27	−21.62
1963	3.50	400	1.16	100		364.18	11.57	3.28
1964	3.53	300	0.86	100		417.86	53.68	14.74
1965	3.64	1,100	3.12	100		491.00	73.14	17.50

Year							
1966	3.78	1,400	3.85	100	549.51	58.51	11.92
1967	3.86	800	2.12	100	430.89	−118.62	−21.59
1968	3.88	200	0.52	100	420.95	−9.94	−2.31
1969	3.93	500	1.29	100	531.20	110.25	26.19
1970	4.10	1,700	4.33	300	681.32	150.12	28.26
1971	4.28	1,800	4.39	300	764.71	83.39	12.24
1972	4.39	1,100	2.57	300	808.73	44.02	5.76
1973	4.43	400	0.91	300	831.11	22.38	2.77
1974	4.51	800	1.81	300	787.72	−43.39	−5.22
1975	4.60	900	2.00	700	889.55	101.83	12.93
1976	4.63	300	0.65	700	840.66	−48.89	−5.50
1977	4.74	1,100	2.38	1,000	953.09	112.43	13.37
1978	4.86	1,200	2.53	1,030	1,101.19	148.10	15.54
				7,630			

Sources: *TJNJ 1989*, p. 396; *TJNJ 1991*, pp. 494, 498; *TJNJ 1998*, p. 544.

Table 3.26 China's railway network and density compared, late 1970s

		Rail length '000 km	km per 1000 population	km per 1000 square km
China	(1981)	53.91	0.05	5.62
Brazil	(1980)	29.66	0.25	3.48
India	(1979)	60.78	0.09	18.50
Japan	(1980)	24.00	0.21	63.55
Korea	(1980)	3.16	0.08	32.07
USA	(1979)	358.90	1.60	38.33
USSR	(1979)	141.10	0.53	6.30

Source: World Bank, *China: The Transport Sector*. Annex 6 to *China: Long-Term Development Issues and Options*, Washington, DC: IBRD/WB, 1985, Table B.29, p. 105. Copyright © 1985 International Bank for Reconstruction and Development/The World Bank. Reprinted by permission of the International Bank for Reconstruction and Development/The World Bank.

(1976–80).[77] The high investment in new railways in the Third Front region between 1964 and 1971 came at the expense of maintenance and expansion of the existing lines elsewhere. It was during this time that serious transportation bottlenecks began to hinder the modernization of the country as a whole.[78]

The technology of the trains was generally several decades behind that in the West. In 1979, 40 per cent of all the steam locomotives had been running for forty years or longer. By that time most countries had long abandoned the use of steam locomotives, which are much less efficient in terms of the amount they can haul per unit of energy used, their ability to function on steep gradients, and their manoeuvrability in the yards. In China at this time, 80.6 per cent of the locomotives were powered by steam, 18.6 per cent by diesel, and only 0.8 per cent by electricity. The first electrification of a line did not occur until 1962, and by 1978 only 1,030 km, or 2.1 per cent of the total trackage was electrified. Only 15.7 per cent of the total operating railways was double-tracked by 1978, and only 12.3 per cent had automatic blocking.[79]

The wooden freight cars, which in 1980 were small by international standards, averaging only 50 tons capacity, were comparable to what had been used in the US in the 1940s. The average freight train in China hauled 2,500 tons in 48–53 cars loaded at full capacity, while in the US 4,500 tons were hauled in 68 cars loaded at only 85 per cent capacity.[80] India had over twice the number of both freight and passenger cars, and 8 per cent more locomotives than China in 1980.[81] Coal used a large proportion of railway capacity in terms both of weight and volume. On average the daily percentage of freight cars devoted to coal was 31.7 in 1952, 46.6 in 1962, and 34.7 in 1978.[82] In Shanxi Province coal occupied an average of 87 per cent of the railway capacity between 1976 and 1981.[83]

The yearly increases in coal production were no where near being matched with similar increases in railway capacity. The negative effects for the coal industry of insufficient investment in the railways were compounded by lack of communication between the Ministries of Coal and Railways. Railway infrastructure plans were set by the SPC. Even if communications between the two Ministries had been good, it was not within the power of the Ministry officials to make substantial, unilateral alterations to the infrastructure allocations and development plans. The management of inter-industry linkages and coordination of infrastructure development were especially poor in central planning. Decision-makers in Beijing were not fully aware of the extent that transport conditions were determining production (*yi yun ding chan*).

Nationally, total coal output increased over nineteen times between 1952 and 1978, but the railway network, in terms of total length of operating rail line, increased only 1.1 times, the number of locomotive engines only 1.35 times, and the number of freight cars only three times.[84] Compared to the US, the USSR, and India, growth in freight train capacity in China lagged the production growth rate of several key commodities, including coal. While China's coal production index was 5.6 in 1978, the railway transportation index was only 5.0. The same indices for the US in 1955 were respectively 0.9 and 1.1; 4.2 and 4.9 for the USSR in 1975; and 1.8 and 2.2 for India in 1978.[85]

Used far beyond its intended capacity, the equipment was constantly breaking down. In 1979 it was found that some 29 per cent of the freight cars sent to seventy-two surveyed mines were defective, having holes and missing doors, panels, and locks.[86] The Ministry had to spend 50 million yuan each year mending them. It was estimated that the loss rate in railway shipments was 3–5 per cent, meaning that over 40 million tons of coal were lost in the yards or along the tracks in 1979 and 1980. The loss rate in waterway shipment varied from 3.85 to 12.15 per cent, with the result that about 4 million tons were lost over the same period. Due to leaking barges and faulty unloading equipment, well over 10,000 tons of coal each year were falling into the river at Shanghai alone. There was also a shortage of warehouses and mechanized loading and unloading equipment both at the railway stations and ports. In the winter, delays were caused by the coal freezing in and onto the cars.

While the industrial centres in the east and south suffered critical shortages, each year enormous piles of coal accumulated at the railway stations in the north, and even at the mine heads, because there was insufficient wagon capacity. In 1980 it was estimated that there were 20 million tons stockpiled throughout the country, half of which were in Shanxi Province alone. Calorific value was lost due to weathering – when not properly stored, bituminous coal can lose up to one-fifth of its heating value and lignite can deteriorate beyond use. Some coal was lost entirely through spontaneous combustion or was washed away by rain. Shanxi alone was losing up to 300,000 tons each year due to exposure to spontaneous combustion and rain.[87] Another problem was pilfering of the coal when loaded wagons sat in station yards. Railway workers and peasants living near the station would help themselves.[88]

Although water transport of heavy cargo had been used for centuries, in the late 1970s it accounted for less than 10 per cent of total coal transport. China has over 3,000 miles of coastline, and in 1978 there were some 136,000 km of navigable inland waterways, but only 42 per cent of these were at least 1 m in depth.[89] Most of the short hauls made by train would have been handled much more efficiently by truck had such vehicles been available.[90] Not only were there not nearly enough trucks being manufactured, there was never sufficient supplies of petrol to operate them. The vehicles were small by international comparison and often loaded and unloaded manually.

When bottlenecks on the main trunk railways became so severe that it was decided to try to transport more coal south down the coast in ships, it became apparent immediately that the loading and unloading capacities of the main harbours in the north and south were mismatched. The southern ports were capable of unloading cargo from much larger ships than the northern ports were capable of loading.

Planners in Beijing organized the transport for all the coal produced by the state mines. The number of linkages was so large that inevitably shipments were frequently late, made via round-about routes, and were often duplicated.[91] For example, at one site all that was needed to transfer the coal from the mine to the nearby thermal plant was a conveyor belt, but the coal was put on a railway that first went 16 km to the east before going by the power plant.[92] In another case, the closest washing plant was 1,188 km away, and the locomotives consumed unnecessarily several extra tons of coal in hauling heavy waste materials with the coal. In 1957, Shanghai's coal requirements were shipped in from thirty-two coal mines, some of which were over 3,000 km away.[93] Inefficient use of railway capacity seems to have been common in many centrally planned economies. In Poland it was estimated that one out of every five railway cars travelled unnecessarily, and in the Soviet Union, misuse of capacity was said to comprise 5–8 per cent of all transportation.[94]

The problems in planning the size and timing of deliveries affected not only coal shipments, but all commodities, as observed by two Chinese economists:

> …poor distribution, planning and management means goods are sent one way along a transport route today, then back along the same route tomorrow, or by railways in one direction and back by water; sometimes goods which are available locally are brought from distant areas; some goods are re-shipped several times; and sometimes goods ideally suited to boat transport are instead shipped by rail, or goods which ought to be transported by truck are put onto boats.[95]

A well-documented consequence of inadequate transportation capacity relative to production in command economies was widespread hoarding, which became a major contributing factor to shortages of all goods, including coal.[96] The shortages of coal in China were such that it was continually bartered and, indeed, coal itself became a form of currency. In frustration with the inability to communicate with each other and the amount of time wasted in making requests through the proper channels, consuming operations and enterprises habitually requested and stocked larger amounts of all supplies than they required in order to hedge against future supply problems, and there was no enforced penalty. Holding large stocks of coal created severe and self-perpetuating imbalances in the production–distribution–consumption cycle. As time went on the planners in Beijing became less able to ascertain accurately which consumers had legitimate needs for coal and which were hoarding extra supplies.

While some enterprises had large unreported stockpiles and railway storage yards were filled to capacity, others continually had urgent need of more.[97] This is exactly the situation Kornai identified in the East European economies: '… shortage and slack are simultaneously present in a resource-constrained economy … the hoarding tendency and, together with this, the almost insatiable demand for inputs for current production represent one of the fundamental processes that 'siphon off' external slack (in this case the seller's output stock) from the economy. In this way they intensively promote permanent reproduction of the shortage in the economy.'[98] Speaking of supply problems in general, one Chinese economist observed:

> Because of the administrative means of managing resources, some departmental comrades are accustomed to devoting their energy towards 'fighting' over resources, with the result that the more the 'fighting', the worse the shortages, and the worse the disputes. Over the past thirty years, the management of materials has been changed several times, but the problems with the centralized administrative way of managing materials have not been solved.[99]

Wan Jing, another economist, noted that the only restriction on hoarding was the inconvenience of providing storage.[100] In China, the fact that not all of the coal produced came under plan added to the confusion. Coal sold on the black market frequently occupied space allocated to other consumers.[101] Transport problems were not unique to China. Much of the railway stock in Russia was old and in poor condition, and the shortage of railway cars was given as one of the official reasons why production in the USSR fell in 1978.[102] Describing the situation in India in 1975, one writer observed: 'It is difficult to think of any sector or aspect of the economy of India in which an improvement in performance is more urgently needed.'[103]

Consumption

Consumption and economic growth

In 1978, 30 per cent of the industrial productive capacity could not function due to a lack of sufficient energy. This is estimated to have caused a loss of over 100 billion yuan, or 22 per cent of gross value of industrial and agricultural output for that year.[104] Most urban areas were put on fixed electricity use schedules, and in order to share the available electricity each sector of the economy was ordered to work without heat and light for one or two separate days, or for three or four consecutive days per week.

In 1979 some 37 per cent of the rural villages still had no electricity at all and had no option but to continue, as for centuries before, to gather whatever burnable material they could find for cooking and heating purposes.[105] Most of the crop residues and dung which ought to have been left on the soil to maintain its nutrients and prevent erosion were used as fuel.[106] The amount of land lost through erosion increased from 1.16 million square kilometres in the early 1950s to 1.5 million in the early 1980s.

One standard technique used to measure the relationship between energy consumption and economic growth is to calculate the energy consumption elasticity coefficient by dividing the energy consumption growth rate for any particular year by the gross domestic product (GDP) growth rate.[107]

A high coefficient is typical of developing countries because, as modernization progresses, the energy consumption rate is usually higher than the growth rate of the economy as a whole, due to the fact that the construction of new industries, infrastructure, commercial development, housing, etc., requires considerable amounts of energy. Furthermore, developing countries tend to have only small amounts of energy-efficient equipment and many fewer management staff trained in the efficient use of energy, if any at all.

Conversely, the energy consumption growth rate in developed countries is generally less than the growth rate of the economy, and hence the coefficient is low. The industrial infrastructure is largely in place, economic activity is usually more consumer- and service-oriented, and more energy-efficient equipment is in use. Per capita energy consumption is high, the annual increases are small, and any extra consumption is by choice, not necessity.

It is difficult to make meaningful international comparisons of energy elasticity and intensity data. For example, when calculating the elasticity coefficient, the consumption of non-commercial fuels is generally excluded, and this exclusion can distort the true energy picture in developing countries where these fuels are the mainstay in rural areas. There are many factors which affect elasticity coefficient data, and many of these are difficult to quantify: climate; prices of domestic and foreign energy; efficiency of energy-consuming equipment; industrial productivity levels; culture; planned economic growth rates; relative availability, ease of distribution, and quality of the various types of energy resources; structure of income, population, and industry; living standards; etc.

Changes in incomes and energy prices are generally considered to be the most significant factors influencing energy consumption. In market economies, consumers are quickly appraised of changes in energy prices, however small or infrequent, and respond immediately, but in centrally planned economies prices and exchange rates were set by state planners and normally altered infrequently.

Between 1971–73 and 1974–77 the energy elasticity coefficients of most countries decreased. In other words, each one per cent increase in GDP required less energy in 1974–77 than in 1971–73. There were two factors at play: first, the decrease in the growth of energy consumption accompanying economic maturity, and second, the sudden effects of the oil shock in 1973 when cutbacks in oil production and shipment embargoes were used as an economic weapon in the

Arab–Israeli conflict. The price of oil more than quadrupled, causing frantic efforts in every importing country to reduce energy consumption and improve consumption efficiency. Some countries managed to reduce the amount of energy consumed per unit of GDP by over half. For others it was impossible to change the structure of their economies and/or fuel mix, or readily improve the efficiency of their equipment.

Table 3.27 shows the tremendous range of China's elasticity coefficients for each of total energy, electricity, and coal consumption between 1954 and 1978. These figures reflect the extreme fluctuations over those years in the growth rates of GDP and energy consumption, which resulted primarily from the political upheavals. As China had not been importing oil and had virtually no trade with countries that did, the effects of the oil crisis were not felt. In fact, isolation meant

Table 3.27 Consumption elasticity coefficients, 1954–78

	Growth rates (%)					Elasticity coefficients		
	Energy consump.	*Electricity consump.*	*Coal consump.*	*GDP*	*Industrial sector*	*Energy*	*Electricity*	*Coal*
1954	15.2	19.6	14.1	4.2	11.8	3.62	4.67	3.36
1955	11.8	11.8	11.2	6.8	2.8	1.74	1.74	1.65
1956	26.3	35.0	26.0	15.0	20.1	1.75	2.33	1.73
1957	9.6	16.3	9.1	5.1	8.0	1.88	3.20	1.78
1958	82.5	42.5	87.1	21.3	53.6	3.87	2.00	4.09
1959	36.0	53.8	36.0	8.8	35.7	4.09	6.11	4.09
1960	26.2	40.4	25.1	−0.3	7.0			
1961	−32.5	−19.2	−34.3	−27.3	−44.2	1.19	0.70	1.26
1962	−18.9	−4.6	−20.7	−5.6	−16.5	3.38	0.82	3.70
1963	−5.9	7.0	−6.2	10.2	14.8		0.69	
1964	6.9	14.3	5.7	18.3	21.7	0.38	0.78	0.31
1965	13.6	20.7	11.6	17.0	23.8	0.80	1.22	0.68
1966	7.2	22.0	7.0	10.7	16.1	0.67	2.06	0.65
1967	−9.6	−6.2	−11.1	−5.7	−14.2	1.68	1.09	1.95
1968	0.4	−7.5	−0.7	−4.1	−7.1		1.83	0.17
1969	23.5	31.3	20.8	16.9	31.0	1.39	1.85	1.23
1970	28.9	23.3	27.2	19.4	27.5	1.49	1.20	1.40
1971	17.8	19.4	15.3	7.0	14.9	2.54	2.77	2.19
1972	8.1	10.1	5.8	3.8	6.5	2.13	2.66	1.53
1973	4.9	9.5	1.3	7.9	8.3	0.62	1.20	0.16
1974	2.6	1.2	−1.1	2.3	−0.6	1.13	0.52	
1975	13.2	16.0	12.7	8.7	14.7	1.52	1.84	1.46
1976	5.3	3.7	2.4	−1.6	1.9			
1977	9.5	10.0	10.0	7.6	11.3	1.25	1.32	1.32
1978	9.2	14.9	9.8	11.7	13.0	0.79	1.27	0.84

Sources: GDP data from *TJNJ 1999*, p. 57; coal consumption data from *NYTJNJ 1991*, p. 136; industry growth data calculated from Beijing Review Press, *The Development of China 1949–1989, China Issues and Ideas*, Beijing: 1989, Table 1.9, p. 22 and Table 1.15, p. 27; price index from *TJNJ 1981*, p. 403.

Note: The GDP data was converted to 1950 constant prices.

that China's consumption efficiency worsened with time and the authorities were hardly aware of it. Whereas the average annual energy conservation rate in some major developed countries in the years immediately after 1973 was above 1.38 per cent, in China it was −2.85.[108]

Between 1954 and 1978 the average yearly increase in GDP was 6.32 per cent and the average yearly increases in energy, electricity, and coal consumption were 11.67, 15.41, and 10.56 per cent, respectively, resulting in overall energy elasticity coefficients of 1.85, 2.44, and 1.67.[109] In other words, each 1 per cent increase in GDP required a 1.85, 2.44, and 1.67 per cent increase respectively in energy, electricity, or coal consumption.

The average yearly coal consumption growth rate was very high at the time of the GLF. In 1958, 1959, and 1960 it grew at 87.1, 36.0, and 25.1 per cent, respectively, but in 1961, 1962, and 1963, the 'bitter years' following the GLF, it fell to −34.3, −20.7, and −6.2 per cent. GDP growth rate over these six years was 21.3, 8.8, −0.3, −27.3, −5.6, and 10.2 per cent, respectively. The worst elasticity coefficients were for the year 1959: 4.09 per cent for total energy and coal consumption, and 6.11 for electricity consumption.

During 'the adjustment years', 1963–65, the average annual growth rate of coal consumption was only 3.7 per cent, but that of GDP in constant terms, 15.2 per cent, with a resulting very low elasticity coefficient of 0.24. This reflects the government's change from emphasizing heavy industry to light industry and agriculture. During the chaotic years of the Cultural Revolution, 1966–76, coal consumption grew at an annual average rate of 7.2 per cent, while GDP grew at an average annual rate of 5.9 per cent, with a resulting coefficient of 1.22. There were two years of negative growth in GDP: 1967 and 1968 during the early years of the Cultural Revolution, and 1976 at the end. However, in 1970 there was an extraordinary 19.4 per cent surge in growth, the highest ever recorded in China's history after the GLF, reflecting a degree of recovery after the worst years of the Cultural Revolution.

Generally speaking, industry grew at a higher rate than the economy as a whole. Looking at the very high figures it is difficult to believe there were energy shortages. However, this enviable growth took place largely at the expense of civilian comfort and convenience. The affects of the GLF and Cultural Revolution on this sector were much greater than economywide growth.

Another method for examining the relationship between coal consumption and economic growth is to calculate the comprehensive coal consumption ratio, derived by dividing total coal consumption by total GDP. Table 3.28 shows that coal consumption per 10,000 yuan of GDP more than doubled over the period. The effects of the GLF are very evident.

The overall picture from these indicators is that the efficiency of coal consumption worsened sharply in the early years of the GLF, recovered slightly, then almost steadily deteriorated after the beginning of the Cultural Revolution.

By international standards China's energy consumption was very inefficient.[110] In 1978 China consumed 24.62 million tons of standard fuel per every

$US10,000 (at 1970 prices) of GNP, while each of the US, the USSR, and UK consumed just under half that amount, at respectively 11.99, 11.29, and 11.88 million tons. Japan consumed less than one-third at 7.07; France about a quarter at 6.10; and West Germany less than one quarter at 5.77.[111] In 1980 China produced just over US$470 of industrial output value per ton of energy consumed, very little relative to the US$1,000 in India; US$1,064 in the USSR; US$1,108 in the US; US$2,334 in the FRG; and US$2,548 in Japan.[112]

Compared to the rest of the world the consumption of electricity, which was generated mostly from coal-fed thermal plants, was also highly inefficient. In

Table 3.28 Comprehensive coal consumption ratios, 1953–78

	tons of coal consumed per 10,000 yuan of GDP
1953	10.03
1954	11.24
1955	11.91
1956	13.29
1957	14.16
1958	21.70
1959	27.05
1960	34.47
1961	31.43
1962	27.45
1963	22.58
1964	19.50
1965	17.95
1966	17.59
1967	16.34
1968	16.71
1969	17.74
1970	19.37
1971	20.58
1972	20.92
1973	19.74
1974	19.15
1975	20.12
1976	21.05
1977	21.72
1978	21.21

Sources: Coal consumption data from *NYTJNJ 1991*, p. 136; price index from *TJNJ 1981*, p. 403; GDP data from *TJNJ 1999*, p. 57.

Note: The GDP data was converted to 1950 constant prices.

1978 consumption at 1.83 kWh per US$90 was second highest in the world after Bulgaria's 1.99 kWh, and four times the world average of 0.45 kWh.[113] The next highest were 1.73 in Zambia and 1.65 in Romania. The average was 0.41 in Africa, 0.31 in Latin America, and 0.39 in Asia, of which it was 0.50 on Taiwan, 0.28 in Hong Kong, and 0.53 in India. It stood at 1.43 in the USSR, 1.00 in non-OECD Europe, and 0.32 in OECD Europe.

According to one estimate, China's energy efficiency in the late 1970s compared to the average for the developed world was less than half in the industrial and private sectors, and about 83 per cent in the thermal electricity generation and transmission sector.[114] The general efficiency for the economy as a whole was about 61 per cent of the average for the developed world.

Breakdown of coal consumption

In 1980 coal accounted for 43 per cent of the energy consumed in the agricultural sector, 45 per cent in construction, 50 per cent in transportation and communications, 66 per cent in the commerce and public catering sector, 15 per cent in non-productive activities, and 91 per cent in the residential sector.[115]

Table 3.29 provides a breakdown of coal consumption from 1953 to 1978. In 1953 the productive: non-productive consumption ratio was 58 : 42. In other words, 58 per cent of the coal was devoted to industry, coke production, power generation, railways, and other industries, while 42 per cent was devoted to home and commercial heating and cooking. By 1978 the ratio had changed drastically to 82 : 18. The proportion of coal devoted to power generation more than doubled to 20 per cent, while that devoted to the railways halved. The proportion to other industries, which includes specific industries such as steel, synthetic ammonia, cement, and coal mining also rose sharply from 30 to 47 per cent.

Obviously, it was the priority of the government that as much of the country's energy resources as possible be allocated to industry. The amount allocated to the civilian sector was minimal, and did not reflect at all what Chinese citizens would have liked. Between 1953 and 1965, total household and commercial consumption increased over 150 per cent, but between 1965 and 1978 it increased only 34 per cent. This shows clearly how the household and commercial sectors were squeezed.

The household sector in China accounted for only 17.2 per cent of *total* energy consumption in 1979, whereas it accounted for 38 per cent in the US, 26 per cent in England, 36.3 per cent in France, and 20.0 per cent in Japan. Industry accounted for 69.4 per cent in China, while it accounted for 36 per cent, 37.9 per cent, 44.5 per cent, and 57.0 per cent in the same four countries respectively, 27.7 per cent in India, and 68.2 per cent in the USSR.[116] While China's total population increased by nearly 70 per cent between 1952 and 1978, per capita energy consumption increased nearly 650 per cent, but at that to only 805 kg of coal equivalent in 1978 (and this was relatively inefficient consumption), whereas it was 11,374 in the US, 7,121 in East Germany, 5,500 in the USSR, 3,825 in Japan,

Table 3.29 Breakdown of coal consumption, 1953–78

	Total consumption	Productive sector	Power generation	Coke production	Railways	Other productive	Residential and commercial
1953 mt	71.49	41.51	6.94	6.27	7.21	21.09	29.98
%	100.0	58.0	9.7	8.8	10.1	29.5	42.0
1957 mt	124.69	70.56	12.23	13.73	9.34	35.26	54.13
%	100.0	56.6	9.8	11.0	7.5	28.3	43.4
1962 mt	206.69	133.70	30.47	18.84	15.15	69.24	72.99
%	100.0	64.7	14.7	9.1	7.3	33.5	35.3
1965 mt	228.84	153.84	38.49	20.38	15.65	79.32	75.00
%	100.0	67.2	16.8	8.9	6.8	34.7	32.8
1978 mt	565.64	465.01	113.44	59.30	26.30	265.97	100.63
%	100.0	82.2	20.1	10.5	4.7	47.0	17.8

Source: Based on *CIYB 1982*, Tables 11 and 12, p. 29.

and 2,461 in Singapore.[117] India's per capita energy consumption was considerably behind China's at 176 kg of coal equivalent.

The efficiency of coal consumption in the industrial sector was particularly poor. Table 3.30, which gives energy consumed in kilograms of coal equivalent per 1980 US dollar's worth of GVIO, shows that China consumed over four times as much as various high-income countries, over twice as much as various middle-income countries, and about the same as India. In steel production, when comparing 1979 statistics for the more efficient Chinese steel mills (Anshan, Wuhan, and Baoshan) with 1977 data for the Nagoya steel mill in Japan, it was found that China consumed 17 per cent more energy in iron-smelting, 12 per cent more in coking, 40 per cent more in sintering, fourteen times more in steel smelting in Bessemer converters, 109 per cent more in pre-rolling, and 166 per cent more in steel-rolling.[118] The efficiency rates at the older, smaller mills were far worse.

Efficiency of consumption in the household sector was, not surprisingly, also low with China consuming over three times the amount of energy that other low-income countries did per unit of GDP, and about five times the average for the high-income countries (see Table 3.31).

Coal in thermal electric power generation

During the First Five Year Plan period about 10 per cent of all the coal output was devoted to power generation, and thereafter the proportion rose slowly to 20 per cent in 1978, compared to an average of about 37 per cent of that in the Pacific countries, 55 in the European, 58 in the OECD, and 63 in the North American in the same year.[119] Over the thirty years from 1949 to 1978 the ratio of thermal to hydroplant generation varied from about 75 : 25 to 90 : 10. However, installed thermal electricity generation capacity as a proportion of total electricity generation capacity decreased from 91.4 per cent in 1949 to 69.7 per cent in 1978

Table 3.30 Comparative energy consumption per unit of gross value industrial output, 1980 (kilograms standard coal equivalent per 1980 $US)

Developing countries		
China	1980	1.06
Brazil	1978	0.40
India	1978	1.04
S. Korea	1980	0.48
Philippines	1979	0.66
Turkey	1979	0.44
Developed countries		
France	1980	0.30
Germany, Fed. Rep.	1980	0.26
Japan	1980	0.30
UK	1980	0.23
US	1980	0.47

Source: Adapted from World Bank, *China: The Energy Sector*. Annex 3 to *China: Long-Term Development Issues and Options*, Washington, DC: IBRD/WB, 1985, Table 1.8, p. 14. Copyright © 1985 International Bank for Reconstruction and Development/The World Bank. Reprinted by permission of the International Bank for Reconstruction and Development/The World Bank.

Note: Includes biomass consumption.

Table 3.31 Comparative energy consumption in the residential/ commercial sector per unit of GDP, 1980 (kilograms standard coal equivalent per 1980 $US)

Developing countries	
China	1.14
Argentina	0.10
Brazil	0.19
Mexico	0.11
India	0.83
Korea, Rep.	0.48
Developed countries	
Canada	0.45
France	0.14
Germany, Fed. Rep.	0.18
Italy	0.16
Japan	0.13
UK	0.22
US	0.35

Source: Adapted from World Bank, *China: The Energy Sector*. Annex 3, *China: Long-Term Development Issues and Options*, Washington, DC: IBRD/WB, 1985, Table 1.7, p. 12. Copyright © 1985 International Bank for Reconstruction and Development/The World Bank. Reprinted by permission of the International Bank for Reconstruction and Development/The World Bank.

Note: Includes biomass consumption.

(see Table 3.32). The fuel used in the thermal plants in 1978 was 73.3 per cent coal, 24.7 per cent oil, and 2.0 per cent gas.[120]

Of the projects with which the Soviets assisted in the First Five Year Plan, 76 were thermal plants. Equipment and technical assistance were also received from Czechoslovakia, East Germany, Romania, and Hungary. Table 3.32 shows that installed thermal generating capacity increased from 1.69 million kW in 1949 to 39.84 in 1978, while the amount of electricity generated rose from 3.6 billion kWh to 212.0. The average growth rates between 1953 and 1978 were 13.3 and 16.2, respectively (see Table 3.33). As with the coal industry, the disruption caused by Mao's campaigns in this period was reflected in irregular annual increments in thermal capacity.

Tables 3.19–3.21 give the amount of investment in thermal and hydro plants in total, and as proportions of total energy and industrial investment. The figures reflect the fact that in the late 1950s it was hoped hydroelectric plants could play a larger role. However, it was found they took too long to build and were too capital intensive, and then in the 1970s hope was transferred to having more oil-fired plants. Domestic oil supplies proved too limited and the plants that had been converted to oil had to be reconverted to coal.

In 1978 China's output of total electricity generated ranked sixth in the world but all of the main cities were forced to ration the available power. Households used electricity almost exclusively for lighting. Per capita consumption of electricity increased from 12.8 kWh in 1952 to 268.4 in 1978.[121] For comparison, in 1978 the average was 159 kWh in Asia, of which it was 1,946 on Taiwan, 1,981 in Hong Kong, and 138 in India. The average was 335 in Africa, 772 in Latin America, 4,175 in the USSR, 2,639 in non-OECD Europe, 3,890 in OECD Europe, and 1,684 in the world.[122]

On the eve of the reform era about 70 per cent of the thermal power was generated by several thousand plants of between 30 and 100 MW, most of which were built in the 1950s, though some were much older – up to sixty years old.[123] The rest was generated by about 1,340 plants with capacities of over 500 MW, and 56 plants in the 250–500 MW range.[124] The first 100 MW unit in the US had been put into operation before 1930, and in the Soviet Union in 1939. However, China did not have one even of this limited capacity until 1968.[125] The largest thermal plant, at Qinghe (Liaoning), was only 1,100 MW, whereas many other countries by this time had 3,000 MW plants and larger.

The thousands of small thermal plants, which were the cheapest and most quickly built sources of electricity available to the rural areas, were highly inefficient and a main source of what in the West would have been unacceptably high pollution levels. Coal consumption in grams of standard coal equivalent per gross kWh at thermal electric plants of 6 MW or larger decreased from 727 in 1952 to 434 in 1978, but this was much higher than the 360 required in the US and 330 in the USSR.[126] The total capacity of the plants was nowhere near enough to meet the demand, and they were used far in excess of their designed capacity. The capacity factor was 0.61 in 1978, and they were operated on average for

Table 3.32 Electricity capacity and generation, 1949–78

	Capacity (million kW)					Generation (billion kWh)				
	Total	Thermal	% Share	Hydro	% Share	Total	Thermal	% Share	Hydro	% Share
1949	1.85	1.69	91.4	0.16	8.6	4.3	3.6	83.7	0.7	16.3
1950	1.87	1.70	90.9	0.17	9.1	4.6	3.8	82.6	0.8	17.4
1951	1.88	1.70	90.4	0.18	9.6	5.7	4.8	84.2	0.9	15.8
1952	1.96	1.77	90.3	0.19	9.7	7.3	6.0	82.2	1.3	17.8
1953	2.35	2.02	86.0	0.33	14.0	9.2	7.7	83.7	1.5	16.3
1954	2.60	2.19	84.2	0.41	15.8	11.0	8.8	80.0	2.2	20.0
1955	3.00	2.50	83.3	0.50	16.7	12.3	9.9	80.5	2.4	19.5
1956	3.83	2.93	76.5	0.90	23.5	16.6	13.1	78.9	3.5	21.1
1957	4.63	3.61	78.0	1.02	22.0	19.3	14.5	75.1	4.8	24.9
1958	6.29	5.07	80.6	1.22	19.4	27.5	23.4	85.1	4.1	14.9
1959	9.54	7.92	83.0	1.62	17.0	42.3	37.9	89.6	4.4	10.4
1960	11.92	9.98	83.7	1.94	16.3	59.4	52.0	87.5	7.4	12.5
1961	12.86	10.53	81.9	2.33	18.1	48.0	40.6	84.6	7.4	15.4
1962	13.04	10.66	81.7	2.38	18.3	45.8	36.8	80.3	9.0	19.7
1963	13.33	10.90	81.8	2.43	18.2	49.0	40.3	82.2	8.7	17.8
1964	14.06	11.38	80.9	2.68	19.1	56.0	45.4	81.1	10.6	18.9
1965	15.08	12.06	80.0	3.02	20.0	67.6	57.2	84.6	10.4	15.4
1966	17.02	13.38	78.6	3.64	21.4	82.5	69.9	84.7	12.6	15.3
1967	17.99	14.15	78.7	3.84	21.3	77.4	64.3	83.1	13.1	16.9
1968	19.16	14.77	77.1	4.39	22.9	71.6	60.1	83.9	11.5	16.1
1969	21.04	15.99	76.0	5.05	24.0	94.0	78.0	83.0	16.0	17.0
1970	23.77	17.54	73.8	6.23	26.2	115.9	95.4	82.3	20.5	17.7
1971	26.28	18.48	70.3	7.80	29.7	138.4	113.3	81.9	25.1	18.1
1972	29.50	20.80	70.5	8.70	29.5	152.4	123.6	81.1	28.8	18.9
1973	33.92	23.62	69.6	10.30	30.4	166.8	127.9	76.7	38.9	23.3
1974	38.11	26.29	69.0	11.82	31.0	168.8	127.4	75.5	41.4	24.5
1975	43.41	29.98	69.1	13.43	30.9	195.8	148.2	75.7	47.6	24.3
1976	47.15	32.50	68.9	14.65	31.1	203.1	157.5	77.5	45.6	22.5
1977	51.45	35.69	69.4	15.76	30.6	223.4	175.8	78.7	47.6	21.3
1978	57.12	39.84	69.7	17.28	30.3	256.6	212.0	82.6	44.6	17.4

Sources: Sinton, Jonathan E. *et al.* (eds), *China Energy Databook*, Berkeley, CA: Ernest Orlando Lawrence Berkeley National Laboratory Report LBL-32822, Rev. 4, Sept. 1996, Tables II.17 and II.19, p. II–60, II–61 and II–64, citing *ZGNY*, various years; *TJNJ*, various years; *China Energy Annual Review*, China Energy Research Society, State Statistical Bureau; and Ministry of Electric Power Industry. Copyright © 1996 Lawrence Berkeley National Laboratory. Reprinted by permission of the Lawrence Berkeley National Laboratory, University of California.

6,018 hours per year, up from 3,457 hours in 1952.[127] Some in Beijing and Shanghai were run for more than 7,000 hours.[128]

The transmission equipment and lines were also of relatively low capacity, and the main grids were not connected. Before 1979 it was not possible to transmit electricity from surplus to deficit areas. The load factors were very high and there were rarely any reserves. In 1949 there were 11,410 km of 110 kV transmission lines. The first 220 kV line was built in 1954, and the first 330 kV line in 1972.[129] By 1979 there were 32 power grids, and 61,000 km of 110 kV lines, 25,000 of 220 kV lines, and 81 of 330 kV lines.[130] No 500 kV lines were built before 1980.

Table 3.33 Electricity capacity and generation growth, 1950–78 (%)

	Capacity			Generation		
	Total	*Thermal*	*Hydro*	*Total*	*Thermal*	*Hydro*
1950	1.1	0.6	6.3	7.0	5.6	14.3
1951	0.5	0.0	5.9	23.9	26.3	12.5
1952	4.3	4.1	5.6	28.1	25.0	44.4
1953	19.9	14.1	73.7	26.0	28.3	15.4
1954	10.6	8.4	24.2	19.6	14.3	46.7
1955	15.4	14.2	22.0	11.8	12.5	9.1
1956	27.7	17.2	80.0	35.0	32.3	45.8
1957	20.9	23.2	13.3	16.3	10.7	37.1
1958	35.9	40.4	19.6	42.5	61.4	−14.6
1959	51.7	56.2	32.8	53.8	62.0	7.3
1960	24.9	26.0	19.8	40.4	37.2	68.2
1961	7.9	5.5	20.1	−19.2	−21.9	0.0
1962	1.4	1.2	2.1	−4.6	−9.4	21.6
1963	2.2	2.3	2.1	7.0	9.5	−3.3
1964	5.5	4.4	10.3	14.3	12.7	21.8
1965	7.3	6.0	12.7	20.7	26.0	−1.9
1966	12.9	10.9	20.5	22.0	22.2	21.2
1967	5.7	5.8	5.5	−6.2	−8.0	4.0
1968	6.5	4.4	14.3	−7.5	−6.5	−12.2
1969	9.8	8.3	15.0	31.3	29.8	39.1
1970	13.0	9.7	23.4	23.3	22.3	28.1
1971	10.6	5.4	25.2	19.4	18.8	22.4
1972	12.3	12.6	11.5	10.1	9.1	14.7
1973	15.0	13.6	18.4	9.4	3.5	35.1
1974	12.4	11.3	14.8	1.2	−0.4	6.4
1975	13.9	14.0	13.6	16.0	16.3	15.0
1976	8.6	8.4	9.1	3.7	6.3	−4.2
1977	9.1	9.8	7.6	10.0	11.6	4.4
1978	11.0	11.6	9.6	14.9	20.6	−6.3

Source: See Table 3.32.

Factors contributing to poor consumption efficiency

A major factor in China's poor energy efficiency was the structure of the economy. Between 1952 and 1978 GDP structure changed from 50.5 per cent primary activities, 20.9 secondary of which industry was 17.6, and tertiary activities 28.6, to 28.1 primary, 48.2 secondary of which industry was 44.3, and tertiary 22.3.[131] Light industry as a proportion of total agricultural and industrial output value increased only slightly from 29.4 per cent in 1953 to 32.4 in 1978, while the proportion of heavy industry more than doubled from 17.5 per cent to 42.8.[132] Average coal consumption at 136,212 tons per 100,000 yuan of heavy industrial output value in 1980 was nearly six times greater than for light industry at 23,458.[133]

In combination with this weighting towards heavy industry was the dominant role of coal. Although the proportion of total energy consumed in the form of coal dropped very significantly between 1952 and 1978, from 94.3 per cent to 70.7, China's reliance on this, the least efficient fossil fuel, was far greater than in most countries (see Table 3.34).[134]

The large proportion of unwashed coal markedly reduced consumption efficiency (see first section of this chapter). Abnormally, large amounts were needed in the industrial boilers and in thermal energy production. Some 18 per cent of the coal supplied to the railways was wasted because it had not been treated. In a shipment of 300 million tons of coal 15 per cent would be waste rock, causing the engines to consume an extra 1.7 million tons of coal.[135]

Possibly up to 65 per cent of the low efficiency of industrial energy consumption was attributable to the mismatch of coal-consuming equipment with available types of coal. Different kinds of coal have different thermal characteristics, and ideally the various coal-consuming industries are supplied with the most appropriate type and form. For example, boilers, kilns, and power generators operate best with steam coal.[136] However, in some provinces, 70–80 per cent of the fuel used was coking coal.[137]

In 1980 some 80 per cent of the coal supplied to the chemical fertilizer industry was of the wrong type. Rather than the preferred medium-sized anthracite, shipments of small-sized unwashed coal and coal dust were supplied, and as a result, 5 million tons of coal were consumed unnecessarily.[138] There could have been energy savings of 15–20 per cent had lump or pellet coal been used by the railways and chemical industry, and of 3–5 per cent had powdered coal been used in electricity generation and civilian consumption.

The lack of recyclable scrap metal also greatly reduced China's energy consumption efficiency. In developed countries, large amounts of steel scrap are available. However, China's scrap usage, though similar to Japan's, was 10–20 per cent lower than in other major steel-producing countries.[139]

Table 3.34 Coal consumption in selected countries as a percentage of total energy consumption, 1978

Australia	46.9	France	19.8
Algeria	4.2	Germany	31.7
Egypt	3.8	Austria	18.1
Nigeria	6.2	Sweden	4.4
Portugal	5.8	Switzerland	1.3
US	20.8	Netherlands	5.6
Brazil	9.5	New Zealand	14.3
Mexico	8.5	Hungary	36.0
Japan	17.4	Poland	82.6
Malaysia	0.4	Romania	21.7
Bangladesh	10.5	USSR	34.7
Pakistan	6.3	East Germany	72.6
Belgium	23.6	Czechoslovakia	70.8

Source: United Nations, Dept. of International Economic and Social Affairs, Statistical Office, *World Energy Supplies*, New York, 1979.

Note: The average for this group of countries is 22.2 per cent.

The technology in China's industrial sector was generally very basic. Of the 26,000 types of products manufactured by the Ministry of Machine Building in the late 1970s, 55 per cent were of the 1940–50s generation, 40 per cent of 1960s, and only 5 per cent of 1970s.[140] The persistent use of such old equipment was due to the planned low rates of depreciation, the lack of capital to buy new equipment, and the fact that China's engineers, isolated from the West, were simply unaware of new energy-saving technologies. Many of the iron and steel plants were of 1950s or earlier vintage and consumed large amounts of coal. About a third of China's steel-making capacity still used open hearth furnaces which in the West had long ago been decommissioned and replaced with the much more efficient electric arc.

Another factor was the size of the plants. According to one estimate, the much more numerous medium- and small-sized iron and steel plants consumed 3.23 more tons of standard coal per ton of steel produced than the larger ones.[141] During the 'Backyard Furnace Campaign' in 1958 there was appalling wastage of coal in steel-making, and much of the iron and steel produced from the thousands of small blast furnaces was of such poor quality it could not be used. Up to five tons of raw coal were required to produce one ton of pig iron in 1958, and consumption in 1960 could have been reduced by about 55 million tons had the 1957 level of thermal efficiency been maintained.[142]

The technical level of China's 180,000 industrial boilers was also basic. In 1980 they had only 55 per cent efficiency, some 20–30 per cent less than in the industrialized world.[143] The efficiency rate for industrial ovens was even worse than for boilers, at 20–30 per cent in China, compared to 50–60 per cent in the West and in Japan.[144]

Old technology also affected the efficiency of the chemical industry. For example, in the early 1980s the chemical industry in Shanghai accounted for one-sixth to one-seventh of the nationwide gross value of production in the industry and generated over one-quarter of the profits. Yet most of the equipment dated to the 1930s and 1940s.[145] The same was true for ammonia production. To produce one ton of ammonia required 1.2 tons of standard coal in Japan, but 2.84 tons in China.[146] The Chinese equipment was of small capacity and low technological standard.

It was explained earlier that coal prices were deliberately kept low. Consequently, consumers of coal in all the socialist countries were quantity takers and they had no incentive to economize in their use of it. As output targets rather than profits were the ultimate gauge of success for them as well, they concentrated on increasing production, almost entirely ignoring the costs. The prevailing attitude was *de bu chang xian* – what one gains cannot offset the losses, or, it is not worth the effort. Industrial consumers had nothing to lose by their wasteful practices and nothing to gain if they spent time and materials trying to improve production efficiency.

The system was such that top level planners were fully occupied trying to ensure that all of the various energy-consuming enterprises reached their output targets and that the products were delivered to the specified ministries. There were no energy conservation targets in the pre-reform years.[147] At the production level, rewards were given to managers for meeting the output targets, but none were given for building improved or specialized types of boilers, railway locomotives, thermal plant equipment, etc., or devising more efficient operation plans. There was no point to doing anything other than to produce more of the same product, and to continue in established work routines rather than experiment with new ideas. In other words, no one at any level would risk the time, effort, and resources to deviate from fulfilling the assigned output targets. This is a recognized problem of command planning. One Soviet specialist put it this way:

> No Soviet producer has to worry about disposing of what he produces, and the problem of the enterprise director is only to fulfill the plan. So unless a pressure for innovation has been built into the plan by people at some higher level, the manager has no incentive to produce a better product... Indeed, he has a vested interest in continuing to produce his present kind of output even if it is technically obsolete.[148]

Besides, there simply was not enough 'slack' in the system for such experimentation. Even if those operating the boilers, thermal plants, etc., had ideas for improving their operations, they could not test them because they were allocated only sufficient capital and materials to fulfil the output targets, nothing more. This tautness is also a characteristic of central planning.

Possibly, 35 per cent of coal wastage was attributable to poor management.[149] A low average level of skills and few trained technicians and engineers are general problems in any developing country. However, in China the situation had been exacerbated by the deliberate demotion of skilled people. During the First Five Year Plan the few fully qualified engineers available were placed in appropriate positions, but later the use of professional advice was shunned, first in the GLF and then more especially during the Cultural Revolution (see Chapter 2). Party personnel intentionally selected to 'manage' the industrial enterprises, and chosen solely on the basis of political criteria, often had little expertise in management. They were satisfied merely to see that their output targets were reached and were either completely unaware that operations could be carried out more efficiently, or were without the knowledge to improve them. The following description of the situation in Hungarian steel mills could have been written about China's steel mills:

> If molten ore poured forth from the furnaces and steel from the Bessemer furnaces, if the rolling plants worked at all, the problem was considered successfully solved; as a matter of fact, nobody paid any attention to the costs.[150]

Annual allocations of coal were largely based on the previous year's allocations and as there was no monitoring of the output value of coal consumption from year to year, the central planners had less and less accurate knowledge of how much was actually needed for any given process. One writer summarized the difficulties: (a) the materials consumption quotas were not standard – some enterprises did not even have quotas, (b) management was ineffectual – for example, cement plants of comparable size used vastly different amounts of coal, and (c) enterprise leadership did not fully understand the importance of consumption quotas.[151] Another said:

> Recently, after various places promoted an economy movement for the use of coal, it was discovered that those units where coal waste is serious, are just those that have excessive coal storage. Because there is plenty of coal, which can be obtained easily, it is used liberally and managed casually. Some units even have no statistics, no quota and no operation regulations. They have no way to find out how much has been wasted.[152]

The lack of contact with the West due to the government's closed door policy compounded all these problems. The Chinese remained oblivious to the progress being made in reducing materials consumption.

Summary

The First Five Year Plan, drawn up with the help of the Soviets, called for large-scale development in the established coal mining areas of the north and northeast. The entire country faced severe coal shortages and, though the Soviet plans for the industry promised enormous production potential ultimately, the demand for coal was immediate and countrywide. Major problems with the Soviet strategy for Mao and his supporters were the lengthy lead-times and the regional concentration. The Soviet-assisted projects generally served only the established urban industrial areas of the north and northeast, doing little to mitigate the demand for coal and economic growth in other regions. The GLF was one of Mao's responses to these shortcomings, and the policies he put forth in 1958 called for decentralized, small-scale production, exactly the opposite of what had been detailed in the First Five Year Plan. The results of the GLF were devastating both in human and economic terms, production fell markedly in the years immediately following, and planning collapsed. It was this disarray that precipitated the departure of the Soviet personnel in 1960.

The economy was just beginning to recover from the GLF when Mao launched the Cultural Revolution, another campaign aimed at reducing the disparities between the urban and rural regions. Apart from a recurrent halting of production and disruption of transport due to factional fighting, the worst effects of the Cultural Revolution on the coal industry, indeed, on the economy as a whole, were the removal to the countryside of the scientists, technicians, and academics, and the government's insistence on pursuing a 'closed-door' policy.

The Third Front Development Programme was equally, if not more disastrous than the Cultural Revolution. The creation of a heavy industry base in a remote area put an enormous strain on the economy. The government allocated huge sums of investment capital to the Programme virtually to the total neglect of the coast. All of the Third Front projects, including the coal mines and railroads, were hastily assembled with little or no preparation and were constantly breaking down. Any spatial rationality that had existed between the good coal production areas and the existing high consumption centres was gone. Production and consumption relationships throughout the country were seriously skewed and the resulting costs of mining poor quality coal and transporting it over treacherous terrain were extremely high.

By the eve of the reform and opening up programme, most of the original leaders of the centralized and decentralized factions had died, and the country faced crippling coal shortages. In the mid-1970s a compromise solution had been set in motion. Providing technical support, the government encouraged the development of LNS mines throughout the country because they had not only proven themselves, but the simple fact was that any coal produced by whatever means was desperately needed. The government decided that the limited investment funds available could best be used in expanding large-scale production in the areas where the reserves were easiest to exploit, namely Shanxi, Shaanxi, Inner Mongolia, and Guizhou, and building the required railway and port infrastructure to transport it to the main consumption centres in the east and south.

Despite many problems associated with central planning, the recurring politically-induced upheavals, general economic backwardness and poverty, and the fact that there was very little opencast mining, China's total production rose to third place in the world in 1978 and production growth over the first thirty years of the PRC had been highest.

Soviet-style command planning, with its paramount emphasis on achieving quantitative targets, failed in the Chinese coal industry as it had in the fuel industries of other command economies. In addition to the inherent motivational weaknesses, it was impossible for the planners to have full knowledge of production, financing, transport, and consumption relationships over so large and diverse a country, and to orchestrate all of the lateral connections among them.

Inadequate investment and the policy of keeping the prices of producer goods very low resulted in the coal industry's having chronic deficits and the recurring necessity to seek 'safety-valve' assistance from the small non-state mines when output was desperately needed. With scarcely enough capital to maintain operations, let alone improve and develop them, heavy government subsidization was required to keep the industry functioning. After the departure of the Soviets no new technologies had been introduced, and the average levels of worker production were very low. Only a small fraction of the coal was washed. Insufficient capital investment contributed to the constantly imbalanced output–capacity ratio.

The Soviet Union and China had the highest railway freight densities in the world, but like the coal industry, the railways also received insufficient capital investment and the price for their services was too low. The two countries probably had the highest losses of coal shipments due to hoarding and equipment malfunction resulting from over-use. Water transport played but a small role in China.

As a proportion of total energy consumption, coal accounted for a larger proportion than in any other country, and consumption efficiency was very low. Coal consuming equipment was old and badly maintained, creating heavy pollution. The capacity of the thermal plants and transmission networks was woefully inadequate to meet the needs of a country trying to modernize quickly.

Prior to 1979, with the exception of the prematurely abandoned First Five Year Plan, the government had no consistent, focused strategy. The amounts of funding given to the industry changed radically from year to year and the directives concerning coal, railway transport, and thermal power plants were too often ad hoc and short-term responses to the crisis of the moment. On numerous occasions the plans were completed late, or were approved and subsequently changed after the beginning of the year in which they were meant to be applied. The coal industry, reflecting the economy as a whole, was in critical need of a comprehensive and integrated long-term strategy. The economy had stalled, the cadres lacked direction, and the masses were cynical. The Chinese leadership realized that they had not only failed to meet their own industrial and modernization targets, but that China was falling behind other Asian developing countries. Deng Xiaoping and his colleagues were keen to address the immediate problems and press on towards the achievement of the Four Modernizations.

4   The industry and Deng's reforms

Introduction

It was at the Third Plenary Session of the Eleventh Central Committee, which began 18 December 1978, that the reform and opening up (*gaige kaifeng*) programme was formally introduced. Included in the Session's communique was the following:

> While we have achieved political stability and unity and are restoring and adhering to the economic policies that proved effective over a long time, we are now, in the light of the new historical conditions and practical experience, adopting a number of major new economic measures, conscientiously transforming the system and methods of economic management, actively expanding economic co-operation on terms of quality and mutual benefit with other countries on the basis of self-reliance, striving to adopt the world's advanced technologies and equipment and greatly strengthening scientific and educational work to meet the needs of modernization.[1]

Only three days earlier, after long negotiations, Beijing and Washington had released a joint communiqué calling for the establishment of normal diplomatic relations on 1 January 1979 and an exchange of ambassadors between Beijing and Washington was to be implemented on 1 March 1979. The Chinese eagerly anticipated a large inflow of American capital and expertise.

In April 1979 the government launched a three-year period of economic 'readjustment, restructuring, consolidating, and improving'. By then Deng Xiaoping had displaced Hua Guofeng as paramount leader of the country, and the targets for the Ten Year Plan (1976–85), which had superseded the Fifth Five Year Plan (1976–80), were dropped because it was all too apparent they were hopelessly unrealistic. A main cause of the economy's plateauing was the persistent shortage of energy, and a major, if not the most critical goal of the economic programme in the following years was to increase energy supply and reduce consumption.[2] The revised 1979 plan stated: 'stress is placed on speeding up the development of agriculture and the light and textile industries and on expanding production and construction in the coal, petroleum, power and building material industries, and in transport services. At the same time, arrangements for production in other industries are to accord with the availability of fuel and power...'[3]

The Chinese Government was very much aware at this time that international oil supplies were again in jeopardy. The reforms in China had hardly been launched when in January 1979 the Shah of Iran was driven into exile and exports of oil from what had been the world's second largest exporter ceased. Panic buying ensued and prices trebled, again causing serious economic repercussions in the rest of the world.

The Sixth Five Year Plan (1981–85) listed many goals for the coal industry, and with the publishing of the first *Zhongguo meitan gongye nianjian* (China Coal Industry Yearbook) in 1983, covering up to and including 1981, a plethora of government plans for reform were announced. Subsequent Yearbooks would include dozens of speeches and reports by Ministry and government officials. Also in these early reform years the academic and professional journals were filled with debate on the appropriate extent and speed of the proposed changes.[4]

The speeches and writings of Coal Minister Gao Yangwen, 1979–85

Gao Yangwen replaced Xiao Han as Minister of the Coal Industry in December 1979 and remained in office until June 1985, when he was succeeded by former Vice-Minister Yu Hongen.[5] It is clear from reading Gao's speeches and writings that he understood the problems in the industry, and that he was candidly critical of what he called the 'blindness' and 'rigidity' of the administration and management. This boldness may have been a consequence of his having close personal ties to Communist Party Secretary – General Hu Yaobang.[6] He summarized the industry's difficulties in the following eight specific points: the planning administration, production line, labour administration, and product distribution were all dead (*guo si*), finances were governed by the central authorities, wages were distributed according to egalitarian principles, the pricing system was irrational, and the management leadership was many-headed and fragmented.[7]

He believed that the industry needed to break through the 'multiple mother-in-law' problem by reducing the number of regulatory bodies. To him the key element of reform was accountability at all levels. He also believed that the significant role of the Party had hampered production, and it was now imperative to remove its manipulative influence on operational decisions. Development of the industry had to take place 'at a relatively steady pace (as against precipitous surges and declines), ... the technical and management quality improved, and intra-industry investment ratios well coordinated...'[8]

Gao saw the impossibility of ever producing sufficient coal to satisfy demand fully when the general consumption efficiency was so low. To meet the country's present and future energy needs, increased production had to be combined with reductions in consumption achieved through technical, planning, and organizational improvements. In one of his key articles he admonished: 'The practice of regarding energy conservation as the central point in the technical transformation and structural reform of the entire national economy is a major and important step toward materializing social modernization.'[9] In a speech in 1982 he noted: 'The [central authorities have] already made it clear that the Ministry of Coal Industry (MCI) should not simply be engaged in coal production, but should also open up a path for the development of coal utilization...'[10] The two means he saw for doing this were to widen the range of coal products available so that users were not restricted to using raw coal, and working in close cooperation with the large consuming industries, for example, thermal plants, chemicals, cement, etc.

Significantly, it was at the Third Plenum of the Chinese Communist Party Central Committee, held in October 1984, that it was decided mandatory planning should give way to guidance planning. In his article published in the principal energy journal in 1985 he raised the following crucial questions:

1. What proportions of the coal should be subject to centralized distribution, guidance planning and market forces?
2. How can command planning be reduced?
3. What means of 'economic leverage' can be used to ensure that plan targets are met?
4. How can coal pricing be changed to ensure that all production costs are covered and a rational profit level is attained but without having a negative effect on the costs of other goods? Should a series of small measures be taken or one major change? Should attention be first given to coal sold at negotiated prices or coal 'under plan'? How can fixed, floating and negotiated prices be integrated?
5. How can coal mine development be funded? How can construction be improved in terms of the amount of time and resources required to build mines and in terms of the quality of construction?
6. How can producers and consumers cooperate?
7. What labour reforms should be carried out to improve safety conditions, stabilize the labour force, raise the quality of the labour force and better link remuneration with output and skills?
8. How can the whole coal industry hierarchy be reformed?[11]

Gao knew that solving these problems would require major structural changes to the economy as a whole.

Production

Overview

The Coal Exploration Corporation was established in March 1982 and a plan for coal production was devised at a national coal planning conference in November that year.[12] In order to meet the target of quadrupling the gross value of industrial output by 2000, it was planned that production should double from 600 million tons in 1980 to 1.2 billion tons (1,200 million tons) in 2000. The planned average annual rate of growth for the entire period was to be 3.4 per cent with an average annual increment of 30 million tons. Growth in the first ten years was to be 2.6 per cent per year with total output in 1985 surpassing 700 million tons and in 1990 exceeding 800 million tons. Emphasis was to be placed on intensifying and expanding development at existing mines in the east region. Between 1990 and 2000 the planned average annual increment was 40 million tons and the annual growth rate was to be 4.1 per cent, with most of the expansion in output taking place in 'the three wests', namely *Shanxi, Shaanxi*, and *Xi Nei Menggu* (western Inner Mongolia), and in Guizhou.[13] For comparison, between 1949 and 1978 the average annual increment had been 20.2 million tons, and the average annual growth rate 12.8. Coal from the three wests was particularly desirable not only because it was easier to mine than coal in other areas, but it had much lower sulphur levels.

There was a two-pronged strategy to reduce the strain on the railways, of which one was to build many more mine-mouth power plants. The electric power would be transmitted along high-tension wires from the remote mining regions to the cities. In addition to the building of thermal power stations it was necessary to integrate the existing fragmented railway network and standardize the gauge of some lines. Shanxi Province was to provide electricity to Jiangsu, Guangdong, and Shandong; Shaanxi and Gansu to Sichuan; and Guizhou to Guangdong, Guangxi, and Hunan.

The other prong was to build power plants near sea ports. Northern coal would be transported by rail from the interior mines to the coast where it would be loaded into large ships for delivery to central and southern port cities. Thermal power stations were to be built in or near coastal cities such as Hangzhou (Jiangsu), Waigaoqiao (Shanghai), Zhenhai (Zhejiang), Fuzhou (Fujian), and Zhuhai and Shenzhen (Guangdong).

A key element of the plans for the coal industry was the use of foreign technology and expertise. In April and May 1980, The People's Republic of China had replaced The Republic of China at the International Monetary Fund and the World Bank, respectively. The government envisaged the same forms of cooperation and assistance in the coal industry as were being considered for other industries: joint ventures, loans, training programmes, cooperative research, exchanges, compensation trade agreements, etc.

To meet the 1.2 billion-ton-target, output from the existing CMAs was to be raised from 344 million tons in 1980 to 400 million in 2000, and from LS and LNS mines from 276 million tons to 500 million, and 400 million tons were to come from new mines. The combined increases equalled 1.3 billion tons. This was planned in order that the target of 1.2 billion tons could be guaranteed.

Of the 400 million tons from new capacity, 200 million tons were to come from opencast mines. The proportion of total output from these mines was to increase from 2.3 per cent in 1980 to 17 per cent in 2000. Thus, the scale of these mines was to be far larger than any previous coal development. The other 200 million tons were to come from medium and small mines. New development was to be focused on ten energy bases, especially on the Shanxi Energy Base.

Existing mines

In order to meet the 2000 target of 1.2 billion tons, it was planned that measures be taken to enable the large mines built before 1980 to maintain high production rates for as long as possible. As noted above, output from the existing CMAs was to be raised from 344 million tons in 1980 to 400 in 2000. In 1983 there were some 554 such mines.[14] There were two ways of increasing this output, through improved mine design and the use of more advanced mining technology. A survey in 1981 revealed that over 130 mines were operating at less than 65 per cent capacity because they had been put into production before construction was complete, had poor management, or incompatible project phases. Thus, it was believed that there was considerable potential in developing these mines, many of which were situated in the east region. The cost per ton of building a new mine was 40–60 per cent higher than renovating or expanding an existing mine.[15] The Production Department of the MCI estimated that nationwide output from these large state mines would, however, be reduced after 2000 by 70 million tons, 40 million due to decommissioning and 30 million due to old age and concomitant lower production.[16]

Opencast mines

Output from opencast mines was to play a much larger role in fuelling the economy. It was well known that production per year at opencast mines was generally twice as high as from underground mines, construction time was at least 25 per cent shorter, construction and operating costs were far less, and that working conditions were much safer. The share of production from this type of mine was to increase from 2.3 per cent of total output in 1980 to 17 per cent in 2000. The use of foreign technology was repeatedly stressed, for this was a situation where the Chinese knew it would be more expeditious to use the latest technology available elsewhere rather than develop their own.[17] The five main opencast mines were Antaibao (Shanxi), and Huolinhe, Jungar, Yiminhe, and Yuanbaoshan (all in Inner Mongolia), all of which were to be developed with foreign assistance.

In May 1979 Dr Armand Hammer, who had made several contracts with the Soviet Union since the 1920s, went to China to try to establish a Sino-American cooperative agreement in either chemicals, oil and gas, or coal.[18] After five years of negotiations and a feasibility study completed in July 1983, the Antaibao Surface Coal Mine Joint Venture agreement was finally concluded in April 1984.[19] The total value of the project was to be about US$649 million.

Initially Hammer's Occidental Petroleum Company had a 50 per cent interest, but in November 1984 Peter Kiewat and Sons, Occidental's 25 per cent equity partner, withdrew and was replaced by the Bank of China. Thus, Occidental ultimately held a 25 per cent stake, the Bank of China Trust and Consultancy Company 25 per cent, the China National Coal Development Corporation (CNCDC) 42.5 per cent, and the China International Trust and Investment Corporation and Shanxi Province together, held 7.5 per cent. Occidental's operating company was

Island Creek Coal Company Ltd. (ICCC), while CNCDC's was the Pingshuo First Coal Company Ltd. (PSF). A US$475 limited recourse loan was syndicated by thirty-nine banks, the main ones being Bank of America, Credit Lyonais, Industrial Bank of Japan, the Royal Bank of Canada, and the Bank of China.

It was the largest joint venture contract with an American company that the Communist government had ever signed and was much publicized.[20] The Vice President of Marketing for ICCC effused:

> the success of Antaibao is of 'supreme importance' to the highest levels of the Chinese government as well as to the joint venture partners ... The government looks to the project 'as a demonstration to the world of the security and health of the China investment environment'. Despite problems and delays, there is a bright future for the mine ... As the closest source of reliable, consistent quality steam coal to many areas of the Pacific Rim, we are confident that, with a well coordinated marketing effort, the mine will enjoy an equally successful and financially rewarding public life.[21]

Located about 9 km north of Pinglu county's township and 23 km south of Shuoxian county's township, the total area of Antaibao was 18.5 km^2 and had close to 500 million tons of low sulphur steaming coal with heating values of up to 7,600 kcal per kg, the highest in the country. Construction began in July 1985 and the opening ceremony was held in September 1987. The contracts for cooperation were to last thirty years. The expected annual production from the first phase of the project was 15.33 million tons, of which 12 million would be washed, and 9 million would be exported, about three times what nationwide exports had been in 1978. It was anticipated that each of the second and third phases would also eventually produce 15 million tons per year. Export sales from both ICCC and PSF over the first twelve years were to be handled exclusively by the China National Coal Import and Export Corporation (CNCIEC). Second and third mines to be built in subsequent phases of the project were to produce another 15 millions tons each.

The foreign side was to contribute about US$200 million worth of foreign equipment and to hire the project design firm. The Chinese side was to provide labour and local materials for the mine site, construct a residential village with housing for 17,000 miners and their families, office buildings, training centres, nurseries, primary schools, cinemas, etc., and to carry out improvements to the railways and Qinhuangdao Port.

For the first twelve years of the project the foreign engineers and experts hired by ICCC were to serve as the managers, while the local PSF men would serve as deputy managers. Over the following eighteen-year period, the roles of the foreign and Chinese would be reversed with the Chinese side in control.

The other four opencast mines, all opened with foreign assistance, are in Inner Mongolia. The Huolinhe mine, in the east, has brown coal. Annual production was to reach 30 million tons, most of which was to be used in the Tongliao power plant and other plants located along the Tonghuo railway line.[22] The Jungar coalfield, 150 km south of Hohhot, has long flame coal with low sulphur content. Annual production was to reach 30 million tons. Also planned for this site were a coal preparation plant and a slurry pipeline to Qinhuangdao.[23] Yiminhe mine, 85 km south of Hailar City, has lignite coal. There was to be a total of seven pits with a combined capacity of 55 million tons, and a large pithead power plant was to be built.[24] Yuanbaoshan mine, located in Chifeng City, has old aged brown coal. Annual production was to reach 8 million tons.[25]

Energy bases

Construction of the Shanxi Energy Base was first proposed in December 1980, and in 1982 the State Council set up the Shanxi Regional Energy Planning Office. The ultimate goal was to make it one of the largest coal operations in the world. It has a total area of about 330,000 km^2 and includes the entire Province of Shanxi, the Jungar region of Inner Mongolia, northern Shaanxi, and western Henan. Some 60–70 per cent of the nation's total reserves are located in this area. The 2000 target set for the entire base was 600 million tons, half of the initial target set for the country as a whole. Of this a minimum of 400 million tons would come from Shanxi Province alone where the reserves are the largest and best quality in China. As of 1983 they amounted to 20.54 billion tons, or one-third of the national total, and covered 37 per cent of the Province's total area.[26] Relatively speaking, the coal has low ash, sulphur, and phosphate content. The coal at Datong is the richest in the country, generating up to 8,000 kcal per kg. The country's largest anthracite deposit is at the Province's Qinshui field. Another advantage of these deposits is that they are generally found in relatively simple geological structures, that is, they are shallow-lying, averaging 200–300 m underground (less than half the national average), the incline is low, there are few faults, and 70–80 per cent are classified as medium to thick. As a result, the coal is relatively easy to extract and the mining costs are far less than the national average.

The base is also rich in other key raw materials such as copper, iron ore, oil, natural gas, mirabilite, salt, soda, and gypsum. Many high energy-consuming industries such as coking, thermal plants, aluminium oxide, electrolytic aluminium, synthetic ammonia, nitrophosphate, synthetic fibres, plastics, rubber, dye, medicine, and explosive plants were planned for the area.

Between 1982 and 2000 ten coal bases were to be established in Hebei, Central Shanxi, Southeast Shanxi, Liaoning, Heilongjiang, Anhui, Shandong, Henan, Shaanxi, and Guizhou.[27] The Shenhua Coal Project was started in 1985 and comprised several sub-projects the main ones of which were: mining of the Shenfu-Dongsheng coal deposit which is on the border of Shaanxi and the Inner Mongolia Autonomous Region and has some 33 per cent of China's verified coal resources; the construction of a second west-east trunk railway line from Shenmu in northern Shaanxi through Baotou to Huanghua Harbour in Hebei Province; expansion of the Huanghua Harbour to an annual handling capacity of 35 million tons; and a large power plant at Fugu, also in northern Shaanxi.[28] It was expected that eventually 100 million tons would be mined each year and shipped to other parts of China, and to world markets.

Local mine development

Output from LS and LNS mines was planned to increase from 276 million tons in 1980 to 500 million tons in 2000. By the late 1970s it had been recognized that the fastest output growth was being achieved by these mines and that their continued and expanded contribution was vital to the economy. Furthermore, as most of the capital was raised locally, basic wages were much lower than those at the large state mines, and as grain rations, housing, sick leave, and medical and retirement benefits were rarely given, the central government was relieved of a major financial burden. The cost of opening or expanding an LS or LNS mine was much less capital intensive than for larger projects. Whereas it cost about 108 yuan per ton to open a large new state mine in Shanxi, it cost only about 40 yuan per ton to build an LS mine and about 16 for an LNS mine.[29] Construction time was generally only two to three years versus six to ten for large mines, and operation costs were some 50 to 80 per cent less. Most importantly, it was thought that locally produced coal could help solve the critical energy shortages problem if it could be made more readily available to consumers, that is, out of plan. In addition, the central and provincial governments hoped the development of local mines would add to the wealth of peasants, reduce rural unemployment, stem rural–urban immigration flows, stimulate the development of rural industry, and halt the ecological damage being caused by the scavenging for firewood and plant stalks.

Between 1980 and 2000 output from local mines was to increase over 80 per cent, whereas output from CMAs was to increase by only 16 per cent. Thus, the local mine sector was immediately given a number of incentives to increase production. For example, the Shanxi provincial government endorsed the establishment of a Local Mine Bureau in 1979 to oversee production from LS, handicraft, and army-run mines, and steps were taken to improve their organization.[30] In December 1981 the State Council approved four financial incentives for local mines: (1) granting of subsidies for mines with deficits, (2) reduction of or exemption from industrial and commercial taxes, (3) increasing investment for capital construction and technical transformation, and (4) increasing the retention of depreciation funds.[31]

There were no restrictions on who could operate local mines. At a discussion on reform in the coal industry, Minister Gao Yangwen stated: '... small mines can be run by communes, brigades and masses ... or run by professional personnel ... cadres, technical personnel and workers can also run small mines individually or through raising funds. In a word, welcome all walks of life to run small mines.'[32] At a conference on industry and communications held in March 1983 it was resolved that:

> Relevant departments should take measures to organize people in localities with better transport facilities to produce as much coal as possible. Localities should be vigorously guided and encouraged to raise funds to build small hydropower stations and small coal mines. Where conditions permit, people may build small thermal power stations.[33]

In April 1983 the State Council approved the *Report on Eight Measures for Accelerating the Development of Small-Scale Coalmines* and ordered all jurisdictions to implement it. In abridged form, the eight measures were as follows:

1. Operation of Mines: Communes, brigades, all trades and industry are encouraged to operate mines, and the masses are encouraged to pool funds to do so. Areas which are short of coal may operate mines in partnership with places having coal.
2. Role of Market: (a) Coal which is to be put under unified distribution and is produced either by communes, brigades or special contracts is to be priced on the basis of quality and should be such as to assure a fair profit. Coal which is not to come under unified distribution may be sold at negotiated prices. (b) The operators of locally produced coal can haul and sell their coal anywhere, but if rail transport is required, the coal will be put under unified distribution. (c) The taxes that small mines pay will be determined on the basis of profits. Mines situated in coal-short areas or where conditions are difficult may apply for discretionary reduction or exemption from taxes. (d) After guaranteeing sufficient funds to continue and expand production, small mines can divide the remainder of their funds for wages and welfare benefits as they choose.
3. Rational Use of Resources: All provincial (or autonomous region) coal bureaus, coal companies and geological departments are to plan the allocations of resources given to operators of small mines to exploit. Small mines can extract resources belonging to state mines that cannot be mined by large mines and within limits, resources that large mines cannot use in the near future.[34] The operators of small mines must abide by the State Council's 'Decisions on Maintaining Regular Production in State-Owned Mining Enterprises'. Reckless or indiscriminatory mining is prohibited and small mines cannot be operated under railroads, highways, reservoirs, dikes, protected places of cultural interest or major structures. Also, in the interests of environmental protection, mining should be done in conjunction with city and rural planning.
4. Multiple Use of Coal Partnership Operations: Processing should be encouraged to produce coal suitable for multiple uses. Small-scale power plants, coal gas plants, construction materials and chemical industries should be encouraged especially in areas lacking convenient communications and transport.
5. Maintain Safety Standards: Small mines should abide by the 'Mine Safety Regulations' and 'Mine Safety Control Regulations' issued by the State Council and the 'Small Safety Rules' issued by the MCI.
6. Assistance: (a) Financial – The operators themselves should provide most of the funds, though where genuine difficulties exist, especially in coal-short areas, the Agricultural Bank of China may issue loans as conditions require. (b) Materials – All materials are to be guaranteed by local bureaus for small

mines whose output is to be put under unified distribution. Explosives are distributed by central materials supply departments. Materials departments and coal management departments are also to assist local mines whose output is not put under unified distribution to procure materials and equipment. (c) Technical advice – Provincial coal bureaus and state mines should organize technical service personnel to help the operators of small mines with development plans, restructuring, safety measures, technical and economic efficiency.

7. Permits: Mining permits are to be issued only to small mines which have been approved by provincial/autonomous region governments or coal departments. Those holding mining permits are also to be issued with business permits by industrial and commercial administration departments. Holders of both permits are protected by national laws. Mines currently in operation must all be checked and under supplemental approval procedures.

8. Local Mine Joint Service Companies: Such companies should be established to assist provinces/autonomous regions share experiences in handling the problems of small mines, to formulate policies, manage funds, and to better coordinate relations.[35]

A ten-year plan for the development of locally-run mines was devised at the National Working Conference of Locally-Run Coal Mines held in August–September 1982.[36] In October 1983 the Local Coal Mine Combined Service Corporation was established to make plans for all local mines in the country – output targets; technical renovation; mechanization of production, road drivage and haulage; safety guidelines; wages; transport; marketing; allocation of reserves and materials – and to liaise between the state mines and all the various types of local mines, to enforce MCI instructions for local mines, and to advise the operators.[37]

Various licensing arrangements were instituted in an attempt to deter haphazard development; to reduce damage to coal seams, and prevent caving in of surface structures such as buildings and railways; and to reduce the large number of accidents caused by indiscriminate digging. In order to improve the productivity of the mines, some of which had a recovery rate of only 10–20 per cent, the government began to levy a tax based on the extent of the resources held and utilized.[38]

Expansion of treatment capacity

By the year 2000, 50 per cent of the total output was to be washed, compared to the 17 per cent washed in 1978.[39] In Shanxi Province the proportion of washed coal was to increase from 5 per cent in 1980 to 45 per cent in 2000.[40] It is apparent that the MCI was now fully aware of how wasteful the consumption of raw coal and the use of the wrong types of coal had been. Specialists wrote numerous articles on the energy savings that could be achieved through increased coal treatment.[41] In November 1980 the MCI held the first ever conference on coal quality inspection and laboratory testing.[42] At the National Working Conference on Coal Preparation held in November 1981, a major agenda item was the need to expand preparation beyond serving just the steel industry.[43] In February 1982 at the National Conference on Comprehensive Utilization of Coal, Minister Gao Yangwen proposed that each mine make a three- or five-year plan for processing.[44] In November of that year the State Council issued a set of very specific instructions entitled *Development of Coal Preparation and Processing and Rational Utilization of Energy*.[45] For example, an outline was given of how various types of coal were best used and specific quality norms, for example, permissible sulphur and ash content levels for defined purposes. Processing facilities were to be built simultaneously with new mines, with priority being given to coal export bases, areas where operations were highly mechanized, where the coal was to be shipped long distances, or where the coal was of inferior quality. Funds for new, expanded, or reconstructed coal washeries and sizing plants, and plants for washing coking coal were to be included in budgets for capital construction, while funds for steam coal washeries and screening plants, and technical transformation of existing coal preparation and screening plants were to be listed as state energy conservation and technical transformation funds.

The State Council further ordered that as of 1983 the big consumers – large iron and steel, fertilizer, power and cement plants; railway locomotives; and large cities with concentrated industries – be supplied with homogeneous shipments of the right sort of coal from designated mines and mining areas, as opposed to diverse sources. A new set of prices according to quality specifications was to be established by the end of 1983, approved by the State Council, and enforced immediately thereafter. As of 1984, cleaned and screened coal were each to be weighed separately, and by the end of 1984 all indigenous coke production was to cease.

The government placed particular emphasis on the need to experiment with the production of briquettes, especially from low-grade coals and/or by-products, with coal gasification and liquefaction, and with the use of slurries for commercial and domestic use.[46] Arrangements for several new foreign-funded coal treatment projects were initiated in the early 1980s.

Mechanization, use of materials, and mine preparation work

In 1978, 32.52 per cent of the state mines were partially mechanized, that is, only extraction and tunnelling were mechanized. In 1984 it was announced that by 1985, 1990, and 2000 the levels were to reach 46, 70, and 85 per cent, respectively, though at an earlier conference on mining machinery held in February 1982, it had been announced that the level of mechanization in state mines would be 50 per cent by 1985 and 95 per cent by 2000. Half of the LS mines were to have 50 per cent mechanization by 2000.[47]

Several measures were taken to raise the quantity and technical sophistication of the machinery used in the mines. The Comprehensive Mining Equipment Service Team, the first national organization to service Chinese-made comprehensive mining equipment, was established in April 1980, and in December of that year the China National Coalmine Equipment Service Corporation was established.[48] The latter was to develop, manufacture and sell new equipment, repair equipment, and maintain service stations in all main coal areas for technical services, purchasing, and marketing.[49] At the National Working Conference on Coal Mining Mechanization held in January 1982, it was advocated that the responsibility system be used in the allocation of equipment, that is, that the chief engineers and team leaders have responsibility for the use and care of it, and that a system of leasing be tried. It was resolved that the supply of spare parts would be improved and that the machinery repair plants of the administration or the manufacturers of the equipment would be responsible for overhauling the fully mechanized face equipment, while routine maintenance would be handled by the repair shops of the mines.[50]

Ten months later the China National Coal Mining Machinery Manufacturing Corporation was established to oversee the manufacturing, supply, and sale of the products made by the MCI's machinery plants, and to cooperate with local coal mine machinery plants and others belonging to ministries which were producing similar sorts of products. In December 1986 this corporation was disbanded and authority was transferred to lower level machinery manufacturing enterprises.[51]

At a conference held the following year, in 1983, it was estimated that the country's major mines would attain a technical level comparable to what advanced countries had in the late 1970s or early 1980s.[52] Ye Qing, one of the coal industry vice-ministers, made several significant statements about the relationship of mechanization and output at a national meeting in April 1984. He noted that the output from fully mechanized faces was many times higher than from partially mechanized faces, and among the serious problems was the machinery industry's production of incompatible and inadequate spare parts.[53]

In its pursuit of accountability the government instituted the contract responsibility system economywide. Materials for production and capital construction under the plan were negotiated in the contracts on the basis of planned output. The bureaux in turn gave quotas to all the mines under their jurisdictions. There were different procedures for goods under the plan and goods not under the plan. Under the terms of the contract, the MCI would guarantee to provide the agreed amount of the materials under the plan. If an enterprise managed to meet the output target without using all of these materials, a cash reward worth 10 to 20 per cent of the value saved would be given. If, on the other hand, the enterprise had to request further supplies, a penalty of 10 to 20 per cent of the value of the extra materials would be levied.[54]

As for materials not under the plan, the enterprises were free to purchase them from the Materials Supply Company of the MCI or from the open market. This Company functioned as a separate economic accounting unit responsible for its own profits and losses. They could also run businesses either independently or jointly with other enterprises that would produce the required materials.[55]

At a National Coal Mine Capital Construction Conference in the spring of 1979, four working groups were established to investigate the relationship between mining and development in state mines. It was discovered that overmining had taken place in ninety-six mines, and in June 1979 at a national mine development conference delegates hoped that the output–capacity ratio problem would be solved within three years.[56]

Experiments in the contracting of capital construction work began in thirteen mines in 1981 and soon after another thirty-five were added to the programme.[57] Contracts were settled on the basis of 'Three Guarantees, Three Assurances', that is, the engineering units would guarantee the engineering, design budget, and lead time, while the MCI would assure the funding, materials and equipment, and labour. The design documents had to be approved by the SPC for projects with a capacity of over 5 million tons. Unlike in the past when the MCI assigned construction crews for all the projects, the enterprises would now be able to call tenders for the work and arrange the sequence of geological and engineering tasks. They would also determine the number of workers and their wages and benefits, as well as set productivity standards. Payment to the chosen crews would be made using investment funds from the state, but the enterprise could retain any remaining funds and could also use funds left over from the previous year. Heretofore all enterprises had to use any funds given them by the state in exactly the way it was planned for them to be used, and any remaining funds had to be returned. Now, if cost-overruns occurred, the state would not make up the difference. These would be first steps in the decentralization of control and hardening of the budget constraint.

Another change was aimed at assuring accountability. The project manager and the contracted workers would now be subject to evaluation, and to dismissal if they did not meet the standards specified in the contract. In the past their work had not been monitored, with the consequence that it was often poorly done and extended over a far too long time. Included in the 1983 plan for the coal industry was the statement: 'It is required that the average lead time of a new mine be no longer than six years.'[58] Wang Senhao, the former chief engineer of the MCI, writing in an article in 1983, suggested that the lead time be brought down from as many as ten years to five in 1990.[59] In September 1983 at the Conference on Coal Industry Capital Construction Work it was resolved that the coal authorities in each province and autonomous region and all mining bureaux should devise a seven-year capital construction projection.

Labour

Well before 1979 it had been recognized that worker productivity in the coal industry was very poor relative to international standards. The main reasons were low pay, no relationship between output and remuneration, and severe overmanning. Soon after the reform programme was launched a series of measures was introduced to increase productivity in terms of OMS. The goal was set in the Sixth Five Year Plan to raise it significantly from 0.912 tons in 1980 to 0.965 in 1985, and to over two tons per day in 2000. In 1979 two conferences were held on reforming wages and bonuses, and experimentation with the conference proposals began at Jinggezhang Mine of the Kailuan Bureau (Hebei) and Xiaonan Mine at the Tiefa Mining Administration in Liaoning.[60] The results of the experiments so pleased the MCI that in 1984 it directed all of the state mines to adopt the changes. Contracts were set with the mining and driving teams and auxiliary groups. A basic wage per ton was to be offered, on top of which was to be added extra wages for above-quota production. According to Shanxi Coal Bureau management representatives, wages for extra output were twice the base wage.[61] For jobs in which the piece-rate system could not be applied, remuneration was to be paid on the basis of responsibility and work points. Bonuses for these workers were to be calculated on the basis of the income generated by the mine from above-quota production or driving footage. Cadres would continue to receive a position allowance.[62] In order to maintain a stable work force, there were several industry-wide increases in average wages, making for a total increase of 54.4 per cent in 1984 over the 1978 level. For comparison, average industrial wages increased by 56.8 per cent over the same period. Wages in the forest industry increased by 39.6 per cent, in the textiles industry by 44.6 per cent, in the paper industry by 49.8 per cent, in the smelting industry by 64.7, and in the electric power industry by 72.1.[63]

Due to the pre-1979 practice of awarding equal wages based on the total wage bill, there had been no incentive for workers to increase their output, and there had been little if any difference in the wages offered to workers with different skills. Minister Gao Yangwen was aware of both these problems. In an article in *Nengyuan* (Energy) he wrote:

> ... we must investigate how to make payment better correspond to the work done in order to stabilize the workforce and raise its quality. We must also widen the gap between the pay grades for surface workers and underground workers and for mental and physical work ... the wages should be such that everyone will be highly envious of skilled workers' wages.[64]

Gao called emphatically for increasing the role of trained scientists and providing them with improved working conditions, and he gave a great deal of attention to increasing the level of education of the miners.[65] In the late 1970s and early 1980s the *China Coal Industry Yearbooks*, *Nengyuan* (Energy), *Meitan kexue jishu* (Coal Science and Technology), and the *Shanxi ribao* (Shanxi Daily) were filled with articles on this subject. One of the twelve development policies of the 1982 comprehensive coal plan was 'to improve technical competency of workers and staff'. Specific targets were: district team leaders or equivalent were to have attended secondary technical school; heads, deputy heads, and department directors were to have college education; and all workers and staff were to have junior middle school by 2000.[66]

Overmanning was a recognized problem and steps were taken to reduce the number of employees at all levels. Between February and June 1982 the industry underwent a major restructuring and many offices were either dropped or merged with others.[67] The number of bureaux went from 16 to 14, departments were reduced from 90 to 63, and the number of posts from 1,043 to 843.[68] Total employment in the industry was reduced by 28.3 per cent. Worker quotas and duty descriptions were defined for each task.[69] Between 1983 and 1985 the work force was further reduced by about half a million.

Financing

Towards accountability

In the pre-reform era the coal industry, as did almost all the state owned industries, operated under soft, if indeed any budget constraint. Financial losses were experienced routinely, and just as routinely the central government was forced to give the industry grants and subsidies to keep it functioning. Inadequate investment allocations, low fixed prices which often did not cover current, far less keep up with rising production costs, and poor management – no accountability at any level, no profit motive, no market dynamic of any kind – kept the industry in a state of financial inertia. Workers of all ranks knew the government would keep the industry going no matter how wasteful and irrational the day-to-day operations.

With the beginning of the reforms, the MCI officials sought to reverse the entrenched and languishing attitudes of its workers by introducing incentives which would not only make each and every worker aware of the costs incurred and the profits/losses made by his unit, but, more importantly, make his earnings directly affected by his unit's financial performance. A contract system came into effect after trials had been carried out in twenty-two mining administrations in the eastern region. Beginning in 1984 the Ministry of Finance made six-year production contracts with the MCI, which in turn made contracts with the provinces, prefectures, departments, and bureaux. These jurisdictions in turn made contracts with the mining administrations, which made contracts with individual mines. Although the contract was set for six years, that is, up to 1990, the administrations and bureaux themselves were to plan the yearly, monthly, and quarterly targets. Each coal producing unit was given an output quota to fulfil by a set date and was to sell the coal to the state at a fixed price for unified distribution.

The two-track pricing system was expanded as an incentive to increased production. Non-state mines had been permitted since about 1972 to sell their coal at market prices. Now, above-quota output that the state mines produced would be sold to the state at 50 per cent (later changed to 70 per cent) above the basic price, or on the open market at a floating price, that is, what the market would bear.[70] Of the profits generated from the coal sales, it was suggested that the enterprises use 50 per cent for maintaining production, 20 per cent for welfare facilities for the staff and workers, 20 per cent for rewards, and 10 per cent for reserve funds.[71] Those units which did not meet the production targets agreed upon in the contract were to make up the difference by buying coal from the markets at the going rate. Besides specifying how much coal would be sold to the state, the contract also set production standards, for example, recovery rates, OMS, and safety requirements.

For its part, the state guaranteed a specified amount of investment funds, materials under unified distribution, and sufficient transportation capacity. If the funds agreed upon in the contract proved inadequate, the enterprises could apply to the MCI for loans at 0.6 per cent. Conversely, if the enterprises did not use all of their allocations, they could lend money to the Ministry at 0.6 per cent.[72] In May 1985, the MCI established its own audit bureau to monitor annually the income and expenditures of all enterprises and institutions under the jurisdiction of the Ministry, and to exert financial discipline on those not under direct MCI control.[73]

The contracts included clauses on losses. In order to make up for the difference between production costs and selling prices, the Ministry of Finance was to subsidize the industry by 300 million yuan per year between 1985 and 1990. If the enterprises incurred losses beyond those expected, they were to make up for these themselves. If, on the other hand, the losses were less than anticipated, the subsidy was to be shared, with 60 per cent going to the enterprises and 40 per cent to the state.[74]

The reforms brought significant changes to the tax system, some of them intended to provide the SOEs with greater incentives and others to regulate the new forms of ownership which were emerging. In 1984–85 profit remittances were changed to taxes and depreciation funds were placed under enterprise control. The relevant Coal Industry Yearbooks indicate that the coal industry received some preferential treatment, for example, the tax rate on coal products was lowered from 8 to 3 per cent,[75] and individual plants within the mining administrations were no longer required to turn over 30 per cent of the basic depreciation funds to the central authorities.[76] What used to be known as the 'industry' and 'commerce' taxes were changed to one 'product' tax, which was to be levied at the rate of 3 per cent. Enterprises making a profit rate of over 12 per cent were to pay a 'resource tax' to reduce the financial discrepancy among enterprises.[77]

The funding allocated to the LNS mine sector was now minimized. It was decided that the limited financial resources available should be spent on a few key large mines and infrastructure projects, that is, undertakings of a magnitude which the peasants could not handle themselves. Operators of local mines were encouraged to open new mines using whatever resources they themselves could collect and, beginning in 1981, the state annually provided loans of 200 million yuan to assist with the opening of new pits or the expansion of old ones. In 1983 this was increased to 470 million. Considerable time at the National Working Conference of Locally-Run Coal Mines held in August–September 1982 was devoted to discussing funding arrangements for local mines.[78] It was decided that sources available to them, apart from the central government were: provincial, prefecture, and county governments; revenue from selling coal directly to other provinces or districts; government subsidies given for transferring coal under the plan; and funds from intra- and inter-provincial joint ventures, bank loans, and foreign funds. It was also decided at the Conference that the coal bureaux at the provincial, prefectural, and county levels must establish their own transport and marketing organizations.

In June 1985 Premier Zhao Ziyang proposed that 'people power' be relied upon for cutting coal in Shanxi Province and that the state concentrate its investment funds on increasing and upgrading transport and communications, aspects which peasants were not technically or financially capable of doing. In rationalizing this plan he said there were three possible strategies for Shanxi:

a.   to rely on state investment to build large mines which would increase production capacity by twenty million tons. The total estimated cost would be forty billion yuan.
b.   to build large, medium, and small mines simultaneously but put most emphasis on the construction of large mines. This would cost about thirty billion yuan.
c.   to build large, medium, and small mines simultaneously but concentrate on the development of township-run mines. The total estimated cost would be twenty billion yuan including nine billion yuan in state loans.[79]

Favouring option (c), he noted that township mines required the least amount of investment, yielded capacity and production increases rapidly, helped economize on the use of state-controlled materials and mining facilities, and helped accelerate the development of the rural economy. The output target for 2000 for Shanxi Province was set at 400 million tons, with 105 million (26.25 per cent) coming from CMAs, 95 million (23.75 per cent) from LS mines, and 200 million (50 per cent) from township mines. He advocated that during the Seventh Five Year Plan no new large state mines be built and that all investment should be put towards the transport and communication sectors. Thus, the resources for the development of the township mines would have to be locally derived. Furthermore, he believed that with the state bearing the entire cost of the construction of the large trunk lines and roads, the masses would be stimulated to build the feeder lines and manage local shipping.

Pricing

The need for a major economywide reform of prices, including a large increase in coal prices, was generally recognized and reflected in the hundreds of articles written in the early 1980s.[80] Prices were regarded as the crucial lever both to raise coal production and reduce consumption. A fundamental issue was the correct timing and rate of change. Since consumers could not immediately improve their consumption performance by acquiring more efficient equipment, the impact of a price increase in the long run would be quite different from what it would be in the short. One viewpoint held that the first step would be to eliminate losses (by making prices more on par with production costs). The second step would be to increase gradually the capital profit level of the industry by raising the average price three or four times so that the average profit level corresponded to that of the heavy industrial sector as a whole.[81]

Most Chinese economists, however, seemed to agree that it would be preferable to make a sizeable increase in the basic price (*zou da bu, yi ci jiejue*, which translated literally means 'take one big step, one decision') rather than to carry out a series of small adjustments (*xiao bu, kuai pao,* meaning 'small steps, quickly run'). Based on the effects of the general price increases made since 1979, they believed that continual change and constant uncertainty of market conditions were leading to economic and political instability.[82] Yet making one sudden, large increase was also seen as potentially harmful. One suggested solution was to raise the average enterprise capital profit level to 5.6 per cent through improving the general efficiency level of the industry and increasing the average state mine price to 48.94 yuan per ton.[83] (In 1978 the average *retail* price had been 30 yuan per ton.)

It was hoped that China's coal prices would eventually become more commensurate with international coal prices. Between 1949 and 1985 the international price rose 400–600 per cent, while in China the average price rose only 239 per cent. If the domestic price were suddenly increased to the international level, the effect on the economy would be catastrophic. Some economists recommended that the price structure should not be altered at all in the near future but rather that the government should opt for a profit redistribution system. They pointed out that while the coal industry had had deficits for some time, the industrial sector as a whole had been making profits. Thus, through taxation or direct subsidization the profits could be redistributed. Most, however, agreed that this would be too controversial, both economically and politically.

A proposal to change the method of calculating profits was also put forward. If profits were calculated on the basis of capital instead of average costs, the older, large raw materials industries such as the coal industry would attain higher output-capital ratios, while those of the smaller, more modern processing industries would be lower. The central government would be able to allocate scarce investment funds better among the various industrial sectors. However, this is contrary to socialist economic theory in that varying production costs within any given industry means that each unit of capital does not yield uniform amounts of profit. As demand and supply fluctuated, so would profit rates, and as these would be constantly changing, investment rates would also have frequently to be adjusted.

In May 1979 *pithead* prices were raised 22.8 per cent from 18 yuan per ton to 22.10 yuan, but by an average of only 2.15 per cent between 1981 and 1984. Another big increase, 17.5 per cent, was effected in 1985 (see Chapter 6).

The purpose of the three-tiered pricing system – a 'national unified price' determined by the state for all coal produced and sold under the plan, a higher price for above-quota production by the state mines, and market determined prices for non-contracted coal produced by local mines – was to maintain the state's ability to allocate directly about half the coal produced to the key sectors and prevent inflation, but at the same time, via the profit motive, stimulate the enterprises and mines at all levels to produce as much as possible using the least amount of materials. Local mine operators were not restricted to selling locally. They were permitted to sell coal to other provinces and regions, and for export. Soon the prices at which local coal was being sold were far higher, up to ten times the fixed price. Legislation was later adopted to prevent this.

Apart from the changes associated with the contract system – selling contract coal at a fixed price and excess production at higher prices – the government widened the price differentials of coal coming from different regions in the country. There were at first two differentials of 5 per cent and 10 per cent for the provinces of Henan, Anhui, Hunan, and Jiangxi. Later the differentials were changed to 10, 15, and 20 per cent and also included Hebei, Shandong, Jiangsu, Beijing, Heilongjiang, Jilin, Liaoning, Sichuan, Guizhou, and three eastern prefectures of Inner Mongolia.[84] Price differentials of 5–10 per cent were also established for the different types of coal and for coals of different quality. Large-consuming industries were encouraged to work out directly with their provincial governments their quality specifications and to negotiate the price.

Trade

Total exports in 1990 were to be 30 million tons, up from 3.1 million in 1978. In July 1982 a new corporation, the CNCIEC was set up under the MCI to manage China's exports and imports of coal, coordinate economic and technical cooperation with other countries, and arrange for the export of coal produced by local mines.[85] In addition, it was to handle compensation trade; the establishment of joint ventures and cooperative coal development initiatives; import foreign technology; and accept investment and loans for development of coal resources from economic organizations or individuals in Hong Kong, Macao, and Taiwan. Branches of CNCIEC were to be established in various locations in China and abroad.

Funding and technical assistance from Japan

A key element of the reforms for the coal industry was the use of foreign technology and expertise. There were several possible arrangements with foreign participants: joint ventures, compensation trade agreements, loans, training programmes, cooperative research, etc. In the late 1970s and early 1980s hundreds of trade protocols, letters of intent, and contracts were signed. As mentioned above, five opencast mines, all of which were to be developed with foreign assistance, were to be a pivotal component of the industry's development plans. In January 1983 the Coal Development Corporation, later renamed the China National Coal Development Corporation, was established to serve as the point of interface with prospective foreign partners.

The Japanese had signed large energy contracts even before China's reform programme had been officially launched. After the 1973 oil crisis the Japanese were desperate to reduce their reliance on Middle East oil. China had vast resources of coal and some oil, but at that time lacked the required infrastructure for exporting large quantities of either. Other motivations were a desire on the part of the Japanese government to encourge and support the reform efforts in China, and also to create job opportunities for domestic manufacturers of plant, equipment, and industrial materials who were hit hard by the two oil crises. In accord with the first Long-Term Trade Agreement (LTTA), signed in February

1978, China was to export crude oil, coking coal, and steam coal to Japan in return for export technology, plants, construction materials, machinery, and parts, in total worth US$10 billion each year.[86] This exchange agreement was initially set for the 1978–85 period, but in March 1979 was extended to 1990 and the yearly value of trade raised to US$20–30 billion. An annual Sino-Japanese Coal Congress was inaugurated in 1981.

The first Resource Bank loans, signed between the Export–Import Bank of Japan and the Bank of China in May 1979, for the establishment of joint oil and coal development ventures, amounted to US$2 billion over 1980–84, repayable at 6.25 per cent over fifteen years. A second group was provided in October 1984, worth US$2.4 billion, repayable at 7.125 per cent over fifteen years. These were devoted mostly to oil projects as Japan was shifting its emphasis away from coal. However, nine coal mining projects were chosen: Baodian and Jiangzhuang (both in Shandong); Xiqu, Malan, Zhenchengdi, Sitaigou, and Dongqu (all in Shanxi); Qianjiaying (Hebei), and Jungar (Inner Mongolia).

Simultaneously, with the Resource Bank Loans, the Japanese provided Official Development Assistance loans for railway and port construction and loading facilities. The Japanese Overseas Economic Cooperation Fund (OECF) administered these loan packages, the first of which was signed in December 1979 for US$1.5 billion, to be given in yearly instalments between 1979 and 1983. The interest rate was a low 3 per cent per annum and the principal was repayable over thirty years following a ten-year grace period. The first four construction projects were: a deep-water wharf extension (phase two) to the Shijiusuo coal port (Shandong); a 300-km railway line connecting the Yanzhou mining area (Shandong) and the Gujiao mining area (Shanxi) to the Shijiusuo port; the 300-km double-tracked, electrified railway section between Beijing and Qinhuangdao (Hebei) of the Datong (Shanxi)-Qinhuangdao railway, which would also take coal to the coast from Kailuan (Hebei); and expansion (phase two) of the Qinhuangdao port. Tenders were placed internationally.

These projects not only facilitated coal exports to Japan, but also provided work for Japanese companies. It was Japanese firms which won the bids for construction of facilities at the Shijiusuo and Qinhuangdao ports, and at many of the sites Japanese industrial materials were used. A second loan was offered in March 1984, for the 1984–90 period, but was devoted entirely to oil projects.

In February 1982 the Japanese New Energy Comprehensive Development Institute signed an agreement for cooperative exploration of Liuzhuang Mining Area in Huainan Coal Field, and in October 1985 construction began on Dongqu Mine in Gujiao Mine Area with a loan of US$200 million equivalent from the Japanese Export and Import Bank.[87]

The Japanese also offered research assistance to the Chinese in improving coal consumption efficiency. In May 1982 China National Coal Development Coal Company and the Briquette Industry Company of Japan carried out a series of experiments on locomotive briquettes, and the Japanese New Energy Comprehensive Development Institute agreed to fund a coal liquefaction pilot plant at the Coal Chemical Research Institute.[88]

Funding and technical assistance from the World Bank

China resumed its seat in the World Bank in 1980. The number of Bank staff assigned to China was third largest after India and Brazil. Over the Sixth Five Year Plan period China received a total of US$2,969.93 million, of which US$74.93 million was specifically designated for the coal industry.[89] The petroleum industry received US$347.98 million, the hydropower (and transmission) industry received US$252.20 million, the gas industry US$24.99 million, the railways US$438.92 million, port development US$67.97 million, and energy saving initiatives US$96.32 million. In all, energy projects amounted to about 27 per cent of the total, and transport, 17 per cent.

The World Bank originally agreed to provide loans for the development of a meagre coal mine at Changcun in Luan Mining Area, and an anthracite mine at Chengzhuang in Jincheng Mining Area, both in Shanxi Province and each with an annual capacity of 4 million tons.[90] Approval was given for the Changcun mine in November 1983.

The United Nations Development Programme commissioned the Montan Consulting Group of West Germany and the Morrison-Knudson International Company of the US to provide technical consulting and management services. Due to a boundary disagreement with a local mine, approval for the Chengzhuang mine was not granted until May 1984. In May 1985 the World Bank signed a loan of US$120.5 million for the project. Two years later, however, this project was dropped in favour of expanding the Kailuan mine and building new mines at Yanzhou in Shandong Province. Development of the Changcun mine went ahead.

In 1985 the World Bank published its seminal *China: Long-Term Development Issues and Options*.[91] Energy and transport were the foci of two of six annexes.[92] The analyses of the difficulties faced in these sectors laid the groundwork for several subsequent specialized coal and railway projects and publications.

Transport

Expansion of transport capacity

The Sixth Five Year Plan gave a very detailed description of the improvements that would be made to railway infrastructure for coal transport between 1981 and 1985 and beyond. Expansion plans were to make it possible for the amount of coal shipped from Shanxi, Western Inner Mongolia, and Ningxia, to increase from 72 to 120 million tons, and of this total increase, shipment capacity to the northeast would increase from 14 to 29 million tons.

In the north, the Datong-Puzhou line from Taiyuan (Shanxi) to Shouxian (Anhui) (210 km) and the Datong-Fengtai (near Beijing) line via Shacheng (Hebei) (377 km) were to be electrified; the Shuoxian-Datong line (122 km) was to be double-tracked; the Datong-Baotou (Inner Mongolia) line was to be partially double-tracked; and a new double-tracked electrified line equipped with special coal cars was to be built from Datong to Qinhuangdao port (Hebei) via Beijing. This heavy-load electrified 653-km line was to be financed by the

Japanese Overseas Economic Cooperation Fund and completed in 1991, and the first phase was begun in 1985. The coal capacity of the Qinhuangdao port was to be expanded from 99.25 million tons to 140 in 2000, Mitsubishi winning the bid to supply most of the equipment. A total of four railways would come from the coal areas to Qinhuangdao. As mentioned earlier, a second west-east trunk railway line from Shenmu (Shaanxi) to Huanghua Harbour (Hebei) was to be built as part of the Shenhua Coal Project.

In the central and northeast regions the plan called for the electrification of the Taiyuan (Shanxi)-Shijiazhuang (Hebei) line (235 km); double-tracking of the Shijiazhuang-Dezhou (Shandong) line (170 km), the Jinan-Lancun (both in Shandong) section of the Jiaoji line (263 km), and part of the Lancun-Qingdao (both in Shandong) line; the construction of a new line from Yangzhou to Shikousuo (both in Shandong) (310 km); and the expediting of a new line from Xinxiang (Henan) to Heze (Shandong).

In the southern region the Changye-Yueshan section (153 km) of the Taiyuan (Shanxi)-Jiazuo (Henan) line was to be electrified. The Taiyuan-Jiazuo and Jiaozuo-Xinxiang lines (90 km), the Zhengzhou (Henan)-Xuzhou (Jiangsu) section (319 km) of the Longhai line, and the Xiuwen-Linfen section (233 km) of the southern Datong-Puzhou (both in Shanxi) line were to be double-tracked.

As for water transport, the plan called for the construction of six coal-loading berths at Qinhuangdao (Hebei), Shijiusuo (Shandong), and Lianyungang (Jiangsu), also funded by the Japanese Overseas Economic Cooperation Fund, enlargement of the coal transfer harbours along the Beijing-Hangzhou (Zhejiang) Canal (the Grand Canal), and the expansion of the Xuzhou (Jiangsu)-Yanzhou (Shandong) section of the Canal.

After the Sixth Five Year Plan was issued, several other rail projects were announced, including technical improvements to the Yangquan-Shijiazhuang (Hebei) line, double-tracking and partial electrification of the Changzhi (Shanxi)-Jiaozuo (Henan) line, and electrification of the Chengzhi-Jincheng (Shanxi) line. The proposals for harbour development were also expanded. For example, the Qingdao (Shandong) and Lianyungang (Jiangsu) ports were to be enlarged, and in order to increase coal handling capacity in southern China, eleven coal terminals were planned for the coastline between Jiangsu and Guangdong Provinces. A completion date of 1990 was set. The terminal at the Ningbo Beilun Power Plant (Zhejiang), capable of handling twelve million tons, was to be the country's largest.[93]

In 1983 the Tongyuan Shipping Company was established for shipping coal by rail from mines in western Henan, southern Shanxi, Shaanxi, and other western provinces to Zhicheng harbour and Wuhan (Hubei), and then via the Huang River to the east coast. Deep water berths were constructed at Nantong (Jiangsu) and piers built at Guixian (Guangxi) to handle coal from Guizhou, bound for Guangdong and Hong Kong via the Xi River.

Transportation capacity in Shanxi Province was to expand severalfold. The rail freight volume in 1983 was 137 million tons, while by 1990 it was to increase 97 per cent to 270 million tons, and between 1990 and 2000, another 48 per cent to 400 million tons. Highway freight volume was to increase 329 per cent from 70 million tons in 1980 to 300 million tons in 2000.[94] Over 75 per cent of the total provincial production was to be shipped out of the Province. Studies were carried out on the feasibility of building slurry lines, but no projects were started at this time.[95]

Unfortunately, the mega plans for expanding the coal transportation infrastructure in Shanxi were begun too late relative to the accelerated pace of mine construction. By late 1983 production growth from the local mines, especially in Shanxi Province, had been so high that it was severely exacerbating the already inadequate transport situation.[96] It frequently happened that those mines which had made out-of-plan contracts, that is, were in a position to make good profits, actually lost money because they had to pay penalties for late deliveries. Thus measures were taken to recentralize authority and reduce production in Shanxi, but to encourage it in other areas. In his speech at the preparatory meeting of the Board of Directors of Local Coal Mine Service Corporation in November 1983, Minister Gao Yangwen stated: 'The subsidies of the State should be mainly used in the provinces (or autonomous regions) where coal is not self sufficient. In the past, Shanxi Province was a focal point. But now, it is already in the saturated condition and its coal cannot be transported out. This is not the area for the energetic development ... It is the coal-short provinces where the energetic development is really needed ...'[97] Indeed Shanxi was still saturated in 1985. Of the 214 million tons mined there, 43.3 million tons or 20.2 per cent could not be shipped out.[98]

Coal transport planning

Insufficient capacity and poor planning had created serious bottlenecks throughout the 1970s. While consumers suffered shortages, coal accumulated at pitheads and railway stations. In December 1980 the MCI held a Symposium on the Work of Reforming the Coal Transport System, at which the groundwork for various reforms was layed.[99] In the past the State Bureau of Supplies had coordinated the sales, allocation, and transport of all coal under plan, but as of October 1981 the MCI was made responsible for these duties.[100] At a National Working Conference on Coal Transport and Sales held the following year, it was announced that a General Transport and Sales Corporation which would function as a separate business entity responsible for its own profits and losses was to be established, and that a task force would be formed to update the storage, loading, and transport facilities available for local mines.[101] Then Li Peng, at a National Conference on Coal Ordering held in the spring of 1984, put forth the idea of 'package' planning of production and delivery for major large consumers such as power plants, railways, and key industries. Producers would arrange with the large consumers five-year contracts in which they would guarantee to meet type, quantity, quality, and delivery schedule specifications. Production and transport were to be put onto one account (*yunding chan, yunding xiao shi xing meitan fenpai, yunshu, jihua*), meaning that coal under plan, no matter what type of mine it came from, would have guaranteed shipment.[102]

The hypergrowth of output from local mines continued to intensify the transport problem. Coal was being brought to the stations far faster than the transport could take it away, forcing the government to curtail output from some of the local mines, especially in Shanxi Province, and to set regional production targets according to available transport.

Consumption

Beginning in 1979, energy saved through conservation measures was referred to as the 'fifth energy resource' and was given prominent attention in the economic 'readjustments'.[103] Indeed, the government was so confident of the gains to be made by reducing energy wastage that the goal was set to quadruple the economy between 1980 and 2000, while coal production was only to double. In June 1980 the Central Committee and State Council announced that the national energy policy should give equal emphasis to production and conservation, but the immediate priority would be energy conservation.[104] In the following years the central government, as well as the governments of most of the provinces and major cities, enacted a series of general energy conservation laws and regulations. At the same time the key industry ministries, such as those for steel, chemicals, smelting, thermal power, etc., published handbooks and innumerable lists of instructions containing highly technical, industry-specific energy consumption efficiency standards for the various stages of production. Together, these comprised a national strategy for mitigating energy shortages through conservation.

Academics and journalists, who before the reform climate took hold were reluctant to write anything critical about the government for fear of losing their standing in the Party, or even their jobs, began commenting openly on the many problems facing the economy, and conspicuous among these was the very poor energy consumption efficiency. The editors of the energy journals, *Nengyuan* (Energy), *Meitan kexue jishu* (Coal Science and Technology), *Shuili fadian* (Water Power Generation), *Tianranqi* (Natural Gas Industry), and *Shiyou kexue yanjiu* (Oil Science Research) began to devote often up to half of the articles in each issue to energy-saving technology and energy management.[105] The general economics journals such as *Jingji yanjiu* (Economic Research), *Jingji guanli* (Economic Management), and *Gongye jingji* (Industrial Economics), as well as engineering and other industry-oriented journals also took up the subject and the daily papers constantly emphasized the need for reducing consumption and making more efficient use of energy, congratulating those enterprises which managed to do so.

A typical article argued that much of the blame for poor consumption efficiency in the steel industry and in the industrial sector as a whole should not be put on old equipment, but rather on poor management procedures.[106] The author contended that tremendous energy savings could be achieved through rationalizing operation procedures. He advocated that equipment be operated steadily in an optimal thermal band, that there be a consistent load factor, little idling, and neither under- nor over-loading. Unnecessary turning on and off were to be prevented, and there should be complete and coordinated equipment systems rather than a collection of components procured from a variety of sources, as found frequently in small and medium enterprises.

The Sixth Five Year Plan called for energy savings of 70–90 million tons standard coal equivalent over the five years.[107] Through energy conservation, the gross social product was to grow at 4 per cent per annum and industrial production also at 4 per cent, while energy production was to increase by only 1.4 per cent. (For comparison, between 1953 and 1978, gross social product grew at an average annual rate of 7.9 per cent, GVIO at 11.4 per cent, and energy production at 12.3 per cent.) Energy consumption per unit of industrial output value was to decrease at an average annual rate of 2.6–3.5 per cent. Detailed measures for accomplishing these conservation targets were given.

There was to be a standardization of energy use measurement; a quota system of energy consumption per unit; monitoring and a system of rewards and punishments; investment of several million yuan in the technical adjustment or expansion of consuming equipment; ongoing substitution of coal for oil; and readjustment of the industrial structure, enterprise structure, and product mix to improve overall industrial consumption. Consumption for steel, electricity, and seventeen other major industrial products would be reduced by 3–12 per cent over the five years. Some specific targets for reductions were: 9.8 per cent in energy consumed per ton of steel, 5.1 per cent in electricity generation, 12.3 per cent per ton of ammonia produced by small fertilizer plants, 12.3 per cent per standard case of plate glass, and 11.5 per cent per ton of processed oil. Urban stoves were to be improved and the proportion of 'formed' coal (as opposed to raw coal) used in the cities was to increase from 18 per cent in 1980 to 46 per cent in 1985. The proportion of gas used by urban residents was to increase from 17 per cent to 26 per cent.

Throughout the period, numerous energy-saving initiatives were launched. In November 1979 Vice Premier Kang Shien announced the start of an annual 'energy conservation month' programme.[108] For one month every year energy conservation would be publicized to make the population aware of the importance of rational energy use and to demonstrate how energy consumption could be reduced. Among the measures he proposed were the formulation and enforcement of consumption quotas for all energy users; the use of coupons or certificates similar to those for the rationing of grain; metered rather than flat rate billing for electricity, gas, and water consumption; technical and organizational energy-awareness training for all workers involved in energy production and consumption; and a system of cash rewards for employees who made a contribution to energy conservation in their enterprises.

A State Energy Commission, established in August 1980, was to allocate energy and monitor its use. However, it ran into organizational difficulties and was absorbed by the State Economic Commission in May 1982. In November 1981 the State Council issued a document entitled *Measures for the Implementation of Price Increases for Fuel Consumption Above the Quota*.[109] As of January 1982 all enterprises that annually consumed 300–600 tons of raw coal, 150–300 tons of fuel oil, or 200,000–350,000 m^3 of gas were to set their own fuel consumption quotas and have these approved by the State Economic Commission. The quotas were to be re-assessed annually by the enterprises themselves in conjunction with the authorities. From then on the charge for all above-quota fuel consumption was to be 50 per cent higher than the regular rate. The extra income earned from the above-quota charges was to go solely towards energy conservation projects.

In July 1982 the State Council issued detailed instructions on the consumption of coal in the country's 280,000 industrial boilers, which were known to be notoriously inefficient.[110] In January 1983 a system of coal rationing for industrial boilers (excluding power station, locomotive, and ship boilers) was instituted whereby coal suppliers would endeavour to provide the quota of the appropriate coal to consumers. Consumers not using all of the quota were entitled to keep the excess, while consumers requiring more were either not given more or were charged a higher rate for extra. Various technical standard specifications and minimum efficiency levels were set for all operating boilers. By 1984 steam heating systems were to be converted to hot water systems; failure to comply would mean reduction of the amount of fuel supplied. By 1990 all old boilers were to be modified to attain a set efficiency level or replaced.

All plans to modify or replace boilers were to be approved by the State Materials Bureau and to be integrated into urban construction plans for central heating and the conversion of oil and gas burners to coal burners. If users did not observe the relevant regulations, fuel supplies would be suspended. New boilers could be manufactured only by certified producers and were to be built strictly according to user requirements and coal type availability. All were to be fitted with standard performance measuring instruments and all boiler operators were to pass examinations on their use. Provisions were included for financial assistance to boiler operators who did not have sufficient funds for bringing their equipment to the new national standards.

5 Central planning in the industry giving way to marketization

Introduction

To understand fully the economics of coal production and distribution in China it is imperative to review the industry's changes in organization and policy within the context of the economywide environment of accelerating but cautious progression towards market socialism. The Chinese government was not following any single development strategy, be it theoretical or tested by other countries. It was very much a process, as Deng Xiaoping expressed it, of '*mokan shitou guo he*' – groping for the stones when crossing the river.[1]

By the late 1980s the consumer and non-grain agricultural sectors had been successfully marketized. However, the state owned industries (SOEs) had changed little, and most were contributing less to the economy with each passing year. Indeed, for the majority, the amount of government support required to keep them operating was increasing at an exponential rate. Introducing market forces in this sector would require fundamental changes in organization, administration, and operations. Managers and workers alike would have to accept a new work ethos which would demand accountability at all levels, even liability.

While the leaders were clear on the desired end results of transforming the SOEs, they were uncertain as to how to achieve these, and to what extent, if any, political reform should accompany economic reform. They watched with great interest and some anxiety the unrest and uprisings in other communist countries: the strike at the Lenin Shipyard in Poland, the rise of Solidarity in the late 1970s and early 1980s, the overthrow of governments in several Eastern European countries in the late 1980s, the fall of the Berlin Wall in October 1989, the collapse of the Soviet Union in the autumn of 1991, and the establishment of the Commonwealth of Independent States in January 1992. They were studying, too, the significant economic progress of state-directed capitalism in Korea, Singapore, Thailand, and Malaysia, as well as the unfolding of the economic reforms in India and Vietnam.

GDP measured in constant prices grew at a remarkable average annual rate of 9.63 per cent in China over the 1979–99 period despite the Asian financial crisis, though it was far from steady growth.[2] In the early years of the reforms the government struggled to prevent the economy from overheating. After 7.6 per cent

GDP growth in 1979 and 7.8 in 1980, it fell to 5.2 per cent in 1981. However, in the years immediately following there was extremely high growth, peaking at 15.2 per cent in 1984. The combination of high industrial production, consumer and investment demand, and wage and bonus hikes, with the decontrolling of prices and increased imports caused soaring inflation.

In 1988 an austerity drive, launched to rectify the overheating, had the effect of slowing growth excessively. By 1990 the economy was in need of stimulus. Furthermore, the demoralizing and adverse economic effects of the June 4th Incident needed to be counteracted.

A great boost came with Deng Xiaoping's highly publicized 'southern tour' to Guangdong in January 1992, when he saw for himself how remarkably success-ful the open-door policy had been in the Province closest to Hong Kong, and par-ticularly in the Shenzhen Special Economic Zone. In June that year twenty-eight inland cities were given preferential treatment comparable to that initially granted the fourteen open coastal cities in 1984. A set of codes and regulations for joint stock companies was also promulgated that month, and in September, price controls on 570 types of production materials were lifted.

These were turning points in the country's economic history, leading up to the momentous announcement in October 1992 at the Third Plenum of the Fourteenth Central Committee of the Chinese Communist Party, that China would work towards becoming a 'socialist market economy'. All references to 'planned economy' in the constitution were changed to 'socialist market economy' and though there were still many unanswered questions, it was clear that Soviet-style central planning would have to be abandoned completely.[3]

In November 1992 the government introduced *Detailed Measures on Changing the Business Mechanism of State-Owned Commercial Enterprises*, the aim of which was to make state owned commercial enterprises more autonomous, profit-driven, and competitive. In April 1993 the National Electronic Trading System (NETS), linking financial and industrial centres by satellite, was put into opera-tion, greatly facilitating trading in stocks and bonds.

However, the rate of economic growth was again too high, and runaway inflation threatened to undermine stability. An economic package, *The 16-Point Programme*, was launched in June 1993 to steady the financial situation. Compared to the measures taken in the late 1980s, this plan was less reactive and the economy experienced a 'soft landing' in 1995–96.

Internationally, the World Trade Organization (WTO) replaced the General Agreement on Tariffs and Trade in January 1995, and in July that year China was given observer status. In the following years the government made several major adjustments to its financial, trade, and investment policies and procedures to meet WTO membership requirements. At the time of writing, membership had not yet been granted.

The Ninth Five Year Plan was launched in conjunction with the publication of another document, *Long-Term Objectives for the Year 2010*. GNP, which was to quadruple the 1980 level by 2000, was set to double again by 2010. It was known that three major historical events would occur in the near future. Deng Xiaoping would pass on, and indeed he did in February 1997. It was a smooth transition to power for his designated successor, Jiang Zemin, who after a time as Shanghai Party Secretary was chosen as General Secretary of the Party before being, in addition, appointed State President. The return of Hong Kong to China in July 1997 was another potentially tense event, but again, political and economic calm prevailed. A few months later, damming of the Changjiang was successfully completed after scientists and engineers around the world had been debating for decades the feasibility and safety of the Three Gorges hydro electric project.

Given that the prices of 80 per cent of producer goods, 85 per cent of agricultural products, and 95 per cent of industrial consumer goods by 1996 were determined by the market, and that their inputs were similarly derived, the Ninth Five Year Plan was described as the first to have been formulated on the basis of the emerging 'socialist market economic system'.[4] Only fourteen items remained under unified distribution that year, including coal, crude oil, oil products, steel products, some ferrous metals, timber, cement, and motor vehicles.[5]

In the economy as a whole in 1998, market forces generally were estimated to influence 70 per cent of labour allocation, 62 per cent of product pricing and distribution, 51 per cent of enterprise management, 23 per cent of land transfers, and 17 per cent of capital distribution.[6] Remarkably, but not surprisingly, in September 1998 it was necessary for the State Statistics Bureau to institute a new economic sector classification system because the one promulgated in 1992 no longer accurately reflected the wider variety of ownership/organization/management arrangements that had come into being.[7]

While there was still a long way to go in marketizing the industrial sector, considerable progress had nonetheless been made towards price reform, incentive-driven management, removing Party and government interference, streamlining work forces, reducing debts, and increasing the revenue base by diversification of activity. The leadership was anxious to continue dismantling command planning by further experimenting with new ownership arrangements, increasing the decentralization of decision-making, and heightening motivation at all levels by introducing more competitive forces. Included in Premier Jiang's report to the Fifteenth National Congress of the Communist Party of China in September 1997 was the following:

> It is a great pioneering undertaking to combine socialism with the market economy. To do this, it is necessary to make active explorations and bold experiments and respect the pioneering initiatives of the masses. It is necessary to deepen the reform and resolve the deep-rooted contradictions and crucial problems that may arise during structural transformation. It is necessary to open China wider to the outside world and absorb and use for reference the advanced technology and managerial expertise of other countries including developed capitalist countries.[8]

He stated that Deng Xiaoping theory required each potential ownership structure to be judged according to the 'three favourables': whether it promotes the growth of productive forces in a socialist society, increases the overall strength of the socialist state, and raises living standards.[9]

Various types of shareholding arrangements aimed at raising capital and productivity would be included in the experimentation. The leadership was vehement that all new forms of 'structural transformation' were to have public ownership as their foundation. How this was to be achieved in practice was not explained, and the leadership's concepts of ownership and property rights have remained vague.

Upon becoming Prime Minister at the NPC in March 1998, Zhu Rongji, a former mayor of Shanghai and vice-head of the Communist Party Central Committee's Leading Group on Finance and Economics, almost immediately made the courageous decision to halt production at thousands of SOEs which had been strangling the economy for decades with hopelessly mounting debts, and pledged his resolve to make the remaining ones profitable in three years. The slogan of the day was *zhuada fangxiao*, meaning grasp the large and release the small. He subsequently restructured and corporatized fifteen government ministries, including the MCI, to hasten marketization, fully aware that millions of workers and bureaucrats would lose their jobs.

In July 1998 it was formally announced that all of the approximately 15,000 commercial operations, including coal mines (which produced about 40 million tons each year), factories, hotels, transport companies, etc., which the People's Liberation Army (PLA) had been operating since the mid-1980s, were to be sold.[10] For some years there had been many occasions when the government had indicated its desire to reduce sharply the size of the PLA and create a much smaller, highly trained, solely military force equipped with the most modern weaponry available.

The main reason given for these profound changes was the need to pave the way for market forces to have full play by differentiating and segregating government and business functions and, in the process, downsizing a mammoth government work force. In addition to cost-effectiveness, this restructuring was aimed at decreasing vulnerability to corruption. When the announcements were made, it was expected that many of the new businesses resulting from the disestablishment of the former ministries would soon be listed on the domestic exchanges. After three years, similar restructuring and cutting were to take place at the provincial level, followed thereafter at the county and township levels.

Making the approximately 118,000 industrial SOEs more competitive in both domestic and foreign markets was recognized as 'the pivotal point' in economic restructuring by the end of year 2000.[11] The government divided them into three groups. About 1,000 of the largest – those producing strategic goods – would be transformed into corporate conglomerates similar to the Korean *chaebol*.[12] These would be largely owned by the state but operated by non-governmental, corporate management. Despite the fact that at precisely this time the Korean economy was in a severe economic recession, and that this particular form of business organization seemed to have been a major contributing factor, the Chinese leaders believed they could learn from both the successes and problems of the *chaebol* structure.

Some 25,000 medium-sized SOEs would be part state, and part cooperative-owned, with no fixed ratio. Most of the over 90,000 small industrial SOEs were to have no state involvement at all and were to become joint stock cooperatives entirely owned and operated by the management and workers, and ideally a foreign partner(s).

The coal industry was in fact specifically identified as one of several industries, including textiles, petroleum, petrochemicals, metallurgy, non-ferrous metals, armaments, cement, glass, and the railways sector in which the government sought to eliminate duplication and over-capacity, backward technology, and/or heavy pollution through restructuring and merging of enterprises. Two other 'main reform' tasks were to create new employment for laid-off workers and to reduce further the role of the government in state enterprises.[13]

In early 1998 the government had identified and prioritized 'three putting into places' and 'five reforms', the former being state enterprise reform, financial reform, and reshuffling of government organizations; and the latter, reform of grain distribution, investment and financing, housing, medical care, and fiscal and taxation systems.[14] Though per capita energy consumption after twenty years of reforms was still very low by international standards, it had ceased being designated a priority problem since about the early 1990s.

In his address to the NPC in March 1996, Li Peng had proposed economic growth of 8 per cent through the end of the century. However, there was no way of predicting the rapid spread of the Asian financial crisis which began in July 1997, striking hard other parts of Asia first, and hitting China in 1998. Economic growth that year was below target at 7.8 per cent, and export growth fell sharply. In 1999 growth fell further to 7.1 per cent. The crisis hit at a bad time for China, shortly after deep structural reforms to the SOEs had been launched.

Jiang acknowledged this in the summer of 1998, and conceded that the plans for transforming the state enterprises would have to be scaled down somewhat. The economy was simply unable to absorb the rising number of workers rendered unemployed with the closing of thousands of medium and small SOEs. Equally serious was the increasing incidence of enterprise workers being forced into buying shares, or in some other way contributing their savings to their 'companies', only to lose them all. Earnest attempts by managers to turn around ailing enterprises by improving the existing line of products or developing new ones inevitably failed due to hopelessly old capital equipment, and the entrenched socialist work patterns and expectations of lower-level workers. Moreover, until such time that market forces were permitted to play a more powerful role in the economy as a whole, management reforms at individual enterprises could go only so far.

Nonetheless the government continued to cherish the hope that per capita income would reach the level of a moderately developed country by mid-century, even though it was necessary at the Ninth NPC in 1999 to reduce the economic growth target to 7 per cent for that year.[15] The annual budget deficit and urban unemployment were both at all-time highs, officially at 150.33 billion yuan and 28.6 million, respectively, though probably actually much higher.[16] Zhu called for '*ji nan, er jin*', meaning 'strive ahead for reforms despite recognized problems', and reaffirmed his goal of making all loss-making SOEs profitable by 2000.

The Ninth NPC elevated Deng Xiaoping theory to the level of Marxism, Leninism and Mao Zedong thought; affirmed 'the principle of ruling the country by law, governing the country according to law and making it a socialist country by law'; changed 'the private economy is a supplement to public ownership' in the Constitution to 'the nonpublic sector, including individual and private businesses, is an important component of the socialist market economy', and assured that 'the State shall protect the legitimate rights and interests of the individual and private enterprises, while exercising guidance, supervision and management over them.'[17] Another major announcement was the granting of permission to foreign banks to set up branches in all major cities.

However, the large numbers of unemployed were restless and disposed to public demonstration. The government was especially nervous about the formation, overt and covert, of labour groups and other disaffected peoples. Without doubt they watched, first with fear and then relief, as a settlement was reached with the 10,000 coal miners in Romania who went on strike in January 1999 demanding a 35 per cent pay raise, and the subsequent reopening of two mines. The authorities, having learned hard lessons from the June 4th Tiananmen Square Incident in 1989, positioned highly trained and equipped security forces in the capital to ensure the passing without incident of the country's fiftieth anniversary and other important political anniversaries in China in 1999 (see Chapter 7).

Further adding to the leadership's anxiety was the fact that relations with the US deteriorated soon after President Clinton's successful visit to China in 1998. The Chinese accused the US negotiators of being too strict with admission requirements to the WTO, were strongly opposed to the NATO strafing of Yugoslavia, and outraged by the bombing of the Chinese Embassy in Belgrade in May 1999. The Americans accused the Chinese of over-protectionism in trade, theft of nuclear warhead and satellite technology, and lack of improvement in human rights. Stability, both inter- and intra-nationally, was being challenged.

The Seventh Five Year Plan: 1986–90

The Seventh Five Year Plan announced a production target for coal of one billion tons for 1990, some 200 million more than what had been called for previously.[18] Emphasis was to be given to technical transformation, renovation, and expansion of existing mines. The share of total production from opencast mines was to reach 10 per cent, the level of mechanization at the CMAs 56 per cent, and exports were to amount to 20 million tons.[19] The northern area was to see the largest amount of new capacity, followed by the northeast and eastern Inner Mongolian areas, the east, southcentral, northwest, and southwest.

An energy base planning office was established in March 1986 to coordinate the development of the ten energy bases (see Chapter 4). Of these the huge Jungar (Inner Mongolia) energy project was given priority. It included eventual production capacity of 12 million tons, with the largest washing and storage facilities ever constructed in China, as well as a 200,000-kW-capacity power plant, and a 216-km electric railway link with the Beijing-Baotou and Datong-Qinhuangdao lines. The 2000 target set for the Shanxi Energy Base was still 600 million tons, but of this, the target set for Shanxi Province specifically was raised from 400 million tons to 440.[20]

Towards the end of the Seventh Five Year Plan, in September 1989, *An Outline for the Development of China's Energy Industry, 1989–2000* was released.[21] It called for the production of 1.4 billion tons in 2000, up from the 1.2 billion set in the late 1970s. Significantly, of the 1.4 billion tons, half was to come from LS and LNS mines, and the other half from CMAs. In 1984 the ratio had actually already reached 50 : 50, and the proportion from the LS and LNS mines had risen steadily thereafter. The government had recognized that the LNS mines were capable of producing the desperately needed coal much faster than the state mines, and, in any event, both state and non-state enterprises were already going through non-state channels to procure it.

Due to the disbanding of the communes beginning in the late 1970s, and the encouragement given to people in rural areas to mine coal, production at the LS and LNS mines had continued to rise sharply in the early 1980s, to the extent that in 1986 Shanxi's total output far outstripped the ability of the railways to ship it out of the Province to where it was needed. Many of the state mines had to operate well below capacity, as well as many local mines. In March 1986 it was announced that:

a. mines producing without a plan or overfulfilling the plan would not be supplied raw materials, loaned funds, or assigned marketing and transport targets,

b. inspection points would be established on all major transport arteries crossing the Province's borders. Officials would examine the shipping documents to determine exactly where shipments were coming from, where they were being sent, and at what price, and,

c. on the basis of criterion included in provincial regulations pertaining to resource use and safety management, production at privately run mines would be curtailed.[22]

In September 1987, the *China Daily* reported: 'Supply of coal will surpass demand on the domestic market this year for the first time in China's history. According to official figures, the output of state owned coal mines is expected to reach 450 million tons this year, while orders have been placed for only 410 million tons.'[23] It was a confused situation. Much more coal was being produced than the railways could handle, but this by no means meant that consumers had sufficient supplies, as the *China Daily* mistakenly seems to imply. Quite the contrary. While direct coal consumption may have been less, shortages of electricity continued to be acute because insufficient coal was being delivered to the thermal electric plants. In 1987 Shanghai was 10 per cent short of its required coal and the East China Power Supply Network had to restrict the power supply to Jiangsu

Province 2,205 times in the first half of the year alone.[24] When these restrictions were in effect, it meant that at least one whole county had its power completely cut off. During the first three months of 1987, Yancheng City (Shanxi) experienced 88 consumption restrictions making for 208 hours without electricity. As in the 1970s, many enterprises had to '*kai san ting si*' (open 3 days, shut for 4) or '*kai si ting san*' (open for 4, shut for 3).[25]

Due to a lack of railway transport in March 1988, coal shipments contracted for by the ministry responsible for electricity were behind by 1.995 million tons, for the ministry responsible for metallurgy by 763,000, for Guangdong Province by 597,000, for Shanghai by 373,000, for Tianjin by 326,000, and for Jiangxi by 310,000. In the summer of that year some 80 per cent of the factories in Jiangsu ceased temporarily to operate due to 55 per cent cuts in electricity, and in November it was reported that Jiangsu was short 16 million tons of coal, that one-quarter of the thermal plants had to stop operating due to a lack of coal, and that Shanghai had almost completely exhausted its stocks.[26]

In January 1989 total steel production nationwide fell by some 10 per cent, the Baoshan Complex (Shanghai) suffering the most from coal shortages. Guangdong General Power Company had to impose power cuts of four days per week, and coal was brought in from Australia and South Africa as a stopgap measure. Six months later the official daily newspaper reported that the country was short over 30 million tons of standard coal equivalent (sce), including 70 billion kWh of electricity, 5 million tons of oil, and 30 million tons of coal. It predicted that the total shortage would increase to 200 million tons sce over the Eighth Five Year Plan period, and 300–400 million tons by the end of the century.[27]

The lack of a sufficient number of railway cars was not the only reason for the power shortages. Some of the coal not under plan was being shipped out of Shanxi illegally. Coal from LNS mines was clandestinely being put onto trains which had been booked to take CMA coal to state industrial enterprises. In 1987 over 9 million tons were shipped out of the Province illegally, displacing 5 million tons of CMA coal, and in 1988 only 74 per cent of the planned amount of coal to be shipped out of Shanxi was actually being moved.[28] In an attempt to reduce the illegal use of railway transport the government announced that no coal could be shipped out of the Province without a permit from the appropriate railway bureau.[29] However, permits were easily 'fiddled' with bribes and the illegal shipments continued.

In 1987, in an attempt to consolidate production, 112 out of the 1,257 counties were designated as 'key coal-producing counties' and given particular government attention. Though at this time the government was pleased that deposits all over the country were being developed, thereby easing the strain on the railways, obviously it could not monitor, far less give assistance to all mines. It was hoped that by 1989 all LNS mines would be registered, have the three required licences obtainable from the local government or Ministry of Agriculture (resource, coal exploitation, and business), meet ten specified safety criteria, and all work according to a set production scale.[30] In January 1987 it was announced that persons who pillaged coal deposits would be fined 3,000 yuan, and those who forged mining certificates 10,000 yuan.[31]

Some 1,500 mines in Shanxi alone were closed for safety reasons in 1987.[32] In the following year there was a nationwide campaign to fill and seal all small pits where operators did not have permits and which had encroached on CMAs.[33] At some of these sites the pit props had become damaged, posing serious hazards to safe production.

With the introduction of the contract system and the gradual coming into play of some market forces associated therewith, the state industrial sector as a whole by 1988 was beginning to operate more rationally, though state enterprises were still not making profits. In the coal industry nearly 75 per cent of the CMAs and over half of the LS mines had deficits in 1987, and in 1988 eighty of the country's eighty-six CMAs (93 per cent) had deficits (discussed in detail in Chapter 6). The alarmingly high inflation rate of over 18 per cent precipitated the launching of an austerity programme in September that year.

In order to facilitate trade and make the industry more financially accountable, the MCI was disolved in April 1988 and put under the jurisdiction of the China National Coal Corporation (CNCC) and Ministry of Energy Resources. The latter also included the former ministries responsible for petroleum, water resources and electric power, and nuclear industry. Yu Hongen, who had been the Minister of Coal Industry, became president of the new Corporation, and Huang Yicheng, who had been Deputy Director of the SPC, became Minister of Energy Resources.[34] The CNCC was to manage only the big CMAs, excluding those in Liaoning, Jilin, Heilongjiang, and Eastern Inner Mongolia, all of which were under the jurisdiction of the North-East and Inner Mongolia United Coal Corporation.

Another change aimed at reducing the role of Party members. Before 1979 all enterprise leaders were assigned by the Ministry, and they were almost always chosen on the basis of political criteria. In order to correct this, enterprise management was given some voice in the selection of persons with demonstrable mining experience or who had specific business skills. In July 1986 *Measures for Simplifying and Strengthening Coal Mine Management* were promulgated. These called for the setting up of cadre quotas per mining administration and stipulated that 'department leaders cannot interfere with enterprise organization matters... Enterprise organization is to be reformed towards stronger business management [and] higher economic benefits...'.[35]

If the industry were ever to make profits it was imperative to reduce the work force at the state mines by thousands. To this end, the CMAs began to assign excess surface workers to 'diversified activities'. Since the 1970s, initially more for the sake of giving employment to these people than to generate extra funds, large mines had assigned some excess labour to growing agricultural produce that could be sold in rural markets, and some to producing a great variety of commodities ranging from cement and automobile parts to artificial leather.

In 1989 discussions began regarding a project run jointly by the United Nations Development Programme and the China International Centre for Economic and

Technical Exchange.[36] Over a period of six years training in safety technology, coal use efficiency, and pollution reduction were given, staff were sent abroad, China's first ergonomics laboratory for the study of the effects on miners of underground conditions – heat and work intensity, etc. – was built, and help was given in establishing a Coal Safety and Health Information Office to collect, process, and disseminate safety data for thirty-nine regional safety training centres.

In late 1987 and early 1988 there were many problems with the Antaibao Mine (Shanxi) joint venture (see Chapter 4). The project had had many difficulties from the beginning, not least of which was the sharp fall in the price of coal on the international markets. When the contract negotiations were begun, the price was nearly US$57, but by 1987 it had fallen to US$35. In February 1988 all mining ceased because the site's stocks of diesel fuel, in short supply nationwide, were not being replenished. Then later that year, in September, all loading of coal for Qinhuangdao Port (Hebei) was suspended following a train crash on the connecting railway. (More details of the problems encountered in this joint venture are given in Chapter 6.)

A great deal has been written about the causes of the June 4th Incident in 1989.[37] There were many factors – the death of Hu Yaobang, the former Party General Secretary who was regarded as a main proponent of political reform, the timing of President Gorbachev's visit, the widening regional economic disparities, the students' frustration with slow progress towards democracy, and their disgust with and sense of helplessness in the face of corruption and nepotism in high government ranks, etc. The reforms had led to rapid economic growth and demand, but also to high inflation. The government had taken radical measures to cool down the economy, including postponing plans to liberalize the prices of more commodities, and re-controlling some which had been allowed to float.

The coal industry, and concomitantly the electric power industry, were seriously affected by the Tiananmen Incident. The immediate effects were disruption of some coal shipments through the Beijing area, and consequent power cuts as plants had to reduce or cease production. For thirteen days power production did not meet its daily target.[38] Eight administrators with Island Creek Coal evacuated their Beijing office, along with fifty family members at the Antaibao mine site. Soon after the killings were broadcast worldwide, much of the capitalist world levied economic sanctions on China, and the World Bank froze a US$780-million loan for infrastructure. Generally speaking, for the remainder of the year, potential investors in the Chinese economy, including the coal industry, were either frightened off or wanted to make a political statement by ignoring China while loudly expressing more interest in investing in other countries. Coal imports in 1989 were over twice as much as they had been in 1988. This was partly due to the disruption of the railways in June.

Japan, slowest to react, postponed the 'Third Yen Loans', which were to be used for new infrastructure development in the early 1990s. However, the period of postponement was not long. By December 1990 another energy agreement with Japan was reached for the 1991–95 period. The CNCIEC was to supply

3.9–5.33 million tons of steaming coal each year from Datong (Shanxi), as well as some coking coal.[39] The relatively small proportion for coal projects, compared to the planned collaboration on oil development, went towards the development of Pingshuo and Anjialing opencast mines (both in Shanxi), and Shenfu and Jiejitu mines (spanning Shaanxi/Inner Mongolia).

At the beginning of the Seventh Five Year Plan the government was still extremely optimistic that coal exports could become a major source of foreign exchange. At a national coal conference held in June 1986 it was announced that total exports over the 1986–90 period should reach 100 million tons, increasing by 5 million tons per year.[40] This was equal to all of the coal exported between 1949 and 1984!

In the spring of 1988, Dr Armand Hammer, Chairman of Occidental Petroleum, began talks with the Taiwanese aimed at bringing an end to the division between the two governments through trade.[41] He had a personal interest in the coal trade in Asia because his Antaibao mine project was looking for export markets. In April he told a news conference that Deng Xiaoping had expressed the desire to export coal to Taiwan as well as to South Korea. The Taiwanese government immediately issued a press statement that such trade was impossible, that indirect trade (i.e. via Hong Kong) for some commodities was permitted, but not raw materials.

Notwithstanding, in June, Hammer announced that Taiwan was considering buying Chinese coal. It was obvious that he had been trying to force a deal. He proposed that Taiwan buy Chinese coal at US$34–36 per ton, and that a counter-trade agreement be made. Taiwan would buy US$500 million worth of coal every year for ten years and pay for it with equivalent value shipments of cement to China. He even suggested that the same ships be used for the transport of both commodities. Should there be any supply problems in China, he would guarantee coal from mines that he owned in the United States. The deal was attractive to large Taiwanese consumers, such as the cement industry in particular, because the US$36 per ton suggested price for Chinese coal was much less than the US$42–46 per ton that they were paying for American and Australian coal. However, the response of the Taiwanese Government was to issue a press statement stating that they had no interest in the scheme.

Soon after, in July, the Taiwanese government issued another statement, this time announcing that state companies such as Taipower would be allowed to purchase coal from China, and another twenty items including coal would be added to the existing list of thirty which the government had allowed to be imported indirectly since 1987. Taiwan would not formally approve the deal until September 1990, and Taipower would be able to purchase Chinese coal only through third parties, and only on the spot market. Chinese Minister of Energy, Huang Yichen, suggested that Taiwan become involved in joint exploration of coal and crude oil, as had Hong Kong and Japan, but the Taiwanese rejected this proposal.[42]

Coal trade discussions were also being carried out with the South Koreans.[43] Despite the absence of diplomatic relations and the fear of upsetting North Korea,

in late 1986 China extended invitations to South Korean businessmen to negotiate trade deals. Indirect trade had been carried out through Hong Kong for many
years, but beginning in the spring of 1987, China began to assign South Korea the
vague title 'port in Southeast Asia' and by the summer several contracts for direct
coal shipments were made. South Korea had for many years been relying on shipments from South Africa and Australia. Now China tried to capture a share of the
market by offering coal at the extremely low price of US$24 FOB.

However, the shipments were highly erratic. While the official export target for
1988 had been set at 17 million tons, by October it was reduced to 13.1. In July,
delays in the deliveries were affecting Korean industries seriously, especially the
Samsung Company. Despite being forced into paying very high spot-market
prices for South African coal in 1987, the Koreans announced they would buy
only half as much from China in 1989.

As of 1 January 1989 control over international coal trade was re-centralized.
Now the CNCIEC and its affiliate, China Coal Import and Export Corporation,
were given exclusive control over the management of coal exports. Communication
among the several trading authorities that had been established at various stages
in the early 1980s had broken down, resulting in severe over commitments in
export shipments. The following companies were affected: China International
Trust and Investment Corporation, China National Metals and Minerals Import
and Export Corporation, the Economic Development Corporation, the All China
Federation of Industry and Commerce, Huaneng International Power Development
Corporation, China Kanghua Development Corporation, and Xinxing Company.

In late 1988 and 1989 the two-track system of pricing was forcing coal-
deficient areas to pay extremely high prices for their coal. Corrupt officials were
buying coal at the low fixed price and selling it at many times this amount. In
March 1990 the State Administration of Commodity Prices, SPC, Ministry of
Energy Resources, and Ministry of Materials issued *Procedures for Setting Price
Ceilings of Coal Outside Plans*.[44] These stipulated that prices for non-quota coal
from the CMAs, LS, and the LNS mines should be between 200 and 350 per cent
of the fixed price. Mine officials caught changing a 'mandatory coal allocation'
to a 'guidance coal allocation', or selling coal under plan at market price incurred
liability for their companies. The Price Inspection Department would fine the
mine where the offending officials were employed and confiscate the unlawful
income.

About 1,300 km of new railway trackage were to be built over the Seventh Five
Year plan period. The second phase of the Datong (Shanxi)-Qinhuangdao (Hebei)
line, 242 km long, connecting Sanhe (Hebei) with Qinhuangdao, and which was
to be a key project for the Eighth Five Year Plan, was begun in 1988, as well as
a line linking the Beijing-Qinhuangdao and Beijing-Shenyang (Liaoning) lines.
In 1988 the 155-km Yongan-Zhangping section of the electrified line between
Yingtan (Jiangxi) and Xiamen (Fujian) was opened, and in 1989 the first 172-km
stretch of the Shenmu (Shaanxi)-Huanghua (Hebei) line was completed to Baotou
(Inner Mongolia), a vital link which eliminated the need for trucking the coal.
Other railway construction projects were single lines linking Hengshui (Shanxi)

and Shangqiu (Henan), Shangqiu and Fuyang (Anhui), Hengyang (Hunan) and Shangqiu, Xian (Shaanxi) and Ankang (Shaanxi), Xuancheng (Anhui) and Hangzhou (Zhejiang), Yanzhou (Shandong) and Shijiugang (Shandong), and Jining and Tongliao (Inner Mongolia); electrification of the Beitongpu (Taiyuan-Datong), Zhengzhou (Henan)-Baoji (Shaanxi) and Zhengzhou-Wuhan (Hubei) lines; double-tracking of the Nantongpu (Jiangsu) (Yuci-Fenglingdu) and Taiyuan (Shanxi)-Puzhou (Shanxi) lines; and construction of double, electrified lines linking Jiaozuo (Henan) and Zhicheng (Sichuan). Phasing out of the steam-powered locomotives began in 1989. Thereafter, only diesel and electric engines were manufactured.

Construction on the 890-km Nankun line, linking Nanning (Guangxi) with Kunming (Yunnan), began in 1990. Due to the rough topography, this was the most difficult and expensive railway line ever built in China. A 58-km line through Panxian county (Guizhou) was to be connected to it. This was of great importance because it would reduce the amount of coal hitherto shipped on the severely taxed north–south trunk railway. The area had excellent coal resources but due to their problematical location a branch line had never been built and production had been limited to what could be shipped out in trucks. Production at this CMA was to reach 80 million tons per year by 2000, of which 20 million tons were to be shipped to other provinces and regions, and 3 million tons to be exported.

The 'Coal Transport Study', begun in 1989, was a collaborative project between the SPC and World Bank to construct an econometric model that would calculate the most efficient ways of delivering coal and electricity over the next fifteen years.[45] The model was to be used in analysing alternative investment strategies and policies. This project also received funding from the Japanese Policy and Human Resources Development Fund and the United Nations Development Programme.

Beginning in the Seventh Five Year Plan period the Ministry of Railways began to divert workers to diversified activities for the same reasons as in the coal industry. A target was set to transfer some 1.4 million railway employees to diversified activities by 2000, and assistance was given to create jobs in other sectors.[46] Another measure to improve the finances of the railways was an increase in freight rates in March 1990 from 0.021 to 0.026 yuan per ton-km. Passenger fares, particularly for the new faster trains, were also raised.

Over the Seventh Five Year Plan period the Grand Canal and various river tributaries were deepened and widened to accommodate ships of over 2,000 tons. A section of the Changjiang was diverted to Jiangsu and Shandong, providing the Canal with a greater flow of water. Coal from Shanxi, Shaanxi, Henan, Anhui, Shandong, and Inner Mongolia was brought by train to Xuzhou (Jiangsu), then taken on the canal to Shanghai and other southern cities. Two big automated ports were built at Xuzhou, facilitating direct transfer of coal from the trains to the ships.

Construction also began in 1988 on a new canal to be completed in the Ninth Five Year Plan, connecting Xiangfan to Wuhan (both in Hubei), then Hanzhong (Shaanxi) to Xiangfan. A new port was built at Yujiahu (Hubei), southeast of

Xiangfan. The Han River, the largest tributary of the Changjiang, was widened and deepened to accommodate 500-dwt ships. Also work on Wangtan, a port 100 km southeast of Tangshan (Hebei) called Jingtang, was begun the following year. It was to handle coal from Shanxi and Hebei.

The Seventh Five Year Plan called for an annual average energy conservation rate of 3 per cent, equivalent to 100 million tons standard coal equivalent over the five years.[47] During this period, the gross value of industrial and agricultural output was to grow at an average rate of 6.7 per cent, industry at an average rate of 7.5 per cent, and energy at 3.4 per cent. Per capita GNP was to quadruple, while energy consumption was only to double. It called for widescale experimentation in energy-saving technologies, equipment, and materials. The plan also called for a 100-ton reduction in the amount of oil burned.

Some economists advocated that the country strive to reduce energy consumption by at least 50 per cent or maintain an energy elasticity coefficient of 0.5, and one forecast that over the Seventh Five Year Plan there would be an equal split between energy conservation achieved through direct means and indirect means.[48] According to another, 80–90 per cent of the total conservation that must be effected was in the industrial sector alone, and of this, 60–70 per cent would be achieved through indirect means.[49] Speaking of the potential for conservation, another quoted an estimate for reducing energy consumption by 57 per cent by 2000, with two-thirds of this coming from indirect conservation.[50]

The State Council issued energy conservation regulations in January 1986 which were to be followed nationwide by all sectors of the economy as of 1 April that year.[51] Large energy-consuming enterprises (those consuming more than 10,000 metric tons standard coal equivalent per year) were to place trained personnel in charge of energy conservation. Smaller enterprises would fall under the supervision of local authorities and departments. The duties of the energy-conservation personnel were to include drawing up and implementing policies for their own jurisdictions, carrying out state and local government policies, and keeping statistics on energy use and submitting these to the State Statistical Bureau. The State Bureau of Standardization was to work with enterprises in the formulation and monitoring of energy-consumption standards. Specific instructions were given concerning the conservation of coal, electricity, and oil.

Construction of new enterprises without guarantees of sufficient energy, or construction of energy-inefficient industry in energy-deficient areas was forbidden. Development of energy-inefficient activities was to take place only in energy-rich areas and with official approval, which was also required to expand boiler capacities. The indigenous method of coking was limited to special situations. Enterprises were encouraged to introduce cogeneration and those that could use coal byproducts and low-grade coal were instructed to do so.[52] The use of honeycomb and smokeless coal and coal-efficient stoves was to be promoted in the civilian sector. Where possible, coal gas was to be provided to urban areas and central heating systems were to replace industrial heating systems.

All large-consuming enterprises were to draw up long-term energy conservation plans and develop energy-conservation projects as part of their technical modernization programmes, and provisions for the funding of such projects were detailed. Low-interest loans, preferential prices, and exemption from product and value-added taxes were to be granted to new energy-saving products for an interim period. In importing technology from abroad, priority would be accorded to energy-efficient equipment. Imports of equipment used for energy conservation would be given import duty exemptions or reductions.

The state would legally protect people who informed local government authorities about energy-wasting activities, and economic benefits were to be awarded to people who made feasible energy-conservation proposals. Penalties in the form of fines, discontinued energy supplies, suspension of business licences or bank loans would be imposed on enterprises that did not follow the regulations. Higher prices were to be charged for above-quota consumption and the income so earned would be used for enforcing energy-conservation measures.

In the spring of 1987 the State Council approved a set of sweeping electricity consumption regulations.[53] Among them was the edict that the fee for above-quota electricity consumption by all sectors would be five to ten times above the regular fee. Arrangements were to be made whereby some enterprises were to operate alternately so as to share available supplies. The charge for electricity used in old inefficient equipment which had previously been declared obsolete would also be five to ten times the normal rate. Such equipment was not to be resold, but immediately destroyed, and persons found re-selling or re-using it were to be fined. As an incentive to enterprises to reduce electricity consumption, those which had surplus electricity were entitled to sell it at market prices.

Tight regulations were put in place for non-productive consumption of electricity. Guesthouses, restaurants (including foreign enterprises), some government offices, stores, and institutions were prohibited from using electricity for heating water or for space heating. Only in tourist hotels, research laboratories, purification rooms, surgical rooms, telecommunications rooms, computer rooms, and theatres, and on special occasions, were air conditioners, electric fans, or heaters to be used, and the electricity so consumed was to be strictly controlled. All above-quota consumption would be charged two to three times the regular rates. Households that used air conditioners, electric fans, heaters, and nonstandard electrical cooking utensils would be charged at five to ten times the basic rate. Cooking with electric stoves was not to be encouraged apart from areas with hydro electric stations or areas where authorization had been given by the SPC. Depending on location, household consumption was generally to be restricted to 60–100 kWh per month except for special occasions, with above-quota consumption charged at five to ten times the basic rate.[54]

At the end of the Seventh Five Year Plan the Energy Data Base of China (EDBC) was established by the Energy Research Institute, Technical Science Department of the Chinese Academy of Social Sciences. Encompassing all forms of energy, it was designed to serve both government and enterprises for economic and technical planning, evaluation, and forecasting.

By the late 1980s the central authorities recognized fully the local and global effects of China's air pollution problem and realized that international assistance was required. Hitherto this subject had received relatively little discussion by the government. Pollution was regarded as a fact of life, as an unavoidable consequence of modernization. However, the concentrations had reached unprecedented levels and were unquestionably the cause of chronic sickness and often death both in China and neighbouring countries. Hundreds of delegations from abroad came to study the causes and many cooperative projects were launched. (see Chapter 6 for wider discussion of the pollution problem).

In 1987 an *Air Pollution Prevention and Control Law* was introduced, and in 1988 the National Environmental Protection Agency (NEPA) was established. The Environmental Protection Law that had been first promulgated in 1979 was revised and re-issued in 1989. During this Plan period most city governments, to reduce pollution levels, called for all smoke/exhaust stacks to be extended such that the emissions would be released higher into the atmosphere, drew up zoning by-laws by which polluting industries were forced to move from residential areas, and expanded district heating networks to obviate the need for highly polluting individual home coal heaters.

The Eighth Five Year Plan: 1991–95

In the Eighth Five Year Plan the Jungar and Shenfu-Dongsheng fields, both in Inner Mongolia, were identified as major construction projects. The latter was to produce 10 million tons annually by 1992, 30 million by 1995, and 60 million by 2000. Production elsewhere in the country was to be consolidated, with the number of key coal-producing counties to be reduced from 112 to 100. The contract system had been introduced in 1985 and was to last for six years. When the six years were coming to an end in 1990, it was decided to extend the contracts for another two years. Initially, the total production target for 1995 had been set at 1,230 million tons, but part way through the Plan period the government sought to stimulate the economy, and it was raised to 1,250 in order to fuel a higher GNP average growth rate target of 8–9 per cent – up from the initially planned 6 per cent – as well as several other new targets.[55] The retrenchment programme launched in 1988 to rectify the overheating had checked growth too severely, and by 1990 the economy needed stimulation. Furthermore, the effects of the June 4th Incident had to be offset by continued visible economic progress. The government leaders were all too aware that their grasp on power depended upon ongoing improvement in living standards.

Interestingly, in the introduction to the *1996 Zhongguo meitan gongye nianjian* (1996 Coal Industry Yearbook), which reported on 1995 activity, it is stated that 'the coal industry conforms with the 14th National Congress of the Chinese Communist Party and the Fifth Plenary Session of the 14th Central Committee of the CCP', both of which had called for the adoption of a socialist market economy. Certainly the industry had already implemented quite a number of market economy principles, but it was by no means clear precisely just how 'public ownership' was to function so as to produce the freedom necessary for market forces to operate maximally.

On the occasion of the launching of the new Ministry in 1993, Vice Premier Zou Jihua announced that the production target for 2000 was to be 1.5 billion tons (not 1.4 as specified in *An Outline for the Development of China's Energy Industry, 1989–2000* which had been released in 1989). Then in December 1994, *Investment Strategies of Coal and Power Industries in China*, put forward by the Economic Research Centre of the SPC, proposed that output reach 1.5–1.6 billion tons in 2000.[56]

Higher production was readily obtainable. Indeed, even before the new targets were announced, output from the LNS mines had been soaring as people from many walks of life became involved in the industry, expecting to make money quickly and easily. The State Council, in an attempt to stem the seemingly uncontrollable increase in the number of illegal small mines, issued several circulars: *Circular of the State Council on Placing Village and Town Coal Mines Under Industry-Wide Management, Circular of the State Council on Approving and Relaying the Opinions of the State Economic Commission and the State Planning Commission on Immediately Overhauling Various Small Coal Pits Operating in the Vicinity of State-Run Coal Mines, Circular of the State Council on Regulating Private Coal Mining*, and *Opinions on Stopping the Indiscriminate Extraction of Small Coal Mines and on Ensuring Safety in Coal Mining*.[57] In all some 13,000 illegal small mines were closed in 1993.[58]

During the previous years the accident rate, at the LNS mines in particular, had been so alarmingly high (see Chapter 6) that the government, in December 1995, warned managers they could face disciplinary action if there were more than one death per million tons of coal mined at the state mines, three deaths at the township mines, and six at the individually-run mines. In December 1994 the MCI drew up a set of formal *Administrative Procedures for Permits on Coal Production*.[59] These specified how to obtain a mining permit, the safety requirements that had to be met, the required training in mine operation, the mining development plan, the penalties incurred if regulations were disobeyed, etc. Instead of being given *minimum* coal targets, the LS mines were given *maximums*. Fines for mining illegally ranged from 20,000 to 50,000 yuan and all the coal so mined was confiscated.[60] In March that year, at an ongoing national work conference of township coal mine managers, it had been noted that over half the 90,000 township mines were not licensed. So much coal was being dug from LNS mines that the government could not keep accurate records of reserves. Significantly, the production figure for 1994 was revised upwards in May 1995 by some 20 million tons after a more thorough survey of the dispersed LNS mines' output had been completed.[61]

At the First Session of the Eighth NPC held in March 1993 it was announced that the CNCC was to be dissolved, and the Ministry which had been disbanded in 1988 as part of the government restructuring programme was to be re-established. The inauguration of the new MCI as an independent entity took place in June 1993.[62] The China Coal Overseas Development Corporation remained unchanged. The 'super ministry' overseeing all the various types of energy had proven simply unmanageable. Moreover, the industry's diversification programme had increased its business responsibilities substantially. The new Ministry was to carry out over-all planning, coordination, supervision, and provision of services, but actual mine management was to be in the hands of the individual enterprise managers.

The new Minister, Wang Senhao, had worked in the Datong Bureau for fifteen years and in the Luan Mining Bureau for eleven, later serving as governor of Shanxi for nearly ten years. Like Gao Yangwen, he had strong opinions about the irrationalities in the industry caused by rigid, top-down central planning.[63] He vigorously supported the establishment of coal exchanges, the liberalization of the price of coal, the opening of the industry to foreign investment, dramatic cuts in the work force, matching production with preparation work, and building mine-mouth power plants to reduce pressure on the railways.

In March 1994 the State Council abolished the Northeast China and Inner Mongolia Coal Industry Unified Corporation as well as the local coal adminis-trations, and established coal industry administration bureaux in Liaoning, Jilin, and Heilongjiang. The specific management functions and responsibilities of the Ministry, provincial government, local governments, Party, and individual enterprise were delimited.[64] Later in the year Beijing compensated the PLA 1.3 billion yuan for turning over army mines to civilian authorities in Shanxi.[65] The military officials involved in the industry were not in it for any strategic reasons, but solely to make profits. In that sense they contributed to the marketization of the industry. However, their dealings generally tended to favour their own *guanxi* (connections), interfering with the full utility of market forces, to say nothing of adding to the already high crime rate.

During the Sixth Five Year Plan fourteen companies (originally bureaux) had been established by the MCI to manage equipment, trade, materials, etc.[66] By the early 1990s it was decided to re-structure these into four 'industrial groups' in an effort to make them even more efficient. The first of these was the China Coal Equipment Group established in 1993, and the others were created in the Ninth Five Year Plan.

The industry's losses nationwide had continued to mount during the late 1980s and peaked at 5.75 billion yuan in 1992. Late that year, the CMAs had difficulty meeting the basic wages of the miners, let alone any extra wages for above-quota production. At the four mining bureaux in Heilongjiang Province the wages had not been paid for some time, production had been suspended, and there had been frequent sit-ins and petitions, and a suicide.[67]

Billions of yuan were owed not only in wages but in pensions to retired employees. Not surprisingly, the men were angry, and were a potential threat to national stability. The Chinese leaders were very aware that it was protests lead by industrial workers which kindled the overthrow of the government in Poland, and that strikes by coal miners in the United Kingdom and Soviet Union had ser-iously imperiled the authority of the governments of those countries.[68] To prevent a potential political disaster, the State Council hastily granted the industry a loan to cover the wages.[69]

A major problem was the inability of large consumers to pay for the coal. Some mine administrations received less than 20 per cent of their payments, meaning they had a fraction of the money needed to run their mine operations. The mines fell seriously behind in basic safety, let alone technical improvements, and again the output–capacity ratio became severely imbalanced. The industry's debt problem had knock-on effects, too. Mining companies owed the railways nearly 200 million yuan in freight charges, and the railways reacted by returning cargo and suspending service. Steel shipments were halted as well as oil supplies. This problem of inter-industry debt, sometimes called triangular debt, or debt chains, had paralysing effects.

Payments in arrears owed by other provinces to Shanxi Province reached nearly 7 billion yuan, crippling the local economy. It was not only the coal industry which suffered, for when it took out large amounts of limited funds from the banks, little was left for the Province's chemical, machine-building, electric power, building materials, and metallurgical industries.

The production capacities of some key state owned coal enterprises dropped as much as 10 per cent and the government was forced to allocate 4 billion yuan in loans to compensate for defaulting consumers. Metallurgical, especially steel, as well as fuel, electric power, and chemical companies owed billions of yuan to the coal industry.

Beginning in 1992 the government decided to take drastic action to reverse the industry's losses and market experiments were initiated to test how the freeing of coal prices would affect both producers and consumers. The results were quite remarkable and it was decided that the proportion of CMAs making profits was to rise from 9.7 per cent in 1992 to 25.7 percent in 1995. The first experiments were carried out in Xuzhou (Jiangsu) and Zaozhuang (Anhui) mining bureaux in July 1992 and later at mines in Shandong. These mines were chosen because they were among the most efficient, and demand for their higher quality products was stable. Also their clients were concentrated in eastern China, where the coal-consuming industrial economy was largely located.

As part of the experiment, all subsidization to the Xuzhou Bureau was withdrawn. By the end of 1992, after using market-determined prices, not only had the Bureau cleared its deficit of 290 million yuan, it had realized a profit of 14 million yuan. In 1993, though production costs increased by 10 yuan per ton and a debt of 100 million yuan had to be paid off, total year-end profits at the Bureau amounted to 35 million yuan.[70] The experiments proved so successful that the authorities were confident enough to begin liberalizing the price of coal nationally (see Chapter 6). In November 1993 the government announced its decision to lift all price controls on coal as of 1 January 1994, one year ahead of schedule.

The decontrolling of coal prices, as well as those of several other commodities, was highly inflationary. The general consumer price index averaged less than 3.3 in 1990 and 1991, but rose to 6.4 in 1992, and 14.7 in 1993. The government launched another austerity programme in mid-1993 to cool down the economy. However, the index continued to rise, peaking at 24.1 in 1994. At the end of the Eighth Five Year Plan it was still far too high at 17.1.

The price of coal sold on the market was two to three times higher than that of coal allocated under the state plan. Consumers, struck over a short period of time with major price increases for a variety of commodities, found themselves unable to cope and slowed down, or in protest actually halted, their purchases of coal, resulting in increased stockpiles.[71] Therefore, in order to help cushion the transition, the subsidies previously earmarked for deficit-ridden mines were used to help key state industries, such as the power, metallurgical, and chemical fertilizer, to buy their supplies at the higher prices.

The expansion of diversified activities was crucial in reducing the industry's debt. It was decided that over the Eighth Five Year Plan period 500,000 workers at the state mines would be transferred to these activities and another 500,000 over the Ninth.[72] Using a government loan and funds it raised itself, the CNCC planned to spend 5,000–10,000 yuan per worker to help them transfer to one of the new endeavours.[73] They were still to be employed by the Ministry, but were to be placed in one of four types of activities: the production of coal-based chemical products, the utilization of mine wastes, the production of coal coexistent materials, and production of all manner of consumer goods and services. Now that these sideline businesses were not only legitimate in the socialist market economy, but strongly encouraged, they multiplied quickly. By 1995 the total output value of the businesses attached to the coal industry as a whole was 41.8 billion yuan, of which those attached to the CMAs alone reached 33.2 billion yuan, taxes and profits were 2 million yuan, and 2 million workers were producing 4,000 products at 20,000 factories.[74]

Other measures were the 4 per cent reduction in March 1994 of the value-added tax rate on coal products from 17 to 13 per cent; the rigorous implementation towards the end of the year of the 'packaging' of purchase, sale, and transport contracts; and the *san bu* (three no's) delivery policy introduced in November 1994 aimed at ending the triangular debts. (The three no's were no shipment without payment, commercial draft, and the clearing of outstanding payments.)[75] At the same time, the actual payment transactions were transferred from the mine administrations to the banks which were able to exert greater pressure on debtors.

The industry's finances were also bolstered – at least initially – by the establishment of several official coal exchanges to help counter the rampant corruption that had arisen primarily due to the two-track pricing system. Coal allocations were still decided by the major coal consumers (the MCI itself, the Ministry of Railways, Ministry of Chemicals, and Ministry of Electric Power) at annual coal conferences. Brokers at the coal exchanges, the first of which began operating in 1992, now matched coal output beyond these key ministries' needs with those of other consumers.

Towards the end of the Eighth Five Year Plan period a major step was taken towards relieving the industry's tremendous burden of maintaining retired miners. At that time 3.5 million workers were supporting 1.06 million retirees at 5 million yuan per year, the industry's total retirement debt being 160 million yuan. Worse, the number of retiring workers was expected to rise sharply in the coming years. In March 1995 the State Council approved social security systems for eleven ministries, among which was the MCI. Under the new pension scheme 3 per cent of each worker's monthly salary would be diverted to build up his own retirement fund. Over the next ten years, the proportion would gradually be increased until it reached 8 per cent. An insurance scheme was also introduced to cover unemployment, medical care, work injuries, and maternity leave.[76]

The final big change in the industry, in what surely had been one of the most extraordinary periods in its history, was the establishment in the summer of 1995 of the Shenhua Group Corporation. It was the first state owned coal enterprise to become a shareholding company with limited liability, a change made possible by the passing of the Company Law in July 1994. Separate subordinate companies were formed to finance the mining of coal and the building of the railway lines, power stations, ships, and ports.

There were some significant events with respect to trade during the Eighth Five Year Plan period, including changes in trading partners. In January 1993, after many years of discussions, an agreement was signed with Israel. In May of the same year, due to prolonged failure of payment, China decided to decrease exports of coal to their old ally, North Korea.[77]

The administration of trade was further consolidated. In October 1994 the CNCIEC became solely responsible for handling the planned export quota set by the state. Separate departments were created to handle the export of power, and coking coals and anthracite, and to manage diversified activities. When the new Ministry had been established in June 1993, Minister Wang Senhao announced that exports would be 50 million tons per year by 2000. However, it was soon realized that this target was wildly optimistic. Over the Eighth Five Year Plan period the average annual increment in exports was only 2.27 million tons, and the total at the end of the Plan period was but 28.62 million tons. Furthermore, the shipments to the Qinhuangdao (Hebei) port were continually late.[78] The results were power shortages in other parts of China, and havoc for international shipping schedules, the government being forced to pay a fortune in demurrage fees. There were also continuing quality problems. In 1995, some 65,000 tons of the 14 million tons processed for export at Qinhuangdao were rejected.

In February 1995, a third coal exporting company, the Shenhua Coal Import Export Corporation, was established to ship coal from the Shenfu fields (Inner Mongolia). (The first and second were the China Coal Import and Export Corporation established in 1982 and the Shanxi Coal Import Export Corporation in 1994). Besides coal, the industry was now exporting mining equipment.

Attempting to attract foreign involvement, the government aggressively advertised the fact that it wanted to form joint and cooperative ventures for the construction of mines, pit-mouth power stations, coal-gasification plants, cement factories, and slurry pipelines. Initially, it sought only loans. Between 1990 and 1994 the Ministry spent US$4 billion in loans (the fourth set) from the Japanese Ex-Im Bank on railway and highway projects, and government loans from Sweden, Italy, and other countries, on mine upgrading. Most foreign funding in the coal industry continued to come from Japan. It was Japanese funding that supported the development of the Shenfu coalfield.

In December 1994 the SPC and Ministry of Foreign Trade and Economic Cooperation released a list of 210 major capital construction and technical renovation projects to be started between 1993 and 2000. The items listed under coal industry were: construction of Huojitu Pit (Dongsheng Mining Area of Shenfu Coal Field) including transportation and water supply facilities, with annual production of 5 million tons; Jining No. 2 Mine, Yanzhou Mining Area (Shandong), with annual production of 4 million tons; Jining No. 3 Mine, Yanzhou Mining Area (Shandong), with annual production of 5 million tons; Xuchang Mine, Yanzhou Mining Area (Shandong), with annual production of 1.5 million tons; Daizhuang Mine, Yanzhou Mining Area (Shandong), with annual production of 1.5 million tons; Fucun Mine, Tengnan Mining Area (Shandong), with annual production of 3 million tons; Anjialing Opencut Mine, Pingshuo Mining Area (Shanxi), with annual production of 15 million tons; Chensilou Mine, Yongcheng Mining Area (Henan), with annual production of 2.4 million tons; and Cheji Mine, Yongcheng Mining Area (Henan), with annual production of 1.8 million tons.[79]

Over 300 joint venture or cooperative coal projects were announced at the 1995 Fujian Investment and Trade Fair, including 16 large mines, 39 power plants, and 180 diversified projects, such as extraction of methane gas and sulphur iron ore, and construction of coking, peat processing, coal treatment, fireproof materials, and coal machinery plants.[80] In conjunction with the announcement of these projects the state relaxed control over joint venture procedures. In order to facilitate negotiations, the MCI and provincial governments were given the authority to examine and approve coal investment projects totalling US$10 million or less, and the Special Economic Zones, open coastal cities, and inland provincial capital cities were authorized to examine and approve projects of US$30 million or less.

The entry in an *Industrial Catalogue Guiding Foreign Investment* released in 1995 read as follows:

1. Design and manufacture of coal extraction and transportation equipment.
2. Design and manufacture of complete sets of gasification equipment.
3. Manufacture of equipment for the production of highly concentrated coal slurry and additives.
4. Comprehensive development and use of low-calorie fuels and associated resources.
5. Comprehensive development and use of coal.[81]

Hammer's death in December 1990 was a blow to the industry's attempts to integrate internationally. In April 1984 he had signed the largest-ever joint venture agreement with China, the Antaibao opencast mine project, and despite all the problems it had encountered, Hammer had been determined to keep it in the company's portfolio. However, soon after he died, the new management of Occidental Petroleum, the parent company, announced that it would divest itself of various loss-making projects, including the Antaibao mine. Production from the mine peaked at 12 million tons, considerably less than the planned 15.33 million tons, and was only about 9 million tons in 1990.[82] Losses of US$31 million were reported for that year. In June 1991 Occidental Petroleum sold its share to the joint venture partner, the Bank of China Trust and Consultancy Company.

Of the new trackage to be layed during the Eighth Five Year Plan period, a key project was the second phase of the 653-km Datong (Shanxi)-Qinhuangdao (Hebei) railway line, along which twelve coal depots with modern loading and unloading equipment were to be installed. Total storage capacity at the port was to be expanded to 18 million tons. Other projects were the electrified double-tracked 252-km Houma (Shanxi)-Yueshan (Henan) line linking up with the Jiazuo (Henan)-Zhicheng (Sichuan) line in central China; the 940-km Jinan (Shandong)-Tongliao (Jilin) line; a third line connecting Qinhuangdao (Hebei) and Shenyang (Liaoning); a 274-km line linking Xian (Shaanxi) and Ankang (Shaanxi); and Inner Mongolia's first electrified line running between Jungar (Inner Mongolia) and Datong (Shanxi), a distance of 264 km, which was to be completed in the summer of 1997.

In 1991 construction of the 810-km Shenmu (Shaanxi)-Huanghua (Hebei) line was begun, followed in 1993 by the 2,553-km Beijing-Kowloon (Hong Kong) line. The latter was to be completed in 1996, and to be used not only as a passenger line, but also as a major coal transport artery. Its routing was planned carefully to minimize distances to power plants along the way. The 1,622-km Lanzhou (Gansu)-Urumqi (Xinjiang) line was ready for operation in January 1995.

At the end of the Eighth Five Year Plan period the railways were facing severe deficits. Not only were operating costs rising by up to 15 per cent per year, but as highway transport improved, ridership on the railways steadily declined. A Railway Construction Fund had been established in 1991 in the form of a surcharge per ton-km on freight, and passenger fares and freight tariffs were raised in October 1995. However, the costs of repairing the railways after extensive flooding, and maintaining the old equipment which had been poorly managed for decades, far exceeded these additional revenues.[83]

In March 1994 the Shanxi Provincial government announced several measures to improve the efficiency of coal transport through economies of scale. Many small railway stations used by local mines were to be closed, while the storage areas of others were to be expanded to a minimum of 300 hundred tons capacity each. Coal depots not on the railway line were to be closed.[84]

In August 1992 and December 1995 the Ministry of Railways issued bonds worth 2 billion yuan and 1.53 billion, respectively. Passenger fares were raised by 50 per cent in October 1995, and freight tariffs by 18 per cent.[85] The government tried to attract foreign financial, technical, and material assistance in expanding and improving the railway network by offering potential investors authority to set prices and railside real estate development privileges. As with coal prices, the government had hitherto not granted pricing authority for fear of inflation. It was announced in December 1995 that the heavily used Guangzhou-Shenzhen (both in Guangdong) line was to be floated on the Hong Kong Stock Exchange and the management given authority to set its own ticket prices.

A very significant transportation development was the US$888.6-million agreement reached in Beijing in August 1994 with a US-led international consortium to build the 600-km 'Yu-Wei' coal pipeline from Yu County, near Yangquan (Shanxi), to Weifang (Shandong).[86] To be completed late 1997, it was to transport 15 million tons of coal per year, and later be extended another 205 km to Qingdao (Shandong). This would make it the longest coal slurry pipeline in the world. The contract called for fifty years of cooperation and it was hoped that the investment would pay for itself within seven years. Five million tons of the coal were to be used annually by a power station in Weifang and 8 million processed tons were to be sold abroad, and the rest on the spot market. The deal included the construction and operation of a coal cleaning plant, an underground slurry line, a dewatering and briquetting plant, and port facilities. The coal would first be crushed and 'deep cleaned', then additives would be injected into the slurry to reduce the sulphur emissions by about 50 per cent and ash by 75 per cent.

As for seaport development during the Eighth Five Year Plan period, the fourth phase of the coal loading project at Qinhuangdao (Hebei) was to be completed, there was to be more development of the Qingdao port (Shandong), and new coal unloading berths at Nantong (Jiangsu), Zhangjiagang (Jiangsu), Zhoushan (Zhejiang), Xiamen (Fujian), Guangzhou (Guangdong), and Shantou (Guangdong). The coal wharves at Shanghai, Hainan, Dalian (Liaoning), Wenzhou (Zhejiang), Meizhou (Fujian), and Guangzhou (Guangdong) were to be renovated.

On the Changjiang some twenty-four berths for coal and phosphorus were to be built, and extensive dredging was to be carried out to enable larger ships to reach cities farther up the river. Ultimately 25,000-ton ships were to be able to berth at Nanjing (Jiangsu), 5,000-ton ships at Wuhan (Hubei), 3,000-ton ships at Yichang (Hubei), 1,000-ton ships at Shuifu (Yunnan), and 300–500-ton ships were to be able to ply up the main tributaries of the Changjiang. In all, a total of 50 million tons of loading capacity were to be added and 30 million tons of unloading capacity.[87] At the end of the Eighth Five Year Plan the government announced plans to consolidate and improve the inland water transportation network. Priority was to be given to the expansion of the Grand Canal and to improvements on the Changjiang and Heilongjiang, as well as the Huai, Xi, and Songhua Rivers.[88]

Several expressways were also to be completed: Beijing-Shijiazhuang (Hebei), Jinan-Qingdao (both in Shandong), Chengdu (Sichuan)- Chongqing, Kaifeng-Luoyang (both in Henan), and Xian-Baoji (both in Shaanxi).

The total energy conservation target for the Eighth Five Year Plan was 167 million tons sce. The amount of coal consumed in mining each 10,000 tons was to be reduced from 305 tons in 1990 to 295 in 1995. The Ministry of Energy Resources also set a goal of replacing thirty-one small electricity generators with larger ones. The small ones would generate a total of 6 million kW and larger ones 7.5 million, resulting in a saving of 3 million tons of raw coal each year.[89] The target for energy conservation was doubled in 1993: 'Great importance will be attached to the practice of economy while accelerating development of energy and raw materials. In the remaining years of the Eighth Five Year Plan, all possible means are to be employed to boost the annual energy conservation rate to 3.7 per cent from the original planned 1.8 per cent.'[90]

A draft set of energy conservation laws was published in late 1993.[91] These set out clearly how energy consumption was to be monitored and what the consequences would be if they were violated. Specifically, a fine of one to five times the value of an enterprise's profits would be levied if it used equipment which did not meet national efficiency norms, refused inspection of its equipment, showed the energy inspectors equipment which it did not actually use, produced products which were labelled as being energy-efficient but actually were not, imported equipment which did not meet national standards, etc. Offending enterprises were liable to pay high prices for energy consumption above set norms, have the profits which were made while using inefficient equipment confiscated, their businesses shut down, and/or their equipment confiscated.

In the early 1990s several foreign energy experts established offices in China to teach energy management in both the public and private sectors. For example, in November 1993 the Beijing Energy Efficiency Centre (BECon) was established. Operated jointly by the Energy Research Institute of the SPC, the Battelle Pacific Northwest Laboratories, and the Lawrence Berkeley Laboratory, the four main tasks of this new organization were to:

1. provide policy advice to central and local government agencies
2. support the development of energy efficiency businesses
3. create and implement public information and technical training programmes for Chinese energy planners and managers
4. design mechanisms for financing energy efficiency, including setting up a disbursement system for loans from multilateral organizations such as the World Bank and Asian Development Bank ...[92]

In June 1993 a Clean Coal Technology Centre was established in Beijing.[93] Of particular importance was research into the 'dry' processing technique, a purification process for raw coal that does not require large quantities of water.

China was one of the first countries to respond to the international goals set at the 1992 Earth Summit held in Rio de Janeiro, by issuing *Ten Strategic Policies for Development and the Environment*. In 1993 the NEPA and the SPC formulated the China Environmental Protection Action Plan (1991–2000), and NEPA devised the *Outline for National Environmental Protection (1993–98)*.[94] An important component of these was a trial programme, aimed at combatting the acid rain problem, of charging coal-burning industries for their emissions of sulphur dioxide. The experiment was implemented in the provinces of Guangdong and Guizhou, as well as nine cities including Chongqing and Yibin (Sichuan), Nanning, Guilin, and Liuzhou (Guangxi), Yichang (Hubei), Qingdao (Shandong), Hangzhou (Zhejiang), and Changsha (Hunan).

In March 1994, the State Council adopted *Agenda 21 – A White Paper on Chinese Population, Environment and Development in the Twenty-first Century*, which included several sections on pollution control and efficient energy consumption.[95] According to an associate project director of the Worldwatch Institute, China has 'crafted one of the most elaborate, ambitious national Agenda 21 Plans'.[96]

In 1995 the Air Pollution Prevention and Control Law which had been introduced in 1987 was amended to include stricter sulphur dioxide control regulations.

The Ninth Five Year Plan: 1996–2000

The Ninth Five Year Plan called for coal production in 2000 to reach 1.45 billion tons (not the 1.5–1.6 which had been proposed before the previous plan), of which 620 million were to come from the CMAs, 220 from the LS mines, and 610 from the LNS.[97] The Shanxi Energy Base would contribute 574 million tons (not 600), and Shanxi Province alone, 400 million tons (not 440).[98] Some 800 existing mines were to be renovated, and 24 new mines were to be built in Shanxi. Production in Guangxi was to increase greatly, from 12.3 million tons in 1995 to 80 in 2000, and of this, 20 million tons were to be shipped to other parts of the country, and 3 million tons exported.[99]

The new underground mine construction priorities were Huainan (Anhui), Huaibei (Anhui), Yongeheng, and Yanzhou (Shandong), and the opencast were Pingshuo (Shanxi), Shenfu-Dongsheng (Inner Mongolia), Jungar (Inner Mongolia), and Panxian (Guizhou). The Jungar Mine, with foreign cooperation, was to go into operation in October 1997 and produce 12 million tons per year for the next fourteen years, all of which would be washed before marketing. Labelled the largest investment project since 1949, besides the mine itself and washeries, it included a thermal plant and a 216-km electrified line to Datong (Shanxi), which would connect with the line to Qinhuangdao (Hebei). Fujian Province, long thought to have virtually no coal, was found to have some thirty-three fields and a 2000 target of 12 million tons was set for the Province.

Partial mechanization was to be installed at 80 per cent of the CMAs, and OMS – 1.217 tons in 1990 – was to reach two tons. The respective levels for the LS mines would be 20 per cent and one ton. Some 100 existing washing plants were to be improved, while another 223 were to be built.[100] This would increase the amount of coal being washed from 202 million tons in 1995 (15.6 per cent of the total output) to 420–450 in 2000 (about 30 per cent, down from the earlier over-optimistic 50 per cent). Prepared capacity at the CMAs was to go from 5.25 to 13 per cent of the developed reserves and at the LS mines from 2.6 to 7 per cent. The 100 counties with the best production potential were each to produce 5 million tons.

Total output value of the diversified activities operated by the coal industry as a whole was to reach 80 billion yuan in 2000, with taxes and profits earning some 6.5 billion. Those operated by CMAs would earn 60 billion yuan, nearly twice what it had been in 1995, and actually surpassing the output value of coal and coal products. They would generate taxes and profits of 5 billion yuan and employ 2.4 million workers. By 2010, the diversified activities were to earn some 150 billion yuan in output value, and 12.8 billion in taxes and profits, and employ about 2.5 million workers.[101]

Approximately 300 coal projects were to be advertised as potential foreign joint ventures, and, in all, the work force was to be reduced by another 500,000 workers. Four 'strategic' areas of research identified for the industry were the construction of small pit-head power plants, using preparation plant refuse and middlings; coal methane gas development; clean coal technology, especially sulphur removal; and slurry transport. Sixty per cent of the miners were to complete senior-middle school.

Considerable emphasis was put on improving safety at the mines. The fatality rate per million tons was to be below one at the CMAs, less than four at the LS mines, and less than eight at the LNS.[102] Not only were the high accident rates receiving unwanted international attention, a major crisis was developing for underground mines. While on the one hand the industry was desperately trying to find jobs for thousands of surplus, often older surface workers, on the other, there was increasing difficulty in finding young men willing to work underground. The present generation of young men in their late teens and early twenties, conceived in the late 1970s, were the first born under the one child policy. Their families were loathe to have their only child take up such a dangerous occupation. In April 1997 it was announced that the head of a state mine where over ten people were killed in accidents in a given year would be fired, and any head of a provincial mine would be dismissed if there were two fatalities in one year.[103] Two years later the goal was set that no more than one, four and seven to eight deaths were to occur per million tons mined at SPMAs (*formerly known as CMAs*), LS and LNS mines, respectively.[104] Total mining fatalities were to be no more than 4,000–4,400 per one billion tons. The *Coal Law* stipulated that as of July 1997 all coal companies had to provide accident insurance for all underground workers. Compensation to families who lost a relative would rise to 50,000 yuan, which was over 20 per cent above the previous level.[105] After accidents at two mines in Henan killed thirty-six people in April 1998 some officials were criminally charged or sued. Others may have been forced to make self-criticisms and/or expelled from the Party.[106]

Wang Senhao, Minister of Coal, made the very significant announcement in April 1996 that the contract system used in the industry since 1985 would be replaced with a new system called the 'Xinji model', named after the three Xinji mines in Huainan (Anhui).[107] Construction of the first mine to employ the model had begun in 1989. The production goals of all three were reached in four years, whereas hitherto most mines had taken at least six to reach their full capacity, and the OMS was a remarkable ten tons. The Xinji mines also had considerable success in bringing down their debt levels, not only by eliminating delays in production from new pits, but by combining the profits of all the various enterprises at the mines to help repay loans.

These mines exemplified the goal for coal enterprises to be responsible entirely for their own management, without any government funding or involvement in operational procedures and decisions. The ultimate goal was *gufenhua*, to make the CMAs limited liability or shareholding companies, and the coal mining bureaux into corporate groups. In January 1997 it was announced that the three mines at Xinji would be listed on the Hong Kong Stock Exchange in the third quarter. This would be the first coal project in China to issue shares abroad.[108]

Under the contract management system it had been found that the great variations across the country in the resources and in market conditions meant that some mines fared much better with less state involvement, while others fell behind, and the differences were widening. In the application of the Xinji model, mines were to be classified into three different groups. The first comprised those that were doing well. They were to be completely responsible for their own profits and losses, and after-tax profits would belong entirely to them for re-investment in such things as pit-head power plants, manufacture of coal methane gas, and their own branch railway lines and/or ports. The second group, less successful but with promise, would be able to obtain from the state some financial support and assistance in changing their operational structures. The third group were those in very serious financial trouble. A special team from the MCI would give them priority investment, loans, subsidies, and technological upgrading.

The first limited liability group, the Zhengzhou Coal Industry Group Limited Liability Company, was established out of the former Zhengzhou Coal Mining Bureau (Henan) in January 1996.[109] It had a deficit of 50 million yuan in 1990, but progressed to making a profit of 80 million by 1995. This was done entirely through changes in management. Support services such as repair workshops, stores, schools, and hospitals, were administratively disassociated from the production work at the pit sites and given complete autonomy. The pits were converted into seventeen companies, each with semi-official status as legal entities. In 1996, the subsidiaries of the Zhengzhou Coal Industry Group became shareholding companies, while the parent corporation continued as a state owned limited liability company. In all, there were six mines, nine supporting companies, and seventeen specialized subsidiaries.

The detailed eighty-one article *Coal Law*, which came into effect in December 1996, was aimed not only at regulating coal development and production, but establishing a rational relationship between the state and individually licensed (LNS) mines.[110] It also sought to ensure efficient use of coal resources, labour, and equipment; to clarify mining rights in ethnic minority areas; and to guarantee the security of coal deliveries. The law was formulated in conjunction with the *Revised Mining Resources Law, the Implementation Ordinances for the Mines Safety Law*, and the *Law on Electric Power*, all issued earlier that year or in 1995. The changes to the *1986 Mineral Resources Law*, released in August, made investment in coal mining much more attractive to foreigners and simplified the approval process. The earlier Law had been ambiguous on the subjects of rights and jurisdictions. Conflict resolution procedures were considerably clarified in the new Law.

In the summer of 1997 the MCI announced plans to establish two enterprise groups out of the Yanzhou (Shandong) and Datong (Shanxi) Coal Mines, and to list four coal companies – the Yanzhou (Shandong), Pingdingshan (Henan), Zhengzhou (Henan), and Antaibao (Shanxi) – on domestic stock markets. It was hoped that many of these would eventually be listed on the Hong Kong Stock Exchange. The Inner Mongolia Yitai Coal Company was launched on the Shanghai Stock Exchange in August 1997.

The first large industrial group, the China Coal Materials and Equipment Group, had been established in 1993. In August 1997 another three were formed: the China Coal Import and Export Group (CCIEG), based on the former CNCIEC, would now supervise domestic trade as well as international; the China Coal Construction and Development Group would manage mine designing, construction, equipment supply, technical consultation, and development; and the China Coal Multi-Utilization Group would oversee the sales of coal-derived chemicals and building materials, clean coal and thermal power, and also supervise sideline service sector businesses.[111] A fifth group, the China Coal Material Group, was to be established later. A notable purchase of the CCIEG was The Holiday Inn Pudong Hotel (Shanghai), situated within the thirty-three storey China Coal Mansion, a property purchased for US$68 million. It was opened in February 1998.[112]

Radical re-organization of the central government affected the coal industry. It was announced in the spring of 1996 that there would be, at some unspecified point in the future, mergers involving the power, metallurgical, electronic, mechanical, chemical, coal, and railway and transportation industries. The forty ministries, departments, and agencies within the State Council would be reduced to twenty-nine, and the SPC and State Economic Commission would also be merged into one agency. In the summer of 1997 the State Council hinted that several of the production ministries would be converted into state corporations in 1998.

The time chosen to carry out this radical re-organization was the NPC held in March 1998. Fifteen ministries and some commissions were either merged or disbanded and four new bodies were created.[113] The MCI was thenceforward re-labelled the *Guojia gongye meitan ju* – 'State Coal Industry Bureau' (SCIB) – sometimes also translated as State Administration of Coal Industry (SACI), and put under the State Economic and Trade Commission (SETC) which had been created in 1993.[114] Other former ministries put under this Commission were those which had been responsible for power, metallurgy, machine building, chemicals, and internal trade. The former national councils for light and textile industries, the former State Grain Reserve Administration, China National Petroleum Corporation, and China National Petrochemical Corporation were also brought under the SETC. Importantly, the environmental protection administration was elevated to ministry status, indicating the priority that reducing pollution had become. Formerly called the NEPA, it was now called the State Environmental Protection Administration (SEPA).

The timing of these stunning announcements was courageous. An end to the Asian financial crisis was nowhere in sight, and China was under considerable pressure to devalue its currency. The level of urban unemployment reached some 13 million by mid-1998, with untold rural unemployed. Though social security systems for eleven ministries, including the MCI, were in the planning stages as early as 1995, by 1998 no comprehensive and nationwide social security system had been implemented. A special task force had been established to try to place the redundant workers in new jobs.

In July 1998 the PLA was ordered to relinquish all business endeavours, including its remaining interests in coal mining, transport and distribution networks all over the country. The State Council in the following month announced the transfer of the ninety-four CMAs, in all accounting at that time for just under 40 per cent of total production, to provincial governments.[115] Thirty of these mines would have auditors assigned them by the State Council, while auditors for the others would be assigned from their local governments. Taxes would go to local governments but profits would stay with the coal enterprises. The challenge to make viable and profitable operations of the newly named 'State Priority Mining Administrations' (SPMAs) was now in the hands of provincial authorities and the mine enterprise managers.

A main goal of the China Coal Industry Association (CCIA), launched in March 1999, was to study ways of integrating the industry into a socialist market economy. Based on the former China Coal Industrial Enterprise Management Association and combining eleven other associations, it was registered with the Ministry of Civil Affairs and under the guidance of the SCIB. Other goals were to promote scientific management, expand foreign cooperation, and draft a series of industrial regulations concerning quality, technology, and management standards.[116]

Preliminary outlines for the Tenth Five Year Plan (2001–2005) were announced in early 1999. The number of sectors given production targets was to be pared down and the Plan would further activate market forces and encourage integration into the global economy. The main priorities were environmental protection, curtailing the disparity between coastal and inland regions, and stimulating consumer spending.[117] At this stage no specific mention of the coal industry was made.

A total of 206.4 billion yuan was budgeted for the coal industry over the Ninth Five Year Plan period, of which it was hoped 20 per cent would be foreign capital – half in the form of loans from foreign governments, and half in the form of foreign direct investment.[118] In June 1996 the State Council announced that foreign companies would be granted not only a greater role in the industry, but actual controlling interests in the mines.[119] In January 1998, another incentive to foreign investors was given in the form of a 50 per cent exemption on import duties for fuel, raw materials, and infrastructural and agricultural products. In March 1999, Zhang Baoming, head of the SCIB boldly indicated that 'a huge amount of foreign funds' would be sought to develop and modernize the industry.[120]

Not long after this announcement the ASEAN Forum on Coal held its inaugural meeting and pledged 'to protect the interests of the region's coal industry and determine how best to exploit a resource that will last much longer than hydrocarbons ... to strengthen bilateral and multilateral cooperation on coal mining and trading among all member countries ... to strengthen institutional and policy frameworks, promote clean coal technology, and enhance private sector investment'.[121] All of the countries in the region were keen to explore the possibilities of reducing their dependence on foreign oil, and to learn how clean coal technologies could produce a cleaner, more efficient, and convenient product. Though China was not a member of ASEAN, the formation of this organization no doubt raised the hopes of the Chinese government that member countries would increase their imports from China.

A formal Memoradum of Policy Understanding (MOPU) was signed between the International Energy Agency (IEA) and China in October 1996. On the 'basis of equality, reciprocity, and mutual benefit' there was to be cooperation in energy supply security, energy and the environment, energy policy, energy statistics and modelling, foreign investment and trading in energy, energy technology, and energy efficiency.[122] The IEA was to work with the State Development Planning Commission, SETC, the SEPA, and National Bureau of Statistics as well as state owned energy enterprises such as the State Power Corporation, China National Petroleum Corporation, and China Energy Conservation Investment Corporation.

At the annual trade fair held in Xiamen (Fujian) in September 1996, 200 coal projects seeking foreign funding were exhibited, and attempts were made to secure investment in coal strip mines, renovating of existing mines, clean coal technology, coal slurry pipelines, coal-bed methane gas exploitation, pit-mouth electricity plants, coking plants, cement plants, coal mining machine factories, and refractory material plants.

The entry in the revised *Industrial Catalogue Guiding Foreign Investment* (see previous section on Eighth Five Year Plan) put into effect as of 1 January 1998, was broadened to include coal dressing and clearly stated that the Chinese partner had to be the majority share holder:

1. Design and manufacture of coal mining, dressing and transportation equipment.
2. The mining, washing and dressing of coal, with special care and rare varieties of coal subject to the Chinese partner holding a controlling share of dominant position.
3. Production of coal slurry and liquefaction of coal.
4. Comprehensive development and use of coal.
5. Comprehensive development and use of low-thermal-value fuels and associated resources.
6. Pipeline transportation of coal.
7. Prospecting and development of coal-seam gas.[123]

In the late 1990s the China National Coal Industry Import and Export Corporation (CNCIIEC) became involved in various international coal development projects. Wholly-owned subsidiaries were set up in Brussels, Tokyo, and Sydney. A 10 per cent share of the Coppabella opencast coal mine in Queensland was purchased in 1998.[124] It was expected that China would be importing more coal from Australia to be used in independent private power projects.[125]

However, while major changes in the industry's organizational structure were taking place, the demand for coal was actually beginning to fall perceptibly, from 1.447 billion tons in 1996 to 1.392 in 1997, and 1.295 in 1998.[126] Not long into the Ninth Five Year Plan it was clear that due to the shutting down of boilers in thousands of bankrupt industrial state enterprises, especially in the northeast, and the facts that the economy had become more light industry-oriented, domestic production of some heavy industrial products was being reduced in favour of imports, more efficient new equipment was consuming smaller amounts of coal, and the urban residential and commercial sectors were converting to gas, there was no need to increase production at that time.

A target of 1,290 million tons had been set for 1996, less than the final figure for 1995, and in February 1997, Wang Senhao, Minister of Coal, announced there was to be no increase in production over the 1996 level because some 100 million tons were stockpiled around the country.[127] This would force many more mines to operate considerably below their intended capacity. A couple of months later Li Peng stated in a report entitled *China's Energy Policy*: 'owing to technological progress, conservation in the use of coal, the development of other types of energy, and the need to protect the environment, the share of coal in the total primary energy structure will decline.'[128]

In October an agreement was reached to control output from Shaanxi, Henan, Hebei, Shanxi, and Inner Mongolia. Wang tried desperately to bring production under control. He called for the closure of all individually licensed (LNS) mines, all mines involved in illegal sale and rental of extraction licences, and mines that infringed on the boundaries of other mines. He encountered difficulty, however, because in the poor rural areas the township government leaders were reluctant to close down what had become lucrative revenue- and job-providing operations.

By the end of 1997 over 14,700 illegal township mines had been closed, including some 8,600 that were unsafe, 3,500 found operating on state mine land without permission, and about 1,400 that were operating beyond their allocated boundaries.[129]

The SPMAs were directed to maintain their 1997 levels of production for 1998, and all mines were ordered shut for twenty-two days at Spring Festival.[130] This was the first time in twenty years that the shutdown was so long. Adding to the factors contributing to the decline in coal consumption at this time was the completion of several gas pipelines. The proportion of city residents with access to tap gas grew from 22.4 per cent in 1985 to 78.8 in 1997.[131]

By late 1997 the government was openly encouraging a reduction in the role of coal, mainly for environmental reasons, though the rate and extent of its replacement with other fuels was not yet known. Since about 1994 government energy officials had been travelling around the world to secure shares in foreign oil developments and to negotiate import agreements. Plans for the construction of domestic ports equipped to handle and refine oil, natural gas, and liquefied natural gas were in progress. These initiatives and their results were not publicized at first.

In June 1998 the production target for the year was reduced by 70 million tons, and in August it was announced that 22,000 mines without licences would be closed by 2000 so as to reduce output by the 200 million tons necessary to bring down supply closer to demand.[132] In November the State Council revised the figures: now 25,800 mines would be closed, of which 2,904 were in Shanxi. Nationwide production was to be reduced by 250 million tons, and of this Shanxi's production would be reduced by 69.8 million tons. All illegal mines were to be closed by the end of 1999.[133]

The government took advantage of the sharp decline in oil prices in March 1997, though the differential between domestic and international prices created a serious diesel smuggling problem.[134] In July 1997 the average domestic market price of oil was US$34.42, while the international price was only US$17. In 1997 imports of crude oil and finished oil products were 33.8 million tons, 140 per cent above the 1996 level.

It was not long before economists writing in the national energy journal, *Zhongguo nengyuan*, were recommending greater use of other forms of energy, and government authorities were stating openly and unequivocally that China's dependence on coal could and must be reduced.[135] In August 1998 officials were quoted predicting that the demand for coal would decrease in excess of 40 per cent over the next seven years.[136] Zhao Xiuguang, a senior official with the China National Offshore Oil Corp (CNOOC) affirmed: '... for a long time the use of coal has caused problems for the environment and transport in coastal areas, and it should be replaced as a matter of some urgency with clean and efficient natural gas to meet the needs of continued economic expansion.'[137] In May 1999 *Xinhua* reported: 'After five years of consideration, the Chinese government is to shift its energy strategy towards the use of more natural gas and away from coal as pressure for environmental protection mounts.'[138] The goal was set to reduce the then current role of coal from about 72 per cent of total energy consumption to about 50 per cent by 2050.[139]

In 1998 electricity consumption in some parts of the country decreased for the first time in two decades, and it was announced in January 1999 that coal output that year was to be less than 1.1 billion tons, down 100 million tons from 1998.[140] This target was further lowered in June to less than 1.03 billion.[141] Stockpiles amounted to an unprecedented 200 million tons at the end of February and there were 500 idle-days of trains of fully loaded cars.[142]

A total of 32 million tons were actually exported in 1998, less than hoped. A goal of 38.5–40 million tons was set for 1999. By this time China was experiencing the full effects of the Asian financial crisis. The yuan was under severe pressure, coal exports dropped nearly 8 per cent in the first quarter of 1999, and economywide foreign direct investment fell about 15 per cent. The coal export target for 2000, initially set to reach 50 million tons, was adjusted down to 40 million in September 1997 when many of the Asian economies were in freefall, but restored to 43–45 million tons in January 2000 when the beginnings of recovery were believed to be in sight, and some months later back to 50 million tons.[143] Coal shipments were resumed to North Korea in July 1999 despite the payment collection problems in previous years.[144] Quantities were the same as before – 2.5 million tons per year for five years. Half would be given in the form of grants and half sold at international prices to be paid partly in foreign currency.

In order to counter the decline in exports the government raised value-added tax rebates to exporters of ships, steel, cement, and coal from 3 to 9 per cent, retroactive from 1 June 1998. Charges for international shipments from Qinhuangdao (Hebei) and Rizhao (Shandong) were also lowered to be on par with those for domestic shipments. In July 1998 it was announced that exports would be increased to draw down stockpiles, and that the Industrial and Commercial Bank of China would offer loans to coal enterprises to facilitate exports.

A group of 150 representatives from the State Economic Commission, the SCIB, and related industry associations was sent to eight coal-producing provinces for three months to enforce the closure of the 25,800 mines. They devoted their attention first to the illegal mines operating on state mine territory. It was necessary for them to confront not only the operators of the mines themselves, but also the local governments and all the services and agencies relating to the production, sale, and transport of coal – banks, dealers, railways, power plants, vendors of equipment, and consumers.[145] In some areas the representatives met formidable resistance because the sale of coal had become a major source of income, especially in areas where there were few options for alternative economic activity. County and township officials were fired for taking bribes from the mine operators and lying to and/or keeping information from the representatives. Often mines were closed in the presence of the representatives, only to be re-opened soon after their departure. Sheng Huaren, chairman of the SETC, admitted that closure of these inefficient, polluting mines with high accident rates was 'not a simple problem, it is a political problem'.[146]

Despite the resistance of local operators and officials to the closure of LNS mines, 23,168 were closed by the beginning of June 1999 and total net production was 406.76 million tons, down 28.35 million tons or 6.52 per cent from that in June 1998. Production from the township mines was down 40.80 million tons or 25.54 per cent.[147] However, due to the problem of excess labour at the SPMAs, production at these mines was actually 11.38 million tons or 5.75 per cent above what it was over the same period the previous year, and at the LS mines 1.07 million tons or 1.38 per cent higher.

By the end of 1999 a total of 31,000 small mines had been closed.[148] In the interests of preventing prices from falling further, the 2000 outputs of coal, steel, and sugar were capped. Coal production was to be kept below 900 million tons, or 72 per cent of the 1998 production level, and 56–60 per cent of the original target of 1.5–1.6 billion tons that had been suggested in 1993–94.[149] In 2000 another 18,900 mines were to be closed and total production was not to exceed 870 million tons.[150] Within the space of five years production had decreased by over 500 million tons to the 1985 level.

The problem of unpaid wages and pensions at the state mines persisted. In March 1998, during the NPC, thousands of miners blocked railways in Heilongjiang in protest over the default on earnings and welfare payments.[151] There were 60,000 unemployed miners in Jixi and 40,000 in Shuangyashan alone. The following year 500 laid-off miners from Dujian state mine held a protest in Chengdu (Sichuan).[152] They had not received basic living expenses for three months. This demonstration received international attention due to the presence of the Hong Kong-based Information Centre for Human Rights and Democratic Movement in China, some members of which travelled to Chengdu to give tacit support to the miners. This organization was again present a year later when 500 miners blocked railway lines between Guiyang (Guizhou) and Kunming (Yunnan) in protest over inadequate provision for 40,000 miners from six mines in Sichuan who were soon to be laid off.[153]

In April 1999 over 4,000 retired miners held up traffic on a railway in Fushun (Liaoning) for 4 hours, in protest over non-payment of pensions.[154] In September 5,000 miners from Xiahuayuan mine (Hebei) blocked the railway between Beijing and Baotou (Inner Mongolia) demanding that the bankruptcy of their mine be declared invalid because it was not approved by their union. They also demanded overdue wages, some nine months in arrears.[155] In February 2000 about 20,000 miners protested the low termination pay granted when their mine near Yangjiazhan (Liaoning) was shut down. They blocked the main roads for three days until a 1,000-member police force was called in to end the violence.

Fourteen SPMAs were forced by the government to declare bankruptcy (a hitherto impossible option) in April 1999, and 40 old LS mines to close within the year. Some 260,000 workers would have to find new employment and another 200,000 would face early retirement.[156] This was in addition to the 400,000 to be moved to diversified activities in that year alone.[157]

It was a hopeless loop. The more coal that was produced, the higher the losses at the state mines. The Ministry had set 'loss-reduction targets' in February 1996: none of the state mines was to have a deficit in 2000, and 60 per cent were to make profits.[158] The slogan at the time was *san nian fangkai meijia, san nian chou hui butie* – in three years liberalize the price of coal, in three years remove all subsidies.[159] This three-year deadline was to apply to the state owned sector as a whole. The industries with the greatest excess capacity and inefficiencies were coal, textiles, paper, petrochemicals, metallurgy, non-ferrous metals, armaments, cement, glass, and the railways.

In the summer of that year, officials from thirty-two large loss-making mines had promised to resign if their operations did not become profitable before the end of 2000.[160] The actual proportion of profitable CMAs was 39 per cent at the end of 1996 (a total of 36 out of 93 at that time), up from 25.7 per cent in 1995. Then in 1997, for the first time for many years, the industry nationwide recorded profits, with sixty-five CMAs registering profits of some 200 million yuan, or so it was reported.[161]

However, early in 1998 it was apparent that the reports of profits were incorrect. It was found that the mine accounts gave the impression that payments had been received for coal purchases when in fact they had not. Losses amounted to a total of 3.71 billion yuan at seventy-six of the CMAs.[162] The mines were faced with large deficits due to critical over-production relative to demand. In January 1999 the targets for profits in year 2000 were necessarily adjusted. The goal now was to have 40 per cent of the SPMAs making profits, and the number of loss-making ones down 50 per cent.[163]

Compounding the problems, it was even more impossible to cover production costs when during the first four months of 1999 the price of coal fell by 15.4 yuan per ton. Total income of the SPMAs at the end of the year was down almost 6 billion yuan and the industry was owed over 40 billion yuan, equivalent to over half the expected sales income (see Chapter 6).

In a scene reminiscent of the coal strikes that had occurred in Britain and Nova Scotia, a miner from the large Longfeng state mine at Fushun (Liaoning) cried plaintively, 'We belong to the enterprise; how can we survive without the mine?'[164] A proposal was submitted to the State Council by the SCIB to pay the auction proceeds from the sale of the fourteen bankrupt and forty closed mines in the form of re-employment funds and pensions to the affected workers.[165] The SPMAs at this point had been under the jurisdiction of the provincial governments for less than a year. Though the energy sector was still classified as strategic and therefore required considerable central government involvement, the provinces now had to shoulder a much greater share of the burden.

It was apparent that the government planned to consolidate coal production radically and focus on improving quality. There was lack of agreement, however, over the rate at which the old mines, many of which had been in operation for a hundred years or more, should be decommissioned. Despite the fact that some of these mines had long since reached their full capacity and the costs incurred of digging deeper could not be rationalized, there was great reluctance on the part of the mine administrations to shut them down. These mines and related services had powered the industrialization of China and had provided a livelihood for millions of families spanning several generations. As in many European countries and Canada the problem was as much a social as an economic one. The main argument for keeping the mines in the east open, besides providing jobs, was the fact that the transport costs, *paid by the consumers*, were considerably less than those for the albeit better quality coal from the central north. After a seminar promoting foreign involvement in the development of China's infrastructure, the following statement was issued in June 1999 for the coal industry:

> Apart from an appropriate increase in aggregate output, the emphasis will be laid on readjusting the product mix and increasing the ratio of clean coal. Meanwhile, priority will be given to building mines in coal-deficient eastern China. The pace of mine construction in the western region will be slowed down. Much effort will be made to improve coal washing technology and to promote the development of the coal seam gas industry.[166]

However, in October that year, the State Development Planning Commission announced the following development strategy:

> China will focus on the development of electric power, and, on the basis of coal, make great efforts to develop oil and natural gas and actively develop nuclear energy and other new and renewable energy sources in the next century... The energy industry will shift its strategic focus westwards to the Shenfu and Dongsheng coal mines in Shaanxi and Shanxi Provinces and the Inner Mongolia Autonomous Region; it should mainly concentrate its efforts on constructing hydropower and pit-mouth thermal power plants and prospecting crude and natural gas in the Xinjiang Uygur Autonomous region and offshore.

China is expected to form major coal production bases in Shanxi, Shaanxi, Heilongjiang, and Guizhou Provinces and western Inner Mongolia; hydropower bases on the upper reaches of the Yellow River, and the lower reaches of the Yangtze River; and crude oil supply centre (*sic*) focusing on Daqing and Shengli Oilfields. In addition the country plans to develop geothermal, wind, solar and tidal energy to change the energy consumption structure and ensure the coordinate development of energy, national economy and environment.[167]

In January 2000 it was reported that production in 1999 had been 'put under effective control' in Beijing, Chongqing, Shanxi, Inner Mongolia, Liaoning, Jilin, Heilongjiang, Jiangsu, Zhejiang, Henan, Hunan, Guangdong, and Sichuan, but that production in Shandong and Anhui still 'exceeded the output quota by a large margin'.[168] It was obvious that the local mining authorities were not cooperating with the provincial governments. No formal statement was issued as to when the old mines in the east would be decommissioned, but it seemed clear that the government was not going to devote the same great effort and expense to keep them going that had been given to the limited and difficult Daqing (Heilongjiang) and Shengli (Shandong) oilfields for already well over a decade.

A total of 8,100 km of railway lines were to be built over the Ninth Five Year Plan period (11,800 had been built in the Eighth), 6,100 km by the central government, and 2,000 by regional governments. Nationwide trackage would amount to 68,000 km by 2000, and 70,000 by 2002, of which 21,000 km or 34 per cent would be double-tracked, and 16,000 km or 27 per cent electrified.[169] Shipment capacity emanating from the 'three wests' was to be increased by a total of 140 million tons, of which 75 million were directed towards the north and 52 towards the south.

Rail cargo capacity was to reach 1.8 billion tons, or 150 million tons more than in 1995. Some of the investment funding would come from bond issues with an interest rate comparable to that of government treasury bonds, and some would come from diversified activities similar to those operated by the coal industry.[170]

Like the coal industry, the railways had to streamline their operations and make them profitable.[171] Transfer of excess workers to diversified activities was ongoing, freight charges were raised frequently, a railway construction fund was in place, and dozens of small railway stations were closed. After months of discussion the State Council finally devolved some of its authority over setting passenger fares and freight rates to the Ministry on 1 April 1996. Beginning in the mid-1990s foreign investors were invited to participate more each year in the development of the railways. In order to attract them it was imperative to give them the authority to set fares and rates.

In 1996 four joint venture railway projects were initiated and by 1998 foreigners were granted permission to develop land adjacent to railways and to participate in both local and trunk railway projects. In January 1999 the hope was expressed that the Ministry of Railways could attract 2 billion yuan worth of foreign development capital every year over the next five years for railway construction, telecommunications, information processing, environmental protection, and railway maintenance machinery.[172] In 1996 the railways experimented with running the trains at higher speeds to provide competitive service for passengers and to permit more freight to be handled. Several lines sought listing on the domestic markets.

A plan for an integrated system of higher speed railways was formally adopted in April 1997. Also in April the Guangshen (Guangzhou-Shenzhen) line became the first to be given pricing and operating autonomy, and the first to be floated on the Hong Kong and New York Stock Exchanges. The government expanded its efforts to recruit foreign investment in the railways and portions of the network, including trunk lines, were to be privatized or leased. In August 1997 the first joint venture line was completed, a 251-km link between Jinhua and Wenzhou (both in Zhejiang), involving a company based in Hong Kong. In late 1997 and early 1998 two private lines were built in Yunnan and Shandong by businessmen who had grown weary of depending on the state-run railways.[173]

Five branches of the railway administration nationwide were closed, 1,000 departments were dissolved, and 68 units at station level as well as 152 small stations were closed or merged. Also like the coal industry, overmanning in the transport sector had been severe. Over the Ninth Five Year Plan period 1.1 million workers out of 2.2 million were to be transferred to diversified activities.[174]

In June 1998 a new five year plan for railway construction and administrative restructuring was announced, along with plans for major highway and rural power grid construction. Not only were these weak sectors of the economy, but it was believed that these infrastructure projects would ease the unemployment problem and help maintain economic growth at the planned 8 per cent for that year.

The railway administration was to be broken into five financially independent firms each responsible for a different aspect of the sector. Two corporations were established, the China National Railway Locomotive and Rolling Stock Industrial Corporation and the China Railway Signalling and Telecommunications Corporation. Losses in 1997 had amounted to about 4.0 billion yuan.[175] These were to be eliminated by 2000. Investment was raised substantially, from 34.9 billion yuan to 45 billion through 2002. Forty-three per cent of this would come from revenues, and the rest from domestic and foreign loans, bonds, and shares. Some would be devoted to accelerating existing projects such as doubling the railway capacity of the southwest region, while the remainder would go towards ten completely new projects also mainly in the southwest region.[176]

Work on the Yu-Wei coal slurry pipeline project, agreed upon in August 1994, began in May 1997. The original targets were greatly reduced. Only 7 million tons of coal–water mixture, instead of the planned 15, would be piped from Shanxi to Shandong, and of this only 1.4 million tons, instead of 8, would be exported.

Considerable port and water transportation improvements were to be made over the Ninth Five Year Plan period. More than 200 berths of 1,000 deadweight tons (dwt) or bigger were to be built, adding an annual handling capacity of 300 million tons. The total capacity increase over the Eighth Five Year Plan period had been 120 million tons. Annual handling capacity at Qinhuangdao port was to be expanded from 60 million tons in 1995 to 100 million tons in 2000, and more wharves were to be built at Tianjin and Huanghua (Hebei).

The capacity at Xiaocao port (Fujian) was to be expanded to 50,000 tons. Coal shipped south from Jintang port (Hebei) was transferred to trains destined for other southern provinces. Also planned were coal trans-shipping and storage bases in east and south China, dredging of deep water channels at the mouths of both the Changjiang and Pearl Rivers, and direct transport links between inland rivers and tributaries. The Han River project, completed in April 1997, enabled ships of 500 dwt to travel from Xiangfan (Hubei) to Wuhan (Hubei).

The Grand Canal was to be further dredged to handle larger coal ships. (Hitherto only the southern section was open, from Xuzhou (Jiangsu) to the Changjiang). The Geographical Research Institute of the Chinese Academy of Social Sciences and the Ministry of Water Resources and Communications, in February 1996, proposed a new project (to be called the Shoutian Canal), which advocated an 867-km channel from the Huang River between northwest Ningxia and Tianjin. They believed that the canal could handle shipments of coal from Shanxi, Shaanxi, Ningxia, and Inner Mongolia, and also ease water shortages in Beijing, Tianjin, Hebei, and Shanxi. However, the project was vehemently vetoed four months later by those who believed that the Huang River would frequently halt canal traffic with either too much or too little flow.

Included in the Ninth Five Year Plan was the target of reducing energy consumption per 10,000 yuan of GDP (1995 prices) from 2.2 tons of sce in 1995 to 1.73–1.78 tons in 2000.[177] Total accumulated power saved would be 330–360 million tons sce, with an average energy-saving rate of 4.4–5 per cent. Coal consumption for power generation was to be reduced from 403 grams per kW in 1995 to 390 in 2000, and the total amount of electricity to be conserved was 1.45 billion kWh. The elasticity of coal consumption was to go from 0.38 to 0.32, and coal consumed in the mining process itself was to be reduced from 295 tons per 10,000 tons of output in 1995 to 285 in 2000.[178] In November 1997 an *Energy Conservation Law* was promulgated which detailed how energy was to be used and management procedures for monitoring use.[179] With a donation of US$35 million from the Global Environment Foundation, the State Economic and Trade Commission established an Energy Conservation Information Dissemination Centre (SECIDC) in 1998.[180]

Among the pollution control targets set for 2000, smoke dust was to increase to only 17.5 million tons from 17.4; industrial fugitive dust was to decrease from 17.3 to 17.0 million tons; and sulphur dioxide emissions were to increase only from 23.7 to 24.6 million tons.[181] Over this period many more international organizations became involved in helping China to reduce pollution and improve energy consumption efficiency (see the 'Quality' section of Chapter 6). In June 1996 the State Council approved the *China Clean Coal Technology Agenda for the Ninth Five Year Plan and Development Outlines for the Year 2010*. The air pollution control laws were revised that year and stipulated that all new power plants had to have desulphurization equipment. NEPA also enacted a new set of regulations and standards.

In the spring and summer of 1997 China's major cities began to issue weekly air quality reports on TV, radio, telephone hotlines, and in newspapers. The general public now had information with which to monitor polluting enterprises and lobby the authorities. In Beijing several measures were introduced to reduce air pollution including banning the use of coal with more than 0.5 per cent sulphur, the establishment of forty coal-free areas, use of higher quality coal in the other areas, increased use of central heating, installation of desulphurized dust removers, testing of vehicle emissions, converting buses to use compressed natural gas, and measures to control dust from construction sites. By mid-February 1999, restaurants and offices in central Beijing were to have switched from coal to gas or other cleaner fuels, and no more coal burning equipment could be installed. By 2002, nine days out of ten would be 'clean air days', up from seven.[182] In August 2000 it was announced that *Shougang* (Capital Iron and Steel), the city's largest polluter, as well as its largest industrial employer, would make plans to move its operations out of the city.[183]

The United Nations Framework Convention on Climate Change was held in Kyoto, December 1997. As China was the second largest contributor to the so-called 'greenhouse gases' problem, there was a lot at stake. Over the course of the convention, the spokesmen for China and other developing countries argued vociferously that the developed world had been polluting for decades and it was unfair now to ask the developing world to turn off instantly all their pollution-creating equipment. (The arguments presented and the final gas-reduction targets agreed upon at the convention are discussed further in Chapter 6.)

In the summer of 1998 another environmental catastrophe related to China's struggle to meet energy demand occurred. Several thousands of lives were lost during the most serious flooding in decades along the Changjiang and Songhua Rivers. Millions of people lost their homes and livelihoods, and millions of yuan worth of infrastructure were destroyed. The denuding of land during the GLF's 'backyard furnace campaign', and continued over-logging and scavenging for burnable material had caused this environmental disaster. In many stretches along these rivers there was no vegetation holding down the soil, and the silt-laden waters washed well beyond their normal channels.

Less than two years later the Beijing area was hit by severe sandstorms. Higher than normal temperatures combined with lower than normal rainfall meant the soil, primarily from Mongolia and the Inner Mongolia Autonomous Region was easily transported by the wind. Experts at the Desertification Prevention and Control Centre concluded that disorderly mining activities, combined with excessive use of water, reclamation of land, grazing, lumbering, and gathering of burnable material were the causes.[184]

6 Economic assessment of the industry after twenty years of reforms

Production

Quantity

By 1989 China had become the world's largest coal producer. Production continued to drop steadily throughout Western Europe, Ukraine, Russia, and Japan during the 1980s and 1990s and, indeed, hundreds of mines in these countries were closed down. China's peak production was 1,374.08 million tons in 1996, compared to the Soviet Union's 771.78 million tons in 1988, the GDR's 312.16 in 1985, and Poland's 206.24 in 1988.[1] China's hard coal production amounted to 37.1 per cent of the world total in 1996, up from 22.1 per cent in 1980.[2] In 1998, following the government's policy to reduce production, it amounted to 33.8 per cent.

Table 6.1, which gives world hard coal production in 1997, shows that there was a wide margin between China and the next largest producer. Targets set in the early 1980s for 1985 and 1990 were easily surpassed, and the original target of 1.2 billion tons for 2000 was met six years in advance and consequently was revised upwards to 1.5 billion tons, to be scaled back later when over-production threatened.

Coke production trebled between 1984 and 1997 from 45.6 million tons to 138, and China became the world's largest exporter.[3] About 40 per cent was produced in Shanxi, some produced using modern methods at steel plants, though most using small brick 'beehive' ovens. Smaller amounts came from Hebei, Liaoning, and Sichuan. Each of these ovens produced 60,000–80,000 tons per year. The increase in production was a direct result of the government's policy in the 1980s to encourage the establishment of rural industries and the potential for earning foreign exchange. Though the quality of the coke was comparatively high, the production process was extremely polluting and in the late 1990s the government tried to close down large numbers of the ovens.

The government would have liked much more assistance in developing opencast mines, not only in terms of capital, but more importantly in technology. At the outset of the reform programme the Chinese mining machinery manufacturers had had almost no experience in building the huge excavators needed for this

Table 6.1 World hard coal production, imports, and exports, 1997 (mt)

	Production	Imports	Exports
Total OECD	1449.8	369.9	303.4
Australia	206.8	—	146.4
Austria	—	3.8	0.0
Belgium	0.4	12.8	1.5
Canada	41.3	14.2	36.5
Czech Republic	16.6	2.3	6.6
Denmark	0.0	13.5	0.1
Finland	—	7.0	—
France	6.3	13.6	0.2
Germany	51.2	20.0	0.5
Greece	—	1.2	0.1
Hungary	—	1.0	—
Ireland	—	3.1	0.0
Italy	—	15.3	—
Japan	4.3	129.5	—
Korea	4.5	49.8	—
Mexico	1.9	1.2	—
Netherlands	—	20.3	3.5
New Zealand	3.1	—	1.2
Poland	137.8	3.2	29.5
Portugal	—	5.8	—
Spain	13.8	11.3	—
Sweden	—	3.3	0.0
Turkey	2.5	9.9	—
United Kingdom	48.5	19.8	1.1
United States	910.4	6.8	76.0
Other OECD	0.4	1.2	0.2
Non-OECD Europe	307.4	17.8	20.1
Former USSR	305.5	3.2	20.1
Former Yugoslavia	0.1	0.3	0.0
Other Non-OECD Europe	1.9	14.3	—
Asia	1767.2	67.7	77.0
China, People's Rep.	1372.8	2.0	30.7
Hong Kong, China	—	5.7	—
India	297.2	9.5	0.1
Indonesia	55.1	0.4	41.5
DPR of Korea	24.1	2.5	0.5
Taipei, Chinese	0.1	36.3	—
Other Asia	17.9	11.4	4.3
Africa and Middle East	228.7	14.6	64.3
Egypt	—	1.5	—
Israel	—	8.1	—
South Africa	220.1	0.4	64.2
Zimbabwe	5.5	0.0	0.0
Other Africa/Middle East	3.1	4.6	0.0

Table 6.1 (Continued)

	Production	Imports	Exports
Latin America	43.2	17.7	32.3
Brazil	5.6	12.3	—
Colombia	30.7	—	26.5
Other Latin America	6.9	5.5	5.9
Total World	3,796.4	487.7	497.1

Source: Adapted from IEA, *Coal Information 1998*, Paris: OECD/IEA, 1998, p. III.11, citing IEA/OECD energy statistics, *Energy Statistics of Non-OECD Countries*, and Secretariat estimates.

Note: The 1372.8 million ton figure for China seems to be an error. Elsewhere in the book, for example, p. I.235, the figure 1,320 millions tons is given.

type of mining. It was hoped that with foreign collaboration 17 per cent, or 200 million tons of the total planned output in 2000 would come from opencast mines. However, by 1998 only 32.55 million tons, or 6.67 per cent of the total came from these mines.

From 1979 to 1996, China's coal production peak year, nationwide output increased 738.54 million tons or 116.2 per cent. Table 6.2 provides total coal production figures with the annual increments and growth rates from 1978. *(So as to provide continuity with the data provided in the tables found in Chapter 3, the beginning date for most tables in this chapter is 1978.)* It shows that though there was an overall upward trend through 1996, it was far from steady.

Production fell by over 15 million tons in 1980 and rose by only 1.5 million tons the following year due to the necessity to divert the miners to mine face preparation, which had been seriously neglected in the late 1970s. Annual growth recovered to an average of over 7 per cent (an average increment of about 46 million tons) over 1982 and 1983, and over 10 per cent (79 million tons) over 1984 and 1985, but slowed to about half this in 1986–88. In 1989 it rebounded to 7.6 per cent (74 million tons), but fell sharply in 1991 to only 0.46 per cent (4.98 million tons). Between 1992 and 1996 annual growth averaged 4.9 per cent (58 million tons), but thereafter dropped steadily, falling 3.6 per cent (49 million tons) in 1997, 7 per cent (93 million tons) in 1998, and 15.3 per cent (189 million tons) in 1999. By the end of 2000, annual output had been reduced to about what it had been fifteen years earlier.

In fact, by 1997 the government was almost frantically taking steps to curtail production. The nature of the nation's 'energy problem' was now completely different from what it had been at the outset of the reforms. In the late 1970s the problem was one of insufficient coal being cut. However, after the government began to encourage the development of LNS mines in the early 1980s, and the imbalance in the output–capacity ratio was rectified at the CMAs, total production increased sharply. By late 1984 the main storage yards of the country were full, but the failure to make commensurate increases in railway capacity – cars,

Table 6.2 Coal production, 1978–2000

	Production (mt)	Increment (mt)	Annual growth rate (%)
1978	617.86	67.18	12.20
1979	635.54	17.68	2.86
1980	620.13	−15.41	−2.27
1981	621.63	1.50	0.24
1982	666.32	44.69	7.19
1983	714.53	48.21	7.24
1984	789.23	74.70	10.45
1985	872.28	83.05	10.52
1986	894.04	21.76	2.49
1987	928.09	34.05	3.81
1988	979.87	51.78	5.58
1989	1,054.15	74.28	7.58
1990	1,079.30	25.15	2.39
1991	1,084.28	4.98	0.46
1992	1,114.55	30.27	2.79
1993	1,151.37	36.82	3.30
1994	1,229.53	78.16	6.79
1995	1,292.18	62.65	5.10
1996	1,374.08	81.90	6.34
1997	1,325.25	−48.83	−3.55
1998	1,232.51	−92.74	−7.00
1999	1,043.63	−188.88	−15.32
2000	880.10	−163.53	−15.67

Sources: *MTGYNJ*, various years; 'China Succeeds in Reducing Coal Output', *MTB*, 17 Jan. 2000, on www.chinaonline.com, accessed 28 June 2000; *China Monthly Statistics*, no. 12, 2000, p. 64.

Note: The figures published in the *MTGYNJ* after 1990 vary with those published in the *TJNJ*.

interchanges at intersections, loading and unloading equipment, etc. – meant that the coal, though available to be transported to the thermal electric plants and to the steel and other large industrial complexes, could not be delivered at the required frequency.

Throughout the late 1980s, coal production consistently exceeded railway capacity. In 1990, production had to be slowed down, indeed halted in many places, because the storage facilities at the mines and railway stations could not cope. The press mistakenly reported that 'the demand for coal was finally being met'.[4] Efforts to expand the railway capacity were made as expeditiously as possible, but the thermal electric plants – many more of which were needed at that time – were still not receiving their required coal supplies. Power was cut on the grid serving Jiangsu and Zhejiang 12,480 times between January and May 1993 for lack of coal, and in late 1995 and early 1996 the governments of Shanghai,

Zhejiang, and Guangdong sent urgent cables to Qinhuangdao requesting coal shipments to keep their generators in operation.[5] There were actually two problems – insufficient railway capacity and local authorities' failure to appreciate the sustained economic growth that the reform measures had generated in their regions. They were underestimating the amount of coal they required. In early 1993 the Shanghai Fuel Corporation seemed to have underestimated its requirements by some 4.5 million tons.[6] A report in the national coal newspaper in 1994 predicted that it would be very difficult for coal supply to meet demand at the turn of the century.[7]

Stockpiles continued to accumulate and again in 1993–94 the press was mistakenly reporting, this time, that the energy problem had been solved. The energy gap had in fact not disappeared, but its causes were changing. The growth rate of coal consumption, and concomitantly production, had begun to slow for a number of reasons. Soon after the price of coal was liberalized, and prices doubled or even tripled depending on location, many consumers tried to play a 'game of chicken' with the government and refused to buy it.[8] They thought by doing so they could force government officials into believing they had erred in marketizing coal (and other key commodities) or, at least, were premature in doing so.

At the same time many insolvent large coal-consuming SOEs were closing down; domestic production of some metals and chemicals was being curtailed in favour of imports; urban residents were enthusiastically switching to towngas, liquefied, or natural gas which were becoming available in their cities; the demand for light industrial goods was growing faster than for heavy; managers of industrial operations were beginning to implement energy-saving techniques including switching to more efficient fuels; new energy-efficient equipment was replacing outdated wasteful equipment; it was becoming imperative for the government to address the local and global consequences of burning large amounts of coal; and following the spectacular growth in exports over the 1980s and 1990s which resulted in greatly expanded foreign exchange reserves, China was finally in a position to import large quantities of foreign oil products.[9]

In the late 1990s the capacity of the railways connecting mines to power plants serving the heavily populated areas of the east and south regions had been substantially increased, and the older, as well as several new and very much larger thermal power plants were for the most part receiving their coal on time. The main 'energy problem', was generally now one of insufficient or low-grade electricity transmission and distribution capacity. There were not enough transformers, the majority of the power lines transmitting the electricity were of only low to medium voltage, and the wiring of most buildings built before the mid-1980s had not been upgraded. In short, very much more electricity was being produced, but it was unavailable to end-consumers. It has been noted several times that China experienced in the coal industry that major problem of all developing economies, the rationalization of the growth rates of the various industries and their infrastructures. The relationship between the production of coal and electric power is discussed in detail in the section on consumption at the end of the chapter.

Structure of production

The evolution of the coal industry mirrored the central government's ongoing efforts to find the right proportion for its direct involvement in the economy as a whole, and in all the various individual sectors. The structure of coal production changed radically over the 1980s and 1990s, eventually coming virtually full circle.[10]

In 1979, 56.3 per cent of the country's total coal output came from the CMAs, 27.0 per cent from the LS mines, and 16.7 from the LNS mines (*see Tables 6.3 and 6.4. Note that after 1994 'individual' and 'other' mines were reported as part of the 'collective mine' category*[11]). In 1984 the CMAs produced exactly half of the total output, and thereafter the combined output from the LS and LNS mines exceeded that of the CMAs. Beginning in 1993, more output came from the LNS mines than from the CMAs. In 1995 CMA production as a proportion of the total was at its lowest at 37.3 per cent, while the proportion of production from the LNS sector was highest at 46.2 per cent. Thereafter, with the closure of thousands of local mines – which proved to be a very difficult task – and the trend towards decentralization, the proportion from the CMAs began to rise and stood at 49.1 per cent in 1999, the level it had been in 1984. The proportion from the LNS mines dropped to less than a third of the total.

In the 1990s the yearly growth in output from the CMAs was very low. Average annual growth between 1979 and 1995 at these mines was only 2.10 per cent and the average annual increment was 7.78 million tons (see Table 6.5 and Figure 6.1). The year 1996 was exceptional with 11.40 per cent growth and a 54.97 million ton increment, likely due to the reclassification of some mines to LS mine status. Of the LS mines, for several years there was negative yearly growth in production at the provincial mines making for an average rate of −0.80 per cent and an increment of −0.64 million tons between 1979 and 1996, while at the prefecture and county level growth was slightly more consistent, averaging 2.17 and 3.47 per cent and 1.10 and 2.52 million tons, respectively.[12] The average rate of growth and average annual increment at the LNS mines was very high at 11.17 per cent and 29.91 million tons. Between 1987 and 1996 the gains in production for the country as a whole were almost solely due to growth in output at the LNS mines. After the government abandoned its policy of energy self-sufficiency in 1996, total production fell 3.6 per cent in 1997, 7.0 in 1998, then a dramatic 15.3 in 1999. Due to the closure of some 31,000 small mines, output at the LNS was down an almost incredible 38.6 per cent.

Many households responded to the government's encouragement in the 1980s to operate small mines. With the disbanding of the communes in the late 1970s family labour not required for growing food sought higher incomes in rural industry and commerce. Production of basic agricultural machinery and consumer goods soared. Family groups became involved in township and village mines (TVMs), or set up their own mining businesses and hired labour as needed. Many military units also became involved, not only in small pits, but in fairly large mines. According to one military analyst annual production by military units had

Table 6.3 Total coal production by mine type, 1978–99 (mt)

Year	Total production	State mine production			Total local	Local mine production							
		Total state	CMA	LS		LS	Components of LS production			Total LNS	Components of LNS production		
							Provincial	Prefectural	County		Collective	Individual	Other
1978	617.86	522.54	341.84	180.70	276.02	180.70				95.30			
1979	635.54	529.23	357.77	171.46	277.77	171.46	69.76	45.18	56.52	106.31			
1980	620.13	506.51	344.39	162.12	275.74	162.12	66.16	42.82	53.14	113.62			
1981	621.63	495.04	335.05	159.99	286.58	159.99	66.89	40.45	52.65	126.59			
1982	666.32	520.25	349.90	170.35	316.42	170.35	70.31	42.52	57.52	146.07			
1983	714.53	544.46	363.12	181.34	351.41	181.34	78.19	44.05	59.10	170.07			
1984	789.23	572.35	394.70	177.65	394.53	177.65	67.00	46.69	63.96	216.88			
1985	872.28	589.04	406.26	182.78	466.02	182.78	61.95	50.92	69.91	283.24	266.71	16.53	0.83
1986	894.04	595.30	413.92	181.38	480.12	181.38	60.47	51.98	68.93	298.74	277.48	20.43	2.07
1987	928.09	601.33	420.20	181.13	507.89	181.13	63.10	49.10	68.93	326.76	296.33	28.36	6.76
1988	979.87	628.34	434.45	193.89	545.42	193.89	66.40	53.41	74.08	351.53	306.73	38.04	10.05
1989	1,054.15	663.75	458.30	205.45	595.85	205.45	68.45	57.39	79.61	390.40	337.27	43.08	4.88
1990	1,079.30	685.31	480.22	205.09	599.08	205.09	66.43	58.71	79.95	393.99	345.80	43.31	4.27
1991	1,084.28	684.15	480.60	203.55	603.68	203.55	64.59	59.28	79.68	400.13	355.93	39.93	3.71
1992	1,114.55	685.46	482.54	202.92	632.11	202.92	57.71	60.68	84.53	429.19	380.71	44.77	4.86
1993	1,151.37	662.06	458.03	204.03	693.34	204.03	55.08	62.91	86.04	489.31	431.52	52.93	6.68
1994	1,229.53	674.63	468.67	205.96	760.86	205.96	56.51	59.11	90.34	554.90	474.63	73.59	3.63
1995	1,292.18	695.63	482.28	213.35	809.90	213.35	57.30	62.06	93.99	596.55	519.63	73.29	
1996	1,374.08	759.31	537.25	222.06	836.83	222.06	58.81	63.81	99.44	614.77		96.58	
1997	1,325.25	754.83	529.16	225.67	796.09	225.67	51.88	66.91	106.88	570.42		84.21	
1998	1,232.51	716.34	503.49	212.85	729.02	212.85	48.08	62.92	101.85	516.17		36.94	
1999	1,043.63	726.63	512.71	213.92	530.92	213.92	44.73	61.81	107.38	317.00		18.73	

Sources: *MTGYNJ* (various years).

Notes: For the years 1978–84 the Yearbooks do not give individual figures for collective, individual, and other mines. The LS mines are together given the title 'Commune and Production Brigade Mines (Including Collective Ownership Mines). From 1995 the output for individual and other mines is included in the collective category. The summing of the 'Total State' and 'Total Local' columns do not equal 'Total Production' because the 'Local State' column has been entered twice, once under the 'State Mine Production' half of the table, and once under the 'Local Production' half of the table. In August 1998 when the CMAs were transferred to provincial governments, they were renamed SPMAs.

Table 6.4 Proportions of total coal production by mine type, 1978–99 (%)

	State mine production			Total local	Local mine production							
	Total state	CMA	LS		LS	Components of LS production			Total LNS	Components of LNS production		
						Provincial	Prefectural	County		Collective	Individual	Other
1978	84.6	55.3	29.2	44.7	29.2				15.4			
1979	83.3	56.3	27.0	43.7	27.0	11.0	7.1	8.9	16.7			
1980	81.7	55.5	26.1	44.5	26.1	10.7	6.9	8.6	18.3			
1981	79.6	53.9	25.7	46.1	25.7	10.8	6.5	8.5	20.4			
1982	78.1	52.5	25.6	47.5	25.6	10.6	6.4	8.6	21.9			
1983	76.2	50.8	25.4	49.2	25.4	10.9	6.2	8.3	23.8			
1984	72.5	50.0	22.5	50.0	22.5	8.5	5.9	8.1	27.5			
1985	67.5	46.6	21.0	53.4	21.0	7.1	5.8	8.0	32.5	30.6	1.9	
1986	66.6	46.3	20.3	53.7	20.3	6.8	5.8	7.7	33.4	31.0	2.3	0.1
1987	64.8	45.3	19.5	54.7	19.5	6.8	5.3	7.4	35.2	31.9	3.1	0.2
1988	64.1	44.3	19.8	55.7	19.8	6.8	5.5	7.6	35.9	31.3	3.9	0.7
1989	63.0	43.5	19.5	56.5	19.5	6.5	5.4	7.6	37.0	32.0	4.1	1.0
1990	63.5	44.5	19.0	55.5	19.0	6.2	5.4	7.4	36.5	32.0	4.0	0.5
1991	63.1	44.3	18.8	55.7	18.8	6.0	5.5	7.3	36.9	32.8	3.7	0.4
1992	61.5	43.3	18.2	56.7	18.2	5.2	5.4	7.6	38.5	34.2	4.0	0.3
1993	57.5	39.8	17.7	60.2	17.7	4.8	5.5	7.5	42.5	37.4	4.6	0.4
1994	54.9	38.1	16.8	61.9	16.8	4.6	4.8	7.3	45.1	38.6	6.0	0.5
1995	53.8	37.3	16.5	62.7	16.5	4.4	4.8	7.3	46.2	40.2	5.7	0.3
1996	55.3	39.1	16.2	60.9	16.2	4.3	4.6	7.2	44.7		7.0	
1997	57.0	39.9	17.0	60.1	17.0	3.9	5.0	8.1	43.0		6.4	
1998	58.2	40.9	17.3	59.1	17.3	3.9	5.1	8.3	41.9		3.0	
1999	68.7	49.1	19.6	50.9	20.5	4.3	5.9	10.3	30.4		1.8	

Sources: Calculated from Table 6.3.

Notes: See notes with Table 6.3.

Table 6.5 Growth rates of coal production by mine type, 1979–99 (%)

	Total production	State mine production			Total local	Local mine production							
							Components of LS production			Total LNS	Components of LNS production		
		Total state	CMA	LS		LS	Provincial	Prefectural	County		Collective	Individual	Other
1979	2.86	1.28	4.66	-5.11	0.63	-5.11	-5.16	-5.22	-5.98	11.55			
1980	-2.42	-4.29	-3.74	-5.45	-0.73	-5.45	1.10	-5.53	-0.92	6.88			
1981	0.24	-2.27	-2.71	-1.31	3.93	-1.31	5.11	5.12	9.25	11.42			
1982	7.19	5.09	4.43	6.48	10.41	6.48	11.21	3.60	2.75	15.39			
1983	7.24	4.65	3.78	6.45	11.06	6.45	-14.31	5.99	8.22	16.43			
1984	10.45	5.12	8.70	-2.03	12.27	-2.03	-7.54	9.06	9.30	27.52			
1985	10.52	2.92	2.93	2.89	18.12	2.89	-2.39	2.08	9.30	30.60			
1986	2.49	1.06	1.89	-0.77	3.03	-0.77	4.35	-5.54	-1.40	5.47	4.04	23.59	
1987	3.81	1.01	1.52	-0.14	5.78	-0.14	5.23	8.78	0.00	9.38	6.79	38.82	2.49
1988	5.58	4.49	3.39	7.04	7.39	7.04	3.09	7.45	7.47	7.58	3.51	34.13	3.27
1989	7.58	5.64	5.49	5.96	9.25	5.96	-2.95	2.30	7.46	11.06	9.96	13.25	48.67
1990	2.39	3.25	4.78	-0.18	0.54	-0.18	-2.77	0.97	0.43	0.92	2.53	0.53	-51.44
1991	0.46	-0.17	0.08	-0.75	0.77	-0.75	-10.65	2.36	-0.34	1.56	2.93	-7.80	-12.50
1992	2.79	0.19	0.40	-0.31	4.71	-0.31	-4.56	3.68	6.09	7.26	6.96	12.12	-13.11
1993	3.30	-3.41	-5.08	0.55	9.69	0.55	2.60	-6.04	1.79	14.01	13.35	18.23	31.00
1994	6.79	1.90	2.32	0.95	9.74	0.95	1.40	4.99	5.00	13.40	9.99	39.03	37.45
1995	5.10	3.11	2.90	3.59	6.45	3.59	2.64	2.82	4.04	7.51	9.48	-0.41	-45.66
1996	6.34	9.15	11.40	4.08	3.33	4.08	-11.78	4.86	5.80	3.05		31.78	
1997	-3.55	-0.59	-1.51	1.63	-4.87	1.63	-7.32	-5.96	7.48	-7.21		-12.81	
1998	-7.00	-5.10	-4.85	-5.68	-8.42	-5.68	-7.00	-1.76	-4.71	-9.51		-56.13	
1999	-15.32	1.44	1.83	0.5	-27.17	0.5			5.43	-38.59		-49.30	

Sources: Calculated from Table 6.3.

Notes: See notes with Table 6.3.

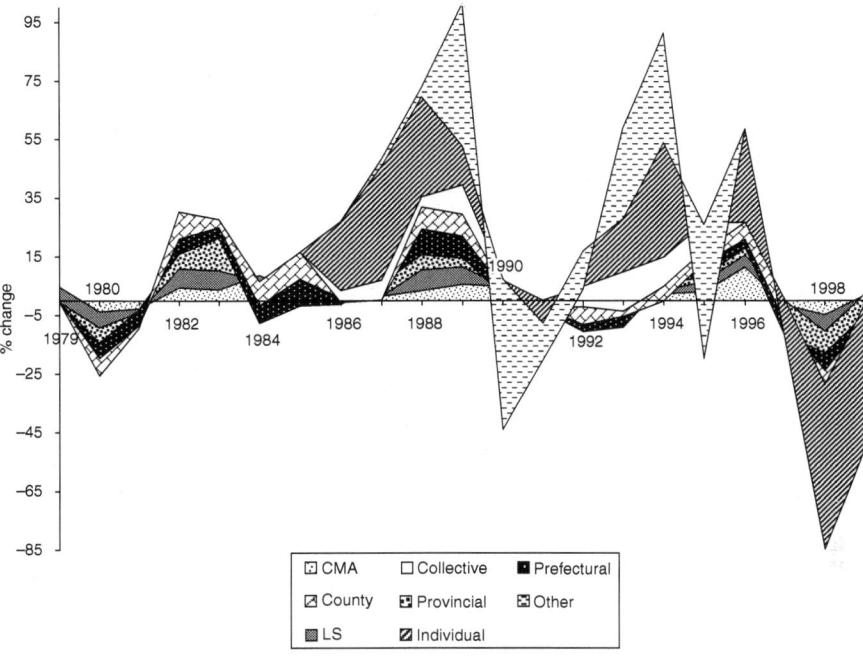

Figure 6.1 Growth rates of coal production by mine type, 1979–99

reached 40 million tons before the government called for the termination of all army commercial activities in 1998.[13]

According to an official from the Ministry of Coal, as of 1998 there were a total of 94 CMAs, about 2,000 LS mines, and some 50,000 LNS mines in operation.[14] It was impossible for the central authorities to know the precise number of LNS mines functioning at any given time. Official sources indicate that at their peak in the mid-1990s, the LNS mines numbered some 75,000.[15] Perhaps the actual number was closer to 85,000. According to the *Township and Village Enterprise Yearbook*, the number of TVMs increased from 20,983 in 1986 to 29,672 in 1996, then dropped to 27,007 in 1997.[16] It was recorded elsewhere in January 1999 that there were 25,640 mines.[17] About 65 per cent of these mines were classified as 'village' mines, the remainder, 'township'. In 2000, when the government was still in the process of shutting down the small mines, the number of LNS mines was said to have decreased from 70,000 to 37,900 and another 18,900 were scheduled to be closed.[18]

From the late 1980s the categories became rather blurred because some of the output from the CMAs and LS mines was sold on the open market, and some from the LNS mines was put under plan. Indeed, in the early *Zhongguo meitan gongye nianjian* (China Coal Industry Yearbooks) the CMAs were called *tongpei meikuang* or 'unified-allocation mines', but in later editions they were re-labelled *guoyou zhongdian meikuang* or 'state priority mines'.[19]

The size/ownership structure of China's coal industry was unique. The number of mines in the other large coal producing countries was a fraction of that in China (see Table 6.6). In the US, the concentration ratio (percentage of total production) of the top fifty mines increased from 66.3 to 75 (estimate) between 1980 and 1993.[20] In Australia there were 121 mines in 1996, of which 108 were privately owned.[21]

After 1979 the allocation process for coal did not change greatly though rampant corruption constantly thwarted official procedures. Each year ministry officials from the main coal consuming sectors – power, steel, chemical fertilizer, and the railways – met at a *meitan dinghua hui* (coal ordering meeting) and negotiated their coal requirements for the following year. One writer referred to these intense bargaining sessions as *malasong dahui* (marathon assemblies).[22] The proportion of total coal production directly allocated by the government (coal under plan) decreased from about 56 per cent in 1979 to 36 per cent in 1989, then rose to 43 per cent in 1991, fell to 39 per cent in 1993, then returned to 43 per cent in 1998.[23] Of the allocated coal, about half went to the above consumers.[24]

It is very telling to compare the extent of the decentralization in coal production in China with that of the entire industrial sector. Table 6.7 gives the breakdown of gross value of coal industry output by ownership and enterprise size, and Table 6.8 gives the percentage breakdowns of gross value of industrial output (GVIO) by ownership structure for the industrial sector as a whole. It is clear that the state's role on average decreased much more in the total industrial sector than in the coal industry. By 1997 state owned units across the industrial spectrum accounted for only about one quarter of total GVIO, whereas in the coal industry

Table 6.6 Comparative structure of production, 1995

	No. of mines	Production (mt)	Average annual output (mt)
China			
CMA	596	439.6	0.7376
LS	1,733	202.2	0.1166
LNS	72,919	596.6	0.0082
Total	79,248	1,238.4	0.0156
US	977	359.4	0.3678
India	355	82.8	0.2332
Russia	196	104.9	0.5352
Poland	67	136.2	2.0328
West Germany	19	53.1	2.7947

Source: *MTGYNJ 1998*, p. 349.

Table 6.7 Breakdown of gross value of coal industry output by ownership and enterprise size, 1979–98 (%)

	1979	1981	1983	1985	1987	1989	1991	1993	1994	1995	1996	1997	1998
State owned (CMAs and LS mines)	62.6	58.1	54.9	51.1	51.4	59.9	53.9	51.3	50.2	49.1	48.9	49.8	50.8
Non-state owned	37.4	41.9	38.5	39.6	39.2	36.4	38.5	42.2	43.8	44.8	45.7	44.8	43.2
of which:													
Collective	—	14.4	15.0	21.4	22.5	23.4	24.7	28.4	29.8	29.9	28.3	27.2	35.5
Coal mine machinery and other plants	—	4.1	6.6	9.2	9.4	8.8	7.6	6.5	6.0	6.1	5.4	5.3	6.0
Large enterprises	—	42.3	40.7	39.9	39.5	41.4	44.9	50.3	49.5	50.7	48.9	52.3	53.1
Medium enterprises	—	17.7	19.5	16.5	16.9	13.5	10.3	6.2	6.3	6.6	6.9	6.6	6.3
Small enterprises	—	40.0	39.8	43.6	43.6	45.0	44.8	43.5	44.2	42.7	44.2	41.0	40.5

Sources: *MTGYNJ*, various years.

Note: No definition is given of large, medium, and small enterprises.

Table 6.8 Breakdown of national gross value of industrial output by ownership, 1978–99 (%)

	1978	1980	1985	1987	1989	1991	1993	1994	1995	1996	1997	1998	1999
State owned	80.7	78.7	64.9	59.7	56.1	56.2	47.0	37.3	34.0	28.5	25.5	28.2	28.2
Collective owned	19.2	20.7	32.1	34.6	35.7	33.0	34.2	37.7	36.6	39.4	38.1	38.4	35.4
Individual owned			1.8	3.7	4.8	4.8	8.0	10.1	12.9	15.5	17.9	17.1	18.2
Other types of ownership		0.6	1.2	2.0	3.4	6.0	10.8	14.9	16.5	16.6	18.5	22.9	26.1

Source: *TJNJ* 1985, p. 306 and *TJNJ 2000*, p. 409.

Note: Included in the notes under Table 13.3, *TJNJ 2000*, p. 409 is the statement 'Adjustment was made on data by different types of owner-ship since 1998 to make them comparable with data of previous years.'

they still accounted for about half. The considerably greater extent of the shift from state to non-state control in the industrial sector as a whole, as compared with that in the coal industry, is all the more remarkable when the percentages of state ownership at the outset of the reforms in 1979 are compared, namely 78.7 per cent versus 62.6. It was clear that coal was still regarded as a 'strategic' industry. In February 1998 rolled steel, timber, and cement were taken out of the plan, but coal, natural gas, and refined oil products remained. Of the state's key 512 enterprises, 32 were coal production enterprises, and of these 30 were CMAs.[25]

As with other industries which underwent drastic restructuring, the coal industry's rapid change from a relatively small number of large enterprises to a large number of small ones had benefits as well as disadvantages. On the positive side, as the share of production from the CMA and LS mines decreased, the central authorities could direct their limited available funds to other critically needing sectors. Moreover, in many rural areas coal sales actually became the main source of income. In some cases the peasants went from barely being able to feed themselves, to making considerable profits from the sale of coal. The growth in industrial activity, especially the production of agriculture-oriented equipment and supplies such as chemical fertilizer, farm tools, machinery, and construction materials, could not have taken place had the locally-derived, albeit poor quality coal not been available. Another indirect positive result of the local coal production was the slowing of soil erosion and the maintenance of soil fertility. Farm wastes formerly used as fuel could be plowed back into the soil. According to one estimate, 40 million tons of coal was equivalent to at least 80 million tons of biomass.[26]

On the negative side, it was difficult for the central authorities to know exactly how much coal would be produced from year to year, and therefore the allocation of materials needed for production, and the assignment of railway capacity became somewhat chaotic.[27] Production in many places was seasonal. Furthermore, the local miners in their ignorance of proper engineering methods often ruined huge coal deposits, and thousands lost their lives in accidents. They removed the coal from the ground using largely improvised equipment and the efficiency of production was so low that the mines were in fact net users. So much sand, rock, and clay was extracted with the coal that it often had only 1,000–2,000 kilocalories per kg.[28] The burning of such low quality coal was extremely damaging to the health of not only the local population and environment, but likely affected global climatic conditions.

The low-technology, low-cost production methods also meant that the local mines severely undercut the state mines in sales. The fact that even the low fixed price at which the state 'sold' allocated coal was above the price at which some local mines sold their output resulted in ever increasing amounts of coal from the state mines being stockpiled. Consumers refused to buy it.

For much of the 1990s the state and non-state mines were working at cross purposes. The lack of clearly defined property rights, managerial functions, and fiscal responsibilities for the various types of ownership/operations provided ripe conditions for corruption. So long as the regulatory structure was vague, and there were minor penalties if any for mining and selling coal illegally, people readily took advantage of opportunities to make extra money. All of these problems are discussed in detail in subsequent sections of this chapter.

Quality

A breakdown of the quantities of the various types of coal produced between 1981 and 1998 is presented in Table 6.9. There were no major fluctuations. The proportion of anthracite remained about 20 per cent of the total, bituminous varieties about 75 per cent, and lignite about 4 per cent. The most notable change was the decrease in the proportion of gas coal (within the bituminous category) from 19.4 to 8.2 per cent. A great deal of progress was made over the reform years in producing coal byproducts such as carbon, methyl alcohol, calcium carbide, ammonia, benzene, naphtha, phenol, potassium, germanium, titanium, and coal tar.

The increase in the tonnage of coal screened and washed nationwide between 1979 and 1997, from 116.1 to 223.42 million tons, was impressive. However, as a proportion of total output, it actually fell from 18.3 to 16.9 per cent and was lower in previous years.[29] For comparison, 100 per cent of total coal production was washed in the UK in 1985, 95 per cent in Japan, 76 per cent in Australia, 49 per cent in Poland, 42 per cent in the USSR, and 42 per cent in the US.[30] In 1980

Table 6.9 Breakdown of coal types, 1981–98 (%)

	1981	*1986*	*1991*	*1996*	*1997*	*1998*
National total	100.0	100.0	100.0	100.0	100.0	100.0
Anthracite	21.1	20.7	19.8	20.8	18.2	18.6
Bituminous	75.1	75.7	76.1	75.2	77.5	77.3
of which:						
Coking:	48.9	46.5	47.8	45.3	48.1	45.0
of which:						
Meagre and lean			1.1	1.1	2.2	2.4
Lean	6.4	6.0	4.3	3.9	3.8	3.6
Coking	11.4	9.1	8.8	8.6	8.2	8.9
Fat	9.1	7.2	7.3	6.0	6.3	6.1
1/3 coking			6.8	7.3	7.7	7.7
Gas and fat			3.8	4.8	5.2	5.1
Gas	19.4	18.1	10.6	8.5	8.5	8.2
Unclassified	2.7	6.1	5.0	5.2	6.1	2.9
General bituminous		29.2	28.2	29.9	29.4	32.4
of which:						
Meagre		4.3	5.1	4.6	4.5	4.8
Weakly caking		6.8	7.3	6.3	6.1	6.2
Non-caking		0.9	1.2	1.3	1.5	2.4
Long flame		6.4	7.9	8.2	8.9	10.1
Unclassified		10.8	6.8	9.4	8.4	8.9
Lignite	3.8	3.6	4.1	4.0	4.3	4.1

Sources: *MTGYNJ*, various years.

some Chinese economists had hoped that 40 per cent of all the coal produced in China would be washed by 2000.[31] In 1998 it did not look as though even the revised target of 30 per cent would be met. In this respect, China's coal industry was similar to India's. In 1989, 147 of 229 mines surveyed in India were never able to supply the thermal plants with the correct grade of coal, the remainder only periodically.[32]

Table 6.10 compares the increases made in production capacity with the increases in washing capacity between 1980 and 1998 at CMA and LS mines. The average ratio was only 1 : 50, that is, for each ton of production capacity, only half a ton of washing capacity were created. Most of the washed coal came from CMAs and was used for coking or smelting, or was exported. Only about 10 per cent of the steam coal was washed. While decentralized production may have first been hastening, then later sustaining the industry's growth, a critical drawback of allowing the LNS mines to overtake the CMAs was that the coal from the former was generally of poorer quality to begin with, and virtually none was washed or treated in any way.

At the end of 1995 there were 226 washing plants with a capacity of about 306 million tons per year operated by the CMAs, 176 plants with a capacity of about

Table 6.10 Comparison of increases in coal production capacity and in coal washing capacity, 1980–98

	CMA and LS mines		
	Productive capacity of mines put into operation (mt)	*Productive capacity of washing plants put into operation* (mt)	*Ratio*
1980	8.29	1.65	1 : 0.20
1981	13.73	2.25	1 : 0.16
1982	8.20	2.55	1 : 0.31
1983	18.52	7.75	1 : 0.42
1984	24.35	5.00	1 : 0.21
1985	16.47	6.21	1 : 0.38
1986	21.05	13.30	1 : 0.63
1987	21.06	11.70	1 : 0.56
1988	33.02	3.90	1 : 0.12
1989	28.05	18.40	1 : 0.66
1990	22.90	12.70	1 : 0.55
1991	28.90	17.25	1 : 0.60
1992	24.19	18.90	1 : 0.78
1993	11.22	10.60	1 : 0.94
1994	7.10	6.70	1 : 0.94
1995	26.24	7.81	1 : 0.30
1996	12.48	11.50	1 : 0.92
1997	26.97	8.80	1 : 0.33
1998	7.04	4.00	1 : 0.57

Source: *MTGYNJ*, various years.

49 million tons per year operated by the LS mines, and some small private plants of unknown capacity.[33] The majority were coke washing plants, and many were in themselves very large consumers of coal and hence heavy polluters. In the early 1990s up to a third did not recycle their water.[34] The recovery rates of the preparation plants for coking and steam coal were not high in 1997 at 58.6 and 79.0 per cent, respectively.[35] The handling of washing plant wastes and ash disposal had improved somewhat since the 1980s, but much more could have been done. Generally, the slag, which does have some potentially useful calorific value, was left unused in large, unsightly mounds frequently prone to spontaneous combustion, and from which liquid sulfuric acid, tar and benzene drained into underground streams.

The average ash level of the coal from the CMAs at 20.21 per cent in 1998 was still some 50 per cent higher than that found in internationally traded steam coals and contributed to a heating value of less than 21 gigajoules per ton compared to the international standard of 29.3.[36] The average ash content of the coal used in power plants was 28 per cent. Nationwide 60 per cent of China's coal has an ash content of 25 to 35 per cent. In the late 1990s coals with an ash content of 40–50 per cent were in use in Jiangxi, and Nanning and Guilin (both in Guangxi).[37] In 1990 the average percentage of ash in Indian coking coal was 20–22 per cent, but in American coal it was only 14.86 per cent.[38] The average sulphur content in Chinese coal, at 1.1 per cent, was 25–50 per cent higher than that found in internationally traded steam coals.[39]

About 10 per cent of total production in 1992, mainly from the southwest and northwest regions, had a sulphur content of over 2 per cent while the average in American coal was 1.04 per cent.[40] (In 1990 the sulphur content of coal burned in power generation was 4.18 per cent in Guangxi, 2.94 per cent in Guizhou, 2.44 in Shaanxi, 2.39 in Sichuan, and 1.95 in Jiangxi.[41]) Coal from the local mines in China had an average level of ash up to 30 per cent, and of sulphur up to 1.5 per cent.[42] Virtually, none of it was either screened or washed and was replete with pieces of rock, wood and other debris.

According to one estimate, if the efficiency with which one billion tons of coal were consumed could be raised by only one per cent, ten million tons of coal could be saved.[43] Another posited that if an additional 200 million tons of steam coal were washed, the amount of unprocessed coal currently being shipped would be reduced by 23 million tons, freeing 6,000 railway cars for other freight.[44] Total system traffic would be 5 per cent lower; sulphur dioxide emissions would be 1.2 million tons less, and particulate emissions 18 million tons less; and there would be 23 million tons less of solid waste per year. Still another estimate was that washing and incremental mining costs excluded, annual savings in transport, ash disposal, and boiler maintenance costs would be about 4 billion yuan per year (in 1993 prices) if the amount of steam coal washed doubled from the 1990 level.[45]

The high capital investment required for washing equipment was a main reason why such a small proportion of the coal was washed. However, the much more difficult, if not impossible problem to solve, is the severe shortage of water. The extraction of 1 ton of coal requires 53–120 litres of water, while 4 tons of water are needed to wash it. China's rainfall distribution is very imbalanced. The northern half of the country, where nearly 60 per cent of the coal is cut, has an arid climate, and in the 1990s the naturally dry conditions were worsened by repeated drought. The southern half, however, lives under a threat of annual flooding. Residents in the north have always been forced to survive with minimal quantities of water, and as more mining was undertaken, and more people moved to the area, the water deficits grew steadily.[46] The average daily per capita consumption of water was a mere 20–30 litres in Taiyuan, the minimum requirement for physiological and personal hygiene needs according to the World Bank. In other parts of the country it was 160 litres.[47]

In the late 1970s the government probably had hopes that the developed world could assist with the water shortage problem. The bottom line, however, was that cheap, unwashed coal, inferior in quality and hence inefficient was usable and could be obtained relatively quickly. At that time the externality of air pollution was not seen as an impediment in the process of modernization. Most mining officials had never travelled to the large coastal cities, let alone abroad, and did not appreciate the health and environmental effects of the burning of raw coal.

When coal prices were finally reformed to reflect quality, mine managers at LS and LNS mines quickly recognized that washed coal would yield considerably higher revenues, but the lack of water prevented them from improving the quality of their output. Despite considerable research by both domestic and international organizations into developing dry processing techniques, there has been to date no major technological breakthrough and the capacity of dry processors remains too small for their application in China.

A great deal of scientific modelling was devoted to determining the best way to divert water from the Changjiang to the coal areas. Progress on this too was minimal. Water was diverted to Jiangsu and Shandong in 1986, facilitating the shipping of coal on the Grand Canal, and was also advantageous to the agriculture and hydroelectric industries, but again did little for expanding coal treatment. Smaller diversions were also carried out in the Qinghai and Gansu areas.

In 1998 the government announced that with partial funding from the World Bank, 1.2 billion m^3 of water would be pumped annually through underground pipes from the Huang River at Wanjiazhai, to Taiyuan in the first phase of the project, then later to Datong and Shuozhou (all in Shanxi).[48] At the time of writing the government is studying three additional south–north water transfer routes.[49] One western geographer is not hopeful stating the plans to ship water north 'will not happen for decades, it at all.'[50]

The pollution problem

From the 1980s the exponential increase in the number of motor vehicles on China's roads added significantly to the carbon monoxide and nitrogen oxide emitted into the air. The indoor and outdoor air pollution created from the burning of coal, however, has accumulated over many decades. Coal combustion has been responsible for about 90 per cent of the emissions of sulphur dioxide,

83 per cent of the carbon dioxide, 70 per cent of the ambient particulates, and 50 per cent of the nitrogen oxide.[51] One western scholar writing in 1997 ventured to say, 'the smoke burden today in major Chinese industrial cities such as Chongqing is significantly worse than it was in [the killer smog] in London in 1952.'[52] The largest consumers of coal, and hence polluters, were the thousands of power stations, 410,000 industrial boilers, and 180,000 furnaces and kilns.[53]

Nine out of ten of the world's cities with the highest air pollution in the mid-1990s were in China.[54] It was estimated that 19 million tons of total suspended particles and 16.22 million tons of sulphur dioxide were emitted into the air annually.[55] The average concentration of suspended particles was 526 micrograms per m^3 in the major northern cities and 318 in the major southern cities, compared to a maximum limit of 60–90 recommended by the WHO.[56] Between 1981 and 1994 ambient particulate emissions decreased at an annual rate of -0.21 per cent, remarkable considering that total coal consumption nearly doubled.[57] It proves that the government did set and successfully enforce measures to reduce these emissions.[58]

China's industrial boilers in the early 1990s, 55 per cent of which were used to heat buildings, and 45 per cent for industrial purposes, were small by international standards. In 1990, 17 per cent of the industrial boilers produced less than one ton of steam per hour (stph), 65 per cent produced 2–4 stph, 16 per cent produced 6–20 stph, and only 2 per cent produced over 35 stph. The thermal efficiency of the four stph categories was 50–55, 63–65, 65–70 and over 70 per cent, respectively.[59] The average thermal efficiency was 31 per cent for thermal power plants, 20–30 per cent for industrial kilns, 15–30 per cent for domestic stoves, and 5–8 per cent for locomotives.[60]

Most large and medium industrial boilers were fitted with basic cyclone ash collectors, though these were 10–20 per cent less efficient than the more modern equipment used in the West. About a third of the industrial boilers had bag houses or electrostatic precipitators.[61] Costing one-quarter to one-third of the price of an entire thermal plant, there are virtually no stack scrubbers in use.[62] Wherever possible, the government ordered boilers to be relocated outside city precincts and stacks be made several feet taller. Households and work unit cafeterias were encouraged first to use the less polluting coal briquettes, then gas. Not surprisingly it was the large, wealthier cities with the most educated populations where the largest reductions in particulates were made.[63] In the medium and small cities the reductions were minimal if any.

Less successful were the measures to reduce emissions of sulphur dioxide, which increased at a rate of 2.2 per cent, roughly paralleling the growth rate of consumption. Total emissions in 1998 were 23.5 million tons, 30 per cent more than in 1993.[64] At the time of writing, clean coal technologies (CCTs), discussed below, were only at the experimental stages in China, and, indeed, in much of the world. The concentration of sulphur dioxide in 1995 was 424 micrograms per m^3 in Guiyang (Guizhou), 338 in Chongqing (Sichuan), 211 in Taiyuan (Shanxi), 132 in Jinan (Shandong), and 105 in Shenyang (Liaoning), compared to the maximum limit of 40–60 recommended by the WHO.[65] By contrast, it was 500 in Tehran, 300 in Jakarta, 70 in New York, 22 in Tokyo, 5 in Los Angeles, and 40 in London.

The problem is most severe in the central and southwestern regions because of higher sulphur content of the coal in the south and inability of the soils in the south, compared to the more alkaline ones in the north, to neutralize some of the sulphur oxide emissions.[66] According to the NEPA, the sulphur dioxide emissions caused acid rain to damage over 30 per cent of the country's land in 1998, up from 18 per cent in 1985. Total economic loss in the early 1990s due to acid rain was estimated to amount to some 16 billion yuan (US$2.77 billion) per year.[67]

In the mid-1990s China accounted for about 14 per cent of the world's carbon dioxide emissions, second highest after the US' 22 per cent.[68] Some scientists claim these are contributing to the phenomenon known as 'global warming' and endangering the planet as a whole through higher temperatures, increased evaporation from the soil, and more intense storm activity. Of the global warming caused by human activities it is estimated that about 57 per cent is due to energy utilization and production, and that some 50 per cent of the greenhouse effect is the result of carbon dioxide emissions.[69] Besides intentional use of coal in China, up to 200 million tons in deposits close to the surface combusted through natural causes through oxidation or forest fires. It was estimated that these uncontrollable flare-ups, which may have been occurring for 300 years in a 200–300 km^2 area in the northern half of the country accounted for 3 per cent of the world's total carbon dioxide emissions.[70] Fires also frequently erupted at the hundreds of mine refuse heaps.

Air pollution is responsible for a very high incidence of respiratory disease. As coal was the dominant urban fuel for decades, Chinese citizens have been exposed to dangerous levels of both indoor and outdoor pollution. The particulates and gases resulting from combustion lead to chronic obstructive pulmonary disease – emphysema, asthma, and chronic bronchitis – which has become the leading cause of death in rural China and the third leading cause in the cities.[71] It is the very small particulates released into the air from small industrial boilers, household stoves, and district heating that are the most harmful to the lungs and difficult to eradicate. Household consumption accounted for only 15 per cent of coal use in urban areas, though contributed to 30 per cent of the urban ground-level pollution.[72]

Despite the daunting geographical proportions of the problem and capital constraints, the government has managed to reduce pollution levels in the large cities considerably and in a relatively short period of time. However, there are still many million peasants (see Consumption section) who still rely completely on the burning of biomass for cooking and heating, and as these people become wealthier they will begin to consume more coal before they ultimately use electricity and gas as do their urban counterparts.[73] It is to be hoped that the backward areas – the countryside, villages, towns, and small cities – will be able to avoid decades of very high emissions by a quantum leap into the new technologies.

Much more can be done to foster compliance with pollution control laws through financial rewards and enforced penalties. The experimentation begun in 1993, with levying charges on sulphur dioxide emissions in two provinces and nine cities was deemed successful and should therefore be extended to cover a larger area. Changing the behaviour of the large state enterprises is a very difficult challenge, as is monitoring hundreds of thousands of rural industries. It was learned from a survey of 573,000 township and village enterprises taken in 1989 that 72 per cent of their boilers were not up to standard, and 86 per cent did not remove dust from their flue gases.[74] According to one estimate, by 2000, rural industries would account for 60–70 per cent of carbon dioxide, 35 per cent of sulphur dioxide, and 48 per cent of fly ash emissions.[75] Well into the 1990s the use of small, inefficient ovens to produce coke continued, in fact, increased, and the proportion of total coke produced from them rose alarmingly from 26.8 per cent in 1979 to 35.5 per cent in 1993.[76]

One difficulty found not only in China, but almost everywhere, even in the developed world, is that of coordinating the plethora of government agencies and NGOs concerned with pollution.[77] The Chinese government is also learning that pollution is an 'externality' in market economies, that is, that market forces in themselves do not prevent or cure severe environmental problems. This is actually one area where the central government must step up its role.

Clean coal technologies (CCTs)

The quantities of impurities released into the air in coal combustion can be reduced through four means: screening and washing, and/or chemical or biological treating of the raw coal prior to combustion; employing of fluidized bed combustors or advanced combustors with additional nitrous oxide controls at the combustion stage; installing sulphur dioxide and nitrous oxide equipment in smoke stacks so as to capture the impurities post combustion; using gasification combined cycle equipment in which the coal is converted to fuel gas that is burned in a turbine, the hot exhaust being used to generate steam for a steam cycle.

There are two sets of CCTs. The combined cycle technologies include Integrated Gasification Combined Cycle (IGCC), Pressurized Fluidized Bed Combustion Combined Cycle (PFBC/CC), and Circulating Fluidized Bed Combustion (CFBC).[78] The steam cycle technologies include Pulverised Fuel Firing (PF-firing) at sub-critical, super-critical or ultrasuper-critical levels, and Atmospheric Fluidised Bed Combustion (AFBC).[79] Both the IGCC and PFBC technologies have high pollution reduction and general efficiency potential but are yet in demonstration phases, mainly in Europe, and are extremely expensive. An advantage of the AFBC technology is that low quality coal can be used as well as other waste materials. Up to 90 per cent of the sulphur dioxide can be removed when limestone is injected into the coal mixture. The most efficient plants in the world are the super-critical PFBC/CC with efficiencies of at least 45 per cent, but they are 'in their commercial infancy', are relatively small at a maximum of only 250 MW, and capital costs are relatively high.[80]

International concern about the environment is such that all countries burning large amounts of coal, even those having no water supply problems, will increasingly be adopting CCTs over the next two decades. The IEA predicts that the main interest in the new coal-fired steam cycle plant will be China, Southeast Asia, and India, and that the particular equipment most in demand will be PF-fired units with sub-critical steam conditions.[81] These can be built at half the capital costs of IGCC and PFBC/CC.[82]

Screening and washing of coal alone does nothing to remove nitrogen or sulphur which is chemically bonded to the coal.[83] They eliminate only the rock, wood, and 30–50 per cent of the pyritic sulphur, while the AFB and PFB technologies can eliminate over 95 per cent of it.[84] These technologies also raise the general efficiency of combustion. For China, a principal advantage is that they bypass the water shortage problem. Other advantages are that the nitrous oxide emissions are reduced, an easier to handle dry solid waste is produced instead of the wet sludge which accumulates after washing, and other materials besides coal can be burned.

At the time of writing China had over 2000 small AFBC plants in operation, some with lime injection equipment.[85] As part of the Coal Utilization Programme, sponsored by the United Nations Development Programme in 1989 and managed by the China International Centre for Economic and Technical Exchange, a CFBC demonstration project was built in Dalian (Liaoning), and design manuals were written.[86] China's Agenda 21 plan, called for the construction of an IGCC demonstration plant, and a Clean Coal Technology Centre was established in June 1993 (see Chapter 5). Unfortunately, funding for the IGCC plant was later blocked by the US Senate.[87]

A key priority in the Ninth Five Year Plan period was to attract foreign assistance for the development of PFBC, AFBC, and IGCC technologies.[88] The installation of CCTs in China's power plants would seem to be imperative, if not inevitable. However, analysts disagree over the expense and viability of the undertaking.[89] Installation of such equipment is further complicated in China by the extremely wide geographical dispersal of the rural, small-scale power plants and boilers. Until the coal gasification techniques are further developed, it would seem that widening the use of AFBC techniques would be the most feasible and appropriate for China.

CAPTURE AND USE OF GASES

Another means of reducing pollution while still relying on coal would be to capture more coalbed methane (CBM) gas. In 1990 about 425–675 million tons of this greenhouse gas were emitted into the air worldwide, of which 24 (4–6 per cent) came from underground mines, and nearly 60 per cent from natural sources.[90] Only about 1 million tons were utilized by industry. The ten largest coal-producing countries captured about 21.82 million tons that year, of which China accounted for 7.7.[91] The average recovery rate of this gas for the ten countries was 19.2 per cent and the utilization rate was 10.1 per cent. In China the respective rates were 13 and 2 per cent. The IEA estimates that China could potentially recover 35–45 per cent and that, as the largest coal producer in the world, easily become the largest producer of CBM.[92] Omnipresent at every mineface, methane had always been considered a menace, causing many lethal explosions every year.

China's total CBM reserves are the largest in the world at some 35,000 billion m^3, equivalent to 45 billion tons of standard coal.[93]

Methane can be used as an input in the production of various chemicals, to power mining equipment, to mix with other fuels for electricity generation, or to create heat for a variety of purposes including cooking. It is non-polluting and smokeless. China has been recovering methane for some years at the CMAs, using relatively low technology, and mainly in the interests of mine safety. One of the largest enterprises, the Taiyuan Coal Gasification (Group) Corporation, was established in 1981. In 1992 the UN embarked on a three-year project to devise a national strategy to harness the gas using techniques that would remove it from the mines before construction. Not only would more lives be saved, but fuller recovery would be possible.[94] If the harnessing and marketing of the gas were successful, the cities would quickly become much cleaner.

In 1996 the China United Coalbed Methane Co. Ltd (CUCBM) was established, and by 1998 there were over fifty coal gas projects in operation and output had expanded to about 400 million m^3.[95] The government continued to promote development of the industry and in January 1998 CUCBM, US Texaco Inc., Arco and Phillips signed a contract to explore methane resources in Anhui and Shanxi. Pilot projects were also established in Liaoning and Hebei. The goal is to produce 10 billion m^3 annually by 2010.[96]

International cooperation

Soon after the 'opening of the door' scientists at the mining institutes and Chinese Academy of Sciences became fully apprised of the fact that China's air pollution was the highest in the world and that carbon dioxide and sulphuric acid emissions carried in clouds could cause irreparable harm thousands of kilometres away from their source. They learned at the same time of the technological advances made over the 1960s and 1970s in the West to lessen pollution. Ambient air quality standards were set for the country which are comparable to those set by the WHO. These are presented in Table 6.11.

Considered a topic of grave international importance not only by environmentalists but also by a wide variety of other interest groups and professions around the world, the reduction of air pollution became the subject of innumerable conferences and publications. Many academics of international repute have devoted a great deal of their attention towards examining China's environmental problems, including air pollution, and have watched very closely the measures taken.[97]

Lending for environmental protection became the fastest growing area of the World Bank's programme in China.[98] Analysts at the Bank have examined in depth the current and future physical effects of air and water pollution and estimated its costs in economic terms.[99] They have also detailed the consequences for China of three alternate action scenarios ('business as usual', a low investment programme, a higher investment programme) based on a set of assumptions about macroeconomic and sectoral growth, and estimated the number of lives that could be saved if China's own 'class 2' pollution standards were met (see Table 6.11).[100]

Table 6.11 Ambient air quality standards (micrograms per cubic metre)

Pollutant	Averaging time	China			WHO
		Class 1	Class 2	Class 3	
Sulphur dioxide	Annual	20	60	100	40–60
Total suspended	Daily	150	300	500	150–230[b]
particulates	Annual[a]	60	120	150	60–90
PM-10	Daily	75	150	250	70
	Annual[a]	20	60	100	
Carbon monoxide	Daily	4	4	6	10[c]
Nitrogen oxides	Daily	50	100	150	150
Ozone	8 hours				100–120
Lead	Annual	0.7			0.5–1.0

Source: World Bank, *Clear Water, Blue Skies: China's Environment in the New Century, China 2020 Series*, Washington: IBRD/WB, 1997, Table 1.1, p. 10. Copyright © 1997 International Bank for Reconstruction and Development/The World Bank. Reprinted by permission of the International Bank for Reconstruction and Development/The World Bank.

World Bank's explanatory notes:
Class 1 are tourist, historic, and conservation areas. Class 2 are residential urban and rural areas. Class 3 are industrial areas and heavy traffic areas.
a. Since China does not have annual standards for total suspended particulates, it has been assumed that the same ratio between daily and annual standards for sulphur dioxide applies to total suspended particulates. Annual standards are needed for comparability with ambient concentration data.
b. Guideline values for combined exposure to sulphur dioxide and total suspended particulates.
c. 8 hours.

The World Bank has provided China a great deal of assistance devoted specifically to the reduction of pollution caused by the mining and burning of coal. The objectives of the 'World Bank's Clean Coal Initiatives', were 'to assist client countries to reform and restructure coal, transportation and conversion sectors, to develop and enforce environmental practices and deploy clean coal technologies'. In June 1996 this group held a Roundtable at which it was decided to launch a study, 'China's Environment in the 21st Century – Evaluation of Emission Reduction Strategies.'[101]

At the United Nations Framework Convention on Climate Change held in Kyoto December 1997, Chinese presenters pleaded for assistance in reducing greenhouse gas emissions, and rebuked the Western countries for insisting upon China's compliance with international air quality standards when the Western countries, in their early and middle stages of economic development, had themselves been very high polluters.[102] The Chinese government argued that reduction targets should be based on total accumulated emissions, not current or estimated future levels. Zhong Shukong, China's senior negotiator at Kyoto appealed: 'Ours [developing nations] are survival emissions. Theirs [developed countries] are luxury emissions. They [only] have two people to a car and yet they want us to give up riding buses.'[103] The Chinese side also pointed out that per capita energy consumption in the developing world was far less than in the developed.

Many of the delegates and demonstrators, and much of the media contingent attending the conference seemed not to realize that the usual means of treating coal were not feasible or of limited use in China due to the water shortage problem. Galling to China and other developing countries was the fact that some nations of the developed world, notably the US, were asking them to reduce emissions drastically *before* these same nations had reduced emissions in their own countries.[104] The US and other developed nations had not yet met the targets they had set for themselves at the 1992 convention held in Rio de Janeiro. Senator Joseph Lieberman, a member of the Senate Climate Change Observers Group was quoted as saying he had told the Chinese 'that without some co-operation from them, the US Senate was unlikely to ratify any treaty drawn up in Kyoto, making any agreement meaningless and leading to China being blamed for derailing the process.'[105]

After several rancorous sessions, it was ultimately agreed that the US would reduce emissions by 7 per cent below the 1990 levels before 2010, the European Union 8 per cent, and Japan by 6 per cent. However, the US held up progress by delaying to sign the protocol until November 1998, and at that, without ratification by the Senate. In 1997 the Senate had passed a resolution urging the President not to sign any treaty that called for reductions of American greenhouse gases without simultaneous reductions in developing countries such as China, India, South Africa, and South Korea.[106]

In 1994 the Asian Development Bank funded a study entitled *National Response Strategy for Global Change: People's Republic of China*, undertaken in collaboration with the Institute for Nuclear Energy Technology in Beijing. They employed modelling techniques to predict energy demand and associated carbon dioxide emissions and concentrations.

In 1996 and 1998 the IEA and State Development Planning Commission held joint conferences on energy efficiency and pollution reducing technologies. In the spring of 1999 it pledged 'references and experiences in energy management and pollution control ...' and would use a thermal plant in Shanxi upon which to demonstrate pollution reduction through technological renovation.[107]

Several individual countries also helped establish special projects. For example, to cite a few, the Guizhou Energy Efficiency Project was launched in 1997 to assist industrial enterprises in Guizhou reduce energy consumption and pollution. Funds were provided by the British government and the British Energy Technology Support Utility (BETSU) offered technical assistance.[108]

In November 1997, a three-day international symposium on CCTs was held in Xiamen (Fujian), sponsored by the Ministry of Energy, the US Department of Energy, and the Energy Administration of the European Commission. The Sino-US Summit was also held that month. Several agreements were signed pertaining to the sale of environmental and energy technology to China. Specifically, the Americans were set to supply equipment to help reduce air pollution caused from the burning of coal in existing equipment, new generators that would burn coal more efficiently, and hardware and software to expand the nuclear energy sector.[109] The following year, when US President Clinton travelled to China in July 1998 the American government further promised technical assistance with air quality monitoring and the development of renewable energy.[110]

A number of universities in the US have created formal research programmes devoted to examining particular aspects of the pollution problem. For instance, the focus of the Harvard University Committee on Environment China Project is the role that energy-derived air pollution and greenhouse gas emissions play in China's international relations.[111] In 1999 the Sino-German Energy Efficiency and Environment Institute was set up in Hangzhou (Zhejiang) by the Zhejiang Provincial Energy Research Institute (ZERI) and Bremen University Environmental Engineering Institute.[112] Part of the fourth tranche of official Japanese Government aid went towards the Sino-Japanese Environmental Cooperation Model City Project for Dalian (Liaoning), Guiyang (Guizhou), and Chongqing. The main objectives of this project were to reduce acid rain and greenhouse gases, and to promote recycling.[113] The European Union also resolved to assist in improving the efficiency and cleanliness of China's coal-fired plants.[114]

A Japan-financed 'Green Aid Plan' was launched in China (and other ASEAN member countries) to carry out demonstration projects aimed at reducing environmental damage caused by the burning of coal.[115] The focus project in China was desulphurization equipment at a power plant in Qingdao (Shandong).

Geographical distribution

By the mid-1990s some 60 per cent of China's 1,257 counties were producing coal, though in 1996 over 50 per cent of the total output came from five provinces alone. Tables 6.12 and 6.13 show the changes in relative distribution of production and consumption from 1981 through the mid- and late 1990s. The north continued to rank first, accounting for 36.5 per cent of total production in 1998, and the east ranked second, accounting for 15.3 per cent. The southwest and southcentral were close behind with 14.1 per cent and 12.6, respectively. Production in Guangxi increased substantially but no where near reached expected levels *(see Ninth Five Year Plan section of Chapter 5)*. The proportion from the northeast, where many of the country's oldest mines were becoming exhausted, showed a marked decline while that from the southwest, where several new mines were opened, increased. The northwest continued to produce the least though its share steadily rose. The rate of production from the Shanxi Energy Base, as well as from the other key mining areas, would have been higher yet in the late 1980s and early 1990s had more transport capacity been available. Table 6.14 lists the largest CMAs in 1996 and these are plotted on Figure 6.2.

Production in Shanxi Province, the largest producer, peaked in 1996 with 349.46 million tons. From 1985 it had accounted for almost exactly one-quarter of total production. After 1996, as a result of government policy to shut down small mines, production dropped to 307.19 million tons in 1998. Henan and Sichuan Provinces were distant second and third largest producers in 1996,

Table 6.12 Distribution of coal production, 1981–98 (mt and %)

	1981	(%)	[Rank]	1985	(%)	[Rank]	1990	(%)	[Rank]	1995	(%)	[Rank]	1996	(%)	[Rank]	1997	(%)	[Rank]	1998	(%)	[Rank]
Total	621.63			872.28			1,079.30			1,292.18			1,374.08			1,325.25			1,232.51		
North	214.58	(34.5)	[1]	316.85	(36.4)	[1]	405.49	(37.5)	[1]	476.72	(37.0)	[1]	506.73	(36.8)	[1]	487.12	(36.7)	[1]	450.33	(36.5)	[1]
Beijing	7.90	(1.3)		9.77	(1.1)		10.03	(0.9)		9.96	(0.8)		10.01	(0.7)		9.80	(0.7)		9.54	(0.8)	
Tianjin	0.00	(0.0)		0.00	(0.0)		0.00	(0.0)		0.00	(0.0)		0.00	(0.0)		0.00	(0.0)		0.00	(0.0)	
Hebei	52.35	(8.4)		60.86	(7.0)		61.91	(5.7)		70.55	(5.5)		74.09	(5.9)		67.86	(5.1)		56.37	(4.6)	
Inner Mongolia	21.79	(3.5)		32.04	(3.7)		47.62	(4.4)		64.45	(5.0)		73.17	(5.2)		79.08	(6.0)		77.23	(6.3)	
Shanxi	132.54	(21.3)		214.18	(24.6)		285.93	(26.5)		331.76	(25.7)		349.46	(25.0)		330.38	(24.9)		307.19	(24.9)	
Northeast	93.51	(15.0)	[4]	131.49	(15.1)	[2]	159.74	(14.8)	[2]	154.79	(12.0)	[5]	167.64	(12.0)	[5]	157.98	(11.9)	[5]	148.57	(12.1)	[5]
Liaoning	33.70	(5.4)		45.91	(5.3)		51.01	(4.7)		52.49	(4.1)		60.41	(4.3)		58.41	(4.4)		56.44	(4.6)	
Jilin	18.07	(2.9)		23.12	(2.6)		26.10	(2.4)		23.79	(1.8)		25.76	(1.9)		24.10	(1.8)		21.23	(1.7)	
Heilongjiang	41.74	(6.7)		62.46	(7.2)		82.63	(7.7)		78.51	(6.1)		81.47	(5.8)		75.47	(5.7)		70.90	(5.8)	
Northwest	49.16	(7.9)	[6]	70.36	(8.0)	[6]	87.54	(8.1)	[6]	105.95	(8.2)	[6]	117.33	(8.3)	[6]	123.00	(9.2)	[6]	115.95	(9.4)	[6]
Shaanxi	18.45	(3.0)		26.93	(3.1)		33.27	(3.1)		39.58	(3.1)		46.13	(3.3)		49.58	(3.7)		44.46	(3.6)	
Gansu	7.88	(1.3)		11.84	(1.3)		15.64	(1.4)		22.09	(1.7)		22.21	(1.6)		22.93	(1.7)		23.16	(1.9)	
Qinghai	1.91	(0.3)		2.77	(0.3)		3.20	(0.3)		2.88	(0.2)		2.97	(0.2)		3.29	(0.2)		3.23	(0.3)	
Xinjiang	11.40	(1.8)		16.68	(1.9)		21.00	(1.9)		26.93	(2.1)		29.86	(2.1)		30.21	(2.3)		29.27	(2.4)	
Ningxia	9.52	(1.5)		12.14	(1.4)		14.43	(1.3)		14.47	(1.1)		16.16	(1.1)		16.99	(1.3)		15.83	(1.3)	
S. Central	95.38	(15.4)	[3]	130.99	(15.0)	[3]	152.45	(14.1)	[3]	188.74	(14.2)	[2]	195.36	(14.7)	[3]	175.07	(13.1)	[4]	155.05	(12.6)	[4]
Henan	58.25	(9.4)		78.57	(9.0)		90.80	(8.4)		101.81	(7.9)		107.86	(7.7)		100.28	(7.6)		86.91	(7.1)	
Hubei	4.34	(0.7)		8.86	(1.0)		9.24	(0.9)		14.37	(1.1)		15.21	(1.1)		15.17	(1.1)		13.26	(1.1)	
Hunan	19.94	(3.2)		29.45	(3.4)		33.71	(3.1)		49.53	(3.8)		50.93	(4.3)		40.23	(3.0)		38.10	(3.1)	
Guangdong	7.24	(1.2)		8.12	(0.9)		8.90	(0.8)		10.69	(0.8)		8.82	(0.7)		8.40	(0.6)		6.81	(0.6)	
Guangxi	5.61	(0.9)		5.99	(0.7)		9.79	(0.9)		12.33	(1.0)		12.52	(0.9)		10.97	(0.8)		9.97	(0.8)	
Hainan							0.01	(0.0)		0.01	(0.0)		0.02	(0.0)		0.02	(0.0)		0.00	(0.0)	
Southwest	65.46	(10.5)	[5]	95.40	(11.0)	[5]	127.06	(11.8)	[5]	180.38	(14.0)	[4]	187.82	(13.5)	[4]	189.03	(14.3)	[3]	174.02	(14.1)	[3]
Sichuan	39.40	(6.3)		55.58	(6.4)		67.84	(6.3)		97.39	(7.5)		95.67	(6.9)		62.22	(4.7)		56.96	(4.6)	
Yunnan	11.90	(1.9)		16.38	(1.9)		22.27	(2.1)		27.89	(2.2)		30.72	(2.2)		32.97	(2.5)		31.03	(2.5)	
Guizhou	14.16	(2.3)		23.44	(2.7)		36.95	(3.4)		55.10	(4.3)		61.43	(4.4)		65.97	(5.0)		65.61	(5.3)	
Chongqing										—			—			27.87	(2.1)		20.42	(1.7)	
East	103.52	(16.7)	[2]	127.16	(14.5)	[4]	146.97	(13.7)	[4]	185.61	(14.4)	[3]	199.27	(14.4)	[2]	193.02	(14.7)	[2]	188.57	(15.3)	[2]
Shanghai	1.65	(0.3)			(0.0)			(0.0)			(0.0)			(0.0)			(0.0)			(0.0)	
Jiangsu	15.71	(2.5)		21.94	(2.5)		24.08	(2.2)		25.49	(2.0)		26.07	(1.9)		24.78	(1.9)		24.81	(2.0)	
Zhejiang	1.33	(0.2)		1.51	(0.2)		1.37	(0.1)		1.13	(0.1)		1.23	(0.1)		1.15	(0.1)		1.09	(0.1)	
Anhui	23.83	(3.8)		29.05	(3.3)		32.05	(3.0)		43.22	(3.3)		46.42	(3.7)		47.69	(3.6)		45.83	(3.7)	
Jiangxi	15.53	(2.5)		19.38	(2.2)		20.27	(1.9)		23.33	(1.8)		24.38	(1.7)		20.64	(1.6)		19.80	(1.6)	
Fujian	4.17	(0.7)		6.06	(0.7)		9.25	(0.9)		8.60	(0.7)		11.68	(0.9)		7.82	(0.6)		7.27	(0.6)	
Shandong	41.30	(6.6)		49.22	(5.6)		59.95	(5.6)		83.84	(6.5)		89.49	(6.4)		90.94	(6.9)		89.77	(7.3)	

Source: 1981–97: MTGYNJ (various years); 1998: 'Zhongguo 98 nian shengchan yuan mei 12.36 yi dun' (China's 1998 Coal Production Is 1,236 Million Tons), Xinhua, 25 Feb. 1999.

Note: Percentages are given in parentheses. In most cases the provincial/regional data given in the MTGYNJ does not sum up exactly to the total given.

Table 6.13 Distribution of coal consumption, 1985–96 (millions of tons and %)

	1985		[Rank]	1990		[Rank]	1996		[Rank]
North	204.30	(24.3)	[1]	236.37	(22.4)	[2]	366.90	(24.5)	[2]
Beijing	20.47	(2.4)		24.27	(2.3)		27.3	(1.8)	
Tianjin	12.65	(1.5)		17.94	(1.7)		22.9	(1.5)	
Hebei	76.56	(9.1)		78.09	(7.4)		113.8	(7.6)	
Innner Mongolia	27.60	(3.3)		39.04	(3.7)		48.9	(3.3)	
Shanxi	67.02	(8.0)		77.03	(7.3)		154.0	(10.3)	
Northeast	151.38	(18.1)	[3]	187.83	(17.8)	[3]	204.40	(13.6)	[4]
Liaoning	69.70	(8.3)		82.31	(7.8)		93.1	(6.2)	
Jilin	30.79	(3.7)		40.10	(3.8)		51.4	(3.4)	
Heilongjiang	50.89	(6.1)		65.42	(6.2)		59.9	(4.0)	
Northwest	60.54	(7.2)	[6]	78.09	(7.4)	[6]	110.90	(7.4)	[6]
Shaanxi	21.86	(2.6)		27.44	(2.6)		41.0	(2.7)	
Gansu	14.55	(1.7)		18.99	(1.8)		26.2	(1.7)	
Qinghai	3.71	(0.4)		5.28	(0.5)		4.9	(0.3)	
Xinjiang	13.88	(1.7)		17.94	(1.7)		27.7	(1.8)	
Ningxia	6.54	(0.8)		8.44	(0.8)		11.1	(0.7)	
S. Central	143.05	(17.0)	[4]	180.44	(17.1)	[4]	271.60	(18.1)	[3]
Henan	59.09	(7.0)		61.20	(5.8)		82.1	(5.5)	
Hubei	22.91	(2.7)		33.77	(3.2)		56.6	(3.8)	
Hunan	34.48	(4.1)		40.10	(3.8)		58.1	(3.9)	
Guangdong	17.16	(2.0)		29.55	(2.8)		50.8	(3.4)	
Guangxi	9.41	(1.1)		15.83	(1.5)		23.5	(1.6)	
Hainan	—			—			0.5	(0.1)	
Southwest	89.64	(10.7)	[5]	116.08	(11.0)	[5]	171.00	(11.4)	[5]
Sichuan	53.66	(6.4)		66.48	(6.3)		92.3	(6.2)	
Yunnan	16.38	(2.0)		22.16	(2.1)		30.4	(2.0)	
Guizhou	19.60	(2.3)		27.44	(2.6)		48.3	(3.2)	
East	190.91	(22.7)	[2]	256.42	(24.3)	[1]	373.40	(24.9)	[1]
Shanghai	16.31	(1.9)		27.44	(2.6)		40.9	(2.7)	
Jiangsu	46.48	(5.5)		62.26	(5.9)		88.3	(5.9)	
Zhejiang	15.80	(1.9)		25.33	(2.4)		45.9	(3.1)	
Anhui	27.72	(3.3)		33.77	(3.2)		52.9	(3.5)	
Jiangxi	19.39	(2.3)		22.16	(2.1)		27.1	(1.8)	
Fujian	8.68	(1.0)		12.66	(1.2)		18.4	(1.2)	
Shandong	56.53	(6.7)		72.81	(6.9)		99.9	(6.7)	
Total	839.82			1,055.23			1,498.2		

Sources: *NYTJNJ 1986*, pp. 150–204; *NYTJNJ 1990*, pp. 145, 171; Ministry of Coal Report.

Note: Percentages are given in parentheses.

with 107.86 and 95.67 million tons, respectively. Shanxi supplied coal to all but six of the twenty-nine provinces in 1996, and the plan for the Province to produce 400 million tons by 2000 would have been easily met had the government not decided to curtail production.[116] Table 6.15 shows that there were major changes in the destinations of the coal shipped from Shanxi by rail. In 1980 almost half was shipped to the Beijing/Tianjin area. By 2000 this was expected to be just over one-third. The east and southcentral received increasingly larger shares of the coal. Table 6.16 gives the destinations of shipments from all

Table 6.14 Output from the largest CMAs
in 1996 (mt)

CMA	Output
Shanxi	
Datong	(31.55)
Pingshou	(12.63)
Yangquan	(16.16)
Xishan	(17.73)
Lu'an	(10.15)
Jincheng	(11.51)
Heilongjiang	
Jixi	(12.08)
Hegang	(13.60)
Shuanyashan	(10.45)
Shandong	
Xinwen	(11.87)
Yanzhou	(18.65)
Anhui	
Huainan	(13.57)
Huaibei	(16.85)
Hebei	
Kailuan	(18.64)
Fengfeng	(10.77)
Henan	
Pingdingshan	(20.71)
Liaoning	
Tiefa	(14.40)
Jiangsu	
Xuzhou	(13.11)

Source: *MTGYNJ 1997*, pp. 174–243.

the coal producing provinces. Each of Shanxi, Henan, Liaoning, Beijing, Heilongjiang, Hebei, Anhui, Jiangsu, Ningxia, Guangdong, Zhejiang, Jilin, and Inner Mongolia shipped out more than half of their total production.

Between 1978 and 1985 there was a substantial geographical shift in consumption away from the east to the north. In 1978 the east had accounted for 35 per cent of total consumption, however by 1985, only about 23 per cent. Greater Shanghai's share alone fell from nearly 12 per cent to just under 2 per cent (compare Table 3.8 with 6.13). Meanwhile the north went from accounting for 15 to 24 per cent with Shanxi and Hebei gaining the most. Consumption in the northeast dropped from third to fourth position, and that in the southcentral region rose from fourth to third. The government-decreed line proscribing central heating south of the Changjiang River (see Chapter 3) remained in place and must be kept in mind when interpreting the Tables. Had central heating been available

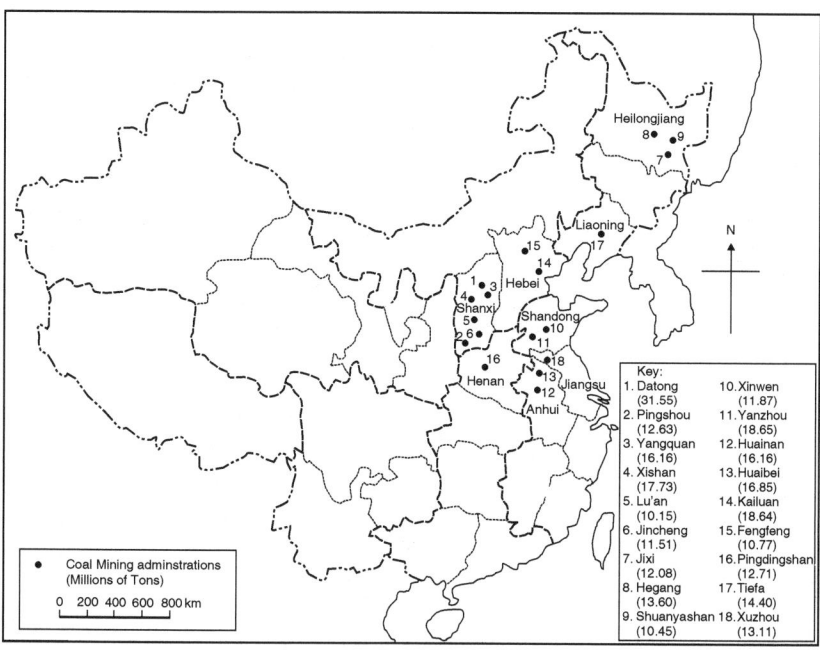

Figure 6.2 Largest coal mining administrations in 1996
Source: *MTGYNJ 1997*, pp. 174–243.

Table 6.15 Distribution of coal output shipped from Shanxi Province by train, 1980–2000 (%)

Shipment destinations	1980	1985	1990	2000
To Beijing/Tianjin area	48.0	41.8	38.0	36.7
To Northeast	10.6	13.9	12.5	8.3
To East	22.8	24.6	32.0	31.7
To South central	14.4	14.8	14.5	20.0
To Other regions	4.1	4.9	3.0	3.3

Source: Calculated from Fu He, 'Tan Guangdong sheng diaoyun meitan de yunshu fangshi he lujing de xuanze' (Discussion on Direct Means of Transporting Coal to Guangdong Province), *NY*, no. 1, 1988, Table 1, p. 8.

Note: The 2000 figures are based on the assumption that total production in Shanxi that year would be 400 million tons.

south of the River, coal consumption would most certainly have been higher there in the winter months.

The geographical production/consumption balance improved slightly over the reform years (see Table 6.17). In 1978 the nationwide spatial deficit had been

Table 6.16 Destinations of coal shipped from each province and autonomous region, 1996 (mt)

Shanxi		Inner Mongolia	
Hebei	(84.36)	Inner Mongolia	(13.42)
Shandong	(22.83)	Heilongjiang	(8.45)
Tianjin	(22.40)	Hebei	(7.79)
Shanxi	(18.80)	Tianjin	(6.00)
Jiangsu	(15.52)	Liaoning	(5.50)
Beijing	(15.12)	Jilin	(3.80)
Liaoning	(12.69)	Ningxia	(0.77)
Henan	(6.91)	Beijing	(0.15)
Hubei	(6.89)	Gansu	(0.09)
Anhui	(3.29)	Qinghai	(0.06)
Hunan	(2.60)	Shanxi	(0.04)
Zhejiang	(2.33)	Shaanxi	(0.01)
Inner Mongolia	(2.07)	Total = 46.07	
Guangxi	(1.30)		
Jiangxi	(1.29)	*Hebei*	
Jilin	(1.16)	Hebei	(27.67)
Guangdong	(1.12)	Beijing	(2.74)
Fujian	(1.05)	Liaoning	(2.62)
Shanghai	(0.65)	Jiangsu	(2.25)
Shaanxi	(0.59)	Tianjin	(2.01)
Heilongjiang	(0.21)	Shandong	(1.75)
Sichuan	(0.20)	Zhejiang	(1.10)
Ningxia	(0.03)	Hubei	(0.66)
Total = 223.41		Inner Mongolia	(0.53)
		Hunan	(0.11)
Heilongjiang		Shanghai	(0.08)
Heilongjiang	(34.60)	Jilin	(0.05)
Liaoning	(15.75)	Anhui	(0.01)
Jilin	(14.19)	Total = 41.57	
Inner Mongolia	(0.50)		
Total = 65.04		*Shandong*	
		Shandong	(27.22)
Henan		Jiangsu	(4.64)
Hubei	(23.94)	Zhejiang	(2.21)
Henan	(15.71)	Anhui	(1.50)
Jiangsu	(8.12)	Shanghai	(0.82)
Anhui	(3.40)	Jiangxi	(0.57)
Hunan	(3.17)	Hubei	(0.34)
Shandong	(2.45)	Hebei	(0.33)
Jiangxi	(2.33)	Hunan	(0.30)
Guangdong	(2.14)	Fujian	(0.20)
Guangxi	(1.57)	Total = 38.13	
Zhejiang	(0.74)		
Shanghai	(0.16)	*Anhui*	
Hebei	(0.15)	Anhui	(21.65)
Liaoning	(0.08)	Jiangsu	(8.37)
Shaanxi	(0.05)	Zhejiang	(2.32)
Tianjin	(0.04)	Shanghai	(1.26)
Jilin	(0.03)	Fujian	(0.87)
Fujian	(0.02)	Jiangxi	(0.29)
Total = 64.10		Shandong	(0.01)
		Total = 34.76	

Table 6.16 (Continued)

Liaoning		Guizhou	
Liaoning	(29.58)	Guizhou	(7.32)
Jilin	(0.72)	Guangxi	(2.14)
Inner Mongolia	(0.31)	Guangdong	(1.67)
Heilongjiang	(0.08)	Yunnan	(1.27)
Hebei	(0.07)	Sichuan	(0.97)
Total = 30.76		Hunan	(0.59)
		Hubei	(0.15)
Sichuan		Jiangxi	(0.12)
Sichuan	(23.99)	Zhejiang	(0.01)
Hubei	(3.39)	Total = 14.24	
Guangxi	(0.96)		
Shaanxi	(0.18)	**Ningxia**	
Guangdong	(0.14)	Ningxia	(4.11)
Hunan	(0.12)	Gansu	(3.90)
Yunnan	(0.12)	Liaoning	(0.98)
Jiangxi	(0.06)	Shaanxi	(0.95)
Total = 28.95		Qinghai	(0.36)
		Tianjin	(0.29)
Shaanxi		Beijing	(0.26)
Shaanxi	(10.16)	Hebei	(0.25)
Jiangsu	(5.19)	Hubei	(0.25)
Hubei	(2.56)	Inner Mongolia	(0.20)
Jiangxi	(0.48)	Shandong	(0.12)
Zhejiang	(0.25)	Xinjiang	(0.12)
Hunan	(0.20)	Jilin	(0.03)
Anhui	(0.18)	Jiangsu	(0.03)
Henan	(0.17)	Henan	(0.03)
Shanghai	(0.03)	Sichuan	(0.03)
Gansu	(0.01)	Shanxi	(0.02)
Total = 19.24		Anhui	(0.02)
		Zhejiang	(0.01)
Jilin		Total = 11.97	
Jilin	(12.46)		
Liaoning	(4.53)	**Jiangxi**	
Inner Mongolia	(0.40)	Jiangxi	(6.95)
Heilongjiang	(0.38)	Zhejiang	(1.51)
Total = 17.77		Fujian	(0.57)
		Hunan	(0.30)
Jiangsu		Guangdong	(0.21)
Jiangsu	(11.73)	Anhui	(0.12)
Zhejiang	(2.43)	Hubei	(0.07)
Shandong	(0.97)	Jiangsu	(0.03)
Shanghai	(0.70)	Guangxi	(0.03)
Anhui	(0.57)	Total = 9.78	
Fujian	(0.10)		
Jiangxi	(0.05)	**Beijing**	
Total = 16.54		Beijing	(2.48)
		Tianjin	(2.39)
Hunan		Hebei	(1.05)
Hunan	(7.89)	Liaoning	(0.40)
Guangdong	(4.41)	Shandong	(0.34)
Guangxi	(1.52)	Jiangsu	(0.12)
Jiangxi	(0.21)	Jilin	(0.02)
Zhejiang	(0.15)	Inner Mongolia	(0.01)
Hubei	(0.10)	Heilongjiang	(0.01)
Total = 14.28		Total = 6.81	

Table 6.16 (Continued)

Yunnan			*Xinjiang*	
Yunnan	(5.60)		Xinjiang	(2.38)
Guangxi	(0.59)		Gansu	(2.36)
Zhejiang	(0.09)		Jiangsu	(0.02)
Jiangxi	(0.09)		Total = 4.76	
Guangdong	(0.09)		*Shanghai*	
Hunan	(0.07)		Shanghai	(0.05)
Sichuan	(0.03)		Zhejiang	(1.13)
Fujian	(0.01)		Total = 1.18	
Total = 6.57			*Qinghai*	
Gansu			Qinghai	(1.11)
Gansu	(4.08)		Jiangsu	(0.02)
Jiangsu	(0.80)		Zhejiang	(0.02)
Qinghai	(0.75)		Total = 1.16	
Zhejiang	(0.35)			
Shaanxi	(0.15)		*Zhejiang*	
Hubei	(0.06)		Zhejiang	(0.75)
Shanghai	(0.04)		Total = 0.75	
Guangxi	(0.03)		*Hubei*	
Shandong	(0.02)		Hubei	(0.16)
Hunan	(0.02)		Fujian	(0.06)
Anhui	(0.01)		Zhejiang	(0.04)
Fujian	(0.01)		Guangdong	(0.02)
Guangdong	(0.01)		Jiangxi	(0.02)
Total = 6.33			Hunan	(0.01)
Guangdong			Guangxi	(0.01)
Guangdong	(5.72)		Total = 0.32	
Total = 5.72			*Tianjin*	
Fujian			Tianjin	(0.13)
Fujian	(4.16)		Hebei	(0.13)
Zhejiang	(0.99)		Hunan	(0.03)
Jiangxi	(0.02)		Hubei	(0.01)
Total = 5.17			Total = 0.30	
Guangxi			*Hainan*	
Guangxi	(3.53)		Hainan	(0.05)
Guangdong	(1.28)		Total = 0.05	
Total = 4.82				

Source: *JTNJ 1997*, pp. 580–1.

nearly 18 per cent. However, this fell to 13.2 after the shift in consumption away from the east, and the official encouragement of construction of small mines throughout the country to increase production and relieve pressure on the railways. By the late 1980s the government was curtailing development of small local mines and in 1990 the deficit of 16.6 per cent was close to what it had been earlier, though it fell to 15.2 in 1996. The east and north each almost equally shared one-quarter of total consumption in 1996.

Table 6.17 Production deficit/surplus for each region, 1985–96 (%)

	E	N	NE	SC	SW	NW
1985						
Production	14.5	36.4	15.1	15.0	11.0	8.0
Consumption	22.7	24.3	18.1	17.0	10.7	7.2
Difference	−8.2	+12.1	−3.0	−2.0	+0.3	+0.8
Total deficit:−13.2						
1990						
Production	13.7	37.5	14.8	14.1	11.8	8.1
Consumption	24.3	22.4	17.8	17.1	11.0	7.4
Difference	−10.6	+15.1	−3.0	−3.0	+0.8	+0.7
Total deficit: −16.6						
1996						
Production	14.7	36.8	12.0	14.7	13.5	8.3
Consumption	24.9	24.5	13.6	18.1	11.4	7.4
Difference	−10.2	+12.3	−1.6	−3.4	+2.1	+0.9
Total deficit:−15.2						

Source: Calculated from Tables 7.8 and 7.9.

As can be seen from Table 6.18, some LS mines in Zhejiang, Sichuan, and Yunnan were reclassified to CMAs in the reform years, while those in Qinghai, Hubei, Guangdong, Guangxi, and Fujian remained entirely within the LS and LNS categories. In several areas most of the production came exclusively from township mines, for example, Hubei, Hunan, Guangdong, Sichuan, Yunnan, and Guizhou.

Since the early 1980s the increases in the amount of coal shipped from mines around the country to the southern provinces were staggering. Table 6.19 gives the net movement of CMA and LS mine coal among the provinces and regions in 1996. Several mine-mouth power plants were built in the north, which helped reduce the strain on the railways, but notwithstanding, coal shipments to Jiangsu increased 212 per cent between 1980 and 1996, and to Guangdong and Shanghai 115 per cent and 91 per cent, respectively.[117] Much of the coal shipped to Shanghai and Jiangsu was transported first by rail to ports, then by water to the south.

Mechanization levels

Over the 1980s and 1990s mechanization at the underground CMAs was increased substantially. Not only were Ministry officials keen to increase productivity, they wanted to improve the image of the industry. At the outset of the reform programme the central government emphasized the crucial role that the coal industry would play in fuelling the country's modernization and

Table 6.18 Geographical distribution of state and local mines, 1981 and 1995–98 (%)

	1981		1995			1996			1997			1998		
	State	Total local	State	Total local:	of which township	State	Total local:	of which township	State	Total local:	of which township	State	Total local:	of which township
National average	53.8	46.2	37.3	61.0	38.6	39.1	61.0	44.7	39.9	60.1	43.0	40.9	59.1	41.9
North														
Beijing	76.3	23.7	49.2	50.8	50.3	49.3	50.7	50.2	50.0	50.0	49.6	48.0	52.0	51.7
Tianjin	0.0	0.0	0.0	0.0	0.0	0.0	0.0	0.0	0.0	0.0	0.0	0.0	0.0	0.0
Hebei	73.4	26.6	54.4	44.4	25.6	53.7	46.3	33.2	57.6	43.6	28.3	64.9	35.1	20.5
Inner Mongolia	67.9	32.1	40.3	57.2	40.4	43.4	56.6	42.9	41.7	58.3	41.8	43.3	56.7	36.2
Shanxi	51.0	49.0	36.3	61.4	47.8	37.5	62.5	50.3	38.2	61.8	46.7	37.3	62.7	48.0
Northeast														
Liaoning	85.7	14.3	66.6	33.4	27.9	68.8	31.2	26.0	72.2	27.8	22.6	71.7	28.3	23.3
Jilin	68.6	31.4	40.9	59.1	40.6	50.5	49.5	31.7	48.8	51.2	32.2	51.3	48.7	29.0
Heilongjiang	79.8	20.2	50.0	43.2	28.5	55.6	44.4	29.4	54.7	45.3	29.9	54.0	46.0	30.2
Northwest														
Shaanxi	63.0	37.0	33.1	62.5	45.6	35.1	64.9	49.4	35.5	64.5	48.7	36.1	63.9	49.5
Gansu	55.5	44.5	27.3	72.7	41.9	35.0	65.0	41.5	35.9	64.1	37.7	34.8	65.2	37.7
Qinghai	0.0	100.0	0.0	100.0	29.9	0.0	100.0	29.3	0.0	100.0	29.5	0.0	100.0	31.9
Xinjiang	26.1	73.9	15.3	81.8	42.5	19.1	32.7	42.5	18.8	81.2	43.2	18.7	81.3	43.0
Ningxia	81.4	18.6	68.4	31.2	11.9	67.3	32.7	13.6	63.3	36.7	16.1	64.2	35.8	13.6

S. Central														
Henan	56.4	43.6	41.6	56.7	34.2	43.4	56.6	40.2	44.1	55.9	38.7	48.0	52.0	35.8
Hubei	0.0	100.0	0.0	100.0	67.3	0.0	100.0	82.1	0.0	100.0	82.9	0.0	100.0	81.7
Hunan	16.4	83.6	13.1	86.7	40.7	12.8	87.2	72.9	14.2	85.8	70.5	13.8	86.2	70.3
Guangdong	0.0	100.0	0.0	100.0	36.1	0.0	100.0	63.7	0.0	100.0	67.6	0.0	100.0	64.5
Guangxi	0.0	100.0	0.0	100.0	34.9	0.0	100.0	52.5	0.0	100.0	50.5	0.0	100.0	44.7
Southwest														
Sichuan	0.0	100.0	18.2	80.3	57.6	21.8	78.2	60.9	17.1	82.9	65.2	15.9	84.1	67.2
Yunnan	0.0	100.0	8.7	91.3	51.5	8.1	91.9	65.7	7.2	92.8	64.2	7.8	92.2	65.6
Guizhou	41.5	58.5	19.1	80.9	34.3	15.4	84.6	79.7	14.2	85.8	60.6	13.1	86.9	82.4
Chongqing	—	—	—	—	—	—	—	—	32.3	67.7	52.1	40.0	60.0	39.8
East														
Shanghai	100.0	0.0	—	—	—	—	—	—	—	—	—	—	—	—
Jiangsu	73.3	26.7	66.7	33.3	16.6	68.0	32.0	17.2	69.8	30.2	16.1	72.5	27.5	14.7
Zhejiang	0.0	100.0	69.9	30.1	10.6	65.9	33.3	23.6	68.7	31.3	23.5	73.4	26.6	19.3
Anhui	92.0	8.0	67.4	32.6	12.3	66.2	33.8	13.5	59.8	40.2	12.4	62.0	38.0	10.9
Jiangxi	39.0	61.0	29.3	70.7	34.4	30.3	69.6	55.0	30.3	69.7	55.6	27.8	72.2	59.3
Fujian	0.0	100.0	0.0	100.0	43.7	0.0	100.0	57.4	0.0	100.0	46.0	0.0	100.0	46.4
Shandong	57.6	42.4	50.4	49.6	23.9	52.2	49.8	21.0	59.6	40.4	21.8	63.3	36.7	16.3

Sources: *CIYB 1982*, p. 32; *MTGYNJ 1997*, p. 25; *MTGYNJ 1998*, p. 20.

Table 6.19 Net coal movement among provinces and regions, 1996 (mt)

Net outflows	
Shanxi	160.2
Henan	22.9
Inner Mongolia	17.7
Heilongjiang	9.7
Shaanxi	7.3
Guizhou	4.3
Ningxia	5.4
Xinjiang	1.9
Anhui	1.3
Sichuan	0.7
Imports/exports and others	10.4
Total	241.8
Net inflows	
Liaoning	31.1
Shanghai	26.5
Jiangsu	24.6
Hubei	22.0
Zhejiang	17.5
Tianjin	16.0
Hebei	15.6
Beijing	11.1
Shandong	11.1
Guangdong	10.1
Jilin	9.6
Hunan	4.9
Fujian	4.8
Gansu	4.0
Guangxi	3.6
Jiangxi	3.5
Hainan	1.0
Qinghai	1.0
Yunnan	0.9
Total	218.9

Source: *MTGYNJ 1997*, p. 28.

Note: CMA and LS mine output only.

took steps to throw off the 'antique, dirty, stupid' image of the state owned coal industry.[118]

The percentage of underground state mine faces with some mechanization more than doubled from 32.34 per cent in 1979 to 73.63 in 1998, and over one-fifth attained complete mechanization. The main types of equipment purchased from foreign companies were shearers, road-headers, hydraulic supports, high-capacity conveyors, and safety monitoring equipment.[119] The first fully mechanized mine was put into operation at Yanzhou (Shandong) in 1986. However, the target to have 70 per cent of the nationwide state mines with some mechanization by 1990 was not met, and at the time of writing the target to have 85 per cent mechanized by 2000 seemed unlikely to be achieved. For context, the proportion of mines in 1991 with mechanization was 83.3 per cent in Russia, 70.8 in Ukraine, 92.5 in Poland, 99.9 in the FRG, 99.9 in the UK, and 83.6 in Japan.[120]

In 1980 there had been an illogical ratio between partially and fully mechanized underground faces. Despite having yields of about only a quarter of those with full mechanization, the partially equipped faces were in far greater abundance than those with no mechanization. By 1983, the share of fully mechanized faces was dominant, indicating that returns on investments in mechanization were much higher than before.

Faces with full mechanization, as opposed to partial or hydraulic mechanization, increased by the largest amount. In 1998, 49.3 per cent of the mine faces were fully mechanized, up from 13.2 in 1980, and the average production rate at these now fully mechanized mines went from 33,607 to 68,889 tons per month per face. The productivity of hydraulic equipment also improved from a low of 15,454 tons per face per month in 1982 to 25,778, though the proportion of faces using this type of equipment remained below 2 per cent.

Some of the problems with the machinery and equipment that had reduced productivity before the reforms had yet to be solved. At many mines machinery dating back to the 1930s was still in use, and at almost all there was a lack of compatibility in equipment. This machinery itself consumed inordinate amounts of electricity. Since 1979 only two new coal machinery factories were built. However, the amount of equipment owned by the Ministry classified as being operational rose from 61 per cent in 1981 to 74.4 per cent in 1997, and that classified as being in good order rose from 84 to 90.4 per cent.[121]

The *Zhongguo meitan gongye nianjian* (China Coal Industry Yearbooks) contain limited data on the LS and LNS mines. In 1993 the level of mechanization at the LS mines was only approximately 25 per cent.[122] They often obtained old or surplus equipment from the CMAs, while the LNS mines often had no machinery at all. Picks, shovels, baskets, ropes, and mules were the main equipment used, and the miners had to rely on natural ventilation. Such operations were described as *laoshu chuan dong* (mouse bores hole).[123]

Somewhat ironically, beginning in 1987 China sold large amounts of basic mining equipment abroad.[124] Roof supports were sold to the US and Australia, and later a great range of equipment was sold especially to various African countries, India, Pakistan, the Philippines, Turkey, and Russia. During the Eighth Five Year Plan period, some US$24 million worth were sold each year.[125] Indeed, China became somewhat of a 'technological leader' in shaft construction.[126] In the early 1990s the CNCIEC began construction on east Africa's largest coal mine, the Kiwira Mine in Tanzania, and contracted the engineering of a shaft at the Jerala Mine in Morocco.[127]

Resource use efficiency

In *Chapter 3* it was noted that ideally half a ton of coal is produced for every ton of prepared capacity, that is, there is an output–capacity ratio of 0.5 : 1, and that there is generally a time lag of two to three years before increased capacity is manifested in increased production. The serious imbalance in the output–capacity ratio in the late 1970s pushed production at the CMAs and LS mines in 1980 down nearly 23 million tons from the 1979 level, and in 1981 there was a further decrease of over 11 million tons, the largest decreases since the Cultural Revolution. It was imperative to divert much of the mining work force from cutting to face construction.

Between 1982 and 1985 the output–capacity ratio averaged a high 1.64 : 1, while between 1986 and 1988 it was what it should be, that is, averaging 0.47 : 1 (see Table 6.20). In the second-half of the 1980s only 67 per cent of the industry's capital construction targets were met due to severe budget cuts.[128] The lag factor noted above meant that production in the late 1980s and early 1990s had to be

Table 6.20 Coal output–capacity ratios, 1979–98

	CMA and LS Mines		
	Annual capacity increases (mt)	*Annual increases in production* (mt)	*Output–capacity ratio*
1978	11.51	50.67	4.40 : 1
1979	13.93	6.69	0.48 : 1
1980	8.29	−22.72	
1981	13.73	−11.47	
1982	8.20	25.21	3.07 : 1
1983	18.52	24.21	1.31 : 1
1984	24.35	27.89	1.15 : 1
1985	16.47	16.69	1.01 : 1
1986	21.05	6.26	0.30 : 1
1987	21.06	6.03	0.29 : 1
1988	33.02	27.01	0.82 : 1
1989	28.05	35.41	1.26 : 1
1990	22.99	21.56	0.94 : 1
1991	28.91	−1.16	
1992	24.19	1.31	0.05 : 1
1993	11.22	−23.40	
1994	7.10	12.57	1.77 : 1
1995	26.24	21.00	0.80 : 1
1996	12.48	63.68	5.10 : 1
1997	26.97	−4.48	
1998	9.69	−38.49	

Sources: *MTGYNJ*, various years; *TJNJ 1985*, p. 446.

Note: The production figures given in the *China Statistical Yearbooks* vary slightly from those in the *China Coal Industry Yearbooks*.

radically curtailed. This coincided with the austerity programme aimed at inflation control and the adverse effects on the economy of the June 4th Incident at Tiananmen Square. Production in 1991 was down 1.16 millions tons from the 1990 level, increased by only 1.31 million tons in 1992 and fell sharply by 23.40 million tons in 1993. The increases in capacity in 1993 and 1994, at only some 11 and 7 million tons, respectively, were considerably less than half what they had been over much over the previous decade, then in 1996 production soared by nearly 64 million tons, an annual increase unparalleled since the GLF. The output–capacity ratio for 1996 was especially high at 5.10 : 1. However, after the government decided to scaleback production drastically in the late 1990s, capacity development continued, making for considerably more rational use of the reserves.

The absence of well-defined, legally defensible property rights resulted in much haphazard and destructive development. Besides the state, provincial, and local government mines there were others operated by rural families and groups, the army, and prisoners. The extraction methods used by groups with no training, and/or the opening of too many pits at one deposit, caused numerous cave-ins which destroyed the seams and ultimately made it impossible to mine by any means.[129] The competition for the resources was such that arguments arose among peasants, and between peasants and Ministry authorities.[130] For example, the World Bank had agreed to provide funds for the development of an anthracite mine at Chengzhuang in Jincheng mining area (Shanxi) in June 1982, but due to a boundary disagreement with a local mine, approval for the project was not granted until May 1984. Two years later, after further wrangling, it was dropped entirely, and the funds were devoted to expansion of the Kailuan mine (Hebei) and construction of new mines at Yanzhou (Shandong).[131] In April 1998 an accident causing fifty-five deaths occurred when miners of one unlicensed mine blocked miners at a rival mine by causing a cave-in which set off explosions at two other nearby mines.[132]

In 1990 the average depth of all types of mines was 330 m, slightly deeper than the average of 300 for the major coal producing countries.[133] However, the recovery rate (proportion of a deposit successfully removed, and ideally about 90 per cent) at the CMAs was not high at 79 per cent in 1996 and only 73 per cent in 1997.[134] At the LS mines it was only 30–40 per cent and a mere 10–20 per cent at the LNS mines.[135] The mining techniques were such that for each ton mined, 5–6 were left unmined and unrecoverable.[136] In north China, the refuse left from the peasants' attempts to cut the coal under a riverbed clogged the natural drainage network, creating latent serious flooding problems.[137]

Use of materials and labour efficiency

The coming into play of market forces, the responsibility system and advances in technology led to reductions in inputs per unit of output required by the CMAs. In 1980, 90.34 m^3 of timber were required on average per ton of raw coal production, but by 1998 this was down to less than 28 (see Table 3.12 for international comparisons).[138] It must be noted, however, that some of the reduction can be explained by the fact that timber was often in short supply, and cement and metal were used instead. Use of timber in the LS mines was still very high at 12.92 m³ per ton in 1985, though this fell to 8.32 in 1994. Not surprisingly, as production became more mechanized over the years, consumption of electricity at the state mines increased from 34.34 kWh per ton in 1980 to 54.88 in 1997.

With the incentives in the 1980s for rural communities to develop local coal deposits, the materials required for mining became in great demand. Just as part of the coal output was sold at fixed low prices and the other at market prices, so too were timber, cement, steel, explosives, etc. The two-track pricing system led to fighting and corruption as the developers of small mines attempted to obtain limited materials as cheaply as possible.

Overmanning at the CMAs and LS mines had been a severe problem before 1979. Thousands of workers were grossly underemployed, many having done little productive work every day for decades and the Ministry began in the early 1990s to make drastic cutbacks.[139] It had finally been recognized that more labour was not a substitute for machinery. Table 6.21 gives a breakdown of total employment at the CMAs. Between 1979 and 1992 it increased from 2.18 to 3.37 million. Thereafter there was an almost steady annual decrease. It stood at 2.63 million in 1998, and by 2000 total workers were to amount to about 3 million of which coal production workers would amount to 1 million, and management, 120,000.[140] Labourers – those with low levels of skill and education, who cut and hauled coal, repaired machinery, monitored safety, carried out rescue work, etc. – decreased as a proportion of the total from 72.0 per cent in 1979 to 55.6 per cent in 1997. The share of service personnel and engineers and technicians made up some of the difference.

Table 6.22 gives the breakdown of coal production labourers, that is those concerned only with cutting and hauling coal. These amounted to 1.17 million in 1979 and peaked at 1.45 million in 1985. By 1998 they were down to 0.94 million. The share of surface workers decreased steadily after 1981 from over 20 per cent to about 15 per cent.

Overmanning had been particularly severe among surface workers and the surfeit was assigned to diversified activities. Generally, the displaced workers were very pleased to make the move to work that offered opportunities to learn new skills and promised higher incomes *(see the Financing section of this Chapter)*. However, in some cases the transfer involved conditions worse than those at the mines. For example, in Heilongjiang the layed off workers were asked to farm in an area some distance away from their families. Not surprisingly, they refused.[141]

Recruitment for underground workers was beginning to become a concern by the mid-1990s. After the introduction of the one-child policy in the late 1970s, parents forbade their only child and son to take up such a hazardous occupation and the industry began to have difficulties finding young men. The normal retirement age was forty-five but many men had to stop working well before reaching that age due to lung disease.[142] In 1983, out of Shanxi's total 165,543 mine

Table 6.21 Employment at CMAs, 1979–98

	Total employees	Labourers	Engineers and technicians	Administrative personnel	Service personnel	Others
1979	2,183,524	1,571,339 (72.0)	34,281 (1.6)	153,065 (7.0)	306,117 (14.0)	106,378 (4.9)
1981	2,417,756	1,699,559 (70.3)	40,667 (1.7)	169,372 (7.0)	360,476 (14.9)	134,494 (5.6)
1982	2,592,139	1,796,224 (69.3)	44,887 (1.7)	175,882 (6.8)	397,221 (15.3)	167,540 (6.5)
1983	2,640,140	1,792,487 (67.9)	48,941 (1.8)	180,172 (6.8)	400,553 (15.2)	208,655 (7.9)
1984	2,826,629	1,861,232 (65.9)	53,487 (1.9)	195,042 (6.9)	426,720 (15.1)	281,355 (9.9)
1985	2,930,815	1,877,138 (64.0)	56,412 (1.9)	197,851 (6.8)	436,911 (14.9)	354,382 (12.1)
1986	2,990,455	1,868,891 (62.5)	57,798 (1.9)	192,057 (6.4)	489,551 (16.4)	374,947 (12.5)
1987	3,035,630	1,883,227 (62.0)	60,503 (2.0)	193,499 (6.4)	521,345 (17.2)	369,438 (12.1)
1988	3,103,217	1,901,301 (61.3)	62,997 (2.1)	194,044 (6.3)	544,655 (17.5)	394,116 (12.7)
1989	3,227,665	1,969,477 (61.0)	69,358 (2.1)	201,540 (6.2)	586,537 (18.2)	395,416 (12.3)
1990	3,297,598	2,006,692 (60.9)	73,507 (2.2)	214,779 (6.5)	624,070 (18.9)	373,966 (11.3)
1991	3,358,083	1,494,088 (44.5)	57,282 (1.7)	164,950 (4.9)	501,059 (14.9)	243,835 (7.3)
1992	3,373,493	1,973,772 (58.5)	77,636 (2.3)	218,172 (6.5)	666,184 (19.8)	432,077 (12.8)
1993	3,267,116	1,866,220 (57.1)	74,521 (2.3)	205,579 (6.3)	627,823 (19.2)	489,652 (15.0)
1994	3,442,127	1,953,874 (56.8)	80,174 (2.3)	223,478 (6.5)	644,877 (18.7)	539,724 (15.7)
1995	3,308,448	1,870,606 (56.5)	77,115 (2.3)	212,199 (6.4)	614,547 (18.6)	533,981 (16.1)
1996	3,220,202	1,805,408 (56.1)	75,366 (2.3)	208,761 (6.5)	594,129 (18.5)	536,538 (16.7)
1997	3,157,066	1,754,321 (55.6)	73,253 (2.3)	201,911 (6.4)	587,827 (18.6)	539,754 (17.1)
1998	2,628,628	1,642,946 (62.5)	69,909 (2.7)	188,329 (7.2)	513,533 (19.5)	213,911 (18.1)

Source: *MTGYNJ* (various years).

Note: The figures in parentheses are percentages of total. No definition is given for 'others'.

employees, 12,871 (7.8 per cent) were found to have pulmonary tuberculosis, and of these 73.3 per cent were underground miners.[143]

Table 6.23 gives the ILOR, which indicates the change in numbers employed at the CMAs per unit increase in output. It shows clearly that productivity was

Table 6.22 Coal production labourers at CMAs, 1979–98

	Total coal production labourers	of which:		
		Working in underground mines	Working at surface	Working at opencast mines
1979	1,172,484	903,579 (77.1)	236,707 (20.2)	26,128 (2.2)
1981	1,301,648	996,393 (76.5)	269,724 (20.7)	28,838 (2.2)
1982	1,373,952	1,055,845 (76.8)	280,320 (20.4)	30,706 (2.2)
1983	1,381,825	1,059,526 (76.7)	284,663 (20.6)	30,198 (2.2)
1984	1,410,371	1,084,334 (76.9)	287,331 (20.4)	30,249 (2.1)
1985	1,447,611	1,125,166 (77.7)	280,366 (19.4)	33,144 (2.3)
1986	1,393,987	1,100,917 (79.0)	253,257 (18.2)	29,794 (2.1)
1987	1,396,997	1,108,348 (79.3)	250,911 (18.0)	28,408 (2.0)
1988	1,394,364	1,120,114 (80.3)	237,207 (17.0)	27,342 (2.0)
1989	1,429,163	1,152,065 (80.6)	236,296 (16.5)	30,326 (2.1)
1990	1,409,240	1,146,731 (81.4)	223,763 (15.9)	28,312 (2.0)
1991	1,389,076	1,126,609 (81.1)	223,164 (16.1)	28,114 (2.0)
1992	1,328,667	1,075,367 (80.9)	213,897 (16.1)	27,934 (2.1)
1993	1,380,334	998,860 (72.4)	192,939 (14.0)	26,117 (1.9)
1994	1,142,988	954,436 (83.5)	154,860 (13.5)	23,891 (2.1)
1995	1,097,658	912,955 (83.2)	150,858 (13.7)	24,494 (2.2)
1996	1,059,770	822,784 (77.6)	145,781 (13.8)	21,570 (2.0)
1997	1,034,388	855,038 (82.7)	146,928 (14.2)	22,007 (2.1)
1998	939,068	771,740 (82.2)	136,127 (14.5)	20,750 (2.2)

Source: Calculated from *MTGYNJ* (various years).

Note: The figures in parentheses are percentages of total coal production labourers. Another category not included in the table performed a variety of other tasks associated with cutting and hauling, such as monitoring safety and carrying out rescue work.

Table 6.23 ILOR 1982–98

	Percentage change		ILOR
	CMA output	CMA labourers	
1982	4.43	5.55	1.25 : 1
1983	3.78	0.57	0.15 : 1
1984	8.70	2.07	0.24 : 1
1985	2.93	2.64	0.90 : 1
1986	1.89	−3.70	−1.96 : 1
1987	1.52	0.22	0.14 : 1
1988	3.39	−0.19	−0.06 : 1
1989	5.49	2.50	0.46 : 1
1990	4.78	−1.39	−0.29 : 1
1991	0.08	−1.43	−17.88 : 1
1992	0.40	−4.35	−10.88 : 1
1993	−5.08	3.89	0.77 : 1
1994	2.32	−17.19	−7.41 : 1
1995	−2.90	−3.97	−1.37 : 1
1996	11.40	−3.45	−0.30 : 1
1997	−1.51	−2.40	1.59 : 1
1998	−4.85	−9.22	1.90 : 1

Source: Calculated from *MTGYNJ*, various years.

increasing. Each unit increase in output required less labour. OMS at the CMAs improved some 139 per cent from 1980 to 98, from 0.912 to 2.180 tons.[144] The output of face workers increased over the same period from 4.129 tons per shift to 9.511, and the average monthly output per coalface increased from 11,220 tons in 1979 to 23,514 in 1998. Yet the overall average output per worker in China's mines compared to that in other countries remained low. OMS in 1994 was 7.17 tons in the UK, 33.53 in Australia, though only 0.55 in India in 1989–90.[145] The official *China Daily* observed: 'In 1994 China employed 7 million people to produce 1.23 billion tons of coal. In the US, 100,000 people mined 923 million tons, and in Germany 145,000 people mined 265 million tons ... The respective employment and production figures remain in sharp contrast.'[146] The 7 million figure for China includes not only CMA and LS mine workers, but also the farmers operating with no mechanization at the LNS mines. However, even when China's CMA output and employment figures alone are compared, as in Table 6.24 with the total figures of other countries, the difference in productivity is still very wide.

OMS at the LS mines was much lower at 0.583 tons in 1985 and 0.795 in 1994. Several thousands of employees were transferred out of these mines to diversified activities as well. Although it was illegal, women and children did work in state mines, and there were also reports that children had been kidnapped to work in LNS mines. One such report was horrific: 'Official media reported this week

Table 6.24 International coal mining productivity data, 1997

	Output (mt)	Labourers	Productivity ('000 tons per manyear)
Australia	217.3	25,900	8.390
Canada	78.8	8,900	8.854
Columbia	15.4	3,800	4.053
China (CMAs only)	529.16	3,157,066	0.168
Germany	46.5	78,100	0.595
India	200.89	548,100	0.378
Poland	139.0	251,200	0.553
South Africa	221.6	57,200	3.874
UK	31	9,300	3.333
US	911	79,400	11.474

Source: *MTGYNJ 1998*, p. 34; International Energy Agency, *Coal Information 1998*, Paris: OECD/IEA, 1999, pp. I-224-5; Sunit Kumar Sarkar and Subhankar Sarkar, *State of Environment and Development in Indian Coalfields*, Calcutta: Oxford and IBH, 1996, p. 8; India, Planning Commission, *Eighth Five Year Plan 1992–97, Vol. II, Sectoral Programmes of Development*, New Delhi: The Commission, 1992, p. 174.

Notes: The Chinese figures are for underground and opencast production combined. The Indian data is for 1989.

more than 100 youngsters were rescued from small mining operations in central China's Hunan Province after they were kidnapped, forced to work and supplied little food or water.'[147]

According to official sources peak employment at the township and village mines was 2,123,722 in 1996.[148] In 1997 there were 1,828,500 employed at such mines, 834,100 at 6,886 township mines and 994,400 at 18,856 village mines.[149] This means there were about 121 per township mine and 53 per village mine.

The average wages paid to state mine workers, compared to those for other industrial workers, decreased over the reform years. Whereas on the eve of the reform era they had been the highest paid work force, they were fourth in 1984, twelfth in 1985, third in 1986, and eleventh in 1987.[150] The average annual wage of underground miners at the CMAs rose from 3,764 yuan in 1992 to 7,900 in 1997, and it was planned that it would reach 10,000 by 2000.[151] The average wage of a surface worker was about half that of a miner's wage.[152]

Compared to the pre-reform years, the proportion of labour costs in total production costs rose substantially. The contract system, which proved to be highly successful when it was introduced, gave the miners a fixed monthly wage and bonuses if targets were surpassed. The differences in pay among the various ranks of workers increased and acted as an incentive for those at the low end of the scale to work towards promotion. In addition, and generally speaking, the contract system facilitated greater communication between upper and lower levels of management.

As noted in the Financing section of this chapter, wages at several mines were repeatedly not paid on time, creating tremendous hardship for mining families. In some cases this was due to triangular debt, in others, to corrupt mine officials. In the case of the former, payment procedures were given to the banks to handle, while in the latter, the Minister of Coal, Wang Senhao, ordered mine managers to take the blame and resign.[153] Groups of unpaid miners demonstrated frequently, and these protests continue at the time of writing.

In many western countries coal miners' unions have been powerful social and political forces capable of virtually paralysing the heavy industry sector. This has never been the case in China, and the guarantee of the right to strike was dropped from the constitution in 1982.[154] The 1993 Regulations for Handling Enterprise Labour Disputes, the 1995 Labour Law, and a new Trade Union Law widened the mandate of unions, but they nonetheless remained under the All-China Federation of Trade Unions, which discouraged campaigning *en masse* for common causes. The new laws aimed at protecting the interests of both the employees and managers. For example, those managers who failed to pay wages and pensions on time would face stiff penalties. However, collective bargaining as practised in the West is still a long way off, if not impossible in the present political environment. This does not prevent frequently recurring local protests. Market forces were playing a greater role in determining wages and employment terms, and union leaders were often enjoying more support from the average worker than were the Party leaders. By comparison, trade unions as well as political parties have been highly vocal in India's coal mining states and in late 1998 were protesting vehemently against restructuring of Coal India Ltd. which monopolizes national production.[155]

The fact that the majority of the miners at the LNS mines had little or no training resulted in appalling accident rates, and the government in the 1980s and early 1990s was understandably not keen to disclose much information. Beginning with the 1996 edition, however, the *Coal Industry Yearbooks* did report accident figures. The Chinese version of the *Yearbooks* give much more information than the English, though it is necessary to combine figures presented in tables to arrive at total figures more accurate than those given in the text. Figures cited by the Chinese and foreign press not only disagree with each other, but with the *Yearbooks*.

According to the *1994 Meitan gongye nianjian* (1994 Coal Industry Yearbook), between 1984 and 1993 a total of 41,000 people died in LNS mines alone, meaning that there was an average of 4,100 deaths per year (at least ten each day), and 12.12 per million tons mined.[156] The deaths at these mines accounted for 63.3 per cent of the total for all mines over that period, and the rate was 4.4 times that at the CMAs. Considerable progress was made in reducing the total death toll for all mines from 10,400 in 1990, to 9,819 in 1991, 9,384 in 1993, 7,271 in 1994, 9,067 in 1995, 6,596 in 1996, 6,155 in 1997, and 6,314 in 1998.[157] However, this rate was still extremely high, and these official figures may greatly understate the true picture. As at least a third of the LNS mines operated illegally, it is possible that dozens of deaths went unreported. The national average fatality rate per million tons mined decreased from 8.17 in 1980 to 4.55 in 1996.[158] In 1985 it was 3.84 at the CMAs, 9.40 at the LS mines, and 11.9 at the LNS.[159] By 1996 the rates were down to 1.168, 4.3, and 8.1, respectively.[160] For comparison, in 1993 the number of fatalities and the fatality rate per million tons were 47 and 0.06 in the US, 2 and 0.28 in Japan, 3 and 0.04 in the UK, 45 and 0.25 in South Africa, 59 and 0.45 in Poland, 295 and 0.96 in Russia, 381 and 3.22 in Ukraine, and 297 and 1.18 in India.[161]

Table 6.25 shows that over 70 per cent of the reported accidents between 1996 and 1998 occurred in the LNS mines, 15 per cent in the LS mines, and 10 per cent in the CMAs. According to the official data, about 54 per cent of the accidents were caused by gas explosions, 25 per cent by roof falls, 7 per cent by flooding, and 7 per cent by runaway wagons. Not readily apparent is the fact that many of the accidents were caused through ignorance and lax safety measures. Many explosions were the result of workers smoking or using inappropriate lighting.

Not only were miners killed cutting, hauling, and hoisting the coal, the illicit trade of explosives that began as a result of peasant involvement in mining led to tragic deaths in urban areas. In February 1996, over 100 people were killed and 400 injured when an explosion occurred in an apartment building in Shaoyang (Hunan).[162]

China's coal officials were aware of international norms with respect to accident rates. Representatives attended the ILO meeting held in Geneva in April 1985 to draw up a code of practice on safety and health in coal mines.[163] The likely reason why the authorities were slow to address the problem in the late 1980s and early 1990s was that implementation and enforcement of mine safety regulations would have increased the start-up and operating costs of the LNS mines, which were not only so vital to total production but were extremely lucrative. It was difficult to enforce what regulations there were, so competitive had the small operations become with each other and with the larger state mines. Encroachment, for example, was a major problem. The small operations – many illegal – would tunnel through to adjacent large mines to take advantage of the

Table 6.25 Breakdown of coal mine fatalities by cause and mine type: average proportions, 1996–98 (%)

	Total	Roof fall	Gas explosion	Electro- mechanical	Runaway wagon	Explosive blast	Flooding	Fire	Other
CMA	10.0	2.6	4.8	0.5	1.3	0.1	0.2	0.1	0.4
LS	14.8	5.5	5.2	0.5	1.6	0.3	0.9	—	0.7
LNS	70.1	16.0	41.0	0.9	3.2	0.7	5.6	0.7	2.0
Total	100.0	25.1	53.7	2.1	6.7	1.2	7.0	0.8	3.3

Source: Calculated from *MTGYNJ 1997–99*.

Note: Totals do not equate to 100 per cent because fatalities that occurred in basic construction, the manufacture of machinery, and geological prospecting are excluded here.

latter's better ventilation, which generally was mechanized. Any measure that added to production costs, for example, better safety equipment, was unwanted.

The high accident rate at the LNS mines became an excuse to extort huge sums from mine owners and managers.[164] Between 1997 and 2000 up to 100 miners were murdered by gangs who first recruited them to work in mines known to have poor safety records. Later, the gangs would kill the men while they were working underground, make it appear as though an explosion or roof fall had occurred, then pose as relatives to collect compensation. They extorted about 18,000 yuan (US$2,300) per worker from the owners and managers by threatening to expose their poor safety procedures to the local authorities.

In the early 1980s it had been hoped that the educational level of the miners at the CMAs could be raised substantially. By 2000, district team leaders or their equivalents are to have secondary technical training; heads, deputy heads, and department directors are to have college education; and all workers and staff are to have junior middle school. Whether or not these targets will be met, it can be said that average technical competence has already increased.[165]

Despite the limited resources with which to buy computers and software, the Coal Industry Ministry, as most of the other government departments, was quick to employ them for the same purposes as in the West, including statistical analysis, scheduling, inventory, etc., and has kept up well with the developed world in obtaining and applying software innovations as they came available. Though the number of terminals relative to users was extremely small, the sophistication of the usage was high.

The China Coal Research Institute in Beijing, founded in 1957, had 6,833 staff in 1999 working at eighteen branches, institutes and centres around the country, including the National Coal Water Mixture Engineering Research Centre, Clean Coal Engineering and Research Centre, National Coal Mine Safety Engineering Research Centre, thirteen quality inspection and test centres, and one national mine safety measurement station. The Institute confers doctorates in three specialties and masters in eight. In recent years it has broadened from strictly technical/ scientific concerns to business management.[166]

The Energy Research Institute (ERI) was established in 1980 and works in conjunction with the SPC and Chinese Academy of Sciences (CAS). Its focus is energy economics including supply and demand forecasting for all energy types, energy conservation, energy and the environment, and the development and application of new energy production and saving technologies.[167]

In 1984 the Chinese Coal Science and Technology Database (CSTD) was created.[168] It indexes articles on coal production, consumption, transport, technology, equipment, mining safety, and environmental protection as published in a wide variety of Chinese journals. In 1980 the ERI initiated the Energy Database of China (EDBC), which contains data on the reserves, production, processing, transportation, distribution, consumption, and conservation of all energy types. An Electricity Abstracts Database (EAD), which indexes 120 periodicals concerned with energy development policy, long-term energy planning, as well as the technical aspects of electricity generation, transmission, and distribution was created in 1987. Dozens of others were set up in the 1990s, and at the time of writing many government and government affiliated agencies involved with coal production have web pages.[169]

Financing

Viability of coal mining

The debt spiral

In the late 1990s most governments around the world were subsidizing their coal industries either through grants, price supports, limitations on coal imports, or long-term agreements between producers and large consumers.[170] Only the Australian, South African, and Colombian coal industries were making sizable profits. In 1997 subsidies per ton of coal equivalent produced were US$125 in Germany, US$139 in Japan, US$79 in Spain, US$82 in Turkey, and US$17 in the UK.[171]

It was announced in 1998 that state subsidization was to end for the entire state sector by 2000. After massive cutbacks and restructuring, 28 of the 40 major industrial sectors, including the textiles, building materials, nonferrous metals, and railway industries all managed to enter 'the black' in 1999, a year ahead of schedule.[172] The coal industry, however, was still in debt. This should not, however, prompt summary negative assessment of the reforms in the industry. Sound appraisals will not be possible until there is greater depth of data. It was only in 1997 that the provinces took over the CMAs – thereafter called 'state priority mining administrations' (SPMAs) – and as recently as early 1999 that many coal enterprises were ordered to declare bankruptcy. Furthermore, like most of the heavy industrial sector, the coal industry is trapped in a web of triangular debt.

In the late 1970s and early 1980s about 60–70 per cent of the CMAs, and about 50 per cent of the LS mines had deficits and the government routinely bailed them out.[173] Table 6.26 indicates that in 1985 it was only the state mines in Shanxi and Hebei that made profits. Jilin had by far the largest losses in total and per ton of output. Nationwide the industry needed a subsidy of 350 million yuan to keep operations going that year.[174] This was far from a good situation, but the mine officials believed, as did the officials of all key industries, that the government would never allow the industry to collapse.

In 1987 an average of 20.3 yuan in profits and taxes was created per 100 yuan of government funding at state owned industries. For the coal industry, however, this figure was -1.91 yuan.[175] That year 72.8 per cent of the CMAs and 53 per cent of the LS mines had deficits, and in the following year 93 per cent of the CMAs had deficits.[176] The total losses at the CMAs fell from their worst, 5.75 billion yuan in 1992, to 610 million in 1996 (see Table 6.27), and it is reasonable to assume that the contract system, the assignment of excess labour to diversified activities, and other streamlining measures, and most importantly price reform, had some part in this. Then in 1997 it was reported that the

Table 6.26 Output, production costs, and losses/profits at CMA and LS mines in each province/area, 1985

Province/ Area	State mines			Local mines		
	Output (mt)	Production cost (yuan/ton)	Loss/ profit (million yuan)	Output (mt)	Production cost (yuan/ton)	Loss/ profit (million yuan)
National Total	40,626.0	29.33	−55,986.5	18,277.7	26.78	−8,048.08
Beijing	597.5	32.81	−2,408.8	14.3	21.22	9.40
Hebei	3,975.5	28.84	1,592.6	971.1	24.73	1,033.78
Shanxi	8,563.5	24.49	26,819.6	3,160.5	16.81	12,632.49
Inner Mongolia	2,065.2	25.69	−552.4	454.6	13.51	469.13
Liaoning	3,411.9			338.1	33.16	−40.20
Jilin	1,177.4	32.15	−41,202.2	370.7	30.72	−3,645.21
Heilongjiang	3,887.0			817.0	26.65	−1,067.30
Jiangsu	1,561.0	29.73	−1,275.6	442.8	47.64	−3,921.27
Zhejiang				134.5	47.94	−69.98
Anhui	2,387.9		−6,592.8	262.7	34.84	−1,053.50
Fujian				397.6	32.20	230.80
Jiangxi	645.7	31.11	−1,025.6	561.3	34.50	456.79
Shandong	2,669.1	33.08	−2,556.3	1,325.6	32.61	−108.00
Henan	3,751.7	28.41	−2,405.1	1,727.9	25.53	4,459.23
Hubei				341.9	35.49	1,580.90
Hunan	285.4	39.44	−5,493.6	1,190.9	38.12	2,937.63
Guangdong				477.6	40.66	2,226.09
Guangxi				450.4	32.21	−3,028.00
Sichuan	1,610.5		−5,304.4	1,551.5	27.64	−5,048.92
Guizhou	716.4	30.00	−9,698.3	307.7	21.95	−506.50
Yunnan				890.3	19.06	−1,562.00
Tibet				3.0		
Shaanxi	1,434.9	28.14	−6,004.1	451.9	15.75	−428.66
Gansu	593.5	30.17	−665.3	312.7	22.12	−186.00
Ningxia	898.2	23.27	−326.4	220.2	21.26	−1,381.23
Qinghai				242.6	19.19	700.80
Xinjiang	393.6	36.67	−38.5	858.3	16.75	750.89

Source: Shanxi shehui kexue yuan, meitan jiage yanjiu keti zu (Shanxi Academy of Social Sciences, Coal Pricing Research Group), *Meitan jiage gaige yu Shanxi meitan jiage* (Coal Pricing Reform and Shanxi Coal Pricing), an unpublished research report prepared by the Academy, Taiyuan, Aug. 1986, p. 53.

industry as a whole was finally able to record profits, 200 million yuan worth, from sixty-five SPMAs.[177] It was hoped that the government finally no longer had to subsidize this industry and could devote more attention to making others self-sustaining.

Table 6.27 Losses at the CMAs, 1992–99

	Losses (billion yuan)	Per cent of CMAs making profits	Location of profit-making CMAs
1992	5.75	9.7	9 all in Shanxi
1993	3.26	19.4	18 in Shanxi, Shandong Anhui, Jiangsu
1994	1.97	26.9	25 in Shanxi, Shandong, Anhui, Jiangsu, Henan, and Guizhou
1995	1.03	28.0	26
1996	0.61	59.0	55
1997	4.20	72.0	
1998	3.71		
1999	2.28		

Sources: See Notes 177–183; Introductions to *MTGYNJ 1995* and *MTGYNJ 1996*; 'Wen shichang jingji tizhi zhuanbian maichu liao xin bufa' (New Forward Steps in Market Economic System, *MTB*, 8 Jan. 1995, p. 1; 'Chuanguo meitan gongye gongzuo huiyi zhaokai' (Summoning of National Coal Industry Work Meeting), *MTB*, 10 Jan. 1995, p. 1; 'Meitan gongye niu kui zeng ying you chengxiao' (The Effects of Loss Reductions in the Coal Industry), *RMRB*, 7 May 1996, p. 2; 'Fuzhong fenjin wangqiang tujin liang ge genben xing zhuanbian shixian niukui zeng ying mubiao' (Carry Heavy Burden, Fight Hard, Push Forward Two Basic Transformations, Achieve Elimination of Losses Target), *MTGYNJ 1997*, p. 68; *GYFZBG 1998*, p. 159; According to Li Peng, 'Zhongguo de nengyuan zhengce' (China Energy Policy), *RMRB*, 29 May 1997, pp. 1–2, losses amounted to 400 million yuan in 1996.

Early in 1998, however, it was found that the mine accounts had been falsified, that payments for coal purchases had not, in fact, been received.[178] Actual debts at the end of 1997 were 4.196 billion yuan and 28 per cent of the SPMAs had losses.[179] In 1998, 81 per cent had losses totalling 3.705 billion yuan.[180] It became impossible to cover production costs when during the first four months of 1999 the price of coal fell by 15.4 yuan per ton. Total income of the SPMA sector at the end of the year was down almost 6 billion yuan.[181] Seventy-six of the SPMAs had deficits amounting in total to 2.28 billion yuan (US$274.6 million).[182] In May 2000, 86 per cent had debts of 1.85 billion yuan.[183]

While some individual operations were apparently financial successes, for example, the Xinji mines (Anhui) (see Chapter 5), the SPMAs as a whole were severely burdened by the need to support over one million retired workers, the total cost of pensions for whom was some 6 billion yuan per year, as well as community facilities and services for the employed.[184] In 1997 the state mines operated 1,268,924 residential buildings, 169,148 collective dormitories, 103,286 education buildings, 15,905 cultural and sports buildings, and 66,533 medical buildings.[185]

The financial picture of the LNS mines was very different. When the ledgers of the state mines were at their bleakest, thousands of small mines were earning huge profits. The low-cost coal was in very high demand, and the taxes and fees levied on the mines helped many previously destitute county governments lift themselves from grinding poverty.[186]

Many rural communities took immediate advantage of the government's incentives to open mines and the coal therefrom quickly became an indispensable input in the phenomenal expansion of rural industrial activity, especially the production of agriculture-oriented equipment and supplies such as chemical fertilizer, farm tools, machinery, and construction materials.[187] Hundreds of thousands of jobs were created in the establishment of mines and industries. In 1978 the average per capita income in Zuoyun County (Yanbei Prefecture, Shanxi Province) was 72 yuan, but with the development of local coal production, by 1984 it had reached 816 yuan. In 1988 coal production accounted for 80 per cent of the county's income.[188] It was estimated that in 1983 the taxes from the local mines in Shanxi Province as a whole provided the local government with 37.5 per cent of its income, and in 1999 that 15 per cent of each farmer's income in the Province came from coal mining.[189]

The income generated was not solely beneficial to townships with coal, but to the rural population as a whole. More of the central government's limited funding could now go to other still backward areas. The benefits of the small mines, however, must be weighed against the problems discussed in other sections of this chapter, such as the thousands of deaths and injuries that took place each year, the subsidence of land, destruction of natural drainage courses, piles of toxic, flammable wastes, and the very low quality of the coal causing dangerous levels of air pollution.

State investment and taxation

Between 1985 and 1999 the coal industry (extraction and processing) received a total of 1,945 billion yuan in state capital construction investment, compared to 3,161 billion for the oil industry (extraction and refining), and 13,443 for the electricity industry (thermal, hydro, and nuclear) (see Table 6.28). Investment in each of the energy sectors fell sharply in 1981 and 1982 due to the general stagnation of the economy at that time.

Initially, the allocations to the coal industry were larger than those to the oil industry. By 1987, however, the government had decided to rely on the LNS mines for an increasing share of total production and state funding to the coal industry decreased relative to that allocated to the oil industry. The margin between allocations to the electricity and coal sectors also grew wider through the 1980s and 1990s. By 1999, investment in electricity was over seventeen times that in coal.[190] There is no doubt that the CMAs, now in direct competition with the LNS mines, had problems adjusting to reduced investment allocations. Long-term capacity development plans had to be changed. Over the Seventh Five

Table 6.28 Capital construction investment, 1976–99 (current million yuan)

	Total industrial investment	Total energy investment	Coal investment	Oil investment	Electricity investment
1976	208.73	69.61	16.54	19.09	33.98
1977	217.36	78.06	22.58	20.76	34.72
1978	273.16	113.83	31.80	31.12	50.91
1979	256.85	109.92	31.86	27.07	50.99
1980	275.61	114.99	33.47	33.38	48.14
Fifth Five Year Plan	1,231.71	486.41	136.25	131.42	218.74
1981	216.01	91.24	23.15	27.95	40.14
1982	260.60	101.38	29.85	25.30	46.23
1983	282.28	126.55	40.07	29.02	57.46
1984	341.59	162.76	55.14	30.63	76.99
1985	446.49	209.88	55.11	33.17	121.60
Sixth Five Year Plan	1,546.97	691.81	203.32	146.07	342.42
1986	531.64	273.94	56.70	38.61	178.63
1987	682.79	348.46	59.60	58.55	230.31
1988	796.09	423.79	63.50	86.45	273.84
1989	822.48	454.42	70.51	93.51	290.40
1990	952.60	565.41	98.82	100.66	365.93
Seventh Five Year Plan	3785.60	2,067.02	349.13	377.78	1,339.11
1991	1,147.21	668.43	116.58	125.91	425.94
1992	1,458.31	840.61	126.85	157.98	555.78
1993	2,004.45	1,152.31	153.05	230.65	768.61
1994	2,761.67	1,578.86	164.98	263.09	1,150.79
1995	3,236.34	1,747.68	189.36	299.83	1,258.49
Eighth Five Year Plan	10,607.98	5,987.89	750.82	1,077.46	4,159.61
1996	3,724.24	2,106.11	216.54	343.60	1,545.97
1997	4,119.51	2,694.71	266.03	489.79	1,938.89
1998	4,169.15	2,774.53	182.01	447.97	2,144.55
1999	3,850.09	2,709.48	124.88	391.42	2,193.18

Sources: *TJNJ*, various years; *NYTJNJ 1996*, p. 27; *GDTZZL*, pp. 79–87.

Note: The oil investment figures include refining. The electricity data was adjusted upwards after 1984.

Year Plan a total of 180 million tons of new capacity was to be developed at the CMAs and 79 million tons at the LS mines, but actual development over the five years was only 111 and 40 million tons respectively.[191]

Government capital construction investment allocations to the industry with no payback requirement of any kind continued, as they had in the pre-reform years, until 1985. In an attempt to make the industry more accountable, the allocations were first changed to loans, then to credit contracts which were signed with the State Development Bank, established in 1993, and with individual enterprises.[192]

Total energy investment as a proportion of total industrial investment rose steadily from 47 per cent in 1985 to 59 per cent in 1990, thereafter decreased to 54 per cent in 1995, but climbed to over 70 per cent by 1999 (see Table 6.29). Investment in the coal industry as a proportion of total energy investment fell almost steadily from 26 per cent in 1985 to less than 5 per cent in 1999 (see Table 6.30). As has been pointed out above, the picture for coal is skewed by the increasingly heavy reliance on the LNS mines. Investment in the oil industry ranged from 14 to 20 per cent over these years, while the proportion devoted to electricity very much dominated total energy investment, increasing its share from 58 per cent in 1985 to over 80 per cent in 1999.

Reflecting the sizable yearly changes in allocations, capital construction investment in constant prices required per ton of output varied considerably over the 1981–99 period (see Table 6.31). The ICORs are also highly uneven. Between 1978 and 1999, the number of years when investment in real terms increased over the previous year is much smaller than the number of years in which it decreased. The years 1982–84, 1990, and 1997 are exceptional in that there were particularly large increases in investment.

Over the 1980s and most of the 1990s the central government went from providing almost all of the coal industry's funding, to providing less than 15 per cent in 1997, and only 2.9 per cent in 1998, by which time control over the key mines had been devolved to the provincial authorities (see Table 6.32). A great many other interests had become involved in the industry after Coal Minister Wang Senhao announced in 1994 his expectation that the largest coal consumers would become principal investors in the coal industry:

> According to 'he who benefits is he who invests' encourage power, metallurgical, chemical, railways, etc., and their subordinate industries, to invest in coal mines; also, areas lacking in coal should invest in coal rich areas ...[193]

The contribution of collective governments increased from only 4.0 per cent in 1979 to over 15 per cent in 1998. The role of local governments in providing equipment and materials for the mines also increased over the 1980 and 1990s. In 1998 over one-fifth of the equipment and materials were procured locally, as opposed to being supplied through central channels.

There were fluctuations in how the capital construction investment funds were used (see Table 6.33). Between 1979 and 1998 the proportion devoted to construction and installation varied from a high of 68 per cent in 1981 to a low of 47 per cent in 1997, while that devoted to equipment, tools and apparatus varied from 17 to 37 per cent. The productive : non-productive investment ratio, reflecting how priorities changed with changes in the availability of funds, varied from 69 : 31 to 90 : 10. Table 6.34, which gives the percentage breakdown of capital construction investment, indicates increases over the 1980s and 1990s in the proportion of funding devoted to the construction of mines, coal washing plants,

Table 6.29 Capital construction investment in energy as a proportion of total investment in industry, 1976–99

	As % of total industry investment			
	Total energy investment	Coal investment	Oil investment	Electricity investment
1976	33.3	7.9	9.1	16.3
1977	35.9	10.4	9.6	16.0
1978	41.7	11.6	11.4	18.6
1979	42.8	12.4	10.5	19.9
1980	41.7	12.1	12.1	17.5
Average Fifth Five Year Plan (1976–80)	39.5	11.1	10.7	17.8
1981	42.2	10.7	12.9	18.6
1982	38.9	11.5	9.7	17.7
1983	44.8	14.2	10.3	20.4
1984	47.6	16.1	9.0	22.5
1985	47.0	12.3	7.4	27.2
Average Sixth Five Year Plan (1981–85)	44.7	13.1	9.4	22.1
1986	51.5	10.7	7.3	33.6
1987	51.0	8.7	8.6	33.7
1988	53.2	8.0	10.9	34.4
1989	55.2	8.6	11.4	35.3
1990	59.4	10.4	10.6	38.4
Average Seventh Five Year Plan (1986–90)	54.6	9.2	10.0	35.4
1991	58.3	10.2	11.0	37.1
1992	57.6	8.7	10.8	38.1
1993	57.5	7.6	11.5	38.3
1994	57.2	6.0	9.5	41.7
1995	54.0	5.9	9.3	38.9
Average Eighth Five Year Plan (1991–95)	56.4	7.1	10.2	39.2
1996	56.6	5.8	9.2	41.5
1997	65.4	6.5	11.9	47.1
1998	66.5	4.4	10.7	51.4
1999	70.4	3.2	10.2	57.0

Source: Calculated from Table 6.28.

Table 6.30 Shares of capital construction investment in energy, 1976–99 (%)

	As % of total energy investment		
	Coal investment	*Oil investment*	*Electricity investment*
1976	23.8	27.4	48.8
1977	28.9	26.6	44.5
1978	27.9	27.3	44.8
1979	29.0	24.6	45.4
1980	29.1	29.0	41.9
Average Fifth Five Year Plan (1976–80)	28.0	27.0	45.0
1981	25.4	30.6	44.0
1982	29.4	25.0	45.6
1983	31.7	22.9	45.4
1984	33.9	18.8	47.3
1985	26.3	15.8	57.9
Average Sixth Five Year Plan (1981–85)	29.4	21.1	49.5
1986	20.7	14.1	65.2
1987	17.1	16.8	66.1
1988	15.0	20.4	64.6
1989	15.5	20.6	63.9
1990	17.5	17.8	64.7
Average Seventh Five Year Plan (1986–90)	16.9	18.3	64.8
1991	17.4	18.8	63.8
1992	15.1	18.8	66.1
1993	13.3	20.0	66.7
1994	10.4	16.7	72.9
1995	10.8	17.2	72.0
Average Eighth Five Year Plan (1991–95)	12.5	18.0	69.5
1996	10.3	16.3	73.4
1997	9.9	18.2	71.9
1998	6.6	16.1	77.3
1999	4.6	14.4	80.9

Source: Calculated from Table 6.28.

Table 6.31 Investment per ton and ICOR, 1978–99

	Investment per ton	Annual per cent change		ICOR
		CMA production	Capital Investment	
1978	93.0	15.77	40.83	2.59
1979	87.3	4.66	−1.76	
1980	92.6	−3.74	−0.90	
1981	62.4	−2.71	−32.46	
1982	75.6	4.43	26.54	5.99
1983	96.4	3.78	32.28	8.54
1984	118.7	8.70	33.86	3.89
1985	105.9	2.93	−8.18	
1986	100.9	1.89	−2.95	
1987	97.4	1.52	−2.01	
1988	84.6	3.39	−10.12	
1989	75.6	5.49	−5.71	
1990	99.1	4.78	37.24	7.79
1991	113.5	0.08	14.65	183.13
1992	116.7	0.40	3.26	8.15
1993	131.1	−5.08	6.59	
1994	113.5	2.32	−11.41	
1995	110.3	2.90	−0.02	
1996	106.7	11.40	7.78	0.68
1997	132.0	−1.51	21.88	
1998	97.5	−4.85	−29.76	
1999	69.2	−0.44	−29.27	

Sources: Calculated from Tables 6.3 and 6.28 using a price index from *TJNJ 2000*, p. 290.

Note: The investment per ton figures were calculated by dividing the annual investment allocations to the state mines by the output of these mines.

and subsidiary plants. The shares of investment devoted to general engineering, including railway spurs from which loaded railway cars from the mine districts could be transferred to the national trunk lines, increased through 1995, but tapered thereafter.

Soon after the reforms were launched experiments were discussed which would replace direct profit transfers of the SOEs to the central government budget with income taxation, and this did eventuate in 1984. By 1988, the majority of SOEs were operating on a contractual basis and thereby had assumed greater responsibility for their own funding. Income tax for all businesses, including all of the new coal corporations was levied at the rate of 33 per cent. In 1989, to help preserve central government revenue, extrabudgetary enterprise funds were subjected to a 10 per cent levy. As of 1994 all mining corporations were subject to the Resource Tax at 0.3–5 yuan per ton, and the Mineral Resource Compensation

Table 6.32 Sources of capital construction investment funds and procurement of equipment and materials, 1979–98 (%)

	Source of funds		Procurement of equipment and materials	
	State	Collective	Items supplied centrally	Items supplied locally
1979	96.0	4.0		
1980	58.4	5.7		
1981	65.3	8.9		
1982	59.9	7.7		
1983	71.3	6.3		
1984	74.2	7.5	86.4	13.6
1985	61.1	9.3	85.7	14.3
1986	70.7	7.9	85.9	14.1
1987	63.9	8.2	88.5	11.6
1988	65.9	8.6	89.4	10.6
1989	60.1	10.0	87.5	12.5
1990	52.1	10.6	89.1	10.9
1991	54.9	8.7	88.5	11.5
1992	37.5	7.2	88.5	11.5
1993	32.5	8.5	86.0	14.0
1994	18.9	8.4	84.1	15.9
1995	21.9	10.9	82.4	17.6
1996	18.7	12.0	78.3	18.9
1997	14.3	10.8	86.4	13.6
1998	2.9	15.1	79.4	20.6

Source: *MTGYNJ*, various years.

Note: Equipment and materials procurement data for the 1979–83 period is unavailable.

Fee equivalent to 1 per cent of the annual sales revenue adjusted according to a mining recovery rate coefficient that factored the variance between actual output and the approved mine design capacity.[194] Tax avoidance by enterprise managers became endemic in the early 1990s, adding to the corruption engendered by the two-track pricing system. In order to help reduce the industry's debt, the value-added tax rate on coal products was reduced from 17 to 13 per cent in March 1994.

Deregulating prices

At the outset of the reforms, a main priority of the government was to transform loss-making industries, including the coal industry, into profitable operations. A crucial step was to bring commodity prices up closer to international levels. The difference between the international coal price and the average price in China was nearly 500 per cent.[195] In the early years, however, out of fear of uncontrollable

Table 6.33 Capital construction investment spending breakdown, 1979–98 (%)

	Construction and installation	Equipment, tools and apparatus	Others	Productive investment	Non-productive investment
1979	59.2	30.5	10.3	80.0	20.0
1980	54.6	36.8	8.6	83.4	16.6
1981	68.0	17.0	14.9	72.5	27.5
1982	67.0	17.0	16.0	68.5	31.5
1983	65.0	18.1	16.9	69.2	30.8
1984	57.4	28.2	14.4	73.8	26.2
1985	62.0	22.0	16.0	71.7	28.3
1986	62.9	22.5	14.6	75.6	24.4
1987	64.2	22.8	13.0	77.0	23.0
1988	65.7	20.2	14.1	78.5	21.5
1989	63.2	22.0	14.8	81.8	18.2
1990	61.7	23.5	14.8	83.7	16.3
1991	60.4	22.9	16.7	82.2	17.8
1992	60.6	22.4	16.9	82.2	17.8
1993	58.9	22.1	18.9	82.6	17.4
1994	58.6	24.2	17.2	79.5	14.5
1995	52.1	20.6	27.3	81.7	18.3
1996	49.8	20.1	30.2	86.9	13.1
1997	47.0	21.7	31.3	85.7	14.3
1998	49.3	18.5	32.2	90.1	9.9

Source: *MTGYNJ*, various years.

Note: Productive capital construction investment refers to investment devoted to mining, while non-productive capital construction investment is for commercial, communications, and residential facilities.

inflation, the government made relatively small increases in the price of distributed coal (under plan).

Chinese economists carefully watched the price reforms of other centrally planned economies. Between 1981 and 1985 Hungary increased the price of coal, gas, and oil by 53 per cent, and in November 1987 Yugoslavia increased the price of coal 62 per cent and of electricity 69 per cent. On 1 April 1988 the price of coal for domestic use in Poland was increased by 200 per cent.[196]

The authorities knew that to avoid uncontrollable inflation, reductions in standards of living, and industrial defaulting, comprehensive adjustment of the *entire* pricing system of the economy would have to be implemented.[197] A key difficulty for the coal industry was the sequence of price reform across the industrial sector. During the 1980s and early 1990s the prices of materials required for mining, such as steel, timber, and cement, half of which were allocated to the coal industry at low fixed prices and the other at market prices, were allowed to rise much more quickly than coal prices. This made it almost impossible for mines built in comparatively hard conditions, such as in the Third Front Region, to function.

Table 6.34 Breakdown of capital construction investment, 1981–98

	1981	1983	1985	1987	1989	1991	1992	1993	1994	1995	1996	1997	1998
Construction at mine sites													
(a) Mines	66.3	67.3	66.4	66.0	66.7	64.5	62.6	62.3	63.8	66.0	71.4	74.4	71.8
(b) Coal washing plants	2.9	6.5	4.3	7.2	8.9	5.3	4.9	5.8	5.1	6.3	6.4	5.1	4.4
(c) Subsidiary plants		2.1	2.8	3.6	5.3	8.0	7.5	6.7	7.9	7.4	6.6	7.0	7.2
(d) General	5.8	5.1	7.9	6.6	7.7	11.4	13.0	12.9	11.2	11.7	8.6	5.3	3.7
Engineering of which: Railway spurs	3.5	2.7	3.4	3.7	4.9	7.7	8.0	7.6	7.4	8.9	6.2	4.1	3.0
(e) Social welfare, cultural, educational, and health facilities	5.2	5.9	6.0	5.1	4.1	4.0	4.9	4.5	3.6	2.7	1.3	1.6	1.6
(f) Others	2.9	3.2	2.7	1.5	0.9	1.8	1.7	1.7	1.5	1.3	1.0	1.1	0.9
Construction of Bases	5.0	3.2	0.9	1.0	1.2	1.0	0.8	0.9	1.5	0.4	0.7	0.7	0.2
Coal mine machinery plants	0.4	0.6	1.2	1.2	0.6	0.5	0.6	0.5	0.4	0.3	0.1	0.2	0.1
Design	0.5	0.3	0.6	0.5	0.3	0.3	0.2	0.2	0.3	0.1	0.1	0.1	0.2
Geology	1.9	1.4	1.3	1.2	0.7	1.2	1.5	1.5	0.3	0.5	0.6	0.6	0.9
Scientific research	0.7	0.8	1.1	1.0	0.5	0.2	0.2	0.2	0.3	0.2	0.4	0.6	0.6
Colleges and schools	2.3	2.0	2.5	2.3	1.4	0.6	0.6	0.6	0.6	0.7	0.7	0.6	1.1
Others	1.9	1.5	2.3	2.8	1.6	1.2	1.5	2.2	3.7	2.4	2.5	2.8	7.5

Source: *MTGYNJ*, various years.

Table 6.35 Average production cost and price of coal, 1978–86 (yuan per ton)

	Production cost	Ex-factory price
1978	16.12	19.27
1979	17.66	23.02
1980	19.87	25.45
1981	21.38	25.67
1982	21.74	25.73
1983	24.09	26.38
1984	25.30	27.21
1985	31.33	31.96
1986	34.33	34.76

Sources: Tian Yuan and Qiao Gong (eds), *Zhongguo jiage gaige yanjiu, 1984–1990* (Research on Price Reform in China, 1984–1990), Beijing: Dianle zhengye chubanshe, 1991, p. 391; Cheng Zhiping (ed.), *Zhongguo wujia wushi nian, 1949–1998* (Fifty Years of Prices in China, 1949–1998), Beijing: Zhongguo wujia chubanshe, 1998, pp. 1086–7.

Note: Beginning in 1983 two yuan were added to the maintenance component of production costs.

Table 6.35 shows that in the mid-1980s the differential between production costs and prices narrowed to almost zero. The average increase in coal production costs between 1985 and 1992 was 15.7 per cent, and 12.7 per cent between 1993 and 1995.[198] The differential between production costs and prices was especially great in Sichuan. In 1985 the average production cost was 27.26 yuan while the ex-factory price was 20.84 yuan. In 1990 the respective figures were 66.29 and 33.34 yuan, and in 1993 they were 125 and 80.61 yuan.[199]

Another major dilemma was the increasing differential between state and non-state mine prices. As the state mines modernized they regularly increased prices to keep them above the production costs. This made them less and less competitive compared to the LNS mines which were proliferating rapidly and had lower production costs because they generally did not invest in the safety and environmental protection equipment that the larger mines had to have by law. As the prices of more commodities were deregulated consumers behaved as in a market economy and opted for the cheapest inputs, including coal. The result was stockpiles of unsold state coal that became unmanageably large.

A significant component of production costs was wages, accounting for over 30 per cent, compared to 5 or 6 per cent for most of the other heavy industries. The several industrial wage increases called for by the government in the 1980s and 1990s in an attempt to keep up with the rapid increases in the cost of living hit the coal industry much harder partly because the coal industry workers were already at the top of the scale in 1979. In 1993 alone they were hiked 14 per cent.

Also contributing to the higher production costs were the increased mechanization, the digging of deeper shafts, and devaluation of the yuan. Much more electricity was needed, half the requirements for which had to be bought at market prices.

In 1992 the average price of coal distributed through state channels was raised by about 20 per cent to 45 yuan. That year the government allowed 22 per cent of distributed coal (73 million tons) to be sold at market prices.[200] The mines at Xuzhou (Jiangsu) and Zaozhuang (Anhui) were the first allowed to deregulate their prices (see section on Eighth Five Year Plan in Chapter 5). Due to the success of marketization in the eastern provinces, it was decided that all major coal mines in the northwest, northeast, and southwest should withdraw from the state allocation system.[201]

Beginning in 1993, the price of coal used in power plants as well as washed coal, and coal used for coking and smelting in the eastern and northeastern regions, and in Hunan, altogether amounting to about 70 per cent of the year's production, were deregulated. This experimentation with price deregulation proved so successful that the authorities were confident enough to begin withdrawing the annual subsidization, which by now amounted to nearly 6 billion yuan, and to liberalize the price of coal nationally.[202] In November the government announced its decision to lift all price controls on coal as of 1 January 1994, one year ahead of schedule.[203]

This pattern of price deregulation in the coal industry was typical of changes occurring in the economy as a whole. Table 6.36 summarizes the radical abandonment of fixed state prices that took place between 1978 and 1993. By 1993 only 5, 10, and 15 per cent of retail commodities, agricultural goods, and capital goods, respectively, were still being sold at state fixed prices.

Table 6.36 Share of goods sold at state fixed prices, 1978–93 (%)

	Retail commodities	Agricultural goods	Capital goods
1978	97	94	100
1992	10	15	20
1993	5	10	15

Source: Nicholas Lardy, *China in the World Economy*, Washington, DC: Institute for International Economics, 1994, p. 11, quoting World Bank, *China Updating Economic Memorandum: Managing Rapid Growth and Transition*, Washington, World Bank, 1993, pp. 49–50, and Niu Genying, 'China's Economic Reform in 1994', *BR*, 37, no. 2 (10–16 Jan. 1994): 10.

Note: Per cent is calculated as a share of transaction volume for each category.

Very significantly, however, there was one exception to the across-the-board coal price deregulation. The government continued to regulate the price of coal sold to electric power plants. In 1996 it stipulated that price increases for coal sold to this sector could not exceed 7 yuan per ton over the November 1995 level for the northern central part of the country excluding eastern Inner Mongolia, 10 yuan for Shaanxi, 9.7 yuan for Heilongjiang, Liaoning and the eastern section of Inner Mongolia, 12 yuan for Jilin, and all other areas 8 yuan.[204] For 1997 they were not to be more than 12 yuan per ton over the 30 September 1996 level, and for 1998 not more than 5 yuan per ton over the October 1997 level.[205]

Figure 6.3 depicts the changes between 1980 and 1999 in ex-factory price indices for the industrial sector as a whole, the coal industry, and a selection of other industries. Apart from 1988 the movements in coal prices closely match those of the industrial sector as a whole, despite the fact that the industry contributed less than 5 per cent to the total gross value of industrial output.

Consumers of coal had always been expected to pay the shipping and handling costs. Table 6.37 gives the average production costs, transportation costs, and

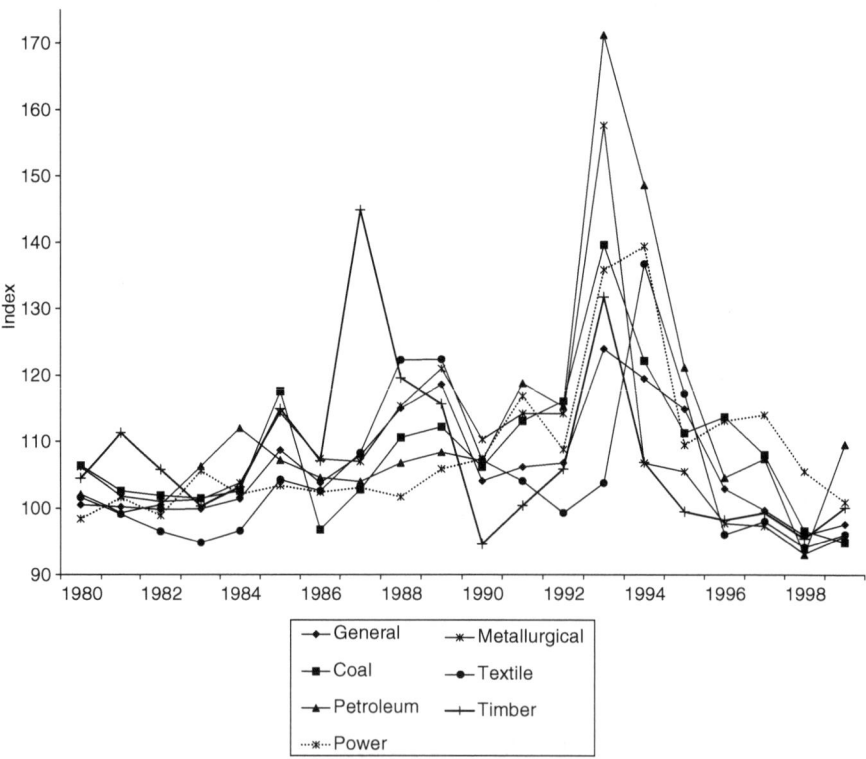

Figure 6.3 Ex-factory price index for coal and other industrial commodities, 1980–99

Source: *TJNJ 2000*, p. 305.

selling prices of coal sold in Shanghai between 1978 and 1986. In the space of less than ten years, production costs more than doubled and average transport fees per ton increased by over ten yuan, making for an increase in average wholesale prices of over 30 yuan per ton.

Some twenty-five different fees were charged *en route* from Datong (Shanxi) to Shanghai and Guangzhou (Guangdong) in 1992. The cost of moving the coal by rail from the interior to the port of Qinhuangdao (Hebei) was 31.19 yuan per ton (see Table 6.38), and of shipping it on the ocean south to Shanghai, an average of about 10 yuan depending on the type of ship. Total costs at Shanghai amounted 142.84–150.59 yuan, and at Guangzhou, 184.8–192.25 yuan. These plus the mine-mouth cost of 77.93 yuan, resulted in Shanghai consumers paying a total of over 220 yuan per ton and Guangzhou consumers, over 262 yuan. The transferring of the coal from one form of transport to another accounted for only 25–30 per cent of the total cost. The other 70–75 per cent was an assortment of fees levied by the railway and port authorities.

Table 6.39 gives the differential in prices between coal under plan and coal not under plan in Shanghai, Nanjing, and Suzhou between 1979 and 1986. In Suzhou the price of coal under plan increased by 47 per cent over this period while that not under plan increased by 93 per cent. The differential between the two prices was 63 per cent in 1986. By December 1993 it was 129 per cent.

By the late 1990s market forces had helped establish different prices for different quality coals. With increasing governmental pressure to reduce pollution, the larger consumers were forced to compete for the best product available. At the low end, it came down to what the coal producers could get to move the coal. Table 6.40 gives the range of average prices charged by the Yanzhou Coal Company for the six months ended 30 June 1999 and 2000. In 1999 the price of the lowest grade coal sold was about 42 per cent that of the highest, and in 2000, 30 per cent.

Table 6.37 Planned production costs and coal prices in Shanghai, 1978–86

	Average production cost (per ton)	Of which: Average mine price (per ton)	As % of production cost	Average transport fee (per ton)	As % of production cost	Average wholesale price
1978	30.73	17.17	55.9	11.48	37.4	33.00
1979	36.35	22.22	61.1	12.02	33.1	37.99
1980	39.69	25.94	65.4	11.99	30.2	41.75
1981	39.56	25.99	65.7	11.96	30.2	42.20
1982	40.73	26.75	65.7	12.44	30.5	42.88
1983	46.60	31.20	67.0	13.31	28.6	46.63
1984	55.20	33.07	59.9	19.60	35.5	57.04
1985	58.41	35.08	60.1	19.69	33.7	63.30
1986	61.61	35.58	57.8	21.50	34.9	63.72

Source: Tian Yuan and Qiao Gong (eds), *Zhongguo jiage gaige yanjiu, 1984–1990* (Research on Price Reform in China, 1984–1990), Beijing: Dianle zhengye chubanshe, 1991, p. 389.

Table 6.38 Cost buildup for Datong lump coal shipped to Shanghai and Guangzhou, March 1992

Cost item	Charge (yuan/metric ton)
A. Minemouth coal cost	77.93
B. Shipping and Handling Costs	
1. Minemouth to railhead loading	20.25
2. Rail shipping cost, Datong to Qinhuangdao	31.19
3. Railroad charges:	
Shipping charge	0.95
Maintenance charge	2.65
Water resource charge	1.00
Car routing charge	0.78
Interest	0.22
Stamp Tax	0.01
Subtotal	5.58
4. Ocean shipping cost, Qinhuangdao to Shanghai (depending on ship type)	6.51–13.21
5. Qinhuangdao port charges:	
Transshipment charge	5.33
Loading charge	23.15
Unloading charge	2.05
Terminal charge	2.05
Port service charge	2.05
Storage charge	0.05
Dust abatement fee	0.30
Port construction fee	1.50
Thawing fee	0.70
Crane charge	16.30
Loan repayment	0.60
Subtotal	55.58
6. Subtotal shipping costs at port of Qinhuangdao	119.11–125.81
To Shanghai	
7. Port of Shanghai port service charge	6.50
8. Port of Shanghai comprehensive charges*	17.23–18.28
9. Subtotal shipping costs at port of Shanghai	142.84–150.59
To Guangzhou	
10. Ocean shipping cost, Qinhuangdao to Guangzhou	37.51
11. Port of Guangzhou port service charge	6.50
12. Port of Guangzhou comprehensive charges*	21.68–22.43
13. Subtotal shipping costs at port of Guangzhou	184.8–192.25
C. Final coal costs at destination	
1. Port of Shanghai	220.77–228.52
2. Port of Guangzhou	262.73–270.18

* Comprehensive port charges include bank interest, profit, management fees, and normal depreciation.

Source: Jonathan E. Sinton *et al.* (eds), *China Energy Databook*, Ernest Orlando Lawrence Berkeley National Laboratory Report LBL-32822, Rev. 4, Sept. 1996; Table VI-2, p. VI-5. Copyright © 1996 Lawrence Berkeley National Laboratory. Reprinted by permission of the Lawrence Berkeley National Laboratory, University of California.

Table 6.39 Comparison of planned coal prices with market prices in Shanghai, Nanjing, and Suzhou, 1979–86 (yuan/ton)

	Shanghai		Nanjing		Suzhou	
	Planned	Market	Planned	Market	Planned	Market
1979	37.99	—	—	—	45.80	56.90
1980	41.75	—	27.71	—	45.80	59.90
1981	42.20	—	29.69	—	45.80	59.90
1982	42.88	—	28.55	34.40	45.80	80.00
1983	46.63	—	31.44	50.85	45.80	87.00
1984	57.04	98.04	29.17	64.55	54.70	100.00
1985	63.30	—	38.36	71.62	62.70	115.00
1986	63.72	64.72	39.54	71.43	67.40	110.00

Source: Tian Yuan and Qiao Gong (eds), *Zhongguo jiage gaige yanjiu, 1984–1990* (Research on Price Reform in China, 1984–1990), Beijing: Dianle zhengye chubanshe, 1991, p. 390.

Table 6.40 Prices charged by Yanzhou Coal Company, 1999 and 2000 (yuan per ton)

	For the six months ended 30 June	
	1999	2000
No. 1 Clean Coal	251.60	222.88
Domestic	193.93	170.60
Exports	193.65	151.30
Screened raw coal	146.65	128.58
Mixed coal and others	105.83	65.75
Average	162.13	135.43

Source: Yanzhou Coal Company Financial Statement, www.yanzhoucoal.com.cn (accessed April 2001).

Average market prices for the month of June each year between 1990 and 2000 are presented in Table 6.41. There were huge increases for the coastal cities especially. For example, the price of coal in Shanghai was raised five times in the first four months of 1993.[206] It was 175 yuan per ton in December 1992, 190 in January 1993, 200 in February, 205 on 3 March, 210 on 25 March, and 215 on 25 April. Within 115 days the price per ton had risen 40 yuan, or 22.9 per cent. The main reason for the hikes was several increases in various 'management fees' and 'service charges' levied by the railway and port authorities. These amounted to 2–15 yuan per ton.

After 1998 the average price, reflecting the over-supply conditions, dropped sharply. By 2000 it was down 11.4 per cent from 251.10 yuan in 1998 to 222.52

Table 6.41 Average market prices for the month of June, 1990–2000 (yuan)

	1990	1991	1992	1993	1994	1995	1996	1997	1998	1999	2000
Average	160.67	144.29	150.73	184.06	226.46	223.38	233.75	251.08	251.10	238.84	222.52
Beijing	141.36			170.00	168.00	196.00	217.00	260.00	260.00	260.00	225.00
Shijiazhuang	85.00			85.00	120.00	170.00	180.00	122.04	122.04	122.04	122.04
Huhhot	75.00				90.00	111.71	115.00	110.00	150.78	150.00	120.00
Shenyang	295.00	139.00	165.00	190.00		247.00	240.00	240.00	240.00	240.00	240.00
Dalian	198.00			170.00		240.00	300.00	270.00	380.00	380.00	260.00
Changchun	158.10		148.00		203.00	203.00	210.00	235.00	245.00	250.00	365.00
Harbin	131.00	134.00	147.00	150.00	175.00	150.00	200.00	245.00	225.00	225.00	225.00
Shanghai	175.00		176.00	215.00	225.00	250.00	280.00	260.00	280.00	280.00	
Xining								176.28	176.28	176.28	176.28
Nanjing	125.00	180.00	190.00	240.00	245.00	280.00	330.00	330.00	350.00	350.00	250.00
Hangzhou	150.00	180.00	195.00	255.00		310.00	390.00	330.00	320.00	320.00	300.00
Ningpo	170.00				290.00	288.15		301.70	310.75	300.58	245.00
Hefei	250.00	136.00			200.00	237.30	300.58	300.58	300.00	300.00	280.00
Fuzhou		151.00	207.50	259.70	344.50	360.00		381.00	378.00	355.00	325.00
Nanchang	145.00	156.00		230.00		260.00	270.00	300.00	280.00	260.00	240.00
Jinan	146.00	139.00							254.00		
Qingdao	163.80		174.30	173.70	209.00	259.00	286.00	336.60	323.18	280.00	285.00
Zhengzhou		106.00	80.00			175.00	180.00		225.00	200.00	150.00

City											
Wuhan	124.20	192.00		245.00	200.00	214.00	260.00	283.63	301.71	270.00	237.00
Changsha	129.00	138.00	120.00	235.00	235.00	260.00	260.00	240.00	235.00	240.00	230.00
Guangzhou	210.00	210.00		285.00	336.96	386.46	361.60	378.55		260.00	
Chengdu	190.00	133.00	140.00	166.00	540.00	199.00	220.00	290.00	350.00	260.00	158.20
Chongqing									158.20	158.20	145.00
Xian	130.00	113.00	120.00	120.00	155.00	155.00		210.00	160.00	150.00	168.00
Lanzhou	117.60	97.80	116.90	129.90	146.00	146.00	167.60	180.00	180.00		150.00
Taiyuan			57.00		112.00	112.00	112.00		76.00	133.24	168.00
Xiamen	262.00		250.00	265.00	330.00	305.00	305.00	355.95	370.64	310.00	290.00
Shenzhen									360.00		360.00
Haikou					135.00	135.00	398.00	150.00			150.00
Nanning	140.00	140.00		86.00	388.00	388.00	150.00	388.00	150.00	170.00	160.00
Kunming					127.00	127.00	146.00				125.00
Yinchuan				61.85	90.00	90.00	108.00	125.00			
Urumqi	54.80		56.25				117.00	112.00		74.10	74.10
Tianjin								248.00	248.00		230.00

Source: *ZGWI*, July issues, 1990–2000.

in 2000. In Shenyang there was no change in the price, but in other cities such as Dalian and Nanjing the fall was as low as 31.6 and 28.6 per cent, respectively. The general price index fell sharply from 115 in 1995 to 1997 in 1999.[207] Consumer demand was low as a result of high unemployment, large surpluses of poor quality goods on the market, and the effects of the Asian financial crisis.

Expansion of the two-tier pricing system was an important first step in reforming prices. On the positive side, the higher prices served as a tremendous incentive to greater production. Many rural mines were extremely lucrative and the availability of more coal enabled rapid expansion of the rural industrial sector (see section on 'Structure'). On the consumption side, they forced consumers to regard coal as a commodity of value, the use of which had to be conserved. Also, consumers and suppliers began to communicate directly with each other, with the result that consumers received the type of coal that more closely met their particular needs.

On the negative side, it led to vicious profiteering by a string of middlemen beginning with corrupt Coal Ministry, army, and railway officials, then by workers at factories, schools, neighbourhood committees, etc., who bought the coal at low fixed prices and re-sold it at greatly increased prices to a variety of consumers. Huge profits were realized.[208]

Another disastrous consequence was triangular debt, sometimes called debt chains. Many state industries had been suffering major losses for years and the deregulating of the price of coal, as well as of other non-substitutable inputs, made it impossible to stay within planned expenditure, if, indeed, to carry on at all. The volatility in prices throughout the 1980s and 1990s made it almost impossible for consuming industries to plan annual budgets.

The dual price system was not able to ease the financial difficulties of the state mines because the government kept the allowed increases in coal prices below the allowed increases in the prices of materials needed for mining.

Initially, the state industries together refused to buy the coal at the higher prices with the result that stockpiles appeared throughout the country. They believed that they could force the government to reconsider.[209] However, within months it was plain that the government was expecting them to continue meeting their production targets and to pay for the coal, and other inputs, through cost-cutting measures of their own. In many cases this was impossible and in the late 1990s this inability of state enterprises to pay for each other's products was leading to the collapse of much of the state industrial sector.

Defaulting consumers owed the Coal Ministry 3.39 billion yuan in 1992, 7.43 billion in 1993, and 13.44 in 1994.[210] Some of the mines received less than 20 per cent of their payments, meaning they had a third of the money needed to run their mine operations. In 1992 the Ministry ordered equipment worth 651 million. However, due to defaults in payments, it was able to order equipment worth only 93 million in 1993.[211]

In 1994 metallurgical companies owed 5.88 billion yuan and accounted for 43.7 per cent of the total amount; fuel companies 5.37 billion yuan or 40 per cent of the total; electric power companies 890 million yuan or 6.6 per cent of the total; and chemical companies 710 million yuan or 5.3 per cent of the total.[212] Among the largest debts were those of the steel industry, itself mired in a partially reformed state, trying to solve critical problems of its own. Of the forty-two steel enterprises, twenty-eight had coal debts, in all amounting to 5.3 billion yuan.[213] As of the end of May 1999 the state coal industry was owed 40.064 billion yuan, equivalent to over half the expected sales income. Of the ninety-four SPMAs about a quarter were each owed 500 million. The Kailuan Administration alone was owed a total of 1.2 billion and the Datong Administration 2.2 billion.[214]

Billions of yuan were owed to Shanxi, the largest coal-producing province, and the frustrated local governors referred to it disparagingly as China's largest interest-free bank. The huge debts reduced its coal industry revenue per ton by 33.4 yuan and its profit by 34.9 yuan. It was not only the coal industry which suffered, for when it took out large amounts of limited funds from the banks, little was left for the Province's chemical, machine-building, electric power, building materials, and metallurgical industries.

The insolvency of the mines created severe hardship for the workers. At the end of 1993, seventy CMAs were withholding 1.56 billion yuan in wages owing, affecting 2.51 million of its 3.21 million workers. A year later fifty owed wages totalling 2.21 billion yuan, affecting 1.58 million workers. Another 300 million yuan were owed in pensions to retired employees.[215] In 1998 a total of 4.32 billion yuan was owed to coal workers nationwide, and in July 1999, 5.27 billion yuan (US$ 634 million).[216]

The rigorous enforcement of the 'three no's' delivery policy, introduced in November 1994, was aimed at ending the triangular debts. The three no's were no shipment without payment, commercial draft, and the clearing of outstanding payments.[217] At the same time, the actual payment transactions were transferred from the mine administrations to the banks. These measures had an immediate positive effect. It was found that some of the enterprises owing money were preferring debt to taking out loans which required interest payments. Some even used the money they owed to purchase capital equipment. However, overdue coal payments at the end of 1997 amounted to 24.9 billion yuan, and the rate of debt collection over the first six months of 1998 faltered at 83.11 per cent.[218]

Triangular debt was not unique to China. It was also a problem in India. There, the Coal Ministry's solution was to deduct outstanding amounts directly from the state's annual plan outlay.[219] It was hoped that by using this method, all debts would be cleared within three years.

Exchanges

The first coal exchange was established in the Beijing area at Zhangshanying town, Yanqing County, in 1992.[220] Later that year, a larger one was opened in Qinhuangdao (Hebei), set up jointly by Qinhuangdao City, the Ministry of Materials and Equipment, and the China General Coal Corporation, and one in

Zhengzhou (Henan).[221] Early in 1993 the CNCC, with the Shanghai municipal authorities, established China's first coal futures market in Shanghai, and the Northeast Coal Trade Centre came into being in Shenyang (Liaoning).[222] In the following months many provincial coal markets were established, including those at Taiyuan (Shanxi), Wuhan (Hubei), Tianjin, and Guangzhou (Guangdong), all of which were operated by the respective provincial governments and the ministries responsible for coal, railways, communications, and internal trade.[223] Others were established at Hangzhou (Zhejiang), Qingdao (Shandong), Chongqing, and Xian (Shaanxi).

The brokers at the exchanges linked coal consumers with coal producers and shippers (rail and water) and earned a 5 per cent commission for every 1,000 yuan of coal traded. For coal under plan, transportation costs were included in the prices, which were not completely free but within specified bands. The trading took place between the 26th and 6th in a monthly cycle. All coal shipments were certified in terms of weight, type, and quality, and brokers managed both spot transactions and forward business, entrusted purchase, and sale and settlements.

When the exchanges opened there was an immediate and very large response. However, within months they were forced to stop operations due to both a lack of railway capacity and means for brokers to book the required transport.[224] The booking problem was solved a year later by having all the brokers book trains through the fully computerized two main exchanges at Shenyang (Liaoning) and Shanghai.[225]

In April 1994 the government ordered a halt in the trading of coal and other key commodity futures in an effort to control inflation. Hundreds of unauthorized individuals and organizations with connections to coal mining and/or railways were trading coal illegally, often completely bypassing the legitimate markets. Throughout the mid-1990s the government investigated hundreds of cases involving the illegal trade of coal and punished the perpetrators, but the problem persisted. In April 1996 some 300–400 illegal traders were operating in Qinhuangdao and the market there was handling only 6 per cent of the local bonafide trade.[226]

Initially, the exchanges experienced a serious cash flow problem. Less than half of the purchases were being paid for punctually.[227] However, as the 'three no's' policy took effect, this situation improved.

The exchange of coal mining goods began in 1995. Mining equipment and constituent parts, mining construction materials, fuel oil, etc., were traded at the first meeting held in Kailuan.

Diversified activities

The profits made from diversified activities were pivotal in ameliorating the coal industry's losses. Over the Eighth and Ninth Five Year Plans, more than two million mine workers were either transferred to existing businesses operated by the mines, or assigned to non- mine related new businesses, including some joint ventures.[228]

In order to assist mines to initiate diversified activities, the central government offered them access to an annual state loan of 3 billion yuan.[229] The industries included everything from all types of agricultural production to construction materials, smelting, machinery, medicines, clothing, shoes, handicrafts, and daily use products. The services included transport, restaurants, hotels, news reporting, schools, and hospitals.[230]

By the end of 1997 over 4,000 different products were being produced by some 1.8 million people at 20,000 factories operated by the CMAs alone, earning an output value of about 45.5 billion yuan, which was 57 per cent of the output value of total coal production. Taxes and profits amounted to about 2 billion yuan.[231] On at least one coal corporation's webpage the information about the diversified activities almost overshadows that for coal production![232]

Joint stock arrangements

In 1994 the Ministry of Coal chose some enterprises to experiment with *gufenhua*, or 'going joint stock', including China Huaneng Fine Coal Corporation, Yanzhou Coal Mining Corporation, Inner Mongolia Yitai Coal, Shenhua Group Corporation, Xingtai, Panjiang, Pingdingshan, Pingshuo, and China Coal Industry Construction Corporation.[233] The listing of these on the domestic and Hong Kong stock markets, and in at least one case on the New York stock market, improved the finances of some individual corporations, but was far from an unqualified success and did little to improve the finances of the industry as a whole.

President Jiang Zemin, who headed the Central Committee's Leading Group on Finance and Economics, had high hopes that this method of financing would be a solution for the ailing state industries. However, rather than yielding the expected immediate positive results, it frequently made things worse for the enterprise employees, often hardly literate, who were cajoled into buying shares. So deep were the problems at many of the enterprises – inexperienced management, obsolete plant, poor quality products, non-existent marketing research – that nothing short of a complete re-appraisal of business targets, if not total plant replacement or dis-establishment was needed. By 1997, Zhu Rongji was publicly downplaying the potential of joint stock arrangements in favour of bankruptcies and mergers where these hopeless situations prevailed.

One of the first coal companies to experiment with *gufenhua* was the China Huaneng Fine Coal Corporation, which was created to manage the Shenfu-Dongsheng coalfield. Shares were sold to the governments of Shaanxi Province, the Inner Mongolian Autonomous Region, some county governments in Shaanxi, and some coal mining bureaux in Shaanxi and Inner Mongolia. People living in the area were encouraged to develop their own mines and sell the coal to the Corporation.[234]

Yanzhou Coal Mining Co. Ltd. mines high quality low-sulphur coal in Shandong Province and sells it to domestic power plants and other consumers mainly in the eastern provinces, as well as to the export market. It had planned to list on the Hong Kong Stock Exchange in November 1997.[235] The US$300 million flotation was sponsored by Bear, Sterns & Co and ABN Amro Rothschild. However, due to the Asian financial crisis the listing was delayed until March 1998 when it listed H shares on the Hong Kong Stock Exchange, American Depository Receipts on the New York Stock Exchange, and B shares on the Shanghai Stock Exchange.[236] Measures aimed in 1997 and 1998 to reduce use of materials and electricity offset the steep decline in coal prices resulting from the fall in demand, and substantial profits were made both years.[237]

The Yitai Coal Company, controlled by Yimeng Coal Group of Inner Mongolia, initially issued US$166 million in B shares on the Shanghai Market at US 40.73 cents in July 1997. Profits fell 64 per cent in first half of 1998 and in August that year the shares valued only 10.8 cents. Yitai managed to produce only 370,000 tons in the first half of 1998, which was but a quarter of the target set for the year, and only 1.87 million tons were shipped, and 1.074 million tons sold, far less than half of the 5 million ton targets set for each for the year.[238]

Shareholding arrangements, besides their perceived potential for raising capital and eliminating debt, were also seen by the central government as an answer to the public ownership dilemma. More importantly, they were expected to act as a hard budget constraint. Managers are, theoretically at least, accountable to the shareholders. However, attempts at ownership reform in the coal industry did not produce clear lines of authority, resulting in near-chaos when the questions to be resolved involved other ministries and sectors.[239] This unsatisfactory situation discouraged foreign investment.

Foreign involvement

Hope and disappointment

When the Chinese opened their door to the world in the late 1970s they hoped that countries from the developed world would be very enthusiastic about investing and collaborating in cooperative projects in China. The Ministry of Coal anticipated that foreign mining companies would be keen to develop China's rich coal deposits, especially for export, and that they would bring their latest technologies. Recognizing the urgent need for large and reliable electricity supply the government also pleaded for foreign investment and technological advice in building power plants.

There was an initial flurry of visiting coal mining and/or machinery company representatives from Australia, Austria, GDR, FRG, Finland, France, Hungary, Japan, Poland, Romania, Spain, Sweden, UK, US, and the Soviet Union.[240] They jostled each other to capture and keep the Chinese mining officials' interest and often hosted lavish inspection tours of mining operations in their home countries. Hundreds of trade protocols and letters of intent were signed involving one or more of mine construction, equipment, and training. Many of the foreign delegations offered the Chinese credit loans to buy machinery and equipment in exchange for the earnings generated from exporting the coal. Others tried to negotiate compensation trade deals.

Investors, especially from the Chinese diaspora, immediately expressed interest in the power sector too, and it was here that some of the first foreign investment agreements were reached. Some investors accepted the less than satisfactory terms initially stipulated by the Chinese government thinking the best strategy was to establish a presence in the market and develop a working relationship with the Chinese over a period of time, anticipating that better terms would be negotiated for subsequent projects. Many others, however, were not prepared to take this gamble.

By 1983, the international investor community began to be somewhat disillusioned with China. Between 1979 and 1981, when the Chinese government was trying to do too much too fast, and oil revenues, which would have quickly and greatly enriched China, did not reach anticipated levels, there were unnerving investment reversals. In all, about 300 agreements in a variety of sectors were withdrawn or postponed. Potential investors feared that payback periods would be repeatedly extended and that there would be obstacles in the repatriation of foreign exchange earnings.

The Chinese market was hard to penetrate. Potential investors often had difficulty even determining where among the multiplicity of government levels, ministries, and planning agencies, the ultimate decision-making authority lay. The Chinese knew little about international business protocols, and the banking, accounting, and legal infrastructure hardly existed. Basic marketing and advertising strategies were virtually unknown, and the fact that the Chinese negotiators frequently demanded western-level wages for unskilled Chinese employees added to the irritation and frustration of would-be investors. Innumerable tariff barriers and a lack of foreign exchange and hard currency guarantees were other deterrents.

In the coal industry, many potential investors were alarmed by Armand Hammer's experience in the development of the Antaibao mine in Shanxi Province (see Chapter 5). There were tremendous difficulties right from the start. The cutting of coal began in 1987, after five years of stormy negotiations, and production thereafter was fraught with problems.[241] The government enacted the 'Provision of the State Council of the PRC for the Encouragement of Foreign Investment' that year, and many other sets of laws and regulations meant to guide foreign investors were promulgated in the years following. These all contributed to the gradual clarification of the rights and roles of the contracting parties, but should have been in place much earlier.

The work of the foreign technicians at the Antaibao mine was seriously frustrated by a lack of translators with good working knowledge of mechanics vocabulary. The technicians had been told they would be assigned seven apprentices each, but in actual fact they were given thirty-five, whom they found relatively unmotivated, unaccustomed to deadlines, and sloppy in their work habits. Worse still, tools, keys, spare parts, and even railway cars ready to be hauled to the port were constantly missing and equipment, especially vehicles, were recklessly handled, with the result that they had to be replaced at a far higher rate than normal. The majority of the Chinese engineers had no experience with computers. The morale of the foreigners was not improved by the fact that they and their families were forced to live an extra eighteen months in the Pingshuo Hotel due to the failure to meet the construction deadline for the planned housing. A steady flow of peasants who attempted to start their own mines at the site had to be forcibly removed.

All mining ceased in February 1988 due to insufficient diesel supplies. After a protest by the foreign workers limited supplies were brought in, but only enough for 4 hours of mining each day until 9 March. Then in September all loading of coal for Qinhuangdao Port (Hebei) was halted following a train crash on the connecting railway. One observer believed the crash was 'too convenient by half' and was 'being offered as an excuse for non-arrival of coal'.[242] The washing plant did not initially function correctly, resulting in product that was too wet. A storage silo collapsed, and when production and washing were finally beginning to reach the anticipated levels, the number of railway cars made available to ship the coal east was only a fraction of what was required. According to one report, the CNCIEC for a period was paying an average of US$20,000 per day in demurrage fees when foreign ships were faced with lengthy delays at the Port.[243]

An atmosphere of mutual suspicion developed when it was discovered that both the ICCC and PSF were overriding the CNCIEC committee composed of members from both sides, and designated to handle all export sales over the first twelve years. It emerged that several contracts were signed that had never been put before the committee.[244]

The Antaibao project was meant to serve as a model to reassure potential investors. Unfortunately, the seemingly unending difficulties frightened them away. Greatly affecting decision-making by both the Ministry of Coal and potential foreign investors was the sharp fall in the international price of coal from a high of US$56.69 per ton in 1982 to US$34.98 in 1987.[245]

Investors in power projects also were troubled. Due to its fear of fuelling inflation, the government unofficially capped the rates of return at 12–15 per cent, somewhat lower than those offered in the rest of Asia, which were usually a minimum of 16 and ranged up to 25 per cent. After examining the risks of constructing and operating a plant in China, most foreign companies believed a minimum of 17 per cent was necessary for their projects to be viable. Many negotiated for months, in some cases years, hoping to persuade the Chinese to match the rates offered elsewhere in Southeast Asia. For most, these efforts were in vain, and many gave up and took their business elsewhere. Another dissuading factor was that, for strategic reasons, foreign companies were not permitted to participate in construction and management of the grid system.[246] The result was that in the late 1980s and early 1990s foreign funding in the power sector as a whole accounted for only 10 per cent of the total.

The many steps required to arrange currency transactions were another hurdle. Initially, foreigners had to use foreign exchange certificates (FECs). After these were phased out, there were at least three different exchange rates, an official rate, an official swap rate, and the black market rate. In 1988 the ratio of these was $100:142:185$.[247] In order to encourage exports, swap markets had been established where yuan could be exchanged for hard currency on demand.

As part of contract negotiations investors had to arrange for the rate at which they could exchange yuan into foreign currency for importing inputs and repatriating profits. Changes in the value of the Chinese yuan *vis-a-vis* the US dollar through the 1980s and 1990s complicated cost–benefit analyses and worried business strategists. The Chinese government was aware that the yuan was overvalued against other major currencies, making exports comparatively unattractive, but feared a major devaluation would trigger and feed inflation. Between 1981 and 1997 the average exchange rate depreciated from 1.71 to the US dollar to 8.62 in 1994, a drop in real terms of 73 per cent. The largest adjustment was made in January 1994 when the yuan was devalued by about 50 per cent. It thereafter steadily appreciated but came under severe pressure during the Asian financial crisis.

The bottlenecks on the railways as well as the two-track pricing of coal and key materials required in mine and power plant construction also dissuaded foreign investors. They were fully aware that corrupt cadres bought at the low prices and resold at much higher prices. It would be impossible to make profits in this uncertain and transitional economic environment.

While Western machinery companies had hoped to sell equipment to China, they became leery about transferring design patents. They feared the availability of cheap labour in China would severely undercut them. Initially, it simply was not known whether or not the Chinese had any understanding of patent law. Some investors took false comfort in the assumption that the Chinese lacked the skills to reproduce foreign equipment, far less improve upon it.

Indeed, the Chinese did set about immediately to make prototypes of what they could. For example, the first heavy-duty coal cutting machine made under a transfer of technology agreement with a UK company was later copied and manufactured in China, and sold in Australia. China could produce certain equipment for 40–60 per cent less than Western companies, though the quality was often inferior and the reliability not proven. Furthermore, the Chinese did not initially appreciate the need for after-sales service, which generally accounts for a large proportion of the selling price in Western countries.

There were many problems in the absorption of technology from the West. In the spring of 1983 the press reported several instances of wasted mining equipment purchases. For example, when equipment from five different countries arrived in 1979 it had to be stored outside because there were no warehouses. Prolonged subjection to the elements drastically shortened its service life. Nor were there repair shops or workers who knew how to operate and maintain the equipment. Toolers at the mines were reluctant to make spare parts due to the time and cost involved, and they dismantled existing equipment for repair requirements. The result was that only 49.2 per cent of the imported equipment was in perfect working order. Millions of yuan had been wasted as a result of this planning failure.[248]

According to an executive of a UK mining machinery company the main problems in the absorption of technology in the late 1980s were the shortage of sufficiently experienced engineers and designers, and the lack of appreciation of the fact that equipment in a tough, hazardous environment requires regular systematic maintenance. Despite these problems, however, he said that the quality of the technological and managerial staff had improved dramatically. The opportunity for many engineers to visit other coal-mining countries had produced a more 'cosmopolitan' engineer.[249]

Between 1978 and March 1999 the coal industry received a total of US$9.34 billion in contracted foreign capital, of which US$4.18 billion had actually been used.[250] Of the total capital construction investment for the industry in 1981 and 1982, over 20 per cent came from foreign sources. Thereafter, however, private foreign investment tapered off.[251] By far the largest contributor to the development of China's coal industry was Japan, in the form of the Long-Term Trade Agreements, Resource Bank Loans, and Official Development Assistance (see Chapter 4). About 20 per cent of the Japanese funding was devoted to railways.[252] It accounted for 52 per cent of all the foreign funding granted to the Ministry of Railways.

China became a member of the World Bank in 1980 and from 1992 was the Bank's largest borrower.[253] China was one of three countries (the others were India and Brazil) to have a country department exclusively devoted to its programme. By 30 June 1999 cumulative lending amounted to about US$33.2 billion, of which US$23.3 billion came from the International Bank for Reconstruction and Development and US$9.9 billion was from the International Development Association. Some 19 per cent of the total lending went to the energy sector, with 24 and 5 per cent to transportation and environmental projects, respectively.[254] Loans designated specifically to the power sector amounted to about US$6.65 billion by late 1998. Another US$1.22 billion in loans was granted to this sector by the Asian Development Bank, and US$3.64 billion by the Japanese government.[255] An additional US$10 billion came from 36 foreign direct investment projects. The largest equipment supplier for coal-fired plants was the German Siemens Group.[256]

In the mid-1990s there were indications that the role of coal in the Chinese economy was beginning to diminish, though there was no doubt that it would continue to be the main fuel for several decades yet. Ministry officials still hoped there would be much more assistance from the West in modernizing the industry.[257] The goal for the Ninth Five Year Plan was to have 20 per cent of the total investment in the industry coming from foreign sources, 10 per cent in the form of government loans, and 10 per cent as direct investment. The share of foreign capital in the power sector was to increase from 20 to 25 per cent. To that end, in 1996, the SPC and Ministry of Power, working with the World Bank, drew up a regulatory framework for Build-Operate-Transfer (BOT) agreements, which gave foreign investors greater flexibility in the financing arrangements and clarified the contract procedures.[258] However, when economic growth and the immediate demand for electricity slowed sharply in the late 1990s the market for new plants almost halted.

In addition to seeking help from international agencies, energy officials began to participate in the Fujian Fair for International Investment and Trade, later renamed the China Fair for International Investment and Trade. Foreign investors were invited to view the entire array of projects related to energy for which the Chinese were seeking collaboration. Over the years, the Chinese participants learned to discuss the intricate details of margins and returns that Western businessmen need in their feasibility analyses, and steadily improved their business competence and presentation skills.

Trade

In the early 1980s the government had great hopes that coal exports would not only help cure the industry's financial problems, but that they would become a major source of foreign exchange. Moreover, the authorities anticipated foreign investors would be keen to establish coal export businesses. However, after examining the viability of such ventures, most of them demurred in the face of problems with transportation, storage and loading, quality, and the lack of satisfactory legal and financial infrastructure. Political uncertainty was another concern.

Coal exports increased by a factor of about eight, from 4.63 million tons to 37.41 between 1979 and 1999, though as a proportion of total coal output the increase was only from 0.73 per cent to 2.61 in 1998. The following year, due to the major reduction in production, the share rose by almost one per cent to 3.58 per cent (see Table 6.42). For perspective, in 1997 China's exports of steam coal accounted for 8.4 per cent of total international trade, hard coal for 6.0 per cent, and coking coal for 2.3 per cent.[259] The steam coal exports of 26.1 million tons in 1997 ranked fifth after Australia's 73.6 million, South Africa's 57.0, Indonesia's 37.6, and the US' 28.8.[260] The coke exports at 4.6 million tons ranked seventh and the hard coal sixth at 30.7 million tons. The main competitors in this area of trade were Australia and Indonesia.

Most of the steam coal exports came from Datong and Pingshuo (both in Shanxi), and Yanzhou (Shandong). Anthracite was exported from Taixi (Shanxi). In 1998, 1999, and 2000 the export price per ton, including cost, insurance and freight (cif), of Yanzhou steaming coal was RMB 211.3 (US$25.52), 170.8 (US$20.63), and 151.3 (US$18.28).[261] Over the same years the average prices of US coal imports were US$32.18, $30.31, and $30.18.[262]

Exports of coke increased sharply in the late 1980s and early 1990s. One economist estimated that by the mid-1990s China accounted for 60 per cent of world coke trade, and believed that China's share of world coke capacity could increase from 29 per cent in 1997 to 40 per cent in 2000.[263] This share is expected to rise as other countries reduce their production.

The earnings from the coal exports were small, often not covering the costs of production and transportation to the ports. In order to gain a niche in the international coal market it was necessary to subsidize the exports of both steam and coking coal. The IEA estimates that the 479,000 tons of steam coal China exported to

Table 6.42 Coal exports and imports, 1979–99

	Total exports (mt)	Value of exports (million current US$)	Exports as % of total production	Exports as % of total foreign exchange earnings	Rank of coal exports' value out of total exports	Total imports (mt)
1979	4.63		0.73			
1980	6.32		1.02			
1981	6.57		1.06			1.94
1982	6.44	33.000	0.97	1.48		
1983	6.56	29.400	0.92	1.32	7	2.13
1984	6.96	31.447	0.88	1.11	10	2.43
1985	7.77	31.447	0.89	1.15	9	2.31
1986	9.82	40.047	1.10	1.29	11	2.47
1987	13.53	47.633	1.46	1.21	11	1.94
1988	15.65	51.495	1.60	1.08	12	1.69
1989	15.34	55.404	1.46	1.05	11	2.29
1990	17.29	65.415	1.60	1.05	12	2.00
1991	20.00	74.848	1.84	1.04	11	1.37
1992	19.70	74.280	1.77	0.87	15	1.23
1993	19.81	70.224	1.72	0.77	15	1.43
1994	24.30	77.075	1.98	0.64	17	1.21
1995	28.62	101.066	2.21	0.68	17	1.61
1996	29.00	110.876	2.11	0.73	16	3.20
1997	30.72	113.253	2.32	0.62	17	2.00
1998	32.29	106.779	2.61	0.58	20	1.58
1999	37.41	108.375	3.58	0.56	23	1.67

Source: *TJNJ*, various years.

Note: Total exports and total imports include both steam and coking coal.

Italy in 1995 were sold at US$8 less than the production cost.[264] India went as far as accusing China of dumping coal onto the world market.[265]

As for their contribution to the country's total foreign exchange earnings, coal exports ranked seventh in 1983, but only twenty-third in 1999 out of some 150 main export categories. Despite the shortages of energy throughout the country, a coal export target of 30 million tons was set for 1990. However, actual exports that year were only 17.29 million tons, miserably short of initial expectations, and the target for 2000 was revised many times. The Ninth Five Year Plan had called for exports of 50 million tons by 2000, though in June 1996 this target was adjusted down to 35, then raised to 40 in January 1997.[266] In 1998 it was expected that exports could be raised even higher, now less in the hope of earning foreign exchange than in an urgent attempt to draw down the over-production and stockpiling. In order to encourage exports the export tax rebate for coal was raised from 3 to 9 per cent in June 1998.[267] However, precisely at this time, most of Asia was suffering the worst effects of the financial crisis. Japan,

South Korea, the Philippines, and Thailand were all unable to import as much as they had in the past. By the summer of 2000, however, the 50 million target was restored.

Over four-fifths of the exports in the late 1990s went to countries in Asia, with Japan taking over two-fifths (see Chapter 4), followed by South Korea, Taiwan, Hong Kong, and Vietnam. Hong Kong was used as a trans-shipment point as well as being a destination. Other importers in Asia after 1979 were North Korea, the Philippines, Malaysia, India, Thailand, Indonesia, Bangladesh, Iran, and Turkey. European destinations were Germany, Belgium, Italy, France, Spain, Denmark, Britain, Ireland, Finland, Austria, the Netherlands, Greece, Romania, and Albania. Coal was also exported some years to the US, Brazil, Israel, and Egypt. As noted in Chapter 5, Taiwan was added to the list with great fanfare in 1990. The mainland government wanted very much for Taiwan not only to buy more coal, but to become involved in the development of PRC mines. There was also considerable jubilation when shipments to Israel were begun in 1993, soon after the establishment of diplomatic relations.

The export trade was poorly administered at the outset of the reforms. In the mid-1980s, for personal gain and fired by the marketization furor, Party members with links to the industry and the railways created coal trading 'companies'. They competed eagerly for the potentially lucrative foreign trade. Very quickly sales were contracted far in excess of the capacities of the railways leading to the ports, storage at the ports, and of the loading and unloading equipment at the docks. Until an umbrella organization, the China National Coal Industry Import and Export (Group) Corporation (CNCIEC), and its affiliate, China Coal Import and Export Corporation (CCIEC), were given exclusive management of the timing of all the international shipments in January 1991, there was chaos at the ports. Many ships queued in the harbour for weeks before they were loaded, and the CNCIEC consequently was obliged to pay vast sums in demurrage fees.

There was also a problem with the quality of the coal reaching Qinhuangdao. In the early stages it did not meet the desired heating value and purity criteria. In 1994, of the 14 million tons processed for export, 65,000 tons were rejected and stringent testing procedures had to be established quickly to certify legally the quantities and qualities.

Figure 6.4 depicts the structure of the CNCIEC. At the time of writing the CNCIEC oversaw shipment of nearly 90 per cent of the coal exports including all coal exported to Japan and Korea.[268] It had more than eighty wholly owned subsidiaries, branches, and representative offices within China, as well as fifteen offices worldwide, either wholly owned subsidiaries or joint ventures, and ranked sixth and seventeenth, in the annual listing published by the Ministry of Foreign Trade and Economic Cooperation, of the top 500 enterprises in terms of export, and combined import and export volume, respectively.[269] In August 1997 four sub-corporations were formed: China Coal Sales and Transportation Corporation, China Local Coal Mines Corporation, China Production and Technology Development Corporation, and Pingshuo Coal Industry Corporation. In 1994 the Shanxi Provincial Coal Import Export Corporation was formed to manage coal

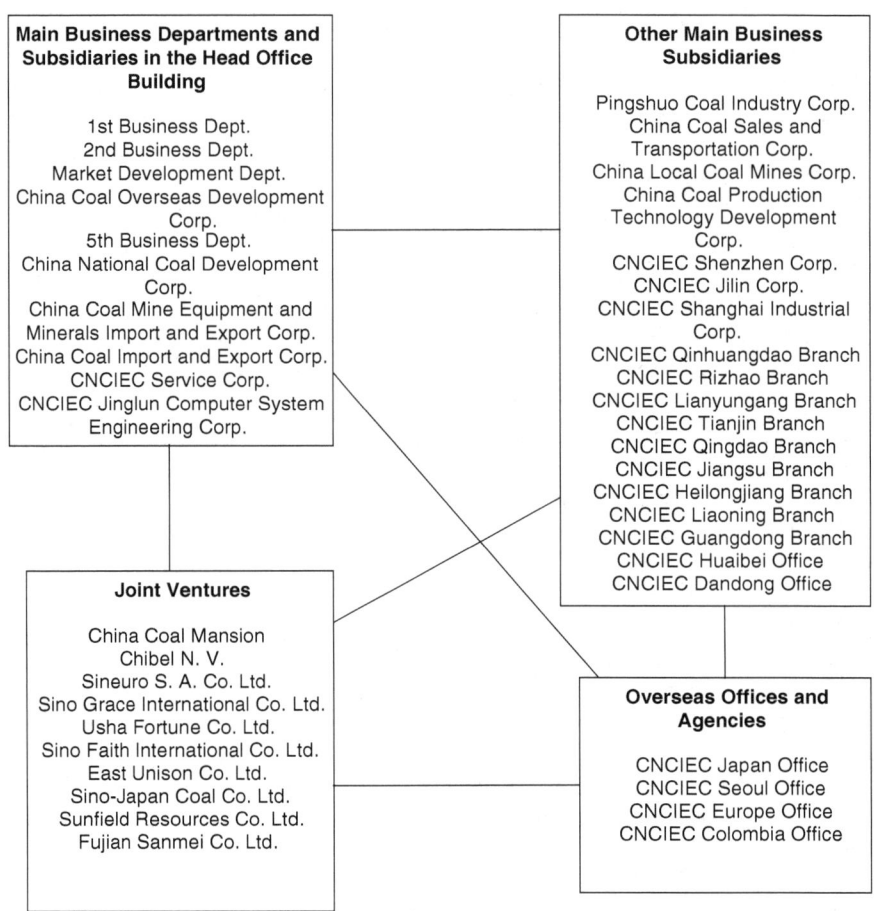

Figure 6.4 Structure of the CNCIEC
Source: *MTGYNJ 1999*, pp. 212–15.

shipments from Shanxi, and in 1995 the Shenhua Coal Import Export Corporation was brought into being to handle export shipments from the Shenfu fields (Inner Mongolia). The other major company approved to handle coal exports is the China National Mineral and Metals Import and Export (Group) Corporation.

Imports of coal before 1999 never accounted for more than 2 per cent of total consumption. Most was steam coal from Australia and South Africa with occasional shipments from Indonesia, Canada, the US, and Russia. Anthracite came from Vietnam, and coking coal was periodically shipped from North Korea, Australia, and New Zealand. There was a sizable but temporary increase in imports after the Tiananmen Incident in 1989. At that time railway shipments were disrupted and some southern consumers were forced into importing coal from South Africa and Canada. Shipments from Indonesia began in late 1993.

Transport

Railways

The year 1997 gave indication of being a turning point in the shipment of coal. Since the 1950s coal had dominated the freight capacity of the trains, but with a fall in the demand, combined with improved coastal shipping infrastructure, this predominance, reflected in several statistical indicators, appeared to be giving away.[270] For example, while coal accounted for about 35 per cent of total freight throughout most of the 1980s, and about 40 per cent in the early 1990s, reaching a peak of 45 per cent in 1996, its share was down to 41.4 per cent in 1999. Similarly, the daily average number of coal freight car loadings rose from 21,627 (35.3 per cent of the total) in 1980 to a peak of 32,720 (42.6 per cent) in 1996, but slipped to 28,987 (39.2 per cent) in 1999. Coal freight in million ton-km increased from 259,593 in 1985 to 404,847 in 1996, but decreased to 354,308 in 1998. The average distance of railway coal shipments also steadily increased between 1978 and 1996 from 368 to 562 km, but was down to 550 in 1999.

Major strides were made over the reform years in expanding and modernizing the railway system. Total trackage increased from 48,618 km in 1978 to 57,923 in 1999 (see Table 6.43). For perspective, India had 63,000 km in 1995.[271] Double-tracking in China rose from 15.7 per cent of the total length in 1978 to 36.1 in 1999, automatic blocking increased from 12.3 per cent to 31.2, and railway stations with electric interlocking increased from 27.7 per cent of the total to 90.1. It is planned that by 2000 the proportion of electrified trackage will have increased from 2.1 in 1979 to 27 per cent of the network, and China's ranking, in terms of total length of electrified railways, will have increased from twenty-fifth to fourth in the world.[272] Indeed, China's expertise in building electric train railway equipment had become such that it began to export components, the first recipient being Iran in September 1998.[273]

Many unit trains – trains composed entirely of specially designed coal cars – were put into service, making for safer and faster transport, loading, and unloading. Several key destination stations were fitted with subways into which the coal could be tipped sideways from the railcars into waiting trucks. The levels of computerized telecommunication and general transportation management were raised considerably.

Despite these achievements, the overall network capacity was still seriously deficient for most of the 1980s and 1990s. Construction slowed down considerably in the late 1980s and early 1990s compared to progress in the pre-reform years. Over the first five five-year plan periods, China had built an average of over 1,000 km of new railways each year, but an average of only 436 over the Sixth,

Table 6.43 Increases and improvements to railway trackage and freight carried, 1979–99

	Length (10,000 km)	Length increment per year, (km)	Length increment per year (% change)	Length electrified (km)	Double-tracking (km)	Coal carried (10,000 tons)	Freight increment (10,000 tons)
1978	48,618	1,200	2.53	1,030	7,630		
1979	49,800	1,182	2.43				
1980	49,940	140	0.28	1,667	8,119		
1981	50,200	260	0.52				
1982	50,500	300	0.60				
1983	51,604	1,104	2.19				
1984	51,741	137	0.27				
1985	52,119	378	0.73	4,151	9,989		
1986	52,487	368	0.71	4,430	10,613	53,106	
1987	52,611	124	0.24	4,643	11,186	54,343	1,237
1988	52,767	156	0.30	5,738	11,771	56,480	2,137
1989	53,187	420	0.80	6,372	12,528	60,891	4,411
1990	53,378	191	0.36	6,941	13,024	62,870	1,979
1991	53,415	37	0.07	7,804	13,380	62,603	−267
1992	53,565	150	0.28	8,434	13,658	64,108	1,505
1993	53,802	237	0.44	8,935	14,315	65,336	1,228
1994	53,992	190	0.35	8,966	15,475	65,943	607
1995	54,616	624	1.16	9,703	16,909	67,357	1,414
1996	56,678	2,062	3.78	10,082	18,423	72,058	4,701
1997	57,566	888	1.57	12,027	19,046	70,345	1,713
1998	57,584	18	0.03	12,984	19,673	64,081	6,264
1999	57,923	339	0.59	14,025	20,925	64,917	836

Sources: *TJNJ*, various years.

about 250 over the Seventh and Eighth (of which only 37 km in 1991), and only 18 in 1998.

Rolling stock and storage capacity were also seriously deficient. In 1993 the daily nationwide demand for coal cars was 120,000, but only 73,000 were available.[274] Though the capacity of specifically designed coal storage yards – covered, guarded, and fitted with sprinkler systems – had been greatly expanded, the time lag between extraction and the availability of transport was such that thousands of tons were left in tall mounds, prone either to being washed away by rain or consumed by combustion. Lack of transport capacity and storage facilities were also serious problems in India.[275]

Pilfering from the railway stations was an ongoing problem, and in 1998 it was reported that tons of coal were being raked off coal trains *en route* to their destinations.[276] In extreme danger to themselves, peasants attached rakes to long poles and waited at points along the route where there were steep embankments upon which they could stand. The chunks of coal fell to the sides of the tracks, were collected in baskets, and used to augment family supplies and/or sold to other peasants.

Rather alarming are the facts that in the mid-1990s about a quarter of the main lines were operated beyond normal life expectancy, and 18.4 per cent of the bridges and 50 per cent of the tunnels had fallen into disrepair.[277] In 1996 it was reported that only about 60 per cent of demand for both freight and passenger railway traffic was being met, and in congested areas, only 40 per cent.[278] Though, due to the fall in demand for coal the amount shipped in 1997 and 1998 was considerably less than in 1996, in 1998 still one-third of the railway bureaux operated above intended maximum limits, with total passenger and freight volume on the five trunk lines exceeding capacity by a factor of 3.4.[279] Total operating track length actually grew by an average of only 0.78 per cent per year between 1986 and 1996, while GDP in constant prices grew at 10.0 per cent, and coal freight, at 3.1.

In 1992 China had 46 km of rail lines per million population, and 6 km per thousand km^2, compared respectively to 808 and 22 km in the US, and 88 and 26 in India. In 1995 China had 21.4 per cent of the world's population and 7.2 per cent of its area, but only 4 per cent of the total mileage of the world's railways.[280]

There were inadequate branch lines, diversion lines, and connectors in the coastal areas particularly, and much of the dispatch equipment in use in the mid-1990s was of 1950s technology.[281] Over half of the marshalling yards were operating over capacity with the result that there were twenty-four bottleneck districts which could handle only 40 per cent of the freight traffic.[282] Railway station platforms were also relatively short, typically 650–850 m in China versus 850–1,050 m in the Soviet Union and 900–2,500 m in the US.[283] This made passing, loading, and unloading time longer.

The amount of coal shipped out of Shanxi from state mines nearly tripled between 1979 and 1996. According to one report, in 1984 a coal train left Shanxi every six minutes, and over 5,000 trucks left each day.[284] Coal occupied 83 per cent of the Province's total rail capacity in 1980, and over 90 after 1985.[285] Yet more coal could have been shipped out had there been more railway cars available.[286] This had serious repercussions for the Provincial economy as a whole. In 1986 it meant that over 700 million tons of metal ores, non-metal ores, chemical materials, and agricultural commodities, together annually worth over 300 million yuan, were also stockpiled.[287] Ideally other goods and materials needed 33.8 per cent of the space.[288] Throughout the late 1980s and early 1990s many mines were forced to operate at less than full capacity due to lack of transport.[289]

The plans for coal exports were seriously affected by the inadequacy of transport capacity. The Long-Term Trade Agreement signed with Japan in 1978 called for the amounts of coal exported to become increasingly larger. However, by 1983 it was apparent that the contracted amounts would have to be reduced somewhat due to lack of railway capacity. Indeed for many of the foreign investors who had proposed giving the Chinese credit loans to buy machinery and equipment in exchange for the earnings generated from the export sales of coal, a critical deterrent, along with the incomplete deregulation of coal prices and corruption, was the inability to guarantee prompt arrival of the coal at the ports. Railway scheduling problems and the lack of rolling stock was among the most serious difficulties that plagued the Antaibao (Shanxi) joint venture, a project that the Chinese government had hoped would serve as a model for other foreign investors.

The average pulling weight of the Chinese trains was considerably less than that in other countries: 2,559 tons in China in 1994 compared to 3,085 tons in the Soviet Union in 1987, and 4,567 tons in the US in 1986.[290] The average speed was 70–90 km per hour in China versus about 160 km in the US and the Soviet Union.[291] The reason was China's continued use of steam engines along with the newer diesel and electric ones, whereas most other countries had completely abandoned steam engines some fifty years ago. Ironically, it was shortages of electricity in the 1980s that necessitated the prolonged use of these half-as-powerful engines. India ceased making steam engines in 1971, while China continued to produce them until 1988, and some were still in use at the time of writing. The number of diesel and electric locomotives in China increased by about fivefold and fourteenfold, respectively over the 1980s and 1990s. Initially, China was importing diesel-electrics from France, Romania, and the US, diesel-hydraulics from West Germany, and electrics from France, East Germany, and Czechoslovakia, but eventually built the latter domestically and actually developed an export market for electric railway equipment.[292]

Inadequate railway capacity was due to four main factors: insufficient investment, irrationally low fare and freight rates, planning failure, and corruption. Between 1980 and 1997, a total of US$5.17 billion of foreign money was spent on improving the railway network, only 4 per cent of the total investment over that period.[293] Most of the foreign funds were in the form of foreign commercial bank and inter-governmental loans from the World Bank and Asian Development Bank. There was virtually no foreign direct investment. The government remained reluctant to allow foreign investment in this sector until the early 1990s. Being the only physical means connecting the widely disparate regions of the country, the railways were regarded as strategic – it was vital to be able to control them in the event of political unrest within the country or foreign attack.

In the early 1990s operating costs rose at some 15 per cent per annum and the erosion of the railways' share in both passenger and freight traffic began to cause serious financial difficulties. The proportion of travellers opting to travel by rail fell from 61 per cent in 1980 to 37 in 1999, while the rail share in freight traffic over the same period fell from 48 to 32 per cent.[294] Profits increased from 3.6 billion yuan in 1983 to 11.31 billion in 1990, the losses beginning in 1994 and most serious in 1995 at 6.4 billion.[295]

In most other large coal producing countries, such as Australia, South Africa, and the US, there are dedicated coal railways which have made high profits. Coal is but one commodity carried on the Chinese railways, and the government, motivated by a combination of socialist economic theory and fear of inflation, as they had been with coal prices, kept passenger fares and freight rates so low that operation costs were often barely met if at all. The rates for shipping coal were so unattractive for the Ministry of Railways, that given a choice, it preferred to ship other commodities for which it could earn higher revenues.[296]

The government had encouraged more and more local coal production in the early 1980s, anticipating that market forces would determine how and where it was used, but failed to provide for the changes in coal transport that this shift in policy necessitated. In theory, the local mines could sell their coal anywhere. However, this was simply impossible due to the lack of transport, as well as the 'tedious' approval procedures for selling coal outside the province of origin.[297] The national railway system was divided into twelve regions, each managed by a separate railway administration, and these in turn oversaw fifty-seven sub-administrations. Local railways often ended at the provincial boundaries.

When it was realized that production far exceeded railway capacity, the government began to set production targets according to available transport – *yi yunding chan, yi xiaoding chan* or *yunding chan, yunding xiao shi xing meitan fenpai, yunshu, jihua* (production according to transport and sales, or no matter what type of mine the coal comes from, if put under plan it will have guaranteed shipment).[298]

The classic problem of developing economies – that of inter-industry linkages and the rationalization of development and reform across the various industrial sectors – was exacerbated by the vast distances between the main producing and consumption centres. Relations between the mines and railway bureaux since the late 1950s were disorganized, sometimes chaotic. In the 1980s and 1990s the government took several major steps to decentralize and marketize the coal industry, and encouraged in a very large way, the development of LNS. However, the marketization of the railways, posing major and complex problems, some security-related, continued to be almost entirely state owned and operated. Now there were hundreds of independent coal producers trying to arrange for shipments with transport authorities, for the most part, still functioning according to central planning. Communications between the two sectors still needed to be greatly improved in order to solve this handling/distribution problem, hasten the success of coal marketization, and eliminate the conditions for rampant corruption.

The uneven progression of the reforms in the two industries was the main factor behind the failure of the coal exchanges, which began operations in 1992, and the structural changes in coal production worsened the already problematic relations between the two industries. When the exchanges opened, consumers and producers alike had high expectations. However, the deals foundered on the inability of the brokers to arrange transport. The effects on the railways of the decentralization and marketization of coal had not been anticipated and the Ministry of Railways was unable to cope with the increased number and complexity of bookings. The solution, established a year later, was to have all the brokers book trains through the fully computerized two main exchanges at Shenyang (Liaoning) and Shanghai.[299]

Poor scheduling meant that railway equipment was not used to the fullest at any given time. For example, in the early 1980s the ratio of full cars leaving Shanxi to empty cars returning was 6 : 1, and there were far too many short trips.[300] In the mid-1980s trips under 20 km accounted for 5–10 per cent of the total coal and non-coal trips, those of 20–50 km for 13–19 per cent, and those of 50–100 km for 23–30 per cent.[301] Furthermore, the work force was often inefficiently deployed. In China 175 men were required to repair and maintain every 100 km of railway, whereas in the US it took only thirteen.[302]

The composition of the annual railway operating costs changed considerably over the reform years. In 1985, wages and salaries accounted for only 14 per cent of the costs, materials accounted for 12, fuel 18, electricity 2, depreciation 45, and other costs 8.[303] In 1998 the share of wages and salaries had risen to 22 per cent, materials accounted for 11, fuel 13, electricity 6, depreciation 15, and others 35.

Throughout the 1980s and 1990s corruption on the part of government officials, mine managers, and military people caused unreliable and tardy delivery. Some used their connections to expropriate railway cars for their own coal-selling interests, bribing the railway workers to falsify cargo reports.[304] They colluded with each other and a variety of middle men, devising means to breach government regulations. Often millions of tons of coal were bought and resold many times before being ultimately used.

However, when Premier Zhu Rongji in 1998 ordered the state sector, including the railways to become profitable within three years, the problems were addressed swiftly and directly. Several measures were introduced immediately to raise profits, including sharply raising passenger ticket prices and freight charges, eliminating short cost-ineffective trips, closing or merging several small stations, reducing the number of departments, laying off some 260,000 workers, enforcing worker accountability, computerizing many applications, raising diesel and electricity consumption efficiencies, installing modern communication signal equipment, improving passenger service, reducing travel time, and adhering strictly to published schedules.[305] Special bond issues, a railway construction fund, more rational taxation, attractive options for foreign investors, greatly increased government investment, and varied shareholding arrangements were aimed at bringing in capital.

The assignment of excess workers to diversified activities had an immediate positive effect on the finances of the railways as it had had in the coal industry. By the end of 1999 some 400,000 railway employees had been moved to these activities. There were over 9,000 enterprises with total assets nearly 62 billion yuan. Profits reached 1.8 billion yuan.[306]

With the fall in the demand for coal transport, requirements from the 'three wests' were finally being met and the proportions of other types of freight carried on the trains were at long last beginning to rise. In early 2000 the government proudly announced that the railways had become profitable one year ahead of schedule despite devastating floods and the Asian financial crisis.

Coastal and inland shipping, highways, and slurry pipelines

The principal means of transporting coal has always been the railways, though as can be seen from Table 6.44, which gives the shares of coal transported by rail, ship, and highway in 1990 and 1995, by which mid-decade ship transport, in terms of ton-kilometres, was accounting for nearly half.

Table 6.44 Modal breakdown of coal transport, 1990 and 1995

	Total	Railway	Water transport	Road transport
1990				
million tons	9.54	6.54	1.30	1.61
	(100.00)	(68.5)	(14.6)	(16.9)
ton-km	5,562.00	3,641.00	1,821.00	100.00
	(100.00)	(65.5)	(32.7)	(1.8)
1995				
million tons	9.41	6.74	1.54	1.13
	(100.00)	(71.6)	(16.4)	(12.0)
ton-km	7,009.05	3,777.20	3,165.74	66.11
	(100.00)	(53.9)	(45.2)	(0.9)

Source: *NYFZBG 1997*, p. 200.

Note: Percentages are given in parentheses.

Coal freight under plan carried on the waterways between 1986 and 1996 increased from 55 million tons to 93, falling to 86 in 1997. Coal freight ton-km rose from 114 million in 1986 to 263 in 1997, and average shipping distance from 2,068 km to 3,064 over the same period. There was still an imbalance between the coal-loading capacity of northern ports and the coal-unloading capacity of southern ports.

The port at Qinhuangdao handled some 70 per cent of all the coal handled by the seaports, and by far the largest amount of the northern coal destined for southern domestic use, followed by Rizhao (Shandong), Qingdao (Shandong), Lianyungang (Jiangsu), Shijiusuo (Shandong), Tianjin, Qingdao (Shandong), Dalian (Liaoning), Zhanjiang (Guangdong), Jinzhou (Liaoning), Suifenhe (Heilongjiang), and Dandong (Liaoning). The main receiving ports were Shanghai, Ningbo (Zhejiang), Huangpu (Greater Shanghai), Yingkou (Liaoning), Guangzhou and Shantou (Guangdong), Xiamen (Fujian), and Hainan (Hainan Island).

As noted above, in the early 1990s there were serious capacity problems on the railway line from Shanxi to Qinhuangdao, and with the storage and loading equipment at the port. Shanghai and other southern cities frequently suffered major electricity shortages due to delayed deliveries from Qinhuangdao. The problems at the port had largely been addressed by the end of the decade and the target of constructing 140 million tons of coal capacity there by 2000 seemed attainable.

The total length of inland navigable waterways decreased sharply over the 1980s and 1990s due to the construction by farmers of sluice gates and dams for irrigation or flood prevention purposes or to serve hydro electric plants, or simply due to a failure to maintain them. In 1978 there were nearly 136,000 km, but this fluctuated between only 109,000 and 110,000 in the 1990s. In 1998 there were 110,263 km and just over 60 per cent had a depth of at least one metre.[307]

The average tonnage of the ships used on the inland waterways in 1994 was 115 tons, a fraction of the size of those used in the US, the Soviet Union, and West Germany in 1980, at 1,400, 1,283, and 1,392 tons, respectively.[308] Most inland ports, apart from several on the Changjiang, had equipment dating from the 1960s, with much of the loading and unloading being done manually on the natural shoreline. In all there were 25 coal ports along the Changjiang, the largest being Nanjing (Jiangsu), Hankou (Hubei), and Zhicheng (Hubei). Hangzhou and Suzhou are the main coal ports on the Grand Canal (Jinghang or Dayunhe Canal) and there are several ports on the Zhujiang (Pearl River) delta.[309] The river and canal systems were poorly linked with each other and with the railways. Considerable dredging work was carried out in the 1990s specifically to enable coal barges to be pulled further inland.[310]

In 1992 China had 902 km of roads, of all types, per million inhabitants, while there were 1,784 in India, 3,335 in the Soviet Union, 9,096 in Japan, and 25,326 in the US. The Soviet Union had 73.7 per cent of its roads paved with asphalt or cement, and India nearly 50 per cent, but only 23.1 per cent of China's were paved, and of these many were not engineered to take heavy trucks.[311] Most of the main highways were carrying traffic far in excess of their capacity, and the average speed of the vehicles was only about half of what was intended.

Over the 1980s and 1990s the Ministry of Communications instituted a National Truck Highway System (NTHS) comprised of interprovincial expressways connecting all provincial capitals and cities with populations greater than 500,000. Foreign loans and credit packages would be used to create some 33,000 km of highways in the form of seven key trunk roads, and 112,000 km of new local feeder roads were to be operational by 2000. A major component of the programme launched in 2000 to open western China was the construction of about 9,000 km of new roads linking all villages to national or regional highway networks. These will certainly accelerate the anticipated switch to coal in remote areas where biomass has been the only available fuel for centuries.

Slurry transport was one of four areas of research designated 'strategic' in the Ninth Five Year Plan. Research had in fact been ongoing since the 1970s, and in 1985 a Pipeline Coal Transportation Experimentation Centre was established in Tangshan (Hebei), but with intense competition for the available water resources, and hardly even enough water for drinking in some areas, the water shortage problem stymied most proposals. In April 1997 work on the Yu-Wei pipeline began, though the amount of coal to be piped from Shanxi to Shandong was considerably less than originally planned.

Consumption

Consumption and economic growth

In 1997 China accounted for 35.9 per cent of all the steam coal, 35.5 per cent of all the hard coal, and 32.9 per cent of all the coking coal consumed in the world.[312] China's total consumption of steam coal was the world's highest at

1,146.4 million tons, with the US a somewhat distant second at 811.7 million tons. For hard and coking coal the figures were respectively 1,332 million tons compared to 839 in the US, and 185.2 compared to 90.9 in the USSR.

Coal changed from accounting for 71.3 per cent of China's total energy consumption in 1979 to a peak in the post-reform years of 76.2 in 1987, 1988, and 1990, and down to 74.7 per cent in 1996. Over 1997 it dropped more than 3.0 percent to 71.5, and another 1.9 per cent over 1998 to 69.6. In 1999 the breakdown of total energy consumption was 67.1 per cent coal, 23.4 oil, 2.8 gas, 6.7 hydro-electric, and less than 1 per cent nuclear power.[313]

For comparison, Table 6.45 gives the breakdown of primary energy consumption for several countries in 1997. South Africa relied on coal to meet 77.2 per cent of its energy needs and was the country closest to China in terms of energy structure. The next most coal-dependent countries were Poland and Czechoslovakia, where coal accounted for 67.8 and 52.1 per cent, respectively.

Throughout the 1980s and early 1990s energy shortages not only affected the economy in terms of several billion yuan of productive value lost each year from domestic capital investment, but were also a key factor discouraging the much needed foreign investors who feared that the unreliable energy supplies would jeopardize the viability of their ventures, and therefore opted to establish their businesses in other Asian countries.

Table 6.45 International comparison of primary energy consumption breakdown, 1997 (%)

	Oil	*Gas*	*Coal*	*Nuclear*	*Hydro*
China	18.6	2.2	71.5	neg.	5.7
Australia	53.2	16.4	41.6	—	1.4
Belgium	42.3	19.7	14.8	21.6	neg.
Canada	34.3	29.8	11.5	9.1	12.7
Czechoslovakia	18.9	19.4	52.1	8.3	0.3
France	35.0	12.8	6.0	41.9	2.2
Germany	40.1	20.9	24.6	12.8	0.5
Hungary	28.2	38.9	16.6	14.6	neg.
India (1994)	31.8	7.5	7.3	0.6	2.8
Ireland	51.7	22.1	24.8	neg.	0.5
Japan	52.8	10.6	16.8	16.3	1.5
South Africa (1994)	19.7	—	77.2	2.7	0.2
South Korea	62.2	7.7	17.8	11.6	0.1
Poland	17.7	9.7	67.8	—	0.1
UK	36.3	33.4	17.7	11.2	0.2
US	38.8	23.5	24.2	8.0	1.4
USSR (1994)	23.1	49.3	21.0	4.5	2.1

Sources: For India, South Africa and USSR: British Petroleum Amoco, *British Petroleum Statistical Review of World Energy*, 1995; all others: International Energy Agency, *Coal Information 1997*, Paris: OECD/IEA, 1998, various pages. (The figures are estimates for 1997.)

In the 1980s production in the metallurgical industries, the building blocks of any economy, was slowed due to a lack of coal. Iron ore could not be refined to western standards, and at Wuhan (Hubei), one of China's major iron and steel centres, there was not even enough power to operate the basic machinery.[314] In 1986 the country's largest aluminium smelter at Fushun (Liaoning) was functioning at only 20–40 per cent of its 100,000-ton annual capacity due to electricity shortages, and it was the same situation in the tin industry.[315]

Ownership of electrical and electronic equipment had soared, creating huge increases in energy demand. For example, between 1981 and 1999, possession of washing machines per 100 average income urban households rose from 6.31 to 91.44, refrigerators from 0.22 to 77.74, electric fans from 42.62 to 171.73, and colour television sets from 0.59 to 111.57.[316] However, the total amount of electricity available to households outside the largest cities in 2000 was still nowhere near meeting the average daily demand, let alone peak demand, and the proud owners of these new pieces of equipment did not enjoy full use of them.[317] Joint venture hotels catering to foreign tourists and businessmen continued to install diesel generators at their own expense.

By 1999 the population had expanded by another 297 million, or 31 per cent over the 1978 level. National per capita energy consumption increased by a factor of 59 per cent from 842.7 kg sce in 1980 to 1,338.6 in 1995.[318] China's official total energy consumption figures belie the magnitude of the shortfall in energy supplies. The government's allocation of as much of the electricity supply as possible to the industrial sector while enforcing low civilian consumption is a key reason why economic growth was so high relative to energy consumption. In addition it must be remembered that much of the rural population was still relying on biomass. In 1996 it was estimated that over 206 million tons of oil equivalent of biomass was consumed which was three times the amount of coal and twenty times the amount of electricity consumed in the residential sector.[319]

At the time of writing, factories in several areas outside the most highly developed conurbations continued to operate below capacity and many urban Chinese, let alone those living in rural areas, lived and worked with minimal electricity, and their supplies were constantly threatened.[320] In most medium and small cities there were still powerouts and dangerous current surges which not only caused frustration but damaged equipment and incurred extra costs in restarting, repairing, and replacement. During the winter months some homes and offices in northern China, due to the notoriously inefficient central heating distribution systems, became so overheated that windows had to be kept open. Most, however, were draughty and occupants wore several layers of clothing all day. Official dates set in the 1950s for turning the central heating on and off at the beginning and end of the winter were still in place. Hot water was available for only a couple of hours each day, and homes and offices had minimal low-wattage lighting. Most indoor lights were turned off during the day, and street lighting, even in the largest cities, was virtually absent apart from main routes. Electricity was insufficient and/or unaffordable, and wiring was too light.

Table 6.46 compares the annual growth rates of total energy consumption, coal and electricity consumption with GDP and industry growth rates. Despite the shortages of electricity and coal, economic growth soared, averaging over 8 per cent between 1978 and 1982, nearly 12 per cent between 1983 and 1988, and again from 1991 to 1999. Industrial growth over the same periods averaged a remarkable 8.8, 16.5, and 18.1 per cent, respectively. The consumption elasticity coefficients after 1979 are vastly improved over those in the pre-reform years (compare this table with Table 3.27) The worst coal elasticity coefficient after the beginning of the reforms occurred in 1989, when economic growth fell sharply due to the disruptions caused by the June 4th Tiananmen Incident. That year economic growth was exactly equal to the growth of coal consumption. The best was in 1992, when GDP growth was particularly high. Between 1993 and 1995 annual economic growth slowed while coal consumption growth accelerated, indicating especially poor consumption efficiency. The growth rate of electricity consumption over the entire period, compared to that of coal consumption, was closer to GDP growth, making for higher coefficients.

Regional energy consumption elasticity coefficients between 1986 and 1990 are given in Table 6.47. The coefficients for less than half of the provinces and

Table 6.46 Consumption elasticity coefficients, 1978–99

	Growth rates (%)					Elasticity coefficients		
	Energy consumption	Coal consumption	Electricity consumption	GDP	Industrial sector	Energy	Coal	Electricity
1978		9.8	14.9	11.7	13.6		0.84	1.27
1979	2.5	3.5	9.9	7.6	8.8	0.33	0.46	1.30
1980	2.9	3.7	6.6	7.8	9.3	0.37	0.47	0.85
1981	−1.4	−0.2	2.9	5.2	4.3	—	—	—
1982	4.4	5.9	5.9	9.1	7.8	0.48	0.65	0.65
1983	6.4	5.2	7.2	10.9	11.2	0.59	0.48	0.66
1984	7.4	10.3	7.4	15.2	16.3	0.49	0.68	0.49
1985	8.1	9.4	9.0	13.5	21.4	0.60	0.70	0.67
1986	5.4	5.9	9.5	8.8	11.7	0.61	0.67	1.08
1987	7.2	6.8	10.6	11.6	17.7	0.62	0.59	0.91
1988	7.3	7.9	9.7	11.3	20.8	0.65	0.70	0.86
1989	4.2	4.1	7.3	4.1	8.5	1.02	1.00	1.78
1990	1.8	2.0	6.2	3.8	7.8	0.47	0.53	1.63
1991	5.1	4.8	9.2	9.2	14.8	0.55	0.52	0.99
1992	5.2	3.3	11.5	14.2	24.7	0.37	0.23	0.81
1993	6.2	5.9	11.0	13.5	27.3	0.21	0.44	0.70
1994	5.8	6.3	9.9	12.6	24.2	0.46	0.50	0.79
1995	6.9	7.1	8.2	10.5	20.3	0.66	0.68	0.78
1996	5.9	5.1	7.4	9.6	16.6	0.62	0.53	0.77
1997	−0.6	−3.8	4.8	8.8	13.1	—	—	0.55
1998	−4.3	−7.0	2.8	7.8	10.8	—	—	0.36
1999	−7.7		6.1	7.1	11.6	—	—	0.86

Sources: *TJNJ 2000*, pp. 55, 239, 243, 409; *TJNJ*, various other years; *NYFZBG 1994*, Table 2-12, pp. 27–8.

Note: The elasticity coefficients are calculated using the GDP data. The GDP and gross value of industrial output data are calculated at comparable prices.

Table 6.47 Regional energy consumption elasticity coefficients, 1986–90

	1986	1987	1988	1989	1990
North					
Beijing	3.01	0.39	0.47	0.59	0.38
Tianjin	0.66	1.37	1.16		3.49
Hebei	2.25	0.84	0.72	0.94	
Inner Mongolia		0.95	0.29	4.20	1.23
Shanxi	1.23	1.47	1.20		
Northeast					
Liaoning	0.63	0.20	0.60	1.02	
Jilin	0.65	0.69	0.51		0.91
Heilongjiang	0.21	0.91	0.48	0.65	0.88
Northwest					
Shaanxi	0.81	0.85	0.39	1.08	0.90
Gansu	0.17	0.70	0.34	0.66	0.41
Qinghai	1.32	0.93	0.87		
Xinjiang	0.06	0.05	1.51	0.84	1.19
Ningxia	0.75	2.10	0.66	1.45	6.87
S. Central					
Henan	0.41	0.43	0.74		0.48
Hubei	1.29	1.03	1.28	2.44	
Hunan	1.03	0.97	0.64		0.39
Guangdong	0.93	1.39	0.66	2.82	0.20
Guangxi	0.19	1.21	0.52	0.88	1.26
Southwest					
Sichuan	0.82	0.62	1.05	1.94	
Yunnan	2.03	0.89	0.33	0.92	0.05
Guizhou	1.16	1.85	1.74	1.89	1.54
East					
Shanghai	2.55	0.70	0.19	0.98	1.23
Jiangsu	0.82	1.15	0.74	5.08	
Zhejiang	0.74	1.63	0.98	4.30	1.55
Anhui	0.58	0.98	0.81	0.58	1.13
Jiangxi	1.79	0.85	0.81	0.11	
Fujian	1.13	0.76	0.76	0.35	0.53
Shandong	1.00	0.80	0.42	1.19	0.75
National Average	0.71	0.70	0.65	1.14	0.36

Source: *NYTJNJ 1991*, p. 19.

regions decreased over these years. Ningxia had by far the highest coefficient in 1990, as well as high consumption per 10,000 yuan of GDP (see Table 6.48) indicating that the rate of coal consumption growth was much higher than that of GDP. In the more developed provinces and regions economic growth was higher than coal consumption growth.

Table 6.48 Coal consumption per 10,000 yuan
of GDP by region, 1990 (tons)

North	
Beijing	4.81
Tianjin	5.76
Hebei	9.42
Inner Mongolia	13.77
Shanxi	19.15
Northeast	
Liaoning	8.54
Jilin	10.20
Heilongjiang	9.89
Northwest	
Shaanxi	7.30
Gansu	7.95
Qinghai	7.09
Xinjiang	7.30
Ningxia	14.57
S. Central	
Henan	6.81
Hubei	4.22
Hunan	5.64
Guangdong	2.03
Guangxi	3.97
Hainan	0.74
Southwest	
Sichuan	5.81
Yunnan	5.53
Guizhou	10.63
East	
Shanghai	3.68
Jiangsu	4.73
Zhejiang	2.98
Anhui	5.66
Jiangxi	5.41
Fujian	2.85
Shandong	5.44

Source: *NYTJNJ 1991*, p. 171; *TJNJ 1994*, pp. 35, 376.

The comprehensive coal consumption ratios given in Table 6.49, indicate that there was steady progress in consumption efficiency.[321] For example, according to these calculations, in 1978 it took over 15 tons of coal to generate 10,000 yuan of GDP, and only 6 in 1998. Indeed, according to the IEA, China's rate of reducing energy intensities during the 1980s, which had been among the highest in the world, was much faster than in any OECD country in a similar period.[322]

Compared to other countries, China's overall energy consumption efficiency in the 1990s was still poor. In 1994, China earned only US$0.7 worth of GDP per kg of energy consumed, whereas it averaged US$1.0 for the low-income

Table 6.49 Comprehensive coal consumption
and coal intensity ratios, 1978–98

	Tons of coal consumed per 10,000 yuan of GDP
1978	15.61
1979	14.78
1980	14.52
1981	13.79
1982	13.66
1983	13.02
1984	12.21
1985	11.63
1986	11.47
1987	11.21
1988	11.49
1989	12.44
1990	11.82
1991	10.93
1992	9.65
1993	8.90
1994	8.53
1995	8.38
1996	8.05
1997	7.12
1998	6.13

Source: *TJNJ 2000*, pp. 53, 242, 290; *NYFZBG 1994*, Table 2-12, pp. 27–8.

Note: GDP data was converted into 1978 constant prices.

economies (including China), US$1.7 for the lower middle income, US$2.8 for the upper middle, and US$4.7 for the high-income economies.[323]

Breakdown of coal consumption

In rural areas coal played a greater role in the 1990s than it had in the 1970s. In 1979 it had accounted for 12.5 per cent of total household energy consumption, with biomass at 85.7 per cent. By 1992 the figures were 24.6 and 71.4 per cent, respectively.[324] However, in the country as a whole, there was a reduction in the role of coal in every sector apart from industry. Table 6.50 shows that the use of other fuels was most noticeable in the construction and transport sectors where the share of coal fell from 44.6 per cent in 1980 to 27.8 per cent in 1990, and from 49.8 per cent to 36.2 per cent, respectively. Coal consumed as a proportion of total energy consumed by the railways also decreased from 89 per cent in 1980 to 55 in 1995.[325] The share of oil products increased from 7 per cent to 25, and electric

Table 6.50 Types of energy used in each economic sector, 1980–90

		Raw coal and products	Crude oil and products	Natural gas	Hydro electricity	Heat power
Agriculture	1980	43.4	27.4	0.0	29.2	0.0
	1985	41.4	27.4	0.0	31.2	0.0
	1990	34.3	31.1	0.0	34.7	0.0
Total industry	1980	60.0	7.7	4.4	25.5	2.4
	1985	53.8	15.4	2.8	25.6	2.4
	1990	51.2	15.1	2.2	28.4	3.1
Light industry	1980	59.8	10.5	1.7	23.5	4.5
	1985	61.9	7.5	1.2	25.7	3.7
	1990	59.3	6.8	0.8	28.6	4.6
Heavy industry	1980	49.1	17.9	5.0	26.0	2.0
	1985	51.8	17.4	3.2	25.6	2.1
	1990	49.2	17.2	2.5	28.4	2.7
Transport and communications	1980	49.8	45.9	0.3	3.8	0.0
	1985	46.7	46.2	0.3	6.8	0.0
	1990	36.2	54.1	0.6	9.2	0.0
Construction	1980	44.6	26.7	8.4	20.2	0.1
	1985	31.2	32.6	14.4	21.8	0.1
	1990	27.8	39.2	11.6	21.1	0.3
Commerce	1980	65.8	8.1	0.0	13.4	12.6
	1985	72.8	7.5	0.0	19.7	0.0
	1990	66.1	9.8	0.0	24.1	0.0
Residential	1980	91.7	1.6	0.3	4.5	1.8
	1985	89.2	2.1	0.4	6.7	1.7
	1990	81.2	3.0	1.6	12.0	2.3

Sources: *NYTJNJ 1986*, pp. 80–4; *NYTJNJ 1991*, pp. 163–8.

power from 4 to 20 per cent. In construction there was much greater use of oil, while in transport there was much greater use of electricity as well as of oil. There was virtually no change in the dependence on coal of the light and heavy industrial sectors.

The breakdowns of total, final, and intermediate coal consumption between 1980 and 1998, are given in Table 6.51.[326] Of the total, the most notable changes over the period are the increase in the share devoted to industry from 71.9 per cent to 88.8, and the decrease in the share consumed by the residential sector from 19.0 to 6.9.[327] Per capita purchase of coal by urban households increased from 240 kg in 1981 to a peak of 271 kg in 1985, but decreased to 115 kg by 1999, while per capita electricity consumption for non-productive purposes rose from 13.4 kWh in 1983 to 106.6 in 1998, per capita liquefied petroleum gas consumption from 0.6 m^3 to 6.2, natural gas from 0.1 m^3 to 1.9, and coal gas from 1.5 m^3 to 6.0.[328] The percentage of urban population with access to gas rose from 22.4 per cent in 1985 to 81.7 per cent in 1999.[329]

Table 6.51 Breakdown of coal consumption, 1980–98

	1980		1985		1990		1995		1996		1997		1998	
	Total (mt)	%	Total (mt)	%	Total (mt)	%	Total (mt)	%	Total (mt)	%	Total (mt)	%	Total (mt)	%
Total	610.10	(100.0)	816.03	(100.0)	1,055.23	(100.0)	1,376.77	(100.0)	1,447.34	(100.0)	1,392.48	(100.0)	1,294.92	(100.0)
Agriculture	15.50	(2.5)	22.09	(2.7)	20.95	(2.0)	18.57	(1.3)	19.17	(1.3)	19.27	(1.4)	19.23	(1.5)
Industry	438.48	(71.9)	586.13	(71.8)	810.91	(76.8)	1,175.71	(85.4)	1,238.86	(85.6)	1,216.71	(87.4)	1,149.52	(88.8)
Heavy industry	—		500.91	(61.4)	695.48	(65.9)	—		—		—		—	
Light industry	—		85.22	(10.4)	115.43	(10.9)	—		—		—		—	
Construction	5.56	(0.9)	5.32	(0.7)	4.38	(0.4)	4.40	(0.3)	4.46	(0.3)	3.83	(0.3)	6.12	(0.5)
Transportation	19.34	(3.2)	23.07	(2.8)	21.61	(2.0)	13.15	(1.0)	11.76	(0.8)	14.31	(1.0)	13.91	(1.1)
Commerce	4.55	(0.7)	7.38	(0.9)	10.58	(1.0)	9.77	(0.7)	10.74	(0.7)	8.63	(0.6)	9.48	(0.7)
Other	10.91	(1.8)	15.80	(1.9)	19.80	(1.9)	19.87	(1.4)	18.35	(1.3)	7.35	(0.5)	7.83	(0.6)
Residential	115.74	(19.0)	156.24	(19.1)	167.00	(15.8)	135.30	(9.8)	143.99	(9.9)	122.38	(8.8)	88.84	(6.9)
Final consumption	388.04	(63.6)	527.04	(64.6)	602.06	(57.1)	661.56	(48.1)	684.54	(47.3)	617.92	(44.4)	563.47	(43.5)
Industry	216.43	(35.5)	297.15	(36.4)	357.74	(33.9)	460.50	(33.4)	676.05	(46.7)	442.14	(31.8)	418.07	(32.3)
Intermediate consumption	194.62	(31.9)	253.97	(31.1)	412.58	(39.1)	694.88	(50.5)	742.12	(51.3)	774.51	(55.6)	731.45	(56.5)
Power generation	126.48	(20.7)	164.41	(20.1)	272.04	(25.8)	444.40	(32.3)	488.09	(33.7)	489.79	(35.2)	494.89	(38.2)
Heating	—	(—)	14.62	(1.8)	29.96	(2.8)	58.87	(4.3)	63.66	(4.4)	62.45	(4.5)	63.20	(4.9)
Coking	66.82	(11.0)	73.04	(9.0)	106.98	(10.1)	183.96	(13.4)	184.56	(12.8)	192.97	(13.9)	156.28	(12.1)
Gas production	1.31	(0.2)	1.91	(0.2)	3.60	(0.3)	7.64	(0.6)	5.82	(0.4)	7.33	(0.5)	6.85	(0.5)
Losses in coal washing and dressing	27.44	(4.5)	35.01	(4.3)	40.59	(3.8)	20.33	(1.5)	20.69	(1.4)	23.11	(1.7)	11.59	(0.9)

Source: *TJNJ*, various years.

Notes: 'Agriculture' is comprised of farming, forestry, animal husbandry, fishery, and water conservancy. 'Transportation' is comprised of transportation, postal, and telecommunication services. 'Commerce' is comprised of commerce, catering services, materials supply, marketing, and storage. Final consumption means direct use of coal, while intermediate means transformation of coal into electricity, heat, coke, or gas.

Intermediate consumption of coal, that is coal which is transformed into another form of energy, increased from 31.9 per cent of the total in 1980 to 56.5 in 1998, with power generation accounting for most of this, rising from 20.7 to 38.2 per cent. Final consumption of coal decreased 20 per cent from 63.6 per cent of the total in 1980 to 43.5 in 1998. Of the final consumption, however, the industrial sector's direct use of coal changed from 35.5 per cent in 1980 to 46.7 in 1996, and then due to the closure of thousands of state owned industries and industry efficiency improvements, down to about 32 per cent in 1998. The proportion of coal lost in processing decreased from 4.5 to 0.9 per cent.

By way of comparison, Table 6.52 gives the breakdown of coal consumption in 1996 for several other countries. These used on average nearly 60 per cent of their coal to generate electricity, with the remainder devoted to fuelling metallurgical industries, or converted into gas or other fuels, or used to manufacture chemicals.

Table 6.52 International comparison of coal consumption, 1996

	China	IEA North America	IEA Europe	IEA Pacific	India	South Africa
Electricity and heat generation	23.8	86.8	59.2	46.5	50.9	46.5
Gas works	0.2	—	—	0.2	—	3.4
Transformation losses[a] of which:	0.7	2.2	6.4	15.9	negl.	0.5
Inputs to coke ovens	10.1	4.3	13.1	29.9	3.8	3.1
Liquefaction	—	—	—	—	—	15.6
Final Consumption[b] of which:	37.6	5.5	16.0	17.4	24.0	17.2
Industry Sector of which:	28.0	5.1	11.9	16.4	23.9	13.2
Iron and Steel	7.3	7.0	6.1	8.7	7.9	4.0
Chemical	6.1	1.0	1.0	0.5	0.6	6.3
Non-metallic minerals	6.9	1.3	2.3	4.2	2.5	0.9
Paper, pulp and print	0.9	0.3	0.3	1.0	0.9	—
Other Sectors[c] of which:	9.6	0.3	3.9	0.9	0.1	4.0
Residential	6.7	0.3	3.2		negl.	1.5
Non-energy use	—	0.1	—	—	—	—

Source: IEA, *Coal Information 1997*, Paris: OECD/IEA, 1998, pp. II-84, 96, 108; III-16, 27, 36.

Notes: Not all the IEA categories are given here, hence the columns do not add up to 100 per cent.
a. Transformation losses refer to the losses in the transformation of coal to secondary products (mainly coke and briquettes) and to tertiary products (mainly blast furnace gas).
b. Includes non-energy use.
c. Includes commercial and public services, agriculture, transport, and residential.

Coal in thermal electric power generation

In 1999 China produced over 1,239 billion kWh of electricity from a capacity of about 299 million kW, which was second highest in the world after the US, in both production and capacity. Of the capacity in China that year, 74.8 per cent was thermal, 24.4 per cent hydro, and 0.8 per cent nuclear (see Table 6.53). For comparison, in India a total of 323.5 billion kWh was generated over 1993–94 from a capacity of 76.7 million kW, of which 76 per cent was thermal, 22 was hydro, and 2 was nuclear.[330]

The annual increases in China's total installed capacity between 1979 and 1999, averaging 8.7 per cent for thermal and 7.2 for hydro, were unparalleled in world history (see Table 6.54). Both thermal and hydro generation more than quadrupled. From 1979, the additional 38.9 million kW of thermal power capacity and the 13.6 million kW of hydro power capacity, completed by 1988 and 1990, respectively, were almost equivalent to the total built of each in the entire pre-reform period.

Although the average annual growth of electricity output was very rapid, it was less than that of GDP. For example, if 1979 is made the base year and given an index number of 100, the figures for GDP in constant terms and electricity production in 1999 were, respectively, 666 and 640.

Throughout the 1980s most urban areas had little more than what would be regarded by developed world standards as basic supplies of electricity. What was available was rationed, with the industrial sector taking precedence over household and commercial consumption. Most foreign joint ventures, including factories and hotels, had no alternative but to install generators to ensure an uninterrupted supply of power.

Generally speaking, after the sharp fall in industrial demand for energy in 1996, Guangdong, Jiangsu, Hunan, Hubei, Jiangxi, and the three northeastern provinces had adequate electricity, except in the summer heat when air conditioners were used. However, the power fluctuated seriously. Brownouts were still occurring in many cities throughout the country, now not due to shortages of generation capacity, but rather to frequent surges in the current and overloading of circuits. The immediate, pressing task in the urban areas was addressing the line capacity constraint, that is, expansion of the capacity of the wiring and fuses, and the installation of more transformers and voltage stabilizers. Simply stated, the electricity was available but it could not be transmitted satisfactorily, if at all, to the point of use. Some power plants even had to operate below capacity due to the high 'failure rate' of the existing distribution network.[331]

By the late 1990s millions of peasants in rural areas had gained access to electricity for the first time in their lives. Electrification of rural households increased between 1987 and 1997 from 74.6 to 95.9 per cent. However, as of the end of 1998 there were still some 2,400 counties without access to electricity, 17 with no power at all, affecting about 65–67 million people. The difference between per capita consumption in urban and rural areas was growing wider.[332] In 1980 nationwide per capita consumption of electricity for household use was 107 kWh

Table 6.53 Electricity capacity and generation, 1978–99

	Capacity (million kW)							Generation (billion kWh)						
	Total	Thermal	% share	Hydro	% share	Nuclear	% share	Total	Thermal	% share	Hydro	% share	Nuclear	% share
1978	57.12	39.84	69.7	17.28	30.3			256.6	212.0	82.6	44.6	17.4		
1979	63.02	43.91	69.7	19.11	30.3			282.0	231.9	82.2	50.1	17.8		
1980	65.87	45.55	69.2	20.32	30.8			300.6	242.4	80.6	58.2	19.4		
1981	69.13	47.20	68.3	21.93	31.7			309.3	243.8	78.8	65.5	21.2		
1982	72.36	49.40	68.3	22.96	31.7			327.7	253.3	77.3	74.4	22.7		
1983	76.44	52.28	68.4	24.16	31.6			351.4	265.0	75.4	86.4	24.6		
1984	80.12	54.52	68.0	25.60	32.0			377.0	290.2	77.0	86.8	23.0		
1985	87.05	60.63	69.6	26.42	30.4			410.7	318.3	77.5	92.4	22.5		
1986	93.82	66.28	70.6	27.54	29.4			449.5	355.0	79.0	94.5	21.0		
1987	102.90	72.70	70.6	30.19	29.4			497.3	397.0	79.9	100.0	20.1		
1988	115.50	82.80	71.7	32.70	28.3			545.2	436.0	80.0	109.2	20.0		
1989	126.64	92.06	72.7	34.58	27.3			584.8	466.5	79.8	118.3	20.2		
1990	137.89	101.84	73.9	36.05	26.1			621.2	494.5	79.6	126.7	20.4		
1991	151.47	113.59	75.0	37.88	25.0			677.5	552.8	81.6	124.7	18.4		
1992	166.53	125.85	76.5	40.68	23.5			753.9	618.2	82.7	130.7	17.3		
1993	182.91	138.32	75.6	44.59	24.4			836.4	685.7	82.0	150.7	18.0		
1994	199.90	148.74	74.4	49.07	24.5	2.09	1.1	927.9	747.0	80.5	166.8	19.5		
1995	217.22	162.94	75.0	52.18	24.0	2.10	1.0	1,007.7	804.3	79.8	190.6	18.9	12.8	1.3
1996	236.54	178.86	75.6	55.58	23.5	2.10	0.9	1,080.0	877.8	81.3	188.0	17.4	14.3	1.3
1997	254.24	192.28	75.6	59.73	23.5	2.23	0.9	1,134.5	924.1	81.5	196.0	17.3	14.4	1.2
1998	277.29	209.88	75.7	65.07	23.5	2.34	0.8	1,166.2	944.1	81.0	208.0	17.8	14.1	1.2
1999	298.77	223.43	74.8	72.97	24.4	2.37	0.8	1,239.3	984.0	79.4	241.2	19.5	14.1	1.1

Sources: TJNJ, various years; Sinton, Jonathan E. et al. (eds), China Energy Databook, Berkeley, CA: Ernest Orlando Lawrence Berkeley National Laboratory Report LBL-32822, Rev. 4, Sept. 1996, Tables II-17 and II-19, pp. II-60, 61, 64; NYFZBG, p. 93; Zhongguo dianli nianjian bianji weiyuan hui (China Electricity Yearbook Editorial Committee), Zhongguo dianli nianjian (China Electricity Yearbook), Beijing: Zhongguo dianli chubanshe, various years.

Table 6.54 Electricity capacity and generation growth (%), 1978–99

	Capacity				Generation			
	Total	Thermal	Hydro	Nuclear	Total	Thermal	Hydro	Nuclear
1978	11.0	11.6	9.6		14.9	20.6	−6.3	
1979	10.3	10.2	10.6		9.9	9.4	12.3	
1980	4.5	3.7	6.3		6.6	4.5	16.2	
1981	4.9	3.6	7.9		2.9	0.6	12.5	
1982	4.7	4.7	4.7		5.9	3.9	13.6	
1983	5.6	5.8	5.2		7.2	4.6	16.1	
1984	4.8	4.3	6.0		7.3	9.5	0.5	
1985	8.6	11.2	3.2		8.9	9.7	6.5	
1986	7.8	9.3	4.2		9.4	11.5	2.3	
1987	9.7	9.7	9.6		10.6	11.8	5.8	
1988	12.2	13.9	8.3		9.6	9.8	9.2	
1989	9.6	11.2	5.7		7.3	7.0	8.3	
1990	8.9	10.6	4.3		6.2	6.0	7.1	
1991	9.8	11.5	5.1		9.1	11.8	−1.6	
1992	9.9	10.8	7.4		11.3	11.8	4.8	
1993	9.8	9.9	9.6		10.9	10.9	15.3	
1994	9.3	7.5	10.0		10.9	8.9	10.7	
1995	8.7	9.5	6.3	0.5	8.6	7.7	14.3	
1996	8.9	9.8	6.5	0.0	7.2	9.1	−1.4	11.7
1997	7.5	7.5	7.5	6.2	5.0	5.3	4.3	0.7
1998	9.1	9.2	8.9	4.9	2.8	2.2	6.1	−2.1
1999	7.7	6.5	12.1	1.3	6.3	4.2	16.0	0.0

Source: Calculated from Table 6.53.

and the average for urban areas was 221 kWh.[333] The average for rural areas at 79 kWh was 36 per cent of that in the urban areas. By 1995 the nationwide average had increased to 829 and the urban average, 1,633 kWh. Though the average for rural areas had risen severalfold to 500 kWh, it was now well under a third of the average for urban areas. The continued burning of dry twigs and crop residues for cooking and heating worsened the country's already severe desertification, soil infertility, and flooding problems. For perspective, about 84 per cent of India's villages were electrified in 1994, but only 27 per cent of rural households had access to electricity.[334]

Nationwide per capita electricity consumption in 1997 at about 846 kWh was only 37 per cent of the world average of 2,258. For comparison, per capita consumption averaged 490 kWh in Africa, 1,453 in Latin America, and 504 in Asia, of which it was 397 in India, 4,959 in Hong Kong, and 6,752 on Taiwan.[335] It stood at 3,003 in non-OECD Europe, 3,716 in the USSR, and 5,409 in OECD Europe.

A task set for the late 1990s and early 2000s was to close down all of the highly inefficient and polluting thermal plants of 30 MW capacity or less, and to ensure that no new plants less than 100 MW be built. Throughout the 1980s and 1990s, debate continued as to what was the most efficient combination of thermal, hydro, and nuclear capacity. Thermal continued to dominate as it had before 1979, ranging from 68 to 77 per cent of the total, with virtually all the remainder being hydro. For perspective, in the OECD countries an average of 69.6 per cent was thermal in 1980, 11.0 per cent hydro, and 19.2 nuclear.[336] In 1997, the percentages were 61.1, 23.4, and 15.0, respectively. In India the thermal–hydro–nuclear ratio was 58.1 : 39.0 : 2.8 per cent in 1980–81, and 69.6 : 27.8 : 2.6 in 1991–92.[337] In both China and India wind, solar, geothermal, wave, and marsh gas power together accounted for less than 1 per cent of the total electricity generated. Nuclear power generation did not begin in China until 1992.[338]

In the late 1970s it was mistakenly believed that China's domestic oil supplies were sufficiently large to warrant converting coal-burning equipment to oil. In 1978 there were 8 million kW worth of oil-consuming power-generation units in operation, of which five had originally been built to consume coal. By the end of 1980, however, 2.7 million kW had been re-converted to coal, and the fuel mix for the thermal plants that year was 72.7 per cent from coal, 7.4 from crude oil, and 18.2 from fuel oil.[339] By 1990, the figures were, respectively, 90.1, 0.9, and 7.2.[340] Between 1980 and 1998 the proportion of total coal consumption devoted to power generation rose substantially from 20.7 to 38.2 per cent. The tonnage consumed annually by power plants nearly quadrupled.

The average size of the thermal plants was still comparatively small. In the late 1990s, of the total installed capacity, units with a generating capacity of 200 MW or larger accounted for only about 45 per cent, while units of 60 MW or less accounted for about 30 per cent. The largest thermal plant in the country with 1,980 MW was in Guangzhou (Guangdong). The situation in India was similar. Of that country's 352 thermal plants in 1994, 175 were 15–100 MW, 82 were 100–150 MW, 86 were 200–210 MW, and 9 were 500 MW.[341]

In the early 1970s hundreds of very small thermal generators of 25 MW or less were built to provide power in the quickest, least expensive way. These did ease shortages temporarily, but in the 1980s it became imperative that they be replaced with larger ones, the main considerations being extreme inefficiency and pollution. Average installed capacity increased from 43.91 GW in 1979 to 137.12 GW in 1993.[342] By the mid-1990s, an added incentive to remove the small plants was the problem of over-production of electricity relative to line capacity and in 1998 the government called for the decommissioning of some 10,000 small plants under 50 MW over the next three years.

Due to the severe, ongoing power shortages in 1989 and 1990 the government had called for accelerated construction of 8,700 MW of thermal electric mine-mouth power plant capacity at openpit bases in Inner Mongolia, and ultimately some 120,000 MW in total. However, the planned proportion of power coming from mine-mouth plants by 2000 was not realized. Part of the problem was the fact that foreign investment in coal development did not materialize (see Foreign Involvement section). More fundamental were the inability of the coal and power industries to agree on a national mine-mouth power strategy, the unavailability of the large amounts of water at the mine sites required for cooling, and difficulties in preventing high transmission losses over great distances. There was also uncertainty over the development of the Three Gorges Project.

China's use of coal to generate electricity was still relatively inefficient. Grams of coal equivalent (gce) consumed per gross kW hour at thermal electric plants of 6 MW or larger decreased from 422 in 1979 to 375 in 1997, and per net kW hour from 457 to 408.[343] It is estimated that each gram decline in China's consumption rate equates to a savings of over 400,000 tons of coal.[344] At some of the small plants consumption could be as high as 1,000 gce. By comparison, 372 grams were consumed in the US in 1992, 331 in Japan, 310 in the former USSR, 344 in France, 310 in West Germany, 323 in Italy, 356 in England, and 351 in Canada.[345] The rate of electricity consumption at China's thermal power plants of 6 MW of installed capacity or above changed from 7.63 per cent in 1979, to a high of 8.22 in 1990, to 7.81 in 1997, while at hydroplants it increased from 0.195 per cent in 1979 to 0.50 in 1997.[346]

Much of the generating equipment was used for many more years than it was intended, and very intensively at that. Insufficient time, if any, was made for regular maintenance, leading to high breakdown and accident rates. Average annual operating hours for thermal plants were 5,114 in 1997, down from the 5,956 in 1979, but much higher than the 4,000 hours per year in most developed countries.[347] In 1997 there were still several provinces and regions operating well over 6,000 hours: Ningxia (6,406), Guizhou (6,206), Gansu (6,053), and Hebei (6,021).[348]

The average demand for electricity as a proportion of peak demand, called the load factor, stood at over 85 per cent at Chinese plants in 1992, compared to 61 per cent at American (1991) and Indian (1993), and 56 per cent at Japanese (1992).[349] Thermal efficiency, that is, output of energy as a percentage of input, at Chinese coal-fired plants in 1997 averaged 27–29 per cent, compared to an average of 38 per cent in the OECD countries, 42.3 per cent in Thailand (1992), 37.5 on Taiwan (1992), and 36.5 in Korea (1993).[350] Chinese plants of 6 MW had efficiencies of only 16 per cent.[351] Losses in generation in 1990 were generally higher than in the rest of Asia at 6.4 per cent, compared to the 3.2 per cent in Korea, for example, but lower than the 7.9 per cent in India.[352]

In 1980 there was a total of 266,843 km of transmission lines in China, of which 64.5 per cent were 35–66 kV, 24.3 per cent 110 kV, 0.2 per cent 154 kV, 10.7 per cent 220 kV, and 0.3 per cent 330 kV.[353] The first 500 kV line was installed in 1981.[354] By 1997, of the 625,871 km of lines, 44.3 per cent were 35 kV, 7.4 per cent were 66 kV, 27.5 per cent were 110 kV, 17.3 per cent were 220 kV, 1.0 per cent were 330 kV, and 2.4 per cent were 500.[355] For comparison, in 1992 India's State Electricity Boards had a total of 4,117,303 km of lines, of which 3.7 per cent were 33 kV, 32.3 per cent were 110 kV, and 64.0 were 400 kV.[356]

In the late 1990s there were five trans-provincial power grids in the northeast, north central, central east, central, and northwest regions (see Table 6.55). In terms of capacity the largest were the Central East and Central. There was also one inter-connected power system linking Guangdong, Guizhou, Yunnan, and Guangxi, and six separate provincial networks in Shandong, Sichuan, Fujian, and Hainan.

Table 6.55 Capacity and output of main electricity networks, 1995

Network name	Installed capacity		Electricity output	
	Total 10,000 GW	Thermal as % of total	Total kWh	Thermal as % of total
Northeast	2,719.7	83.9	1,308.9	90.1
North Central	2,994.3	96.0	1,531.6	98.5
Central East	3,702.9	93.7	1,806.9	96.1
Central	3,039.9	62.6	1,458.4	70.0
Northwest	1,256.4	60.2	630.8	73.5
Shandong Province	1,229.2	99.6	734.6	99.9
Fujian Province	610.0	40.6	251.1	42.9
Guangdong Province	2,271.7	79.5	821.1	84.0
Guangxi Province	519.4	41.9	206.0	38.3
Sichuan Province	1,067.7	59.3	517.7	62.1
Yunnan Province	443.9	29.0	187.3	33.4
Guizhou Province	365.9	57.8	182.1	65.3
Hainan Province	151.8	66.6	31.9	64.1
Total	20,372.8		9,398.4	

Source: Adapted from Table 5.4, *NYFZBG 1997*, p. 94.

Until the late 1980s many eastern and central power networks continued to be overloaded all year round, necessitating rationing of peak-time electricity use. In 1999 there still was rationing during the peak summer season in some rural areas. Over the 1990s considerable progress was made in connecting surplus electricity areas with deficit areas. Five electric power groups, together controlling 70 per cent of national capacity, were established in January 1993 corresponding to the five main grids. Each power group was to operate as an independent business enterprise. In June 1996 The National Power Grid Development Corporation was established to improve the coordination of local and trans-provincial grids, and it was hoped that an integrated national grid would be completed between 2010 and 2020, after the full realization of the Three Gorges Hydro Project.

Transmission and distribution losses in 1997 amounted to 8.20 per cent in China, down from 9.24 per cent in 1979, more than twice as much as the 1990 figures of 3.4 per cent in the Philippines and Singapore, but much less than the 23 per cent in India, 16 per cent in Indonesia, and 11 per cent in Malaysia.[357]

Energy conservation

Beginning with the Sixth Five Year Plan, conserved energy was regarded as an additional energy resource, and detailed energy conservation targets and measures were included in that Plan and all the plans thereafter.[358] The government recognized the double role of coal conservation in improving the supply/distribution situation on the one hand, and lessening pollution on the other.

In general terms the conservation efforts were focused on improving boilers and kilns, using excess heat from electricity production, replacing old equipment with energy-saving equipment, replacing small equipment with big, avoiding the use of low-quality coal, and perfecting industrial production techniques and management. The results over the 1980s and 1990s, while all commendable, varied from sector to sector. Table 6.56 provides the coal conservation rates by region for 1992. The wide range of figures indicates that the conservation measures introduced since 1980 were not adopted uniformly across the country. In general, the wealthier, more populated coastal areas had better conservation rates than the less populated remote ones. The former likely had newer, more efficient, and better maintained consuming equipment.

According to various Chinese researchers consumption of coal per ton of steel decreased from 2.04 tons (sce) in 1980 to 1.52 in 1995.[359] This was still very much higher than the 0.76 tons it required in Japan in 1986, and the 0.8 tons in West Germany and France in the early 1980s.[360] Key cement-producing enterprises in China required 191.2 kg (sce) to heat one kilogram of cement mixture while Japan required only 121.1. The average consumption of coal (sce) per ton of steam generated by industrial furnaces was 118.4 kg in advanced countries, while it was 155.9 kg in China.

There are a variety of factors which can contribute to improved energy intensity, such as higher imports of energy-intensive products like steel and fertilizer, changes in the structure of the economy, replacement of old equipment with new, changes in the fuel mix, changes in equipment operation procedures, etc.

Table 6.57 shows the contribution of each of the constituent sectors making up GDP from 1980 to 1999. The top half shows that the share of primary industry dropped about 12 per cent from over 30 per cent to about 18 in 1999, the 2 per cent loss and gain in the manufacturing and construction industries respectively cancelled each other out, and that the share of tertiary industry increased from about 21 to 33 per cent. China's economy was still relatively oriented towards energy-intensive heavy industry. For comparative purposes, the tertiary sector accounted for an average of 55 per cent of the OECD member economies.

The growth rates of primary industry and the wholesale, retail, and catering trade sector approximately halved between the Sixth and Eighth Five Year Plan periods. Growth in secondary industry, including manufacturing and construction, was very high at over 17 per cent over the Eighth Five Year Plan period, but dropped back to about 9 per cent in the late 1990s. The highest period of growth for the tertiary sector was over the Sixth Five Year Plan period at nearly 14 per cent. It slowed to about 8 per cent over the Ninth. There was little variation in the growth rate of the transportation and communication sector.

Various economists have formulated econometric models in an attempt to isolate the determining factors in China's success in conserving energy in recent years. There is no consensus. One Chinese economist estimated that in general terms 50 per cent of the successes as of the late 1980s were a result of changes in China's economic structure, 40 per cent from improved energy management and technical renovation, and 10 per cent from the importing of products whose production is energy-intensive.[361]

Table 6.56 Coal conservation rates by region, 1992 (%)

National average	8.45
North	
Beijing	10.91
Tianjin	4.12
Hebei	8.33
Inner Mongolia	0.49
Shanxi	3.70
Northeast	
Liaoning	3.85
Jilin	4.11
Heilongjiang	2.94
Northwest	
Shaanxi	2.88
Gansu	−0.66
Qinghai	−17.00
Xinjiang	−1.08
Ningxia	4.19
S. Central	
Henan	6.88
Hubei	8.27
Hunan	7.86
Guangdong	9.52
Guangxi	8.92
Southwest	
Sichuan	5.00
Yunnan	0.92
Guizhou	12.23
East	
Shanghai	3.73
Jiangsu	14.64
Zhejiang	8.65
Anhui	5.09
Jiangxi	11.85
Fujian	15.92
Shandong	4.90

Source: '1992 nian jieneng jiang hao he ziyuan zong heli yong chengxiao xianzhu', (Remarkable Progress in Resource Conservation and Use in 1992), *ZGNY*, no. 9 (1993): 24.

Economists at the Energy Research Institute in Beijing concluded that about 70 per cent of the reduction in industrial energy intensity over the 1980s was due to shifts in the structure of the economy, specifically a shift away from agriculture and industry to commerce and, within the industrial sector, a shift away from heavy to light.[362] Soon afterwards two Americans argued that it was shifts

Table 6.57 Breakdown of GDP, 1980–99

	1980	1985	1990	1995	1996	1997	1998	1999
A. Percentage breakdown								
Total GDP	100.0	100.0	100.0	100.0	100.0	100.0	100.0	100.0
Primary industry	30.1	28.4	27.1	20.5	20.4	19.1	18.6	17.7
Secondary industry	48.5	43.1	41.6	48.8	49.5	50.0	49.3	49.3
Manufacturing	44.2	38.5	37.0	42.3	42.8	43.5	42.6	42.7
Construction	4.3	4.7	4.6	6.5	6.7	6.5	6.7	6.6
Tertiary industry	21.4	28.5	31.3	30.7	30.1	30.9	32.1	33.0
Transportation and communications	4.5	4.5	6.2	5.2	5.1	5.1	5.3	5.4
Wholesale, retail and catering trade	4.7	9.8	7.7	8.4	8.2	8.3	8.4	8.4

	6th FYP (1981–85)	7th FYP (1986–90)	8th FYP (1991–95)	First years of 9th FYP (1996–98)	1998–99
B. Average growth in constant prices					
Total GDP	10.4	7.9	12.0	8.7	7.1
Primary industry	8.3	4.2	4.2	4.0	2.8
Secondary industry	10.2	9.1	17.5	10.6	8.1
Manufacturing	10.1	9.3	17.7	10.9	8.5
Construction	11.4	6.9	14.9	7.6	4.3
Tertiary industry	13.6	9.5	10.0	8.2	7.5
Transportation and communications	10.4	9.9	11.1	10.1	9.3
Wholesale, retail and catering trade	15.1	5.1	7.6	7.2	7.2

Source: *TJNJ 2000*, pp. 54–5.

between energy intensive subsectors which accounted for most of the reduction in total industrial energy intensity.[363]

According to a World Bank study about two-thirds of the reduction in energy intensity was the result of shifts in final and intermediate demand for goods and services, effected mainly through changing the product mix. Efficiency gains in industrial production through technical changes, and the slow growth of energy consumption in the residential sector were also factors.[364]

Two Hong Kong economists concluded it was not due to changes in the structure of the economy, but rather to major improvements in the efficiency of energy consumption, especially in heavy industry and in the transport and communications sectors.[365] Similar results were obtained by yet another Chinese economist who believed that the declines in energy intensity between 1981 and 1987 were due to energy conservation programmes, improvements in macroeconomic performance, and increases in energy prices.[366]

Economists at the IEA contend that the energy intensity figures derived from official data do not actually reflect the true situation because the total GDP figures are underestimated and the GDP growth rate figures are overestimated.[367] Three economists working at Harvard University also found the data problematic, and concluded that the fall in the energy–output ratio between 1987 and 1992 was due mainly to technical change within sectors.[368] They believed that the import of some energy-intensive products also contributed to the decline, while structural change actually increased the use of energy.

The fact that many of the Chinese industrial workers, at all levels, still did not appreciate the economic potential of saving energy also pulled down average energy efficiency rates. Having not received professional education, nor travelled to developed countries and acquired knowledge of international best practices, they were completely unaware of standards elsewhere. There was also great difficulty in the enforcing of conservation measures. Some government officials rightly complained that the discussion and research about conservation greatly surpassed consistent and effective action.[369] Throughout the 1980s and early 1990s energy prices were still too low to influence consumption behaviour. However, this began to change in the mid-1990s when marketization of the economy was progressing rapidly.

Ideally, more investment should have been devoted to replacing relatively small, inefficient boilers in particular. It has been agreed by energy economists that improving the efficiency of these would reduce coal consumption by at least 50 million tons per year, possibly up to 90.[370] Economists at the World Bank estimated that in all some 247 million tons of coal, or 20 per cent of total coal consumption in 1996, could have been saved through the adoption of 'best-practice technologies', including the most efficient industrial boiler, power plant, steel, cement, fan, pump, and electric motor-making equipment available.[371] The savings from improved boilers alone would amount to over 28 per cent of this total.

Plate 1 Pile of unprocessed coal from an LNS mine near Shenyang (Liaoning)

Plate 2 Coal briquettes stacked outside the main entrance of an old building in Shenyang (Liaoning)

Plate 3 Outdoor cooking using coal briquettes in Shenyang (Liaoning)

Plate 4 New tunnel construction at Yanzhou CMA (Shandong)
(Photo: Courtesy of *The Beijing Review*)

Plate 5 Coal cutting in Nantun Mine, Yanzhou CMA (Shandong)
(Photo: Courtesy of *The Beijing Review*)

Plate 6 Operation of comprehensive mining equipment at Yongdingzhuang Mine at
Datong CMA (Shanxi) (Photo: Courtesy of *The Beijing Review*)

Plate 7 Coal miners at the end of their shift, Yanzhou CMA (Shandong)
(Photo: Courtesy of *The Beijing Review*)

Plate 8 Coal miners and staff attending university night classes, Yanzhou CMA
(Shandong) (Photo: Courtesy of *The Beijing Review*)

Plate 9 Railway cars loaded mechanically in Zhengzhou (Henan)

Plate 10 Railway cars loaded with coal traversing Henan Province

Plate 11 Trucks loaded with clean coal parked outside the Zhengzhou Heat and Power Plant (Henan)

Plate 12 Coal waiting to be transferred to ships at Qinhuangdao (Hebei)

Plate 13 Coal being loaded onto a ship at Qinhuangdao (Hebei)

Plate 14 Coal ship docked at Qinhuangdao (Hebei)

Plate 15 Coal barges in Shanghai Harbour

Plate 16 Camel teams transporting coal from Baidong Mine at Datong CMA (Shanxi) to hillside residents (Photo: Courtesy of *The Beijing Review*)

Plate 17 Coal pile at the back of a hotel in Hangzhou (Zhejiang)

Plate 18 Stoking the boiler at a hotel in Hangzhou (Zhejiang)

Plate 19 Central heating and hot water pipes leading from a boiler to a building (Beijing)

Plate 20 Coal transport system operations room at the Zhengzhou Heat and Power Plant (Henan)

Plate 21 Zhengzhou Heat and Power Plant (Henan) cooling tower and chimney; railway cars loaded with coal in the foreground

Plate 22 Power lines emanating from the Zhengzhou Heat and Power Plant (Henan)

7 Retrospect and prospect

Retrospect

China's modern coal industry progressed from relatively primitive beginnings in the mid-nineteenth century to that of the world's largest producer, and this position is not likely to be challenged. The path to this achievement was not consistently upward, and the success was qualified. While production goals for 2000 were achieved before the end of the century, those for quality, technical upgrading, resource use, labour efficiency, exports, foreign investment, and profitability for the industry as a whole, were not.

The period from the opening of the treaty ports in the 1840s to the fall of the Qing dynasty in 1911 was one of foreign commercial enterprise. The mainly European capitalists needed coal to fuel their trading ships, powered by the steam engine which had revolutionized industry and transportation alike. Despite constant warlord activity, the largest mines and trunk railways were constructed before 1911 by foreign companies which, in the course of serving their own requirements for coal, laid the foundation for China's industrialization. With an abundance of cheap labour at hand, it was inevitable that mine mechanization remained at a low level. Owners employed contractors who were exacting of gain and often cruel.

The Japanese gained ascendancy in 1905 when they defeated Russia, the principal competitor for the coal resources which were so vital to Japanese industrial development and imperialist goals. Victory in the Sino-Japanese War and the setting up of the Manchuguo puppet state in 1932 assured this ascendancy, and production from Japanese controlled mines thereafter exceeded that from the Chinese mines and the remaining few with European interests. To a large extent they insisted upon using their own people as managers, engineers, and technicians. During the Second World War some of the railways and largest mines were bombed. After the Japanese surrendered in 1945, looting and plundering by Russian forces reduced the plant and equipment capacity of the industry by some 80–90 per cent and further damage was incurred in 1947–48, when the Nationalist–Communist conflict resumed. In 1949 China's total coal production was ninth largest in the world. Quality was poor. There was no national plan for development.

The coal industry, like the economy as a whole, saw tremendous gains punctuated by severe setbacks under Mao. Coal output had been restored to the pre-1949 peak within the first three years of the establishment of the PRC. Soviet-style central planning had been chosen as the economic model and the First Five Year Plan set the coal industry on a rational course. Not long into this Plan, however, Mao grew impatient with what he perceived to be an unnecessarily slow pace and urban bias, and responded by launching a series of campaigns aimed at stimulating much more rapid growth by capturing the enthusiasm and energy of the vast rural masses. The GLF had disastrous consequences for the entire country. Millions died of starvation and the reckless extraction of coal in the backyard furnace campaign set the coal industry back by almost a decade. The Soviet government angrily withdrew all assistance in 1960.

Production and capacity had nearly been restored by 1965, the end of a three-year period of readjustment, when Mao launched the Cultural Revolution, another 'quick-fix' campaign aimed this time at equalizing what he perceived to be a resurgence of inequalities in wealth and education. The coal industry was badly affected in the mid- and late 1960s by the mysterious death of the Minister of Coal, the sabotaging or monopolizing of key railroads by the Red Guards, and by the Third Front Development Programme.

Growth in industrial production in the 1970s was very high despite the tumultuous political climate and the use of equipment far beyond its planned life expectancy. However, as many of the country's top scientists and intellectuals had either been killed or banished to the countryside, the cumulative neglect of research combined with a virtually total absence of communication with the outside world resulting from Mao's closed door policy, led to China's falling seriously behind in its goals of surpassing other developing countries and catching up with the developed world. At the time of Mao's death in 1976 the industrial sector was in a state of stagnation with a demoralized work force, and, in the case of the coal industry specifically, crippling imbalances in output–capacity ratio and spatial rationalization, and fundamental structural problems.

After a brief period of turmoil caused by the Gang of Four and an unsteady short interval under the chairmanship of Hua Guofeng, Deng Xiaoping emerged as the paramount leader in 1978. He recognized the imperative need of raising living standards, opening the country to foreign investment, and seeking the technical assistance of the developed world. Over a three-year period of 'readjustment, restructuring, consolidating, and improving', scientists and intellectuals resumed their places in the research institutes and were encouraged to discuss and write freely about the developmental problems of the country and to devise plans for solving them. In the case of the coal industry mining engineers and workers with technical training were sought out and their skills appropriately utilized.

Deng realized that incentives and accountability had to be injected into the work environment at every level and considerable decision-making power was devolved from Beijing to local governments. Mandatory planning gave way to guidance planning. The central government department responsible for coal was re-organized several times.

The Minister of Coal at that time, Gao Yangwen, fully espoused Deng's goals for efficiency. The contract-responsibility system was implemented in an attempt to realize the greatest productivity from both labour and materials. There were rewards for above-target performance and penalties for sub-performance. Many of the reforms in the coal industry were taking place in the economy as a whole, though at different rates. However, both the coal industry and the railways were classified as strategic and both had huge labour forces, factors which in significant measure accounted for the government's especially cautious approach to their reform. A large number of competing interests had to be rationalized as energy policy cut across the whole economy.

The development of the coal industry was one, if not the most important goal, of the Sixth Five Year Plan (1981–85). Coal was needed not only to sustain the planned modernization of the existing industrial sector, but to fuel the new and burgeoning rural industrial sector. The central government set both short- and long-term targets for production at the existing and new CMA, LS, and LNS underground mines. Strong encouragement was given to the peasants to develop small local mines. It was recognized that nationwide production from these could increase total output most quickly and cheaply. Ambitious development plans for new opencast mines were also set, as well as targets for improved coal quality, and higher coal consumption efficiency. It was hoped that the industry would attract considerable foreign funding, and that coal exports would become a major source of foreign exchange. The planned expansion of railway and water transport of coal, and of thermal power plant and transmission capacity was by far the largest ever planned anywhere.

Not long after the reforms had begun, however, the economy as a whole and the coal industry especially, began to experience the classic problem of developing economies – the uneven rate of progress in the various sectors and the resultant complications in inter-industry linkages. When the peasants from the now disbanded communes began to operate their own mines as town and village enterprises, the annual increments in coal production in China reached levels unmatched in history. Though the increases in transport capacity had been unprecedented, and may have well served the anticipated production from the state mines, the rapid production from the non-state mines quickly rendered transport critically inadequate.

Another crucial problem involving timing and interface was price reform. As more and more coal was sold at market prices, other sectors of the economy in which prices had not been deregulated and which required coal as an input, were seriously disadvantaged, and the coal suppliers were left without cash flow. Inter-industry debt was paralysing, many CMAs fell several months behind in wage and pension payments, and miners all over the country staged sit-ins. The key advantage of the two-track pricing system was its stimulating effect on production, but it was quickly found to give corrupt mine and railway officials ample opportunity to abuse their powers and commandeer shipments which were meant to be under plan.

These problems – transportation inadequacies, incomplete deregulation of prices, and corruption combined with inadequate legal and business infrastructure – were pivotal in deterring the hundreds of potential foreign investors who flocked to China in the early 1980s to examine the feasibility of coal mine construction or equipment contracts.

The Tiananmen Square Incident on 4 June 1989 was a major, though as it turned out, short-lived setback for the coal industry as it was for the entire economy. It occurred about a year into a planned period of retrenchment necessitated by double-digit inflation. Party and government alike were sharply divided as to the wisdom of pushing on with reform.

Deng's tour of south China in early 1992 was the turning point. Buoyed by the very visible economic impact that the economic reforms and open door policy had had in the south he called for higher economic growth nationwide, and the replacement of all references to 'planned economy' in the country's constitution with 'socialist market economy'. Soviet-style central planning had definitely come to an end. The raised coal production targets in 1993 were easily obtainable. However, the already serious transportation problems worsened. Coal production was necessarily being determined by transport capacity.

During the Eighth (1991–95) and Ninth (1996–2000) Five Year Plan periods under Jiang Zemin and Zhu Rongji, the focus was on industrial restructuring, and experiments, including various shareholding and cooperative ownership systems, were carried out. The government wanted public ownership to remain the foundation of the economy, but sought to introduce market principles which would foster sound financing, efficiency, competitiveness, and flexibility. All state industries, including the coal industry were to pay off their debts by the year 2000.

However, the state coal mines were encumbered with heavy welfare obligations to their bloated workforces and their families, and to millions of retired workers. They had already experienced difficulty in keeping up with wage payments, and when restructuring and the streamlining threw millions of workers out of jobs, the state faced an explosive situation. Unpaid workers joined forces with the unemployed and staged repeated protests.

The central government's response to the industry's continuing mounting losses was to concentrate large state mine production under a re-organized Ministry; give the small mines production maximums, and prosecute illegal mines; introduce coal exchanges; convert some of the CMAs to shareholding limited liability companies which would list on the stock exchanges; greatly expand diversified activities; entice foreign investors with the promise of controlling interests; and build more mine-mouth thermal power plants.

Radical central government re-organization came in 1998, when production ministries, including that for coal, were converted into state corporations. Premier Zhu Rongji halted production at thousands of SOEs. The ninety-four CMAs were transferred to provincial governments, henceforward to be called 'state priority mines'.

The decreasing demand for coal began to show up in statistical reports in 1996. The government had closed, and would continue to close, hundreds of insolvent SOEs across the entire economy; gas for cooking was becoming available in the cities; modernization was now fostering a greater need for more light industry than heavy; new efficient coal consumption equipment was replacing the old in the urban centres; some sectors of the economy were being forced to use more efficient non-coal-burning equipment to be internationally competitive; the need to address the domestic and global consequences of burning so much coal was becoming urgent; and the rapid growth of exports through the 1980s and 1990s created sufficient reserves of foreign exchange with which to begin importing large quantities of foreign oil products.

Coal had been identified as the only feasible energy option in the early years of the PRC, and its primacy was confirmed on the threshold of the reforms, other alternatives not yet being practicable. By the mid-1990s, however, the Chinese government saw both the necessity and the opportunity to end the strategy of energy self-reliance.

The water shortage was a main factor preventing the reduction of pollution. The Chinese knew how to build coal washing equipment, and had the materials to do so, but no plan had been agreed upon to divert the abundant water supplies in the centre of the country to the north. As in the other coal-producing countries the official agencies responsible for the environment were watching closely the global efforts in developing clean coal technologies. A Clean Coal Technology Centre was established in Beijing in 1994. Strong national and provincial government support was given to the international organizations which were carrying out demonstration projects within China.

The government was anxiously trying to reduce output from the LNS mines not only because they were in serious competition with the state priority mines, but because the coal they produced was of poorer quality. It was difficult to persuade the operators of these mines to close because the profits they had earned from selling coal could never be matched in any agricultural pursuit. The severe climates of the main mining regions, combined with the lack of water, made growing cash crops impossible.

At the time of writing the deliveries of coal were generally meeting the quantity and timing requirements of the existing thermal power plants. Often the electricity could not be transmitted to consumers for lack of sufficiently heavy long distance transmission and urban distribution lines, and voltage stabilizers. While other forms of energy were rapidly replacing the use of coal in the railway and urban residential sectors, and in much of the industrial sector, its use in electric power production was expected to increase. The Chinese government also hoped that the use of coal-derived fuels could be expanded. Though coal would not continue to account for over 70 per cent of total energy consumption, it would certainly remain the dominant fuel for at least another half century. The government also had high hopes for exports.

Over the 1990s major strides were made in raising the efficiency of coal consumption and reducing pollution levels in the large cities, but in the smaller urban centres and rural areas the serious situation remained unchanged, in many regions actually worsening. The government's regulatory frameworks for handling these problems were yet immature. It was generally the enterprises themselves which, in the interests of cutting production costs, converted or replaced small-scale, outdated boilers and kilns. There were no incentives, for example, in the form of major tax breaks, nor were any penalties for not doing so sufficiently large to motivate enterprises to act. Enterprises which did reduce their emissions were not rewarded, and the level of emission charges was yet too low to have significant effect.

In the one hundred and fifty years under review in this book, China moved from a closed and isolated economy to one of the world's most powerful and competitive players. Throughout the past two decades of reform the world has been increasingly affected and, indeed, awed by China's rapidly expanding economy. Between 1978 and 1998 the industrial sector grew at an average annual rate of about 12 per cent, accomplishing what had required centuries in other countries. The per capita 1980 GNP quadrupled by 1996, well in advance of the year 2000 target date. Between 1979 and 1996 total GDP rose from twenty-fifth position in the world to seventh, and the average annual growth rate of the economy was 9.9 per cent, surpassed only by that of South Korea.[1]

In the 1990s China became the world's largest producer of coal, steel, cement, cotton yarn, cotton cloth, silk fabrics, clothing, knitwear, bicycles, washing machines, and television sets, as well as many food crops, and the second largest holder of foreign exchange reserves after Japan. The country's contribution to total international trade ranked thirty-second in 1978.[2] By 1998 the value of total exports ranked ninth in the world, and imports tenth.[3]

In several respects China's economy, by the end of the twentieth century, had taken on the appearance of a developed country, and indeed some nations of the developing world chafed at China's continuing receipt of a large proportion of aid and credit from international agencies. In other ways, however, China was still very much a developing country. Average living standards, though greatly improved, were still very low, and per capita GDP was only one-third of the average for what the World Bank classifies as 'middle-income' economies'. In 1997 there were yet 65 million citizens below the internationally recognized poverty line.[4]

The abundant reserves of coal were one of the few comparative advantages that the country possessed when it set out in 1949 to modernize and become a world power. However, the politically-fraught trial and error process of switching from large-scale, centralized production to small-scale, decentralized production and back, combined with continuously low consumption efficiency, was costly and wasteful of resources. By allocating most of the available coal and electricity to the industrial sector, and forcing several generations of Chinese people to endure great discomfort and inconvenience, the government was able to keep the economy growing. In the coal industry, growth would probably have been greater had political conflict not interfered with policy and planning among the coal, railway, and power sectors. However, a qualifier should be added to this speculation. The Soviet-style planning was causing major problems – output–capacity ratio, distribution, hoarding, etc. – in the coal industries of Russia and the Eastern European states even before it was applied to the Chinese coal industry. One may ask the question how these problems would have affected the Chinese coal industry in the long term, that is, had not Deng begun the cautious introduction of market forces.

Prospect

Issues to be resolved in the coal industry

In the immediate future, the four main issues facing China's coal industry are managing the effects of industry restructuring in conjunction with the decline in demand, achieving and maintaining the solvency of the SPMAs, accelerating opencast mining, and reducing air pollution.

Managing the effects of industry restructuring and the decline in demand

In the past five years the demand for coal dropped sharply by nearly 25 per cent from over 1.3 billion tons in 1996 to less than one billion in 2000. The need no longer to produce and distribute incomparable quantities of coal solves many problems but creates others, notably unemployment and social dislocation. The use of non-coal forms of energy will soon result in no demand for coal in the railway and urban residential sectors. It will therefore be possible, as has been the case for some time in most other countries, to plan production exclusively for thermal power production and heavy industry. This will permit economies of scale, promote more intensive use of inputs, ease distribution problems, and prevent some environmental destruction.

The decline in coal demand is contributing to alarmingly high national unemployment and concomitantly encumbering the coal enterprises with astronomical welfare costs in the form of severance packages. Most affected are the legions of peasants who had worked in the LNS mines, but there are also many thousands of large state mine workers who have been displaced by downsizing and labour contracting. The number of workers who will be laid off from the large, century-old mines in the northeast will increase in the coming years as these operations are gradually decommissioned. The exact timing of the closures has become highly political, but the escalating costs and dangers involved in mining these deposits render closing them imperative in favour of the expansion and construction of new underground and opencast mines in the central north and west.

The problem of surplus labour in the coal industry is especially severe because the workers generally have very little education and most of the mining areas are environmentally barren. There is often little basis from which new employment opportunities can be developed. So severe is the problem that keeping the small mines closed is proving difficult; the workers find ways of re-opening them. The coal industry cannot alone shoulder the long-term welfare of its laid-off workers. A nascent system of countrywide social security, including unemployment insurance, will require considerable time to become functional. Standardized legislation must be in place and it must be made impossible for managers to divert employee contributions in any way. Besides caring for the unemployed, there are the ongoing heavy obligations in the form of pensions to the millions who retired before the restructuring.

Achieving and maintaining solvency of the SPMAs

In August 1998 the State Council announced the transfer of the ninety-four CMAs to their respective provincial governments. Henceforward they would be identified as SPMAs. Thirty of the mines would have auditors assigned them by the State Council, while auditors for the others would be assigned from their local governments. The profits would stay with the coal enterprises, but the taxes would go to the local governments.

The onus for transforming the SPMAs into profitable operations is now on the provincial authorities and mine enterprise managers, but during this period of massive downsizing, providing basic assistance to the laid-off workers must be shared among all levels of government. However, the relative jurisdictions of these under the new organizational structure are as yet not entirely clear. Pensions and severance payouts are not the only welfare-related sources of debt. The SPMAs must be divested of financial responsibility for support services such as schools and hospitals. Indeed, this streamlining began in 1996 with the Xinji model.

Although some mining administrations have become profitable, the industry as a whole is still in debt. Many mine operations have no means to pay regular wages and are borrowing heavily from the banks to meet commitments. While wages for all workers must keep up with the cost of living, those for the underground workers must be especially attractive. The one child policy has already made recruiting for the most dangerous jobs difficult. Coal miners have traditionally been among the highest paid in the industrial sector, a fact which they will not forget when sectorwide wage hikes come up for discussion.

At the outset of the reforms the mining officials believed that huge revenues could be amassed through exports. Such trade, however, was found to be impossible due to a lack of transport capacity and quality problems. Now that domestic coal demand is lower and transport capacity is less burdened, the authorities are again hoping that export sales can bolster the industry.

A certain increase in exports can be expected. The developing countries in Southeast Asia are the fastest growing markets for energy and China's coal is nearby and relatively cheap. China's coal industry faces stiff competition from Australia and South Africa, however, for both can supply higher quality coal and reliably. Coal enterprises in China need to learn a great deal more about participation in spot markets, which demand quick response to short-term demand. Development of opencast mining, too, could improve the export picture.

Fundamental to achieving and maintaining solvency is economywide price reform. As a basic input, coal must be priced such that on the one hand production costs are covered, but on the other, consumers can afford it. Many countries have faced the same dilemma and found the national economic destabilization caused by raising the price of a basic input politically hazardous. The solution is often cross-subsidization. In this case, profits generated by production of other goods and services within the company are used to subsidize loss-making coal. China's coal industry has already benefited greatly from the expansion of diversified activities in the 1990s, and the revenues so earned by many administrations are sustaining the mining operations.

The central government has made it clear that it will not offer the industry further support. The provinces and mine managers are experimenting with various schemes for capital accumulation. Employee shareholding arrangements tried thus far do not seem to have been successful. In many cases disgruntled workers have complained their savings are being used to support state debt. Managers must be made responsible to shareholders. Beijing set up four loan-recovery agencies in 1999 and new debt workout schemes are under discussion. The 'debt for equity swap' put forward by the Zhengzhou Coal Industry Group has Beijing's approval in principle.[5]

In recent years the opportunities for foreign involvement in the industry have been greatly expanded, and when China is admitted into the WTO it can be expected that the scope will be widened yet further, and that prices and wages, as well as standards throughout the entire production-marketing continuum will converge towards international norms. The industry must continue to pursue foreign investment and aid aggressively to further computerize existing operations, accelerate the development of opencast mines, and carry out CCT research.

Accelerating opencast mining

At the beginning of the reforms, there had been high hopes that the share of total output from opencast mines in Shanxi, Inner Mongolia, and Shaanxi could within two decades, increase from a negligible amount to about a fifth of total production. The cost, safety, and environmental advantages of this type of mining, compared to underground mining were well known and it was expected that foreign investors, immediately upon learning of the export potential of the superior quality coal at these deposits would hasten to become involved.

However, the many contractual and transportation difficulties encountered by Occidental Petroleum at the Antaibao Mine (Shanxi) in the mid-1980s quickly discouraged further investment in the sector. Despite this, Chinese companies, on their own, did manage to complete by 2000 Phase I of most of the originally planned mines. Output was a fraction of what had been planned for by this date.

Another factor affecting the development of these mines was the unforeseen rapid rate of production from the LNS mines. There was not the railway capacity to transport coal from the opencast operations in the remote north and northwest.

Now that the LNS mine sector has been drastically curtailed, the transport problems eased, and the government envisages a greatly consolidated industry comprised of a relatively few large operations, the focus has returned to the opencast mines. It is hoped that foreign investors will invest in construction and machinery in return for export earnings. The foreign investment community has observed that the Chinese government is considerably more knowledgeable now about international contractual arrangements and pricing than it was in the mid-1980s, and is beginning to show a renewed interest in mining projects, including integrated thermal power projects.

Reducing air pollution

The pollution problem is in large part a water shortage problem. Since the early 1980s the government has fully understood the local and international ramifications of burning huge quantities of unwashed coal. China has the technology to produce washing equipment, but the lack of water is a physical constraint which does not seem to have a solution. Geologists and environmentalists disagree as to the feasibility of diverting the Changjiang River northwards.

When the large, old mines in the east are eventually closed, a much larger share of total production will have to come from the arid, northwest and northcentral areas. Unless there is a breakthrough in dry purifying techniques, no attempt will be made to reduce the pollution at the production stage. All the coal will be loaded onto the trains without being washed, and the pollution reduction efforts will necessarily occur at the consumption stage. Water for the miners and their families may eventually have to be trucked in. The authorities are now reconciled to the fact that the water problem is such that almost no other industrial activities can take place in these areas, and that the number of sites where it is possible to build mine-mouth power plants is relatively few.

The Chinese government is not struggling alone with the pollution problem. Though many of the countries party to the Kyoto agreement are falling behind in their commitments, none are unaware of their ultimate responsibility to their own people and to the world to reduce, if not eliminate emissions through the use of CCTs and cleaner fuels. Since the early 1990s many governments and international organizations have been cooperating in expanding the use of existing CCTs and developing new ones. Large amounts of funding and scientific expertise have been made available and Chinese laboratories are at the forefront of the research.

The incentive mechanisms, including penalties, for raising national energy consumption efficiency and pollution standards in China, are as yet very weak. The Chinese government must examine its tax base and decide which sectors can and should pay higher taxes to provide the revenue with which to purchase pollution control and monitoring equipment. It must be decided how much of the burden of controlling pollution should be borne by the industrial community, and how much by the service and residential sectors. The system of offering rewards, in the form of tax breaks, to enterprises which of their own accord improve their coal consumption efficiency must also be broadened.

Expected development of hydro and nuclear power,
oil, gas, and renewable energy

China has some 14 per cent of the world's hydroelectric power reserves, but of these only about 14 per cent was tapped as of 1998. Total installed hydropower capacity, ranking third after that in the US and Canada, is to more than double by 2010, at which time about 33 per cent of the resources would be harnessed. Since 1979 hydropower has accounted for about 20 per cent of total electricity generated. By 2010 this proportion is to be raised to 30 per cent.[6]

Though there is considerable Chinese expertise in constructing small hydroplants, foreign assistance is needed to build the necessary large-scale turbines, transformers, and transmission lines. Most of the consumption is still in the eastern provinces, thousands of kilometres away from the best sites for hydropower development in the southwest of the country. The immediate practicability of building more large hydropower projects is limited also by the problems of seasonality and silting, the conflicting navigation and irrigation needs for the water, and the huge expense and long lead times.

The highly controversial Three Gorges Project (Hubei) on the Changjiang River, when completed in 2009, will have a capacity of 18,200 MW, the world's largest output from any single facility, and will help flood control and allow large barges to be pulled as far as Chongqing. However, the 84.7 billion kWh of electricity it will generate per year is equivalent to a mere 7.8 per cent of the combined hydro, thermal, and nuclear power generated in 1996, and savings of coal will amount to only 40 million tons.[7]

In 1999 nuclear energy accounted for less than 1 per cent of total electricity generation. There are two plants, one in Qinshan (Zhejiang) with 300 MW capacity, which began producing electricity in 1992, and one at Daya Bay (Guangdong) with two 900 MW reactors, which began operation in 1994. The US had placed sanctions on nuclear energy equipment and expertise in 1985 in an attempt to prevent China's exporting them to other countries. By January 1998 China met the non-proliferation conditions set by signatories to the Non-Proliferation Treaty, and the sanctions were lifted.[8]

Originally, it had been hoped that 20,000 MW of nuclear energy could be installed by 2000. This wildly optimistic target was substantially reduced, however, when many delays in the construction of the plant at Daya Bay and funding problems at the Qinshan plant were encountered. Discussions with Russian experts regarding the building of a plant in Dalian were cancelled when the Soviet Union collapsed. The huge capital investment and protracted time required to build nuclear power plants make them a long-term option.

As of late 1998 it was planned that there would be 8,500 MW by 2005 (which would mean that nuclear power would account for about 3 per cent of total electricity generated), 20,000 by 2010, and 40,000 by 2020 (5 per cent of the total).[9] Construction of two 1,000 MW plants, to be built with Russian assistance at Lianyungang (Jiangsu), began in 1999. One is to be completed in 2004, and the other in 2005. Qinshan Phase II called for two 600 MW plants to begin operation in 2003, and Phase III, a turn-key project contracted with the Canadian Atomic Energy Corporation, for two 700 MW plants, in 2004. Two 900 MW plants were to be built at Lingao (Guangdong) before 2005, and feasibility studies were under way for plants in Fujian, Shandong, Jiangxi, Zhejiang, and Guangdong. By April 1999, however, due to the decrease in demand for electricity largely due to the downturn in the economy, the government announced that no new nuclear projects would be started for three years.[10]

Elsewhere in the world, for example, Canada, US, UK, Germany, Hong Kong, and Taiwan, the increasing lack of public acceptability for nuclear developments is affecting energy policy. Despite the problems of radioactive waste and decommissioning of old plants, the Chinese government seems eager to increase the role of this particular form of energy. According to Qian Jihui, deputy secretary of the International Atomic Energy Agency:

> China has good prospects for developing nuclear power because there is no antinuclear lobby. The chief reason for the brake on nuclear power development in the Western countries ... is lower public acceptance, partly due to misleading and unscientific media reports that make the public believe accidents such as Chernobyl can happen any time. But in China, public acceptance is not a problem. China has a correct understanding and clear policy concerning nuclear power that leads the world.[11]

Total production of crude oil in 2000 stood at 162.6 million tons. China has been the world's fifth largest petroleum producer and fourth largest refiner since 1996, though the country has only 2.3 and 1.2 per cent of the world's proven recoverable reserves of petroleum and natural gas, respectively.[12] China's expected future production of oil and gas are not expected to expand at nearly the pace that imports of these will.

The largest hydrocarbon fields, those at Daqing (Heilongjiang) and Shengli (Shandong), have reached their peak production. The Tarim (Xinjiang), Turpan-Hami, and Jungar Basins are to replace the eastern oil fields early in the next century. Though large deposits, the oil is deep and the desert mining conditions make for very high production costs. The Chinese government did not grant permission to Western companies to survey the area until 1993, but now it is very actively seeking investment and technical assistance. A major component of the goal to supply eastern China with cleaner energy is the current renewed exploration for offshore oil and gas in the South China, Bohai, and East China Seas after a reprieve in the late 1980s, when a major switch away from coal was considered impossible.[13]

Confirmed reserves of natural gas stand at 3 trillion m^3 and production is to more than quadruple from 23 billion m^3 in 1998 to 30 in 2000, 70–80 by 2010, and 100–110 by 2020.[14] The planned construction of thousands of kilometres of gas pipelines from Sichuan, Shanxi, Gansu, Ningxia, Qinghai, and Xinjiang will enable millions of homes to convert from using coal for cooking and heating to gas. Also planned is the production of 10 billion m^3 of coalbed methane gas by 2010.

Research into the use of solar, wind, wave, and geothermal energy is ongoing to try to mitigate shortages in rural areas, though it is realized that these methods will continue to play only a very minor role. The goal for 2015 is to have 2 per cent of China's total energy consumption come from renewable sources.[15] China's production and consumption of solar energy ranks first in the world.[16] Exploitable wind power nationwide is estimated to be 253 GW, and in 1997 a total of 217 MW (0.1 per cent) were harnessed.[17] This is expected to reach 1,000 MW by 2000, and 3,000 MW by 2010. Of the estimated 3.5 GW of geothermal energy resources that can be used for electricity generation, by 1990 a total of 21 MW (or 0.6 per cent) were in use, and of the potential 21.7 GW of tidal power capacity, total installed capacity stood at 6.12 MW (or 0.03 per cent) in 1996.[18]

The role of coal in China's foreseeable economic future

The phrases 'energy demand' and 'energy consumption' are often used interchangeably and this can lead to misunderstanding. For a number of reasons, the total amount of energy actually consumed, whatever the particular mix of types, especially in developing countries, is usually considerably less than what citizens, individual, and corporate, would consume in the absence of physical and/or policy constraints. 'Energy consumption' should be used to refer to what is actually consumed, and 'energy demand' for the amount that all the various sectors – industrial, transport, residential, and commercial – would like to have at their disposal.

It is a simple matter to estimate China's future energy demand. It can be done by multiplying the developed world's average per capita energy consumption, for example, the average per capita energy consumption for the OECD countries, by China's population. As more Chinese travel abroad, and watch Western television and films, they will demand more electricity to power the conveniences taken for granted in the West, whether they be truly necessary to well-being or not.[19]

Obviously, it will not be possible for China's per capita energy consumption level to reach that of the OECD countries for some time. The Chinese government is working towards this level over the long term and setting minimum targets over the short. When establishing these targets the Chinese energy authorities, as those in any other country, must take into consideration such factors as: anticipated changes in the growth rate and structure of the domestic economy, population, and trade; expectations for living standards; per capita income; the rate of inflation; residential densities and the pace of urbanization; and national regulations specifying the efficiency of consuming equipment.

Specific to energy consumption planning in China is the imperative of furthering and fine-tuning the economic reforms. Political and social stability depend on the government's continually improving the average standard of living for the entire population. Per capita income must continue to increase.

The government has called for economic growth of 7 per cent per annum up to the year 2010, with inflation less than 5 per cent. Between 2000 and 2010 GNP per capita is to double, and between 2000 and 2050 to quadruple. China's leaders are aiming at making 2020 the year when the country will be the world's second power after the US in science and technology, and believe that by the middle of the century 'the modernization programme will have been accomplished by and large, and China will have become a prosperous, strong, democratic and culturally advanced socialist country'.[20] Living standards by then are to be similar to those of a middle-developed, comparatively wealthy country.

The government is vigorously encouraging economic growth in the west and interior regions of the country, mainly in response to the widening economic disparity, but also to mitigate ethic tensions. The expansion of existing cities in these regions and the construction of new ones, requiring huge amounts of building materials, will be very energy-intensive.

In 1999 coal accounted for 68.2 per cent of total energy production, oil for 20.9, natural gas 3.1, hydro 7.8, and nuclear less than 1 per cent.[21] On the consumption side, coal accounted for 67.1 per cent of total energy consumed, oil for 23.4, natural gas 2.8, hydro 6.7, and nuclear less than 1 per cent. In 1996 the policy of energy self-sufficiency was abandoned. The authorities made it clear that henceforward there must be greater use of fuels which have higher calorific value and are cleaner. Li Hongxun, Deputy Director of the Transport and Energy Department of the SPC remarked succinctly in early 1998: 'The lack of *high-quality* energy will be the main problem facing China's future energy supply.'[22] (author's italics)

Any estimates of China's future energy consumption structure carried out before 1996 are invalid now because they were based on the assumption that coal would continue to dominate the energy mix at the then current rate, that is, account for at least 70 per cent of total energy consumption. For example, in 1996 economists at the IEA predicted that China would account for 50 per cent of the total increase in global coal demand.[23] This is now rather more debatable.

The government of Guangdong has announced its intention not to build any more coal-fired thermal plants. The acid rain caused by burning coal with a high sulphur content is most severe in the relatively wealthy south of the country, and the logistics of shipping coal here have always been problematic.[24] For the rest of the country, however, coal will continue to be the mainstay of electricity generation for the next couple of decades at least.

It is not easy to predict how much coal will be imported in the future, depending as it does on how quickly the quality of domestic coal can be improved. The IEA estimates that imports could increase to 50 million tons by 2010, up from 1.58 in 1998.[25] What can be said with certainty is that the structure of coal consumption will change quite dramatically. The direct use of coal in the transport and urban residential sectors will cease and virtually all output will be used for power generation and in heavy industry.

Changes in China's energy consumption structure will be affected by: the growth potential of China's foreign exchange earnings; energy security concerns; relative prices of fuel types and required infrastructure; national and worldwide environmental imperatives; relative efficiencies of fuel types; economies of scale; costs of research; and costs of conversion from one fuel type to another.

The external factors which will affect China's ability to earn foreign exchange, and concomitantly the country's energy imports, include the length of time the Asian region requires to recover fully from the financial crisis that hit the area after June 1997, worldwide competition in China's main export sectors such as electric equipment and accessories and textile materials and products, Japan's economic restructuring, other countries' willingness to maintain large trade surpluses with China, and international and trade agreements such as NAFTA and the WTO.

Beginning in 1994 the government began to play a very assertive role in international oil and gas prospecting and exploitation. The China National Petroleum (Group) Corporation (CNPC) and the China Petroleum (Group) Corporation (Sinopec) have held extensive negotiations with officials in Azerbaijan, Bangladesh, Canada, India, Indonesia, Iran, Iraq, Kazakhstan, Kuwait, Malaysia, Mongolia, Pakistan, Papua New Guinea, Peru, Siberia, Sudan, Thailand, Turkmenistan, and Venezuela. In some cases China is buying shares in oil companies, in others actually buying oilfields, or offering assistance with port, pipeline, and refinery construction. The interaction with Iran and Iraq irritates the West, but the Chinese government is undeterred and is building the necessary infrastructure at several home ports to handle large shipments of foreign oil and gas. There is disagreement over whether or not there are the quantities of oil available in Kazakhstan to merit the construction of a crude oil pipeline to Xinjiang, and discussions are ongoing over the construction of a pipeline from the Irkutskaya oblast of Eastern Siberia.

In order to raise capital for the development of foreign oil and gas, the CNPC and China National Offshore Oil Corporation were promoted to ministerial level and placed under the SETC. They were also to be listed on the Hong Kong Stock Exchange and issue bonds both domestically and overseas.

Net imports of oil had been steadily rising since 1994 to account for over 30 per cent of total consumption by 2000. They had dropped in 1998 due to the smuggling problem, but soared in 1999 and 2000 despite a tripling in international prices.[26] In its 'business as usual' projection, the IEA predicts that total oil imports could increase from 163.9 million tons of oil equivalent in 1995 to 505.7 in 2020.[27] China also plans to import large quantities of liquid natural gas after 2005 when a dedicated terminal is completed at Shenzhen (Guangdong).

There could also be imports of electricity. In June 1997 an agreement was signed for the transmission of 20,000 gWh of electricity from Siberia.[28] The transmission of larger quantities from this source depends upon resolving the terms of trade, and on the ability of the Chinese grids and urban networks, in terms of capacity, to handle larger volumes of current.

Table 7.1 presents some scenarios for China's energy consumption structure up to 2050 put forward by several Chinese energy analysts. There is clearly no agreement as to the exact rate at which coal will be replaced by other forms of energy. However, as the proportion of coal in total energy consumption had already fallen to 68.2 per cent by 1999, it seems many of these analysts were too conservative in their estimates. Interviewed by the author in August 1998, officials from the Ministry of Coal conjectured that by 2005 coal, as a percentage of total energy consumption, could fall to 60 per cent. They foresaw much higher imports of oil and gas from the CIS and the Middle East.[29] In the spring of 2000 Xinhua reported that the government planned to reduce the country's reliance on coal to 50 per cent by 2050.[30] As for the growth rate of coal demand in China between 1995 and 2020, the IEA predicted it would be the slowest of all the main fuel types, at 3.1 per cent compared to 9.6 for nuclear power, 6.5 for gas, 5.5 for hydro, and 4.6 for oil.[31]

It was in the 1950s and 1960s that most developed countries, largely for reasons of efficiency and cleanness, began reducing with the expectation in many cases, of virtually eliminating the use of coal. China would have liked to have taken this step too, on the threshold of the reform programme. In the 1970s two defining factors came into play on the world energy scene, the Middle East conflict, and the growing intolerance of the pollution caused by the burning of carbon-based fuels.

Reliance on a single fuel, regardless of source, is no longer advisable, if indeed possible, for any industrialized country. What is needed is an energy mix which is diversified as to both types and sources, thereby providing flexibility and security in the face of geopolitical and/or technical change. The individual components in the mix must be efficient, clean, reliable, and safe. It is this policy which China is pursuing.

In China the change in energy consumption structure is taking place while the country is in a state of profound economic, social, and political transformation. The state industrial sector, in which the coal industry is a crucial player, is the last production component of the economy to be reformed. The restructuring of the coal industry was especially slow because of its classification as strategic, complicated multi-tiered production structure, wide spatial dispersal, huge labour force, endemic corruption, and low prestige.

Globally, it is expected that coal will prevail for several decades as the most commonly used fuel for electricity generation and input for a variety of heavy industries. Many energy economists believe, however, that coal use could rise dramatically in the coming decades once the technical and financial difficulties associated with the CCTs are solved.

For China, the huge coal reserves remain one of the country's few comparative advantages. While coal is quickly being replaced in the railway and urban residential sectors, its use in the rural areas is only beginning. The government

Table 7.1 Forecasts of future energy consumption structures (%)

	A			B		C		D				E	F		
	2000	2010	2020	2010	2050	2000	2010	2000	2010	2030	2050	2010	1995	2010	2020
Coal	72.5–68.6	64.6–64.0	60.5–57.5	65.0	50.0		57.3	70.0	66.7	61.1	58.6	64.1	76.9	69.7	67.4
Oil	18.6–19.8	20.3–20.6	21.5–21.8	18.7			22.9						19.0	22.8	24.1
Gas	2.6–3.0	5.9–6.2	8.5–9.7	5.6			8.4						2.0	3.7	3.9
Nuclear				10.7			9.4						0.3	1.2	1.6
Hydro	6.3–8.6	9.2	10				2.0						1.9	2.5	3.0
Other renewables														0.1	0.1

Sources

A Chen Shutong and Zhu Yuezhong, 'Woguo meilai nengyuan jiegou bianhua qushi' (Developmental Tendency of China's Energy Structure in the Future), *GYLJ*, no. 12 (1997): 62.

B Xie Heping, President of China Mining University, cited in 'China's Energy Resources to be Diversified in Next Century', *Asia Pulse*, 28 August 1998.

C Yang Zhaohong, 'Liyong guoji youqi ziyuan, gaige woguo nengyuan jiegou' (Use National Oil and Gas Resources, Reform China's Energy Structure), *ZGNY*, no. 11 (1998): 14.

D Lui Deshun and Shi Zulin, 'A Strategy to Promote the Use of Clean Coal Technologies', in Preeti Soni (ed.), *Energy and Environmental Challenges in Central Asia and the Caucasus: Windows for Co-operation*, New Delhi: Tata Energy Research Institute, 1996, p. 43, citing Qiu Daxing, Ma Yuqing and Shi Zulin, *China' Energy Strategy Research: Energy Demand Analysis in China for 2050*, Working Paper of ITEESA/Institute of Nuclear Energy Technology, Tsinghua University, Beijing, 1993.

E Wu Yin, 'Meitan chanye jingzhong jiegou fenxi' (Analysis of the Competition Structure of the Coal Industry), *ZGNY*, no. 1 (1999): 18.

F IEA, *World Energy Outlook 1997*, Paris: OECD/IEA, 1997, Table 15.4, p. 281.

has high hopes that exports can be raised to account for a much larger share of international coal trade. Massive amounts of capital are needed to put into place the infrastructure necessary to import oil and gas and develop nuclear electricity. Coal will definitely dominate the energy mix for the foreseeable future. Progress in opencast mining and CCT technology could extend this future.

Epilogue

As this book was going to press, the Chinese economy and living standards generally, were still greatly affected by power shortages. While some areas in the east had enjoyed surplus power for some time, they anticipated future power deficits unless more new projects were approved soon. The authorities in Guangdong, were, in fact, expecting to face a 25 per cent power shortage over the next five years.[1] Millions of peasants in outlying areas had no access to electricity at all and the planned development of the western region pivoted on the construction of several large power projects. Though steadily being replaced with other cleaner fuels in all but thermal electric power generation and a few heavy industries, there was no doubt that in the overall consumption structure coal would continue to be the main form of energy for some decades.

The coal industry was still suffering from over-capacity, though the demand outlook seemed slightly better than at the end of 2000. Output during the first three months of 2001 was up 14.2 per cent over the previous year's figure.[2] The quantity of coal that was washed during these months was up 16.8 per cent. Industry officials were still travelling around the country trying to close down small illegal mines, only to find them re-opened by the miners shortly thereafter. Horrific accidents continued to be reported, seemingly at a higher frequency than ever.

Coal prices remained depressed. In March 2000 the average price of a ton of steam coal was over 243 yuan, but only 218 from August through October, and in the following months averaged about 223 yuan.[3] Serious financial problems persisted, but the government's plans to corporatize the coal industry in a way similar to that of the oil industry were nevertheless taking shape. By 2005 there were to be only about eight large coal corporations, and foreign capital would be sought through international stock offerings.

The government decided to go ahead with the long-discussed plan to divert water over 1,000 km from the Yangzi River to the arid north and northwest regions. It was listed as one of the key items in the Tenth Five Year Plan. When the project is finished, 45 billion m^3 of water will be transferred each year via separate eastern, central, and western routes. The main priority is to meet basic human consumption and hygiene requirements which are expected to grow steadily until population growth peaks in 2030. Attempting to meet the water needs of industries in these regions is necessarily of lower precedence.

Indeed, it would be impossible to transfer the huge amounts of water required by the main coal producing regions of Shanxi, Shaanxi, and Inner Mongolia from the south. Given the considerably smaller amount of coal being produced compared to the mid-1990s, however, it is finally possible to transport the best coal south for washing. The government has approved plans for the construction of several new coal distribution centres around the country, and among the targets for 2005 included in the Tenth Five Year Plan is the almost unbelievable goal of washing 95 per cent of the coal entering the market.

As for coal trade, total exports reached an unprecedented high of 55.05 million tons in 2000, then in the first three months of 2001 were up to 101 per cent over the same months in 2000.[4] Admission to the WTO was expected to be delayed until 2002. The agreed reduction or abolition of import tariffs on coal was expected to trigger a sharp increase of imports into China.

The authorities were hopeful that several coal liquefaction plants to be built over the next decade would become viable.[5] The CCT transforms coal into gasoline, diesel, LPG, and other oil products. Though the technology has been around for some decades, it has been insufficiently cost-effective to warrant undertaking on a large scale. At a model project launched in Yunnan it was estimated that a ton of oil could be produced from local coal at 1,500 yuan, while the cost of importing a ton, including duty, was 1,900 yuan. It was also estimated that at current oil prices the plant's construction costs could be recovered within six years.[6]

Impressive progress continued in the reduction of air pollution in the country's major cities, but was very slow in the rural areas. According to estimates of the SEPA, China's continued heavy reliance on coal was causing environmental problems that eroded 10 per cent of the nation's GDP.[7]

The Bush administration's decision not to ratify the Kyoto Protocol was met with disappointment and anger around the world, not least of all by China.[8] The reaction might have been less severe had a plausible alternative plan for dealing with the problem of global warming been presented. Without one, however, anti-American sentiment was inflamed. The administration claimed that the set targets were impossible to meet and it was a waste of time even to discuss the current dialogue. Furthermore, the US objected to the exclusion of developing countries such as China, India, Mexico, Brazil, and South Korea, from making reductions in their greenhouse gas emissions.

It would have to be said that the debates following the announcement were often more emotional than informed. The enduring lack of agreement among the scientific community as to the true causes and consequences of global warming, and even if it exists at all, left the general public confused and frightened.[9]

To the Bush administration, the problem was the length of time required to restructure the national electricity industry. In 1999 over 52 per cent of the US' total electricity was generated by coal-fired thermal plants.[10] Oil, gas, hydro, and nuclear plants accounted for 3.1, 15.2, 7.2, and 20.0 per cent of electricity generation, respectively. There were brief periods before and after the first oil crisis during which the industry increased its use of oil, but the second oil crisis was so alarming as to force indefinite postponement of greater reliance on oil.

Given that proposing more nuclear power plants had been virtually political suicide for governments in most parts of the world, and that the renewable forms of energy such as solar, wind, wave, etc., are suitable just for niche applications, the only way to meet the emission reduction targets in the US was to increase hydro and gas plants. Even if the necessary resources were readily available, harnessing them usually required a decade or more. The political and physical obstacles involved in replacing a large proportion of the coal plants with cleaner units, or building new hydro or gas powered plants, simply could not be surmounted before 2012. Attempts to reduce oil consumption in the meantime, it was argued, would lead to job losses, the transferal of jobs to other countries, and reductions in living standards generally.

The fact that California was suddenly faced with crippling shortages of electricity when the new administration was just finding its feet did not bode well for the signing of the Kyoto Protocol. Due to the chaotic de-regulation of the industry over the previous decade, and atypical demand/supply conditions, the State in 2001 was forced to draw down its budget surplus by millions of dollars each day to pay for electricity, some obtained from out-of-state suppliers. The utility companies were bankrupt, and the situation was certain to worsen as summer approached. In such circumstances it was impossible for the administration to ask the public to reduce consumption voluntarily. Nor was it surprising that Vice President Cheney's energy plan, announced in May 2001, not only called for a continuing role for coal, but included new nuclear initiatives as well as looking northward to Alaska for oil and gas.

The developing world for its part pointed out that, on a per capita basis, its contribution to global emissions was minute. India argued that the US' emissions and consequent contribution to global warming would raise sea levels, affect the monsoon, lead to the disappearance of the Himalayan glaciers, and result in water scarcity in tropical Asia. The American emissions are more than twice those of China where there are still millions of households with no access to electricity. The developing world accuses the Bush administration, of asking immature economies to reduce their 'survival emissions', while doing little to reduce the American 'luxury emissions'.

To many people in the rest of the world, the concept of carbon sequestrian was appalling. That public officials – some of them elected on green agendas – could seriously consider the barter of pristine areas to be used as 'sinks' in exchange for 'carbon credits' was almost unbelievable.

It was clear that China's policy of energy self-sufficiency had truly ended. Imports of oil continued to grow rapidly. In 2000 they were 92 per cent higher than in 1999, despite international prices increasing eight times over the year, by a total of 37 per cent.[11] Over the first three months of 2001 imports of crude were up 262 per cent over the previous year's figure.[12] Both of China's two newly consolidated, vertically integrated global oil firms, the China Petrochemical Corporation (Sinopec) and China National Petroleum Company (CNPC), as well as the China National Offshore Oil Corporation (CNOOC), gained greater international stature in 2000. They continued to be very assertively making deals around the world to purchase oil fields or buy shares in them, and/or construct oil infrastructure. Net imports of oil amounted to 28 per cent of total oil consumption in 2000. International energy analysts were expecting China soon to account for half of Asia's crude oil demand and become the world's third largest oil importer after the US and Japan. For China, the increased reliance on foreign energy meant increased vulnerability to the capriciousness of international markets and a surrendering of some control over its economy. For the world, it meant a significant, and likely irreversible impact on world energy supplies.

The consumption of gas was expected to grow very rapidly in the coming years. The future of the coal industry depends to a large extent on what gas resources come available. Given that gas is a more efficient fuel and burns more cleanly, it was anticipated that if its cost-effectiveness were proven to surpass substantially that of coal, more and more coal-burning equipment would be converted.

The completion of a gas pipeline from southwest Xinjiang to Shanghai would alter the energy consumption structure of the coastal provinces significantly. Though there seemed to be little doubt that the project had approval in principle, at the time of writing it was being delayed due to disagreement over the size of the gas deposit relative to the length of time it would take to recoup the construction investment. Estimating the costs of feeding gas from other deposits, such as a newly discovered gasfield in Ordos Basin, Inner Mongolia, into this pipeline were also prolonging the contracting process. Other lines in the planning stages were from Shanxi to Beijing, northeast Hebei, and Shandong; from Sichuan to Hubei and Hunan; and from Qaidam (Qinghai) to Lanzhou (Gansu). A line designed to deliver gas from Irkutsk in Siberia to China and South Korea was also under consideration. The completion planned for 2006 of the liquid natural gas port terminal in Shenzhen (Guangdong), and later two others up the coast, would also lead to major changes in energy consumption patterns.

All plans for new nuclear plants were on hold. A three-year moratorium on new nuclear plants had been imposed in 1999, and no plants were listed as part of the Tenth Five Year Plan. However, Zhejiang, Fujian, Guangdong, and Shandong were all lobbying the central government for approval of plants in their jurisdictions.[13] Though the Chinese government had always been very positive about the potential use of more nuclear energy, it could hardly have failed to notice that since the early 1990s, or even earlier, governments around the world that proposed building nuclear plants faced fierce objections from their citizens.

An accident at a nuclear plant in September 1999 in Tokaimura, 120 km northeast of Tokyo caught world attention, as did the campaign promise of the Democratic Progressive Party in Taiwan to suspend building of the Island's fourth nuclear plant. The project had been approved in 1994 under the previous Kuomintang government and was to be completed in 2004. Construction was duly halted soon after the new leader, Chen Shui-bian, came to power. However, with the plant already a third completed and billions of dollars in contracts at stake, he was forced to rescind the decision to abort the project after parliament threatened to impeach him. The US' recently announced intention to resume nuclear power initiatives may change the stance of other countries and will be closely observed internationally.

The downing of the American surveillance plane and death of a Chinese pilot, as well as the apprehension of several Chinese–American scholars strained Sino-American political relations, but economic ties were not affected. Indeed, some Chinese industries, including the energy industry, were steadily becoming more deeply integrated into international business networks. The largest enterprises were beginning to be identified with the international business establishment, which around the world was increasingly under attack from a very vocal, and socially and environmentally aware, global public. The world's largest consumers of coal – China, the US, the CIS, and India, as a group, find themselves unable to ignore the continuous calls to strengthen their collaborative efforts and achieve results in the area of CCT research.

Notes

Introduction

1 For example, C. H. Chai, *China: Transition to a Market Economy*, Oxford: Clarendon, 1997; Fan Qimiao and Peter Nolan, *China's Economic Reforms: The Costs and Benefits of Incrementalism*, New York: St Martin's, 1994; Donald Hay, Derek Morris, Guy Liu and Shujie Yao, *Economic Reform and State-Owned Enterprises in China, 1979–1987*, Oxford: Clarendon, 1994; Y. Y. Kueh, Joseph C. H. Chai, and Gang Fan (eds), *Industrial Reforms and Macroeconomic Instability in China*, Studies in Contemporary China, Oxford: Oxford University Press, 1998; Nicholas Lardy, *China's Unfinished Economic Revolution*, Washington, DC: Brookings Institute, 1998; Barry Naughton, *Growing Out of the Plan: Chinese Economic Reform, 1978–1993*, New York: Cambridge University Press, 1996; and Edward Steinfeld, *Forging Reform in China: The Fate of State-Owned Industry*, Cambridge: Cambridge University Press, 1998.
2 Tim Wright, *Coal Mining in China's Economy and Society 1895–1937*, Cambridge Studies in Chinese History, Literature and Institutions, Cambridge: Cambridge University Press, 1984; Alexander B. Ikonnikov, *The Coal Industry of China*, Canberra: Research School of Pacific Studies, Australian National University, 1977.
3 Abbreviated as *CIYB* (English version) or *MTGYNJ* (Chinese version). As the author did not have access to a complete set in either language, she used both English and Chinese versions to constitute a complete run. The Chinese version generally contained much more material than the English and therefore was preferred when available.
4 Fang Weizhong (ed.), *Zhonghua renmin gongheguo jingji da shiji 1949–1980* (Major Economic Events in Communist China 1949–1980), Beijing: Zhongguo shehui kexue chubanshe, 1984.
5 Liu Suinian and Wu Qungan (eds), *China's Socialist Economy: An Outline History 1949–1984*, Beijing: Beijing Review, 1986.
6 For example, *Zhongguo nengyuan* (China's Energy), *Zhongguo meitan bao* (China Coal Newspaper), *Zhongguo gongye jingji* (China Industrial Economics), *Jingji yanjiu* (Economics Research), *Jiage lilun yu shijian* (Pricing Theory and Practice), *Jingji guanli* (Economic Management), *Wuzi guanli* (Materials Management), *Jingji wenti* (Economic Problems), and *Tiedao yunshu yu jingji* (Railway Transportation and Economics).

1 The industry before 1949

1 This chapter is based almost entirely on Tim Wright, *Coal Mining in China's Economy and Society 1895–1937*, Cambridge Studies in Chinese History, Literature and Institutions, Cambridge: Cambridge University Press, 1984 and Alexander B. Ikonnikov, *The Coal Industry of China*, Canberra: Research School of Pacific Studies, Australian National University, 1977. Other sources employed were: C. Y. Hsieh and M. C. Chu, 'Foreign Interest in the Mining Industry', in China Institute of Pacific Relations, *Economic Trends and Problems in the Early Republican Period*, Preliminary Paper Prepared for the 4th Biennial Conference of the Institute of Pacific Relations, Hangzhou,

21 Oct.–4 Nov. 1931, Shanghai: The Institute, July 1931; Ralph W. Huenemann, *The Dragon and the Iron Horse: The Economics of Railroads in China 1876–1937*, Cambridge, MA: Council on East Asian Studies, Harvard University, 1984; Wang Kung-ping, *Controlling Factors in the Future Development of the Chinese Coal Industry*, Morningside Heights, NY: King's Crown Press, 1947; and *Zhonghua renmin gongheguo guojia tongji ju gongye tongji sibian* ([China] State Statistical Bureau, Industrial Statistical Department), *Wo guo gangtie, dianli, meitan, jixie, fangzhi, zaozhi gongye de jinxi* (The Past and Present State of China's Steel, Electric Power, Coal, Machinery, Textile, and Paper Industries), Beijing: Tongji chubanshe, 1958.

 2 Wright, *Coal Mining in China's Economy*, p. 52, citing Huang Jiamo, *Jiawu zhan qian zhi Taiwan meiwu*, Taibei: 1961.

 3 Ikonnikov, *The Coal Industry of China*, pp. 19–20, citing The Geological Survey of China, *Chungkuo k'uangyeh chiyao, no. 7 (7th General Statement on the Mining Industry of China)*, Chungking: 1945, pp. 646, 676; The Geological Survey of China, *Chungkuo k'uangyeh chiyao, no. 2 (2nd General Statement on the Mining Industry of China)*, Peking: 1926, pp. 32, 37, 42; and Young Ta Chin, *Chintai Chungkuo shihyeh t'ungchih (Modern Industries of China)*, Nanking: 1933, vol. 2, pp. 26, 55.

 4 Wright, *Coal Mining in China's Economy*, p. 121, using data from Wang Jingyu, *Zhongguo jindai gongye shi ziliao, dierji, 1895–1914 nian*, Beijing: 1957, pp. 34–5.

 5 Hsieh and Chu, 'Foreign Interest in the Mining Industry', p. 51.

 6 Ibid., p. 1.

 7 Wright, *Coal Mining in China's Economy*, p. 118.

 8 Ibid., p. 132.

 9 Hsieh and Chu, 'Foreign Interest in the Mining Industry', p. 49.

10 All references to 'total' coal output or production comprise anthracite, bituminous, and brown coal.

11 Wang Kung-ping, *Controlling Factors*, p. 52.

12 Ibid.

13 Wright, *Coal Mining in China's Economy*, pp. 32–3 and 156, citing *Kuangye zhoubao* (Mining Weekly), 60 (28 Aug. 1929): 179–80; 67 (21 Oct. 1929): 289–90; 36 (28 Feb. 1929): 591; 381 (7 May 1936): 1194; and 391 (21 July 1936): 97.

14 Wang Kung-ping, *Controlling Factors*, p. 50, citing The *World Coal Mining Industry*, Geneva: International Labour Office, 1938, pp. 54–7.

15 Wu Yuan-li and H. C. Ling, *Economic Development and the Use of Energy Resources in Communist China*, New York: Praeger, 1963, pp. 238–9, citing *General Statement on the Mining Industry*, Seventh Issue [Beijing: 193?], pp. 49–50.

16 Neil K. Buxton, *The Economic Development of the British Coal Industry*, London: B.T. Batsford, 1978, p. 85.

17 Wu Yuan-li and Ling, *Economic Development and the Use of Energy Resources*, p. 239, citing Edwin S. Pauley, *A Report on Japanese Assets in Manchuria to the President of the United States*, Washington, DC: 1946, p. 74.

18 Robert Carin, *Power Industry in Communist China*, Communist China Problem Research Series EC44, Hong Kong: Union Research Institute, 1969, p. 30, using data from Yu-Kuei Cheng, *Foreign Trade and Industrial Development of China: An Historical and Integrated Analysis Through 1948*, Washington, DC: University of Washington Press, 1956, p. 267.

19 John K. Chang, *Industrial Development in Pre-Communist China: A Quantitative Analysis*, Edinburgh: University of Edinburgh Press, 1969, p. 82, citing Chinese Association for the United Nations, *A Report on Russian Destruction of Our Industries in the North-Eastern Provinces*, Taipei: 1952, pp. 3–4. According to another source, about one-quarter of the country's coal mines had been completely put out of production. Ikonnikov, p. 175, citing *Jen-min shou-tse* (People's Handbook), Shanghai, Tientsin and Peking: 1951, p. VIII–10.

20 Zhonghua renmin gongheguo guojia tongji ju, p.101; International Labour Organization, Coal Mines Committee [Report on the 4th Session], *Productivity in Coal Mines*, Item 3, Geneva: ILO,1951, pp. 22–3, using data from [United Nations] Economic Commission for Europe, Coal Division, *Monthly Bulletin of Coal Statistics* and United Nations, Statistical Office, *Monthly Bulletin of Statistics*.

21 Wright, *Coal Mining in China's Economy*, p. 78.

22 Ibid., p. 37, citing Minami Manshu tetsudo kabushiki (South Manchurian Railway Company) shomubu chosaka, *Kairan tanko chosa shiryo*, Dalian: 1929, p. 109; and Kailan Mining Administration, *Kaiping Coal*, Tianjin: c. 1920, p. 10.

23 Ibid., p. 82, using data from Hou Defeng, *Disici Zhongguo kuangye jiyao* (Fourth China Mining Industry Summary)1932, pp. 91–2.

24 Ibid., pp. 89–90.

25 Ibid., pp. 86–7, citing *Nongshang gongbao* (Agricultural Workers' Newspaper), 9 Apr. 1918, xuanzai 5, and Donald B. Gillin, *Warlord: Yen Hsi-shan in Shansi Province, 1911–1949*, Princeton: 1967, p. 88.

26 Ibid., pp. 87–8, citing Quan Hansheng, 'Shanxi meikuang ziyuan yu jindai Zhongguo gongyehua de guanxi', in Quan Hansheng, *Zhongguo jingjishi luncong*, Hong Kong: 1972, vol. 2, p. 765. Note that the portion of the Changjiang from Yangzhou to the river mouth used to be known as the Yangtze. China Handbook Editorial Committee, *Geography*, China Handbook Series, Beijing: Foreign Languages Press, p. 108.

27 Wu Yuan-li, *An Economic Survey of Communist China*, New York: Twayne, 1956, p. 283.

28 Robert P. Taylor, *Rural Energy Development in China*, Washington, DC: Resources for the Future, 1981, p. 149, citing Vaclav Smil, 'China's Energetics: A System Analysis', in *Chinese Economy, Post-Mao*, vol. 1, *Policy and Performance*, JEC Report, 1978, p. 368.

29 Wright, *Coal Mining in China's Economy*, p. 37, citing Manshikai, *Manshu kaihatsu yonju nen shi*, Tokyo: 1964–5, vol. 2, p. 66; *Nongkuang gongbao* (Rural Mine Newspaper), 20 Jan. 1930, p. 209; and *Kuangye zhoubao* (Mining Weekly), no. 109 (7 Sept. 1930): 196.

30 The room and pillar method involves working the coal to the edge of the mining concession, leaving pillars of coal to support the roof. About half of the coal is left in place in the form of these pillars and therefore this method can be quite wasteful. In some cases, however, the cutting retreats back to the shaft, and the pillars themselves are mined causing the abandoned areas to cave in. In areas where the seams are especially thick, the coal can be worked a second time once the debris is sufficiently compacted. In the longwall method, cutting begins at the outer edge of the concession and retreats to the centre shaft or main road. A large section remains to hold up the shaft. It does not require nearly as much timber, and offers more leeway for the easier use of machinery.

31 Wright, *Coal Mining in China's Economy*, p. 38.

32 CIYB 1982, p. 27; Neil K. Buxton, *The Economic Development of the British Coal Industry*, p. 179, calculated from data provided by [Gt. Brit.] Royal Commission on the Coal Industry, 1925, *Minutes of Evidence*, Cmnd. 2600, vol. III (1926): League of Nations, *Problem of the Coal Industry*, Memorandum on Coal I, pp. 39–40; International Labour Office, *The World Coal-Mining Industry, I, Economic Conditions 1938*, pp. 106–10; Political and Economic Planning, *Report of the British Coal Industry*, 1936, pp. 153–4; and [Gt. Brit.] Ministry of Fuel and Power, Coal Mining: Report of the Technical Advisory Committee, Cd. 6610 (1945) (Better known as the *Reid Report*).

33 Wright, *Coal Mining in China's Economy*, p. 41, citing Alan R. Griffin, *Coalmining*, London: 1971, p. 61; Masao Tezuka, *Shina jukogyo hattatsu shi*, Kyoto: 1944, p. 314; and Konwakai Kozan, *Nihon kogyo hattatsu shi*, Tokyo: 1932, vol. 2, pp. 325–7.

34 Wang Kung-ping, *Controlling Factors*, p. 98.

35 Wright, *Coal Mining in China's Economy*, p. 221, citing Masao Kanda, Shisen sho soran, Tokyo: 1936, p. 605.

36 Ikonnikov, *The Coal Industry of China*, calculated from Table 12, p. 103.
37 International Labour Organization, Coal Mines Committee [Report on the 4th session] *Productivity in Coal Mines*, p. 28.
38 Wright, *Coal Mining in China's Economy*, calculated from Table 39, p. 172 which cites A. T. Shurick, *The Coal Industry*, London, 1924, p. 359, Bal Raj Seth, *Labour in the Indian Coal Industry*, Bombay, 1940, pp. 282–3, and several Chinese and Japanese sources. The fatality rate in the US in 1910 was 6 per million tons. William S. Peirce, *Economics of the Energy Industries*, 2nd edn, Westport, CN: Praeger, 1996, p. 119.
39 Wang Kung-ping, *Controlling Factors*, p. 120.
40 Ibid., p. 121, citing *China Handbook, 1937–1943* , New York: MacMillan, 1943, pp. 464–6.
41 Ibid., p. 157, citing J. S. Burgess, *Guilds of Peiping*, New York: Columbia University Press, 1928.
42 Ikonnikov, *The Coal Industry of China*, pp. 4, 151–3.
43 Wang Kung-ping, *Controlling Factors*, pp. 190–1.
44 Table 1.2 and Buxton, *The Economic Development of the British Coal Industry*, p. 166, using data from [Gt. Brit.] Secretary for Mines. Annual reports, 1921–; and B.R. Mitchell and P. Deane, *Abstract of British Historical Statistics*, 1971, pp. 115–19.
45 Wright, *Coal Mining in China's Economy*, p. 72.
46 Ibid., p. 44.
47 Ibid., p. 45.
48 Heunemann, *The Dragon and the Iron Horse*, p. 2, citing Li Kuo-ch'i, *Chung-kuo tsao-ch'i ti t'ieh-lu ching-ying*, Taipei: 1961, p. 201.
49 Wright, *Coal Mining in China's Economy*, pp. 29, 44, and 83, citing Ellsworth C. Carlson, *The Kaiping Mines (1877–1912)*, 2nd edn, Cambridge, MA: 1971, pp. 19–22.
50 Thomas G. Rawski, *Economic Growth in Prewar China*, Berkeley: University of California Press, 1989, p. 208, citing Rhoads Murphey, *The Outsiders: The Western Experience in India and China*, Ann Arbor: University of Michigan Press, 1977, p. 110.
51 Huenemann, *The Dragon and the Iron Horse*, calculated from Table 11, p. 124, using data from China, Ministry of Railways, *Statistics of Chinese National Railways for the Year Ended 31 December1933*, Nanking: 1935, Tables IX and XXXVI.
52 Wright, *Coal Mining in China's Economy*, p. 159.
53 Wang Kung-ping, *Controlling Factors*, p. 125, citing D. N. Rowe, *China Among the Powers*, New York: Harcourt Brace, 1945, p. 105.
54 Ibid., p. 11.
55 Buxton, *The Economic Development of the British Coal Industry*, p. 166, using data from [Gt. Brit.] Secretary for Mines. Annual reports, 1921–; and B. R. Mitchell and P. Deane, *Abstract of British Historical Statistics*, 1971, pp. 115–19.
56 Wright, *Coal Mining in China's Economy*, p. 51.
57 Ibid., pp. 53–4.
58 Buxton, *The Economic Development of the British Coal Industry*, p. 174, using data from [Gt. Brit.] Secretary for Mines. Annual reports, 1921-.
59 Wright, *Coal Mining in China's Economy*, p. 55.
60 Ibid., p. 76, citing data from International Labour Organization, *The World Coalmining Industry*, 2 vols., Geneva: 1938, p. 69.
61 Li Jingwen, 'Energy Economics in Building a Modern China', in Michael B. McElroy, Chris R. Nielson and Peter Lydon (eds), *Energizing China: Reconciling Environmental Protection and Economic Growth*, Cambridge, MA: Harvard University Committee on the Environment, 1998, p. 352.
62 *CIYB 1982*, p. 29. This 1953 figure is the only one available.
63 Preeti Soni and Vishal Narain, 'India: Energy and Environment Situation', in Preeti Soni (ed.), *Energy and Environmental Challenges in Central Asia and the Caucasus: Windows for Co-Operation*, New Delhi: Tata Energy Research Institute, 1996, p. 167.

64 *Zhonghua renmin gongheguo guojia tongji ju*, pp. 43–4.
65 Vaclav Smil, *Energy in China's Modernization: Advances and Limitations*, Armonk, NY: M. E. Sharpe, 1988, p. 118; According to Carin, *Power Industry*, p. 111, there was a total of 11,410 km.
66 Carin, *Power Industry*, p. 117, citing Tang Shih, 'What the Soviet Experts Have Contributed to China's Power Industry in Past 8 Years', *Jen-min tien-yeh* (People's Power Industry), no. 31 (5 Nov. 1957).
67 *Zhonghua renmin gongheguo guojia tongji ju*, p. 42.
68 Ibid., p. 49.

2 Mao's adaptation of Soviet central planning: effects on the industry

1 Chi-ming Hou, 'Manpower, Employment and Unemployment', in Alexander Eckstein, Walter Galenson, and Ta-Chung Liu (eds), *Economic Trends in Communist China*, Edinburgh: Edinburgh University Press, 1968, p. 369.
2 Eric Hills Carlsen, *Soviet Aid and the Development of Communist China's Coal Industry, 1949–1960*, PhD diss., Cornell University, 1969, p. 6.
3 Barry Naughton, *Growing Out of the Plan: Chinese Economic Reform, 1978–1993*, Cambridge: Cambridge University Press, 1996, pp. 41–2; Cheng Chu-yuan, *China's Economic Development: Growth and Structural Change*, Boulder, CO: Westview, 1982, p. 181; Christine P. W. Wong, 'Ownership and Control in Chinese Industry: The Maoist Legacy and Prospects for the 1980's', in *China's Economy Looks Toward the Year 2000, Vol. I, The Four Modernizations*, JEC Report, 1986, Table 7, p. 603. The three sources were not in agreement.
4 The main sources for this section were: Lei Jihua (ed.), *Shengchan jihua guanli: meikuang qiye guanli jichu duwu* (Production Planning Management: Coal Mine Enterprise Management Basic Reading), Beijing: Meitan gongye chubanshe, 1986; and Wang Maolin (ed.), *Meitan gongye qiye guanli* (Coal Industry Enterprise Management), Taiyuan: Shanxi renmin chubanshe, 1982.
5 Above-norm investment projects were large and state-funded, in contradistinction to below-norm projects which were small and financed by an industry's own funds.
6 Morris Bornstein, 'Soviet Price Theory and Policy', in *New Directions in the Soviet Economy*, JEC Report, 1966, pp. 65–7.
7 The profit margin for any industrial good is calculated on the basis of average production costs.
8 Ye Ruixiang, 'Wo guo de nengyuan jiage yanbian ji qi gaige' (Our Country's Energy Pricing Development), in Wang Zhenzhi (ed.), *Jiage gaige yu jiage guanli* (Pricing Reform and Pricing Management), Beijing: Zhongguo wuzi chubanshe, 1987, p. 236; Zhang Mengceng, 'Guoying meitan gongye chanpin jiage wenti' (Problems with the Prices of Products from the State Owned Coal Industry), *CZ*, no. 3 (1958): 15.
9 Li Shiyi, 'Meitan jiage gaige de chubu shexiang' (Preliminary Ideas on Coal Price Reform), *JDYJ*, no. 22 (1986): 1.
10 For the sake of comparison, throughout the period the price of a ton of coal was less than the price of one hundred bars of soap, or one ton of sugarcane or sugarbeet. *TJNJ 1985*, 544–7.
11 In 1979 (the earliest data available) coal washed for smelting was 57 yuan/ton and other washed coal was 44 yuan/ton. Xu Shoubo, *Nengyuan jishu jingji xue* (Energy Technology Economics), Changsha: Hunan renmin chubanshe, 1981, p. 519.
12 Alexander B. Ikonnikov, *The Coal Industry of China*, Canberra: Research School of Pacific Studies, Australian National University, 1977, p. 137, Table 26, citing I. I. Bazhenov *et al.*, *Ugol'naya Promyshlennost Kitayskoy Narodnoy Respubliki* (Coal Industry of the People's Republic of China), Moscow: 1959, p. 453. The 1978 breakdown was materials at 35.4 per cent, wages at 29.2, depreciation at 12.6, electricity at

9.7, and other at 13.1. Zhang Liqing, 'Shanxi meitan gongye zai chuanguo de diwei' (The Position of Shanxi's Coal Industry in the Whole Country), *SXJDYJ*, p. 411.

13 'First National Coal Mines Conference, November 1949', *CIYB 1983*, p. 49.

14 Liu Suinian and Wu Qungan (eds), *China's Socialist Economy: An Outline History 1949–1984*, Beijing: Beijing Review, 1986, p. 109.

15 Ibid., p. 83.

16 Fang Weizhong (ed.), *Zhonghua renmin gongheguo jingji da shiji 1949–1980* (Major Economic Events in Communist China 1949–1980), Beijing: Zhongguo shehui kexue chubanshe, 1984, pp. 4, 97.

17 Wu Yuan-li and H. C. Ling, *Economic Development and the Use of Energy Resources in Communist China*, New York: Praeger, 1963, p. 45, citing Wu Yuan-li, *An Economic Survey of Communist China*, New York: Twayne, 1956, pp. 284–5.

18 Leung Chi-keung, *China: Railway Patterns and National Goals*, University of Chicago, Dept. of Geography Research Paper no. 195 (1980): 111.

19 *CIYB 1983*, p. 52.

20 Carlsen, *Soviet Aid*, p. 73, quoting a Soviet source.

21 China, State Planning Commission, *First Five Year Plan for the Development of the National Economy of the People's Republic of China in 1953–1957*, Peking: Foreign Languages Press, 1956, p. 41.

22 Ibid., pp. 68–70.

23 'Joint Instruction from the Ministry of Fuel Industry and National Coal Mine Trade Union', *CIYB 1983*, p. 51.

24 Carlsen, *Soviet Aid*, p. 85, quoting *RMRB*, 30 Oct. 1953 and 27 Nov. 1953.

25 T. J. Hughes and D. E. T. Luard, *The Economic Development of Communist China*, London: Oxford University Press, 1961, p. 54.

26 For example, Zhou Enlai, 'Zengchan meitan he jieyue yong mei tong shi bingju' (Promote the Simultaneous Increased Production and Saving of Coal), *RMRB*, 2 Nov. 1957, p. 1.

27 'Report on the Organization of the Coal Market for 1957, 18 March 1957', *Xinhua banyue kan* (New China Semi-Monthly), 9 (10 May 1957): 110. Extracts published in Chao Kuo-chun, *Economic Planning and Organization in Mainland China: A Documentary Study, 1949–1957*, Cambridge, MA: East Asian Research Centre, Harvard University, 1963, vol. 2, pp. 41–2.

28 Liu Suinian and Wu Qungan (eds), *China's Socialist Economy*, p. 171.

29 Construction of the following new lines was to begin during the First Five Year Plan period: Lanzhou (Gansu)-Xinjiang; Baoji (Shaanxi)-Chengdu (Sichuan); Jining-Erenhot (both in Inner Mongolia); Baotou (Inner Mongolia)-Lanzhou (Gansu); Yingtan (Jiangxi)-Xiamen (Fujian); Litang (Guangxi)-Chanchiang (Guangdong); Shacheng-Fengtai (both in Hebei); Beijing-Chengde (Hebei).

30 *First Five Year Plan*, pp. 142–7.

31 He Baisha, 'Fahui difang jijixing, da li fazhan difang meikuang' (Bring into Play the Positive Aspects of the Development of Local Coal Mines), *JJJH*, no. 2 (1958): 13,25 'Proposals of the Eighth National Party Congress for the Second Five Year Plan for Development of the National Economy, 1958–1962', (adopted 27 Sept. 1956), in Stuart Kirby (ed.), *Contemporary China, Vol. II, 1956–1957*, Oxford: Oxford University Press, 1958, p. 146.

32 Christopher Howe and Kenneth R. Walker, 'Mao the Economist', in Dick Wilson (ed.), *Mao Tse-Tung in the Scales of History*, Cambridge: Cambridge University Press, 1977, p. 196.

33 'The Stipulations on Improving the Administrative System of Enterprises and Institutes under the Ministry of Coal Industry', *CIYB 1983*, p. 53.

34 *CIYB 1983*, p. 53.

35 Telephone Conference, October 1958, reported in *CIYB 1983*, p. 53; He Baisha, 'Zai meikuang jianshe zhong yao guanche zhixing da zhong xiao xing xiang jiehe de

fangzhen' (In Coal Mine Construction Implement Principle of Integrating Big, Medium and Small Mines), *RMRB*, 3 Mar. 1958, p. 3.

36 Wang Jinlin, 'Meitan gongye zhixing di yige wu nian jihua de zhuyao jingyan jiaoxun' (Principal Learning Experiences of the Coal Industry During the First Five Year Plan), *MTGY*, no. 19 (1957): 5.

37 'Report on the Organization of the Coal Market for 1957', p. 42.

38 Carlsen, *Soviet Aid*, p. 214.

39 Liu Suinian and Wu Qungan (eds), *China's Socialist Economy*, p. 234.

40 'Gu qi geming ganjin shixian shengchan jianshe da yaojin' (Revolutionary Energy to Achieve Great Leap Forward in Production and Construction), *MTGY*, no. 4 (1958): 4.

41 Liu Shaoqi, 'Report on the Work of the Central Committee of the Communist Party of China', delivered at the Second Session of the 8th NPC, 5 May 1958, *PR*, no. 14 (1958): 10.

42 Joseph G. Crosfield and Sons Ltd, *The Chinese Coal Industry*, Warrington, UK: 1961–62, part 6, p. 33.

43 Liu Suinian and Wu Qungan (eds), *China's Socialist Economy*, p. 251.

44 Ikonnikov, *The Coal Industry of China*, p. 71, citing *MTGY*, no. 6 (1959): 6–7.

45 Lui Suinian and Wu Qungan (eds), *China's Socialist Economy*, p. 252.

46 Ibid., pp. 252–3.

47 Ibid., p. 259. Various agricultural targets were also cut.

48 Ibid., pp. 260–1.

49 Fang Weizhong (ed.), *Zhonghua renmin gongheguo*, p. 271; 'Guanyu 1960 nian guomin jingji jihua caoan de baogao' (Report on the Draft Plan for 1960), *RMRB*, 31 Mar. 1960, p. 2.

50 *CIYB 1983*, p. 54.

51 Liu Suinian and Wu Qungan (eds), *China's Socialist Economy*, p. 268.

52 There were disagreements on borders and the sharing of weapons technology. China was also frustrated by Moscow's lack of support in relations with India and Taiwan.

53 Communist Party of China, Central Committee Approves Report of Leading Group of the Ministry of Coal Industry (MCI), January 1961, *CIYB 1983*, p. 55.

54 Telephone Conference of the CPC Leading Group of the MCI, May 1961, *CIYB 1983*, p. 55.

55 *CIYB 1983*, p. 55.

56 'National Coal Cadres Meeting, September 1961', *CIYB 1983*, p. 55.

57 Fang Weizhong (ed.), *Zhonghua renmin gongheguo*, p. 313; Chen Yun, 'Talk at the Forum on Coal Work (Oct. 1961)', in Nicholas R. Lardy and Kenneth Lieberthal (eds), *Chen Yun's Strategy for China's Development: A Non-Maoist Alternative*, Armonk, NY: M. E. Sharpe, 1983, pp. 174–5.

58 Liu Suinian and Wu Qungan (eds), *China's Socialist Economy*, pp. 280–1.

59 The Ministry of Food and the MCI jointly issued the *Notice on Increasing Food Rations for Coal Miners of Different Provinces and Autonomous Regions – January 1961, CIYB 1983*, p. 56.

60 Communist Party of China, Central Committee, *Regulation on the Commodity Supply to Miners of State Controlled Mines, October 1961, CIYB 1983*, p. 55.

61 Premier Zhou Enlai's instructions came after he had requested the State Economic Commission to call a meeting of the Industrial Secretaries of six important coal-producing provinces. The Commission was to report to him and the Central Committee (April 1962, *CIYB 1983*, p. 56).

62 Forum of Directors of Ten Provincial or Regional Bureaux called by the Ministry June 1962, *CIYB 1983*, p. 57.

63 *CIYB 1983*, p. 59.

64 Liu Suinian and Wu Qungan (eds), *China's Socialist Economy*, pp. 283–4. The target for total industrial and agricultural output value was reduced from 140 billion yuan to

130 billion, for total agricultural output value from 45 billion to 42, and for total industrial output value from 95 billion to 88 billion.

65 The Ministry established a Long-Term Planning Committee, which was to be chaired by Vice Minister Zhong Ziyun, February 1963, *CIYB 1983*, p. 57. No details are given as to what planning periods were discussed.

66 Working Conference on Coal Mining Technology, convened by the Ministry, *CIYB 1983*, p. 58.

67 Ibid., p. 60.

68 Barry Naughton, 'The Third Front: Defence Industrialization in the Chinese Interior', *CQ*, 115 (1988): 353, citing Fang Weizhong (ed.), *Zhonghua renmin gongheguo*, p. 379.

69 Liu Suinian and Wu Qungan (eds), *China's Socialist Economy*, p. 339. The Plan was not published.

70 Cyril Chiren Lin, 'The Reinstatement of Economics in China Today', *CQ*, 85 (Mar. 1981): 3.

71 For example, Qi Sisi, 'Shengchan danwei jieyue yong mei da you ke wei' (Encourage Production Work Units to Save in the Use of Coal), *RMRB*, 9 Feb. 1968, p. 4.

72 *GYJJNJ 1998*, p. 58.

73 *CIYB 1983*, p. 62 reports that he 'died by persecution', but according to Lieberthal and Oksenburg, he apparently committed suicide. Kenneth Lieberthal and Michel Oksenberg, *Policy Making in China: Leaders, Structures and Processes*, Princeton: Princeton University Press, 1988, p. 90.

74 Telegram sent by CPC Central Committee May 1968, *CIYB 1983*, p. 63.

75 For example, 'Bixu jiuzheng hushi chanpin zhiliang de pianxiang' (Must Correct the Tendency to Neglect Product Quality), *MTGY*, no. 10 (1956): 10–11; Tong Boyin, 'Jige langfei yuan cailiao de qiye' (Enterprises that Waste Materials), *JHJJ*, no. 3 (1956): 27–8.

76 Joint Conference of the Ministries of Coal Industry, Petroleum, Railway, Communications, Metallurgy, and the State Bureau of Supplies and State Planning Commission, October 1969, *CIYB 1983*, p. 63.

77 Liu Suinian and Wu Qungan (eds), *China's Socialist Economy*, p. 358.

78 Mei Gong, 'Jiangnan meitian "meiyou kaicai jiazhi ma?"' (Do Coalfields South of the River Have 'No Value'?), *RMRB*, 10 Aug. 1970, p. 3.

79 Chen Dongsheng, 'Tiaozheng jingji jiegou, jiasu meitan nengyuan jidi jianshe' (Transform Economic Structure, Increase Speed of Coal Base Construction), *SXJDYJ*, p. 261.

80 Zhou Shulian and Pei Shuping, *Zhongguo gongye fazhan zhanlue wenti yanjiu* (Research Concerning Problems in China's Industrial Development Strategy), Tianjin: Tianjin renmin chubanshe, 1981, p. 110.

81 Transfers of jurisdictions, which required the approval of the State Council, occurred throughout the early 1970s, *CIYB 1983*, pp. 63–4. This decentralization occurred in all sectors of the economy as a result of discussion in the Forum of National Planning which issued *Rudimentary Ideas Conceived by the Ministries Under the Central Government on Transferring Enterprise Management to the Lower Levels*, in Liu Suinian and Wu Qungan (eds), *China's Socialist Economy*, p. 363.

82 Liu Suinian and Wu Qungan (eds), *China's Socialist Economy*, p. 351.

83 Like the Third Five Year Plan, it was not published.

84 Liu Suinian and Wu Qungan (eds), *China's Socialist Economy*, p. 370.

85 Ibid., p. 372.

86 Mao had designated Lin Biao to be his successor. However, the two men began to differ in 1970. Afraid of being officially removed, Lin Biao planned a military coup, but his plot was leaked. It is believed he died in a plane crash while trying to escape China. Historians are not in agreement as to the manner of death.

87 For example, Field Conference in Xuzhou (Jiangsu), June 1972; National Coal Mine Capital Construction Conference in Tangshan (Hebei), July/August 1972; a national

meeting called by the Ministry in September 1972; and National Planning Conference on Capital Construction and Geology in Coal Mines in Zouxian (Shandong), October 1972, *CIYB 1983*, pp. 65–6.

88 Fang Weizhong (ed.), *Zhonghua renmin gongheguo*, p. 507.
89 Liu Suinian and Wu Qungan (eds), *China's Socialist Economy*, p. 377.
90 Ibid., p. 385. Generally speaking, the main consumption centres, such as the thermal stations, and the large urban distribution centres strove to maintain a three-month reserve supply of coal on site.
91 Fang Weizhong (ed.), *Zhonghua renmin gongheguo*, p. 518.
92 This was a movement against traditional Confucian values, some of which were believed to be holding back the progress of the country, and was also an indirect attack on Zhou Enlai and other economic 'pragmatists', including Deng Xiaoping.
93 Liu Suinian and Wu Qungan (eds), *China's Socialist Economy*, p. 392.
94 Fang Weizhong (ed.), *Zhonghua renmin gongheguo*, p. 525.
95 'Emergency Meeting of all Key Coal Mines', June 1975, *CIYB 1983*, p. 67.
96 This line was completed in October 1973, but was not used to full capacity until much later.
97 Liu Suinian and Wu Qungan (eds), *China's Socialist Economy*, pp. 388–9.
98 Ibid., p. 395.
99 'Zhengfu gongzuo baogao' (Government Work Report), *RMRB*, 21 Jan. 1975, pp. 1–2.
100 Liu Suinian and Wu Qungan (eds), *China's Socialist Economy*, pp. 393–4 and 404–5; Fang Weizhong (ed.), *Zhonghua renmin gongheguo*, p. 541. The 1976 target was set at a National Planning Conference in December 1975 and January 1976. Also discussed were the *Outline Programme for the Ten-Year Planning of the Development of the National Economy (1976–1985)* and the Fifth Five Year Plan (1976–1980). The official output figure for 1975 was 480 mt. It seems the planners assumed production was somewhat less than this. Nevertheless, this shows that little capacity was available.
101 Liu Suinian and Wu Qungan (eds), *China's Socialist Economy*, pp. 404–5.
102 *CIYB 1983*, p. 66.
103 *Measures for Running Small Coal Mines and Provincial Regulations on Technical Transformation*, both adopted at a conference held in September 1973, *CIYB 1983*, p. 66.
104 Ibid., p. 578.
105 The Gang of Four was a radical group from Shanghai who tried to gain national power after the death of Mao Zedong in 1976.
106 Deng Xiaoping was elected Vice-Chairman of the Central Committee of the Chinese Communist Party and member of the Political Bureau Standing Committee at the Second Plenary Session of the Tenth Party Central Commitee held January 8–10, 1975. He was purged at the time of Zhou Enlai's death in January 1976 but was rehabilitated in the spring of 1977.
107 Deng Xiaoping in August and October 1975, *CIYB 1983*, p. 67.
108 IMF, *International Financial Statistics Yearbook*, Washington, DC, 30, 2 (Feb. 1977): 38–9.
109 Instructions given by Deng were conveyed at a National Symposium on Coal Mine Mechanization – November 1977, *CIYB 1983*, p. 69.
110 *CIYB 1983*, pp. 69–70.
111 Emergency Meeting of the State Council – July 1977, *CIYB 1983*, p. 70.
112 *CIYB 1983*, p. 70.
113 'Tuanjie qilai, wei jianshe shehui zhuyi de xiandaihua qiangguo er fendou' (Unite, Struggle to Build a Strong, Modern Socialist Nation), *RMRB*, 7 Mar. 1978, p. 3.
114 This project came virtually to epitomise all the difficulties that the Chinese government faced on the eve of the reforms. See Chae-Jin Lee, *China and Japan: New Economic Diplomacy*, Stanford, CA: Hoover Institution Press, 1984, pp. 30–75.

115 'Ruhe gaosu du fazhan meitan gongye?' (How to Speed up Coal Production?), *RMRB*, 4 Jan. 1978, p. 2; Fang Weizhong (ed.), *Zhonghua renmin gongheguo*, p. 588.
116 *See* Zhang Shihong, 'Qie shi jiejue ranliao langfei wenti' (We Must Be Practical in Solving the Fuel Waste Problem), *RMRB*, 11 May 1978, p. 2.
117 *CIYB 1983*, p. 72.
118 Ibid., p. 71.
119 Deng Xiaoping when visiting Tangshan Colliery – 19 September 1978, *CIYB 1983*, p. 71.
120 *CIYB 1983*, p. 71.
121 Chao Yu-shen, 'The Thirty-Point Decision on Industry: An Analysis', *Issues and Studies*, 15, 1 (Jan. 1979): 60, citing New China News Agency, Beijing, 1 May 1978.
122 For example, Huang Zaiyao, 'Dui wo guo jinnian lai nengyuan taolun zhong jige wenti de shangtao' (Recent Discussions about Energy Problems), *GYJJGLCK*, no. 1 (1981): I–41. In his speech on the occasion of the First Nationwide Energy Conservation Month Radio Broadcast and Television Rally, Vice Premier Kang Shien placed blame squarely on the Gang: 'Due to the long period of disturbance and damage caused by Lin Biao and the "Gang of Four", energy administration is in chaos ...'; 'Jieyue nengyuan shi guomin jingji fazhan zhong yi ge tuchu er jinpo de renwu: Kang Shien tongzhi zai chuanguo di yici 'jie neng yue' guangbo dianshi da hui shang de jinhua [zhaiyao]' (Conservation of Energy Is an Important, Urgent Task in the Development of the National Economy: Summary of Vice Premier Kang Shien's Speech for the First National Energy Conservation Month Radio and Television Broadcast), *GRRB*, 2 Nov. 1979, p. 1–2.
123 Hua Guofeng, 'Report on the Work of the Government', delivered 26 Feb. 1978, *PR* (10 Mar. 1978): 12.
124 The official government function of maintaining and publishing statistics had been provided for in the early days of the PRC, but ceased with the GLF.

3 Economic assessment of the industry on the eve of the reform and opening up programme

1 George Philip, *The Political Economy of International Oil*, Edinburgh: Edinburgh University Press, 1994, p. 122, citing J. Darmstadter and H. Landsberg, 'The European Background', *Daedalus*, 104 (Autumn 1975).
2 *MTGYNY 1997*, p. 321.
3 Alex Toohey, 'Country Experiences and Lessons Learned', in IEA, *Energy Efficiency Improvements in China: Policy Measures, Innovative Finance and Technology Deployment*, Proceedings of a Conference held in Beijing 3–4 Dec. 1996, Paris: OECD/IEA, 1997, pp. 73–4. Toohey was CEO of the World Coal Institute at the time. The Conference was jointly sponsored by the IEA, the Chinese State Planning Commission, and the Chinese National People's Congress, Environmental and Resources Protection Committee.
4 Alexander B. Ikonnikov, *The Coal Industry of China*, Canberra: Research School of Pacific Studies, Australian National University, 1977, p. 109, citing *MTGY*, no. 19 (1957): 3, 7.
5 'Opencast', 'strip', or 'surface' mining procedures are used when the deposits are lying in strata less than about 150 feet below the surface. It is far less expensive than underground mining because higher capacity machinery can be used and equipment for ventilation and for the prevention of flooding are not needed. Labour costs are lower because conditions are less difficult and dangerous. On the other hand, deposits near the surface, by their very nature, are of poorer quality due to greater oxidation. Opencast mining creates larger amounts of waste rock and is more disruptive in terms of the land used, imbalances to the water table and dust pollution. Also, depending on the climate, opencast mining may or may not be possible all year round. The World

Bank estimated that in 1978 average underground mining costs in the world were US$20–30 per ton and for opencast, US$10–15 per ton. World Bank, *Energy Options and Policy Issues in Developing Countries*, Staff Working Paper 350, Washington, DC: August 1979, p. 40.

6 United States, Congress, House, Committee on Interstate and Foreign Commerce, Subcommittee on Energy and Power, *Energy Factbook: Data on Energy Resources, Reserves, Production, Consumption, Prices, Processing, and Industry Structure*, Washington, DC: 1980, p. 544; Leslie Dienes and Theodore Shabad, *The Soviet Energy System: Resource Use and Policies*, Washington, DC: V. H. Winston, 1979, Table 25, p. 106; Peter James, *The Future of Coal*, 2nd edn, London: Macmillan, 1984, p. 217. Though this chapter pertains to the 1949–78 period, some 1979 or early 1980s data for China or for other countries has been used when no earlier data was available. It was assumed that, unless otherwise specified, statistics given in Chinese articles written in the early 1980s were late 1970s figures.

7 Dienes and Shabad, *The Soviet Energy System*, p. 264, citing G. G. Khorev, *Ekonomika otkrytoy dobychi uglya* (Moscow: *Nedra*, 1974), p. 5 and *Narodnoye Khozyaystvo SSSR za 60 let* (Moscow: *Statistika*, 1977), p. 206. Dienes and Shabad's note: 'Data refer to 1972, but there does not seem to have been any change in the degree of concentration since', Iain F. Elliot, *The Soviet Energy Balance: Natural Gas, Other Fossil Fuels and Alternative Power Sources*, New York: Praeger, 1974, p. 164.

8 Calculated from William S. Peirce, *Economics of the Energy Industries*, 2nd edn, Westport, CN: Praeger, 1996, citing Keystone Coal Industry Manual, 1995, p. S-4 and James, *The Future of Coal*, p. 107.

9 *The Energy Factbook*, p. 623.

10 Wallace E. Tyner, *Energy Resources and Economic Development in India*, Leiden: Martinus Nijhoff, 1978, p. 41.

11 P. D. Henderson, *India: The Energy Sector*, Delhi: Oxford University Press, 1975, pp. 38, 49–50.

12 The term 'processing' used in a general way, refers to the removal of various impurities such as rock, wood, clay, and up to half of the the pyritic sulphur, but does not remove the organic, chemically bonded sulphur and nitrogen. It may include physical separation, which mechanically distinguishes the relative gravities and/or surface properties of the coal, or magnetic or chemical separation. At a later stage the lumps are crushed and screened into various standard sizes. In the jiggery method of processing, coal is fed onto a mesh screen. Water is forced slowly upwards through the screen causing the light pieces of clean coal to float above the rock, clay, wood, etc. The clean product is drawn off in one direction while the debris is conveyed in another. The small, but heavier pieces of coal that go through the mesh land into a hutch below and are ideally processed again to remove pyritic material.

13 Liu Zaixing, 'Shanxi meitan nengyuan jidi jianshe ruogan wenti tantao' (Investigation into Certain Problems about Shanxi Coal Energy Base Construction), *JJWT*, no. 3 (1980): 2.

14 Jonathan E. Sinton *et al.* (eds), *China Energy Databook*, Berkeley, CA: Ernest Orlando Lawrence Berkeley National Laboratory Report LBL-32822 Rev. 4, Sept. 1996, Table IX.24, p. IX-51, using data from The Carbon Dioxide Information Analysis Centre, *Trends '93: A Compendium of Data on Global Change*, Oak Ridge, TN: Oak Ridge National Laboratory, 1994. This measurement of carbon dioxide includes emissions from fossil fuel combustion, cement production, and gas flaring.

15 Ibid., Table II.9, p. II-38, using data from *NYTJNJ* (various years), *TJNJ* (various years), *ZGNY*, no. 2 (1995), and data obtained by Sinton from the Energy Research Institute in Beijing, 1991.

16 *CIYB 1982*, p. 8.

17 Ibid.

18 For example, 71.8 per cent of the workers in Ukraine were using handtools in 1969. Elliot, *The Soviet Energy Balance*, pp. 165–6.

19 Henderson, *India: The Energy Sector*, p. 40.
20 United States, Central Intelligence Agency, *Chinese Coal Industry: Prospects Over the Next Decade*, ER 79-10092, Washington, DC: National Foreign Assessment Centre, 1979, p. 2.
21 Calculated from *CIYB 1982*, p. 36, the earliest data available; Guy Doyle made this point using 1984 data: *China's Potential in International Coal Trade*, London: International Energy Agency Coal Research, 1987, p. 30.
22 Hydraulic machinery greatly reduces the costs of shaft-building because the amount of tunnelling required is reduced by some 30–40 per cent and total construction time by about 20 per cent. Consumption of prop timber is far less, and fewer men need to be at the face underground. Joseph G. Crosfield and Sons Ltd, *The Chinese Coal Industry*, Warrington, UK: 1961–62, part 5, section viii, p. 37.
23 *CIYB 1982*, p. 31; B. O. Szuprowicz, 'Coal: China's Economic Backbone', *Coal Age*, 82, 6 (June 1977): 114.
24 Herbert S. Levine, 'Possible Causes of the Deterioration of Soviet Productivity Growth in the Period 1976–1980', in *Soviet Economy in the 1980s: Problems and Prospects: Part 1*, JEC Report, 1982, p. 158, citing CIA data.
25 Gr. Brit., House of Commons, Energy Committee, Session 1986–87, First Report, *The Coal Industry*, London: HMSO, 28 January, 1987, Table III.5, p. 36.
26 *CIYB 1983*, p. 141.
27 Robert Leggett, 'Soviet Investment Policy in the Eleventh Five Year Plan', in *Soviet Economy in the 1980s: Problems and Prospects, Part 1*, JEC Report, 1982, footnote 19, p. 143.
28 Gt. Brit. National Coal Board, *The Coal Industry of the USSR: A Report by the Technical Mission of the NCB*, London: HMSO, 1957, p. 62.
29 Wang Maolin, 'Zhuajin xian you qiye jishu gaizao jiakuai Shanxi meitan fazhan' (Must Modernize and Improve Enterprise Technology in Order to Raise Shanxi's Coal Production Quickly), in *SXJDYJ*, 1984, pp. 496–7.
30 In 1979 there was a mismatch between hoisting and extraction equipment in 76 mines. Martin Weil, 'China's Troubled Coal Sector' *CBR*, (Mar.–Apr. 1982): 24.
31 Gr. Brit. National Coal Board, *The Coal Industry of the USSR*, pp. 42, 47.
32 For example, 'Speed Preparatory Work in Donets Basin Mines', *Izvestia*, 23 May 1952, p. 2, in *CDSP*, iv, 21 (5 July 1952): 20–1; 'Donets Basin Miners in Struggle for New Five Year Plan', *Izvestia*, 1 Nov. 1952, p. 2, in *CDSP*, iv, 44 (13 Dec. 1952): 31–2; 'Let Us Make Fuller Use of Coal Mine Reserves', *Pravda*, 19 Mar. 1953, p. 2, in *CDSP*, v, 11 (25 April 1953): 20–1; 'Improve Direction of Coal Mines', *Pravda*, 17 Feb. 1954, p. 1, in *CDSP*, vi, 7 (31 Mar. 1954): 26–7; 'For Increased Coal', *Pravda*, 10 Oct. 1954, p. 1, in *CDSP*, vi, 41 (24 Nov. 1954): 15–16; 'Bureaucratic Planning and its Results', *Pravda*, 16 Oct. 1955, p. 2, in *CDSP*, vii, 42 (30 Nov. 1955): 26–7.
33 'Vital Problems in the Development of Coal Industry', *Pravda*, 2 June 1953, p. 2, in *CDSP*, v, 22 (11 July 1953): 26–7.
34 'Why Mine Construction Lags', *Pravda*, 12 July 1954, p. 2, in *CDSP*, vi, 28 (1954): 17–18. Similar articles appeared the following year: 'Develop Production Capacity of Coal Industry More Rapidly', *Pravda*, 28 Nov. 1955, in *CDSP*, vi, 48 (12 Jan. 1956): 23–24; 'Make Full Use of Donets Basin Reserves', *Pravda*, 19 Mar. 1955, in *CDSP*, vii, 11 (27 Apr. 1955): 26.
35 Ed A. Hewett, *Energy, Economics and Foreign Policy in the Soviet Union*, Washington, DC: Brookings Institute, 1984, p. 89, 92.
36 This data is not ideal in that the mine capacity data is for CMA and LS mines while the coal production data is for CMAs only. It does, however, reflect the main trends.
37 *GDTZZL*, p. 161.
38 Zhang Benlian, 'Yingxiang guoying meitan gongye yingli shuiping tigao de jige wenti' (Some Questions Influencing Raising of the Profit Level in the State Mining Coal Industry', *CZ*, no. 7 (1957): 30.

39 Choe Boum-Jong, *A Model of World Energy Markets and OPEC Pricing*, World Bank Staff Working Paper, no. 633, Washington, DC: International Bank for Reconstruction and Development/World Bank, 1984, p. 72, citing Martin B. Zimmerman, *The US Coal Industry*, Boston: MIT Press, 1981; Colin Robinson and E. Marshall, *What Future for British Coal*, London: Institute of Economic Affairs, 1981, p. 51; and other World Bank sources.

40 *GYJJNJ 1998*, pp. 57–8.

41 Chen Dianmo, 'Fahui Shanxi jingji youshi de genben tujing' (Bring into Play Shanxi's Superiority), in *SXJDYJ*, p. 251.

42 Doyle, *China's Potential in International Coal Trade*, p. 30.

43 Gt. Brit., National Coal Board [Report on the 1976 Delegation's Visit to China] [London: 1976], paragraph 2.6.

44 *CIYB 1983*, p. 70.

45 United Nations, *Industrial Statistics Yearbook*, New York: Department. of Economic and Social Affairs, State Statistical Office of the United Nations, various years; James, *The Future of Coal*, p. 217.

46 This refers to the miners at Shanxi's CMAs. Zhang Shouyi, 'Gaige guanli tizhi jiakuai Shanxi nengyuan jidi jianshe de bufa' (Measures to Hasten the Reform Management System of Construction for the Shanxi Energy Base), in *SXJDYJ*, pp. 287–8.

47 Ikonnikov, *The Coal Industry of China*, pp. 48–9, citing *MTGY*, no. 5 (1957): 22, 47; 10 (1957): 10.

48 *CIYB 1982*, p. 30.

49 Ibid., p. 31.

50 Xia Guocai, 'Accelerating the Development of an Energy-Heavy-Chemical Industrial Base in Shanxi Is a Major Strategic Decision in China's Modernization Programme', *JJRB*, 3 (15 Sept. 1983): 28, in *JPRS*, 14 (1984): 3.

51 'Improve Direction of the Coal Mines', *Pravda*, 17 Feb. 1954, p. 1, in *CDSP*, vi, 7 (31 Mar. 1954): 26–7.

52 Chinese miners were always being encouraged to strive for 'the red banner', and in the Soviet Union, at mines where the quota had been met, the miners were 'permitted to light up a red star at the head of [the] mine – a symbol of labour and victory'. 'Reserves for Increasing Labour Productivity in Krivoi Rog Mines', *Izvestia*, 7 Aug. 1954, p. 2, in *CDSP*, vi, 32 (22 Sept. 1954): 25.

53 *LDGZZL*, p. 159.

54 'State Owned Mines Boost Output', *Xinhua*, 21 Sept. 1985, in *JPRS*, 96 (1985): 44 and Conference on Coal Industry Capital Construction Work, Sept. 1983, *CIYB 1984*, p. 140.

55 'What Disturbs Miners', *Izvestia*, 13 Aug. 1954, p. 2, in *CDSP*, vi, 32 (22 Sept. 1954): 25–6.

56 Zhang Hongshun, 'Building Shanxi Energy Base Hinges on Upgrading Existing Technology', *Jishu jingji yu guanli yanjiu* (Technology Economics and Management Research), 4 (31 Dec. 1983): 44–7, in *JPRS*, 32 (1984): 40.

57 'The Challenges and Opportunities Faced by China's Coal Industry', *SJMTJS*, 2 (Feb. 1985): 3–6, in *JPRS*, 2 (Feb. 1985): 59.

58 Thane Gustafson, *Crisis Amid Plenty: The Politics of Soviet Energy under Brezhnev and Gorbachev*, Princeton, NJ: Princeton University Press, 1989, pp. 23–4, 47, 73–6, 100–1; Steven Rath, *The Hungarian Coal Industry*, National Committee for a Free Europe, Mid-European Studies Centre, Research Document no. 280, Washington: Photoduplication Service, Library of Congress, 1955, pp. 145, 181.

59 James, *The Future of Coal*, pp. 200, 222; Robert P. Greene and J. Michael Gallagher (eds), *Future Coal Prospects: Country and Regional Assessments*, World Coal Study, Cambridge, MA: Ballinger, 1980.

60 IEA, *Coal Prospects and Policies in IEA Countries, 1987 Review*, Paris: OECD/IEA, 1988, Table V.3, p. 61.

61 Barry Naughton, 'The Third Front: Defence Industrialization in the Chinese Interior', *CQ*, 115 (1988): 381, 351.
62 Ibid., p. 365.
63 Ibid., p. 366.
64 Ibid., p. 379.
65 Zhang Mengceng, 'Guoying meitan gongye chanpin jiage wenti' (Problems with the Prices of Products from the State Owned Coal Industry), *CZ*, no. 3 (1958): 14; Shanxi shehui kexue yuan, meitan jiage yanjiu keti zu (Shanxi Academy of Social Sciences, Coal Pricing Research Group), *Meitan jiage gaige yu Shanxi meitan jiage* (Coal Pricing Reform and Shanxi Coal Pricing), an unpublished research report prepared by the Academy, Taiyuan: 1986, p. 51.
66 The author interviewed Ye Ruixiang in September 1987.
67 Originally only the production cost profit rate, adopted from the Soviet pricing procedure during the First Five Year Plan period, was used. It is derived by dividing the total revenues from any commodity by the total production costs. The concept of 'capital profit rate' was later devised by the Chinese. For any commodity or group of commodities, an average rate of profit on capital funds is set such that the more capital used, the higher the expected rate of profit.
68 Calculated from *TJNJ 1981*, p. 207.
69 Wang Zhenzhi and Wang Yongshi, 'Epilogue: Prices in China', in Wei Lin and Arnold Chao (eds), *China's Economic Reforms*, Philadelphia: University of Pennsylvania Press, 1982, p. 223.
70 Morris Bornstein, 'Soviet Price Policy in the 1970s', in *Soviet Economy in a New Perspective*, JEC Report, 1976, p. 23; Stanley H. Cohn, 'Sources of Low Productivity in Soviet Capital Investment', in *Soviet Economy in the 1980s: Problems and Prospects, Part I*, JEC Report, 1982, p. 190. Interestingly, in 1988 the price of coal in Poland was increased 200 per cent. *FT*, 1 Feb. 1988, p. 2.
71 Kenneth Lieberthal and Michel Oksenberg, *Policy Making in China: Leaders, Structures and Processes*, Princeton: Princeton University Press, 1988, pp. 165–6; Doyle, *China's Potential in International Coal Trade*, p. 39.
72 Sinton *et al.* (eds), *China Energy Databook*, Table VII.1, p. VII-4.
73 Liu Song, 'Japanese Loans Pooled to Build Railways', *BR*, 13 Sept. 1999, p. 27.
74 Holland Hunter, 'Transport in Soviet and Chinese Development', *Economic Development and Cultural Change*, 14 (Oct. 1965): 75.
75 Ralph W. Huenemann and Nicholas H. Ludlow, 'China's Railroads', *CBR*, 4, 2 (1977): 32.
76 World Bank, *China: The Transport Sector. Annex 6 to China: Long Term Development Issues and Options* , Washington, DC: 1985, p. 33 citing George W. Wilson, *Economic Analysis of Intercity Freight Transportation*, Bloomington, IN: Indiana University Press, 1980, p. 275; *GDTZZL*, pp. 45–9, 101.
77 *TJNJ 1981*, pp. 296–7.
78 Naughton, 'The Third Front', p. 380, citing [China] State Economic Commission, Comprehensive Transport Research Institute, 'How Transport Became a Weak Link in National Economic Development', *GYJJGLCK*, no. 4 (1981): 1–3.
79 *TJNJ 1998*, p. 544.
80 Albert S. Peterson, 'China: Transportation Developments, 1971–80', in *China Under the Four Modernizations*, JEC Report, 1982, pp. 144, 147.
81 World Bank, *China: The Transport Sector. Annex 6*, Table 4.5, p. 34.
82 *TJNJ 1988*, p. 510.
83 Wu Runtao, 'Jin mei yunshu yu qita wuzi yunshu de bili guanxi' (Relationship Between Coal Transport and Other Materials Transport), *SXJDYJ*, p. 643.
84 Sun Shangqing, 'Nengyuan jiegou' (Energy Structure), in Ma Hong and Sun Shangqing (eds), *Zhongguo jingji jiegou wenti yanjiu* (Problems in China's Economic Structure), Beijing: 1980, p. 266.

85 Thomas P. Lyons, *Economic Integration and Planning in Maoist China*, New York: Columbia University Press, 1987, pp. 143–8.
86 'Meitan yunshu sunhao jingren' (Astonishing Loss of Coal During Shipment), *RMRB*, 14 Sept. 1981, p. 2; Hu Benzhe, 'Guanyu Shanxi nengyuan jidi meitan shusong fangshi de tantao' (Investigation into Methods of Coal Transportation in Shanxi Energy Base), *NY*, no. 4 (1983): 2.
87 Sun Shangqing, 'Nengyuan jiegou' (Energy Structure), p. 266; Wu Runtao, 'Jin mei yunshu', p. 644.
88 'Meitan yunshu sunhao jingren' (Astonishing Loss of Coal During Shipment), *RMRB*, 14 Sept. 1981, p. 2.
89 *TJNJ 1998*, p. 538.
90 Wu Runtao, 'Jin mei yunshu', p. 647.
91 'Gaige meitan yunxiao gongzuo de jidian yijian' (Some Views On Reforming Coal Marketing Work), *MTJJYJ*, no. 9 (1984): 7.
92 Ta M. Li and K. P. Wang, 'China's Emerging Mining Industry: A Long March Toward Modernization', *Mining Engineering*, 32, 3 (Mar. 1980): 277. Another example is given in Liu Guoguang and Wang Riusan, 'Restructuring of the Economy', in Yu Guangyuan (ed.), *China's Socialist Modernization*, Beijing: Foreign Languages Press, 1984, p. 98.
93 Christopher Howe, *China's Economy: A Basic Guide*, London: Granada, 1978, p. 108, citing two sources: He Baisha, 'Fahui difang jijixing, dali fazhan difang meikuang' (Bring into Play the Positive Aspects of the Development of Local Mines), *JJJH*, no. 2 (1958): 12–14, and Qin Yuzheng, 'Jianshe xin mei jing yingdang zhuyi diqu chan xu pingheng' (In Building New Coal Mines, Attention Must Be Paid to the Regional Balance of Production), *JJJH*, no. 6 (1958): 10–11.
94 Bogdan Mieczkowski, *Transportation in Eastern Europe: Empirical Findings*, New York: East European Quarterly, 1978, pp. 148, 147. See also Holland Hunter, 'Transportation in Soviet Development', in Gary Fromm (ed.), *Transport Investment and Economic Development*, Washington, DC: Brookings Institution, 1965, p. 130.
95 Lei Ding and Liang Kuangbai, 'Jiaotong yunshuye zai wo guo guomin jingji zhong de zuoyong' (The Role of Communications and Transportation in the National Economy), *JJYJ*, no. 2 (1965): 41.
96 For example, Julia Kwong, *The Political Economy of Corruption in China*, Studies on Contemporary China, Armonk, NY: M. E. Sharpe, 1997, pp. 51–78; Joseph Berliner, *Soviet Industry from Stalin to Gorbachev: Essays on Management and Innovation*, Ithaca, NY: Cornell University Press, 1988, pp. 75–6; David Granick, *Management of the Industrial Firm in the USSR: A Study in Soviet Economic Planning*, New York: Columbia University Press, 1954, pp. 144–5; Alexander Gerschenkron, *Economic Backwardness in Historical Perspective*, New York: Frederick A. Praeger, 1962, p. 286.
97 For example, the vice party secretary of Henan was accused of under-reporting coal production by 5.3 million tons, hiding nearly two million tons of cement and not reporting thousands of machines, tractors and cars in the mid-1970s. Kwong, *The Political Economy of Corruption in China*, p. 67, citing *RMRB*, 13 Nov. 1978, p. 1.
98 Janos Kornai, *Economics of Shortage*, vol. A, Amsterdam: North-Holland, 1980, pp. 119, 102.
99 Li Yinglu, 'Wuzi guanli tizhi bixu gaige' (Materials Management System Needs Reform), *JJGL*, no. 10 (1979): 24.
100 Wan Jing, 'Wajue qiye qianli, jiasu wuzi zhouzhuan' (Tap Enterprise Potential, Speed up Materials Turnover), *JJGL*, no. 8 (1979): 27.
101 'Gaige meitan yunxiao gongzuo de jidian yijian' (Some Views On Reforming Coal Marketing Work), p. 7.

102 James, *The Future of Coal*, p. 138.

103 Henderson, *India: The Energy Sector*, p. 156.

104 Gong Guangyu, 'Wo guo nengyuan de xianzhuang he weilai' (China's Energy Present and Future), in 2000 nian yanjiu xiao zuzhu (Year 2000 Research Group), *Gongyuan 2000 nian de Zhongguo* (China in the Year 2000), Beijing: Kexue jishu wenxian chubanshe, 1984, p. 33.

105 Sinton *et al.* (eds), Table II.28, p. II-79.

106 In 1979 the breakdown of rural energy consumption was 12.7 per cent from firewood, 34.9 per cent from dried stalks, 24.6 per cent from dung, 14.2 per cent from small hydro electricity plants, and 13.6 per cent from small coal mines. Sun Shangqing and Zhai Ligong, *Zhongguo nengyuan jiegou yanjiu* (Research into China's Energy Structure), Taiyuan: Shanxi renmin chubanshe/Zhongguo shehui kexue chubanshe, 1987, pp. 199–202. Sun Shangqing, 'Nengyuan jiegou' (Energy Structure), p. 277.

107 The author is aware there are problems with respect to the accuracy and comparability of the official GDP figures. See Ren Ruoen, *China's Economic Performance in an International Perspective*, Paris: OECD Development Centre, 1997, Chapters 1 and 2.

108 Jiang Zhenping, 'Energy Efficiency in China', a paper presented at the *Industrial Ecology and Global Change*, conference sponsored by the Global Change Institute, held in Snowmass, CO, July 1992, p. 1. Dr Jiang, a senior research fellow at the Energy Research Institute, State Planning Commission, sent the author a copy of his paper in advance of publication of the conference proceedings by the Chinese University Press in 1994.

109 These coefficients obviously overlap. Coal was the dominant form of energy and main fuel used in the generation of electricity.

110 It is impossible to make meaningful international comparisons of *coal* consumption because very few economies relied on coal to the extent that China did (70–92 per cent of total energy consumed over the period in question). As the breakdown of coal consumption is very similar to that for total energy consumption, international comparisons of *energy* consumption are substituted.

111 Wang Jiacheng, 'Cong wo guo nengyuan de xiaofei goucheng kan jieneng de qianli he tujing' (Discussion of the Means and Potential for Energy Saving from Observing the Consumption Structure of Energy in Our Country), GYJJGLCK, no. 3 (1981): 2.

112 Fang Rukang, 'Wo guo nengyuan ziyuan de xianzhuang ji heli liyong wenti' (Rationalizing China's Current Use of Energy Resources), DLYJ, no. 3 (Dec. 1984): 33, citing 'Guoqing yu zhanlue, quanmian anpai nengyuan de kaifa he jieyue' (National Conditions and Strategy: Overall Arrangements for Energy Production and Conservation), LW, 11, 1982.

113 IEA, *Energy Balances of Non-OECD Countries, 1996–1997*, Paris: OECD/IEA, 1999, pp. II484–489.

114 *JJNJ 1982*, p. v–98.

115 *NYTJNJ 1986*, pp. 80–5.

116 Wang Jiacheng, 'Cong wo guo nengyuan', p. 3.

117 World Bank, *World Development Report 1980*, Washington: DC, 1980, p. 122–3.

118 Meng Renlun and Shen Changzhi, 'Jinhou gangtie gongye de jieneng fangxiang' (Future Directions in Iron and Steel Industry Energy Conservation), NY, no. 5 (1983): 1.

119 IEA, *Coal Information 1985*, Paris: OECD/IEA, 1985, pp. 155, 167, 203, 247.

120 Sun Shangqing and Zhai Ligong, *Zhongguo nengyuan jiegou yanjiu*, p. 113.

121 *TJNJ 1999*, p. 36. These are per capita output figures which are equivalent to per capita consumption.

122 IEA, *Energy Balances of Non-OECD Countries, 1996–1997*, pp. II490–95.

123 Gong Guangyu, 'Wo guo nengyuan de xianzhuang he weilai', p. 35.

124 Seiichi Nakajima, 'China's Energy Problems: Present and Future', *Developing Economies*, 20, 4 (1982): 489.

125 Vaclav Smil, 'Energy' in Leo A. Orleans and Caroline Davidson (eds), *Science in Contemporary China*, Stanford: Stanford University Press, 1980, p. 422, citing I. Fan, 'Industries in Mainland China', in R. Carin (ed.), *Communist China 1969*, Hong Kong: Union Research Institute, 1970, p. 312.

126 *GYJJNJ 1998*, p. 59.

127 Ibid.; Sinton *et al.* (eds), *China Energy Databook*, Table II.21, p. II-70.

128 Nakajima, 'China's Energy Problems', p. 490.

129 Robert Carin, *Power Industry in Communist China*, Communist China Problem Research Series EC44, Hong Kong: Union Research Institute, 1969, p. 111, citing 'Construction of High Voltage Transmission Line in Northeast', NCNA (New China News Agency) English Release, Mukden, 9 Nov. 1953; 'The Northeast's New High-Tension Transmission Line', *People's China*, 16 (16 Aug. 1954): 26–7; and Huang Zhiming, 'The Development of High Voltage Overhead Transmission Line Technology in China', *Electricity* (Chinese Society for Electrical Engineering), 10, 1 (1999): 39.

130 Nakajima, 'China's Energy Problems', p. 490.

131 *TJNJ 1998*, p. 38.

132 Calculated from Beijing Review Press, *The Development of China 1949–1989*, China Issues and Ideas, Beijing: 1989, Table 1.15, p. 27.

133 *NYTJNJ 1986*, p. 17.

134 To convert a unit of coal, coke, crude oil, gasoline, diesel oil, or natural gas to standard coal equivalent, the coefficients are 0.7143, 0.9714, 1.4286, 1.4714, 1.4571, and 1.3300 respectively. *NYTJNJ 1991*, p. 413.

135 Mei Yan, 'Fazhan meitan xixuan jiagong shi jieneng de you xiao tujing' (Development of Coal Washing and Dressing Is an Effective Method of Energy Conservation), *JJGL*, 12 (1981): 29.

136 Steam coals are assessed on the basis of ash, volatile matter, sulphur, and moisture content, in combination with heating value and grindability.

137 Xu Shoubo, *Nengyuan jishu jingji xue* (Energy Technology Economics), Changsha: Hunan renmin chubanshe 1981, p. 232.

138 Mei Yan, 'Fazhan meitan xixuan', pp. 28–9.

139 Meng Renlun and Shen Changzhi, 'Jinhou gangtie', p. 1.

140 Gong Guangyu, 'Wo guo nengyuan de xianzhuang he weilai', p. 35.

141 Lin Senmu, Zhou Shulian, and Qi Mingchen, 'Industry and Transport', in Yu Guangyuan (ed.), *China's Socialist Modernization*, Beijing: Foreign Languages Press, p. 334.

142 Wu Yuan-li and H. C. Ling, *Economic Development and the Use of Energy Resources in Communist China*, New York: Praeger, 1963, pp. 110, 125.

143 Sun Shangqing, 'Nengyuan jiegou', p. 285; Fang Rukang, 'Wo guo nengyuan ziyuan', p. 34.

144 *JJNJ 1982*, p. v–98.

145 Wang Xianglin, 'Jieyue nengyuan shi tigao jingji xiaoyi de zhongyao tujing' (Saving Energy Is an Important Way of Raising Economic Results), *NY*, no. 3 (1984): 6.

146 'Statistics of Energy Utilization in China', *Economic Reporter*, Feb. 1981, p. 9.

147 Jiang Zhenping, 'Energy Efficiency in China', p. 1.

148 Robert W. Campbell, *Soviet Economic Power: Its Organization, Growth and Challenge*, London: Stevens, 1960, p. 181.

149 Mei Yan, 'Fazhan meitan xixuan', p. 28.

150 Ferenc Janossy, 'The Origins of Contradictions in Our Economy and the Path to Their Solution', translation of *Közgazdasagri Szemle*, 6, 1969 in *Eastern European Economics*, 8, 4 (Summer 1970): 378.

151 Liu Kaimo, 'Wuzi xiaohao dinge zhong cunzai de jige wenti' (Problems Concerning Materials Consumption Quotas), *JHJJ*, no. 1 (1956): 34.

152 'Is It Good to Have Excess Stored Materials?', *Dagongbao*, 19 Mar. 1966, p. 3, in *JPRS*, 262, 29 Apr. 1966, pp. 1–5.

4 The industry and Deng's reforms

1 Communique of the Third Plenary Session of the 11th Central Committee of the Communist Party of China', *PR*, 52 (29 Dec. 1978), p. 11.

2 Not all observers of the Chinese economy believe that it was in a state of exhaustion in the late 1970s. Some hold the view that the stronger impetus of the reform movement was the desire by Deng Xiaoping and his supporters to win political popularity by raising living standards. See Barry Naughton, *Growing Out of the Plan: Chinese Economic Reform, 1978–1993*, Cambridge, Cambridge University Press, 1996, pp. 60–1. Whatever the true state of the economy at that time, a large and reliable supply of energy was required to move it forward.

3 Yu Qiuli, 'Arrangements for the 1979 National Economic Plan', *PR*, 20 July 1979, p. 7.

4 For example, *Nengyuan* (Energy), *Meitan kexue jishu* (Coal Science and Technology), *Jiage gaige* (Price Reform), *Jingji guanli* (Economic Management), *Jingji yanjiu* (Economic Research), *Gongye jingji* (Industrial Economics), *Wuzi guanli* (Materials Management), and *Jingji wenti* (Economic Problems).

5 Yu Hongen was a graduate of Beijing Mining Engineering Institute and had been Deputy Chief of Heilongjiang Provincial Coal Administrative Bureau.

6 Martin Weil, 'Coal's Promises and Problems', *CBR*, Mar.–Apr.1984, p. 40.

7 Gao Yangwen, 'Chuang li you Zhongguo tese de chongman shengji he huoli de meitan gongye jingji tizhi' (Creating a Distinctly Chinese-Style and Vigorous Coal Industry Economic System), *NY*, no. 2 (1985): 1.

8 Gao Yangwen, *CIYB 1982*, Introduction.

9 Gao Yangwen, 'On the Issue of China's Energy Policy', *MTKXJS*, no. 4 (Apr. 1981): 4–8, in *JPRS*, 166 (1981): 15.

10 Gao Yangwen, Speech at the National Conference on Comprehensive Utilization of Coal, February 1982, *CIYB 1983*, p. 98.

11 Gao Yangwen, 'Chuang li you Zhongguo tese', pp. 5–6.

12 'Strive for the Modernization of the Coal Industry in the Spirit of the CPC 12th National Congress', *CIYB 1983*, pp. 91–5.

13 'xi' means 'west'.

14 China, Ministry of Coal Industry, Production Department, 'Speed Up Technical Transformation of Existing Mines – How Existing Mines May Produce 400 Million Tons by the End of This Century', *CIYB 1983*, p. 184.

15 Liu Zaixing, '1990 nian san yi dun de chanliang shi nan dadao de' (The 1990 Target of 300 Million Tons Is Difficult to Reach), *SXJDYJ*, p. 221. It cost about 108 yuan per ton to open a new state mine in Shanxi, whereas it cost only 60 yuan on average to renovate or expand one. Gao Huimin, 'Jianshe Shanxi nengyuan jidi touzi guimo yuce' (The Scale of Investment in Shanxi Energy Base), *SXJDYJ*, p. 394.

16 'Strive for the Modernization of the Coal Industry in the Spirit of the CPC 12th National Congress', *CIYB 1983*, p. 92; 'Speed Up Technical Transformation of Existing Mines', *CIYB 1983*, p. 184; Gao Yangwen, 'Quanli kaichuang meitan gongye xiandai hua jianshe de xin jumian' (Considerable Effort Must Be Devoted to Modernizing the Coal Industry), *NY*, no. 2 (1983): 4.

17 At that time the Ministry had virtually no opencast mining equipment. Gao Yangwen, 'Quanli kaichuang meitan gongye', p. 5; Wang Senhao, 'Some Preliminary Conceptions of Technical Progress in China's Coal Industry During the Next Eighteen Years', *SJMTJS*, no. 5 (1983): 2–6, in *JPRS*, 377 (1983): 53, 55. Wang Senhao was the former chief engineer in the Ministry of Coal Industry.

338   *Notes*

18 Armand Hammer with Neil Lyndon, *Hammer Witness to History*, London: Simon and Schuster, 1987, p. 460; Li Zhengping, 'Pingshuo Antaibao lutian meikuang jianshe juece jishi' (Strategic Policy for Construction of Pingshuo Antaibao Opencast Coal Mine), *JJRB*, 20 Oct. 1987, p. 1.

19 *A Brief Introduction to Antaibao Surface Mine*, a pamphlet published by the Joint Control Committee of Antaibao Surface Mine which was given to visitors at the site.

20 Wu Zhen, 'Qilai ba, Pingshuo' (Rise Up Pingshuo), *SXRB*, 8 Sept. 1987, p. 2; 'Antaibao lutian meikuang touchan' (Antaibao Opencast Coal Mine Begins Production), *SXRB*, 11 Sept. 1987, p. 1; Xu Yuanchao, 'Giant Coal Mine Shows Sino-Foreign Cooperation', *CD*, 11 Sept. 1987.

21 'Antaibao Mine Partners Work to Overcome Development Obstacles', *CWI*, 24 Feb. 1988.

22 *CIYB 1983*, p. 218.

23 *CIYB 1983*, p. 218; *CIYB 1984*, p. 225.

24 *CIYB 1983*, p. 219; *CIYB 1985*, p. 169.

25 *CIYB 1983*, p. 218.

26 'Brief Introduction to Coal Industry in Shanxi Province', *CIYB 1985*, pp. 177–8.

27 'Broad Prospects for China's Coal Industry', *GRRB*, 25 Oct. 1982, translated in *CIYB 1983*, pp. 120–1.

28 'Shenhua' is formed from the names 'Shenmu' and 'Huanghua'.

29 Gao Huimin, 'Jianshe Shanxi nengyuan', *SXJDYJ*, p. 394.

30 *SXJJNJ 1985*, p. 187.

31 'Economic Policies Concerning Locally Administered State Coal Mines', *CIYB 1982*, p. 86.

32 'The Coal Industry Decides to Relax the Restrictions and Encourage the Masses to Run Small-Sized Mines', *RMRB*, 10 Mar. 1983, translated in *CIYB 1984*, p. 114.

33 'Measures to Improve Energy and Transport', *Peking Home Service*, 1200 GMT, 24 Mar. 1983, *SWB Part 3, The Far East, C*, China: FE/7305/C/1,

34 Small mine operators were permitted to use abandoned areas of large administrations, and shallow or structurally damaged seams. Huang Zaiyao, 'Dui Shanxi sheng fazhan shedui meikuang jige wenti de shangtao' (Discussion about Problems with Commune Coal Mine Production in Shanxi Province), *JJWT*, no. 5 (1984): 20.

35 'Liberalization of Policies for Accelerated Small Coal Mine Development, State Council Approves and Forwards Ministry of Coal Industry's "Report on Eight Measures for Accelerating the Development of Small Coal Mines", Requiring All Jurisdictions to Implement It', *MTB*, 11 May 1983, p. 1, in *JPRS*, 389 (1983): 34–7; 'The State Council Approves the Coal Ministry's Report and Notice Is Dispatched to Relax the Restrictions and Rapidly Develop Small Mines', *JJCKB*, 21 May 1983, translated in *CIYB 1984*, p. 114.

36 'National Working Conference of Locally-Run Coal Mines', *CIYB 1983*, pp. 169–70.

37 *CIYB 1984*, p. 96.

38 Speech by Li Peng in Nov. 1984, *CIYB 1985*, p. 127.

39 'Speed Up Technical Transformation of Existing Mines', *CIYB 1983*, p. 190.

40 *SXJJNJ 1985*, p. 189.

41 For example, Mei Yan, 'Fazhan meitan xixuan jiagong shi jieneng de you xiao tujing' (Development of Coal Washing and Dressing Is an Effective Method of Energy Conservation), *JJGL*, no. 12 (1981): 28–30; Chen Zhu, 'Da li fazhan meitan xixuan jieyue nengyuan' (Increase Coal Processing to Save Energy), *NY*, no. 4 (1984): 27–8; and Shao Tonghan, 'Ti gao meitan zhiliang jieyue nengyuan' (Raise Coal Quality, Save Energy), *NY*, no. 3 (1984): 32–3.

42 *CIYB 1982*, p. 17.

43 Ibid., p. 20.

44 Gao Yangwen, Speech at the National Conference on Comprehensive Utilization of Coal, February 1982, *CIYB 1983*, p. 99.

45 'Instructions of the State Council on Development of Coal Preparation and Processing and Rational Utilization of Energy', 12 Nov. 1982, *CIYB 1983*, pp. 86–8.

46 Wang Senhao, 'Some Preliminary Conceptions', p. 58; *CIYB 1983*, p. 190.

47 Ye Qing, Speech at the National Coal Mining Mechanization Working Meeting, April 1984, *CIYB 1985*, pp. 42–7; National Conference on Mining Machinery, *CIYB 1983*, pp. 154–5.

48 *CIYB 1982*, pp. 14–15, 17.

49 'Main Coal Enterprises and Institutions', *CIYB 1982*, p. 262.

50 'National Working Conference on Coal Mining Mechanization', *CIYB 1983*, pp. 151–2.

51 'Gaige meikuang jixie zhizao qiye guanli tizhi chexiao zhongguo meikuang jixie zhizao gongsi' (Reform Coal Mine Machinery Manufacturing Management System, Abolish China Coal Mine Machinery Manufacturing Company), *MTGYNJ 1987*, pp. 75–6.

52 'National Working Conference on Manufacturing of Coal Machinery', *CIYB 1984*, p. 131.

53 Ye Qing, Speech at the National Coal Mining Mechanization Working Meeting, April 1984, *CIYB 1985*, pp. 42–7.

54 Zhou Yimin and Zhang Du, 'Jianchi gaige gaohao meitan gongye de wuzi gongying guanli' (Persist in Reforming Coal Industry's Materials Supply Management), *WZGL*, no. 4 (1986): 15.

55 Ye Qing, 'Explanations on the General Contract for State-Controlled Mines', *CIYB 1985*, p. 146.

56 *CIYB 1983*, pp. 73–4. At a national planning conference in November 1982 it was announced that conditions in 87 of the 96 mines had been 'alleviated', *CIYB 1983*, p. 92.

57 'Regulations Regarding Contracting Capital Construction Projects for the Coal Industry' were made public in December 1980, *CIYB 1982*, pp. 80, 109.

58 'Make an All-Round Start and Push Forward in Unison: Key Points of Work of the Coal Industry in 1983', *CIYB 1983*, p. 95.

59 Wang Senhao, 'Some Preliminary Conceptions', p. 54.

60 'Chronicle of Events in the Coal Industry Since Founding of the PRC 1949–1979', *CIYB 1983*, pp. 73–4; 'The Fixed Wage Per Ton Coal System Will Be Exercised in State-Controlled Mines Country-Wide', *MTB*, 11 July 1984, translated in *CIYB 1985*, pp. 65–6.

61 Author's interview with representatives of the Shanxi Mining Bureau in 1997.

62 China, Ministry of Coal Industry, Policy Research Office, 'New Wage Contract System: The Wage/Tonnage System Implemented on a Trial Basis in the Jinggezhuang Coal Mine', *CIYB 1984*, pp. 163–5.

63 *LDGZZL*, p. 159.

64 Gao Yangwen, 'Chuang li you Zhongguo tese', p. 6.

65 Gao Yangwen, 'Relying on the Advance of Science and Technology to Open Up a New Prospect for the Realization of Doubling Coal Production', *CIYB 1984*, pp. 173–96.

66 'Strive for the Modernization of the Coal Industry in the Spirit of the CPC 12th National Congress', *CIYB 1983*, p. 95.

67 'Restructuring of the Ministry of Coal Industry', *CIYB 1983*, pp. 200–1.

68 These were the departments of Production, Planning, Capital Construction, Technology Development, Finance, Labour and Wages, Education, and Personnel; bureaux for Product Allocation, Transportation and Sales, Coal Processing and Utilization, and Veteran Cadres; the General and Policy Research Offices, and Safety Inspectorate.

69 Luo Hongda, 'Reform of Financial System Taking an Encouraging Step', *CIYB 1984*, p. 152; 'Meeting on Consolidation of Key Enterprises' (July 1982), *CIYB 1983*, pp. 89–91; 'National Working Conference on the Management of Labour and Wages

in State-Controlled Mines' (Apr. 1982), *CIYB 1983*, pp. 160–1; and 'Forum on Consolidation of Major Coal-Mining Enterprises' (July 1982), *CIYB 1983*, p. 162.

70 Hu Fuguo (Vice-Minister of Coal Industry), Speech at Meeting for Management Work in State-Run Coal Mines, 20 Aug. 1984, *CIYB 1985*, pp. 47–52.

71 Ye Qing, 'Explanations on the General Contract for State-Controlled Mines', *CIYB 1985*, p. 148.

72 Luo Hongda, 'Reform of Financial System Taking an Encouraging Step', *CIYB 1984*, p. 154.

73 *CIYB 1985*, p. 53.

74 Luo Hongda, 'Reform of Financial System', p. 152; Ye Qing, 'Explanations on the General Contract for State-Controlled Mines', *CIYB 1985*, p. 145.

75 Ye Qing, 'Explanations on the General Contract for State-Controlled Mines', *CIYB 1985*, p. 146.

76 Luo Hongda, 'Accelerate the Tempo of Financial Reform and Implement Various Economic Policies Specified by the General Contract System', *CIYB 1985*, p. 108.

77 Hu Fuguo, 'Speech at Meeting for Management Work in State-Run Coal Mines', *CIYB 1985*, pp. 51–2.

78 'National Working Conference of Locally-Run Coal Mines', *CIYB 1983*, pp. 169–70.

79 Zhang Qinwen and Ren Boping, 'The Choice of Strategy for the Construction of the Shanxi Energy Base', *Shehui jingji daobao* (Social Economic Herald), 4 June 1985, p. 5, in *JPRS*, 69 (1985): 1–4.

80 For example, Gao Yangwen, 'Chuang li you Zhongguo', pp. 1–6; Gao Yangwen, 'Minister Unveils Grand Strategy to Revamp Coal Industry', *MTKXJS*, 10 (1982): 2–4, in *JPRS*, 336 (1983): 82–7; Ye Ruixiang, 'Wo guo nengyuan jiage wenti de yanjiu' (Research into China's Energy Pricing Problems), in Zhongguo nengyuan jingji yanjiu weiyuanhui (China Energy Economics Research Committee), *Zhongguo nengyuan jingji wenti wenji* (China Energy Economics Problems: Collected Works), Beijing: Nengyuan chubanshe, 1981, pp. 24–9; Meitan jiage gaige xueshu taolun hui (Coal Price Reform Study Group), 'Shenru tantao meitan jiage gaige de lilun he shijian wenti' (Thorough Investigation into the Problems and Practice of Coal Pricing Reform), *JDYJ*, no. 22 (1986): 1–8; Mei Yan, 'Tiaozheng meitan jiage shi zai bi xing' (Transforming Coal Prices Is Imperative), *GYJJGLCK*, no. 11 (1983): 55–6, 45; Xu Jingyi, 'Cong meitan ziyuan zengjia jiage xiajiang kan jiazhi guilu de zuoyong' (The Effect of Raising Coal Resource Prices on Law of Value), *WZGL*, no. 2 (1985): 2–3; Wang Jiachun and Li Minxin, 'Guanyu wo guo nengyuan zhengce yu jiage de maodun' (China's Energy Policies and Price Contradictions), *JJWT*, no. 4 (1984): 17–21; Li Shiyi, 'Meitan jiage gaige de chubu shexiang' (Preliminary Ideas on Coal Price Reform), *JDYJ*, no. 22 (1986): 1–8; Qin Meng, 'Meitan jiage gaige yu Shanxi meitan jiage guandian zhaiyao' (A Point of View on Coal Price Reform and Coal Prices in Shanxi), *JDYJ*, no. 19 (1986): 3–5.

81 Meitan jiage gaige xueshu taolun hui (Coal Price Reform Study Group), 'Shenru tantao', p. 5.

82 In 1979 the prices of several basic agricultural products were raised, while in the early 1980s prices of many products sold in the cities were decontrolled. The goal was to decentralize pricing control and gradually eliminate the state monopoly over purchasing and marketing. As a result of the price reforms the quantity and quality of some products was better than ever before. However, price increases also led to high inflation and social divisions. Some farmers' incomes were greatly increased, while urban residents' spending power was often actually decreased, and state enterprises' debts spiralled. The hoped-for immediate increases in productivity generally were not realized.

83 Meitan jiage gaige xueshu taolun hui (Coal Price Reform Study Group), 'Shenru tantao', p. 5.

84 Ye Qing, 'Explanations on the General Contract for State-Controlled Mines', *CIYB 1985*, p. 145.

85 'Zhongguo meitan jinchukou zong gongsi chengli' (China National Coal Import Export Corporation Established), *RMRB*, 23 June 1982, p. 1. The Corporation is the subject of several articles in the *MTGYNJ*.

86 All information here pertaining to the Japanese funding is from Erika Platte, 'The Role of China in Japan's Quest for Energy Security', *Hitotsubashi Journal of Economics*, 37 (1996): 69–86; Martin Weil, 'China's Troubled Coal Sector', *CBR*, Mar.–Apr. 1982, pp. 23–34; Dori Jones, 'The Dawning of Coal's "Second Age"', *CBR*, May–June 1980, pp. 38–46; Kenji Hattori, 'Trends in China's Coal Development and Japan-China Coal Trade', *China Newsletter*, no. 43 (Mar.–Apr. 1983): 14–21, and Shuichi Ono, *Sino-Japanese Economic Relationships: Trade, Direct Investment, and Future Strategy*, World Bank Discussion Paper No. 146, China and Mongolia Dept. Series, Washington, DC: IBRD/WB, 1992. Many of the details in these sources did not agree.

87 *CIYB 1984*, p. 228; *CIYB 1986*, p. 62.

88 *CIYB 1983*, p. 223; Weil (1982), p. 34.

89 Calculated by the author from The World Bank's website in June 2000: www.worldbank.org.

90 *CIYB 1983*, pp. 220–1; *CIYB 1984*, p. 226; *CIYB 1985*, p. 171; *CIYB 1986*, p. 62.

91 World Bank, *China: Long-Term Development Issues and Options*, Baltimore: John Hopkins University Press, 1985.

92 The other four were Issues and Prospects in Education, Agriculture to the Year 2000, Economic Model and Projections, and Economic Structure in International Perspective.

93 Qi Hongmin, 'New Docks to Send Extra Coal South', *CD* (17 Oct. 1987): 2.

94 Jing Wei, 'Large-Scale Development Mapped Out', *BR* (17 Dec. 1984): 26.

95 Wu Jing, 'Prospects of China's Coal Industry', *BR* (12 Sept. 1983): 14; 'Pre-Feasibility Study on "Changjiang" Slurry Pipeline', *CIYB 1985*, p. 231.

96 For example, see Kou Zongji and Li Sufang, 'Xietong nuli, jinkuai jiejue Huoxian kuang wuji meitan chaochu jiya wenti' (Strive Towards Solving As Soon As Possible the Problem of Overstocked Coal at Huoxian Mine), *JJWT*, no. 7 (1985): 44.

97 Gao Yangwen, 'Relaxing the Policy and Speeding Up the Development of Small Coal Mines', *CIYB 1984*, pp. 89–92.

98 'Meitan yao shixing yi xiao dingchan de fangzhen' (The Coal Industry Must Follow Production Plans Based on Consumption Levels Guideline), *SXRB*, 4 Mar. 1986, p. 1.

99 *CIYB 1984*, p. 7

100 *CIYB 1982*, p. 145.

101 *CIYB 1983*, p. 171.

102 *CIYB 1985*, p. 81; Li Guangchun, 'Gonggu he fazhan dangqian meitan gongqiu de da hao xingshi' (Strengthen and Develop the Current Good Demand and Supply Situation in the Coal Industry), *WZGL*, no. 2 (20 Feb. 1986): 8–10; 'Meitan chanyun xiao sanben zhang hecheng yi ben hao' (Combining Coal Production, Transport and Consumption on One Account), *RMRB*, 11 Sept. 1984, p. 1; *CIYB 1984*, p. 25.

103 Sun Shangqing, 'Nengyuan jiegou' (Energy Structure), in Ma Hong and Sun Shangqing (eds), *Zhongguo jingji jiegou wenti yanjiu* (Problems in China's Economic Structure), Beijing: s.n., 1980, p. 284; Luo Genji and Fan Lianfen, 'Chongfen fahui jingji ganggan de zuoyong cujin jin jieneng' (Abundantly Bring into Play Economic Levers to Promote the Conservation of Energy), *NY*, no. 6 (1985): 9–10, 13; Chen Zhicheng, 'Dui dangqian qiye nengyuan guanli gongzuo de jidian kanfa' (Some Points About Current Enterprise Energy Management Work), *NY*, no. 5 (1982): 12–14; Chen Heping, 'Jieneng gongzuo gaige chutan' (An Initial Look at Reforming Energy Conservation Work), *NY*, no. 2 (1985): 7–9; Xu Shoubo and Wu Jiapei, 'Jiaqiang nengyuan kaifa, dali jieyue neng hao' (Strengthen Energy Exploitation, Vigorously Economize on Energy Consumption), *RMRB*, 21 Jan. 1983, p. 5.

104 'Chronicle of Events in Coal Industry', *CIYB 1982*, p.15.
105 The journal *Nengyuan* (Energy) was retitled *Zhongguo nengyuan* (Energy of China) in 1989.
106 Sun Hongzheng, 'Zuzhi jingji yunxing shi jieneng de zhongyao huanjie', (Organizing Economic Systems Is an Important Link in Energy Conservation), *NY*, no. 5 (1983): 38–9.
107 'Zhonghua renmin gongheguo guomin jingji he shehui fazhan di liuge wu nian jihua' (The People's Republic of China Economic and Socialist Development Sixth Five Year Plan), *Zhonghua renmin gongheguo guowuyuan gongbao* (Bulletin of the State Council of the People's Republic of China), 9 (10 June 1983): 340–1.
108 Kang Shien, 'Jieyue nengyuan shi guomin jingji fazhan zhong yi ge tuchu er jinpo de renwu: Kang Shien tongzhi zai chuanguo di yici "jie neng yue" guangbo dianshi da hui shang de jinhua [zhaiyao]' (Conservation of Energy Is an Important, Urgent Task in the Development of the National Economy: Summary of Vice Premier Kang Shien's Speech for the First National Energy Conservation Month Radio and Television Broadcast), *GRRB*, 2 Nov. 1979, pp. 1–2.
109 'Measures for Implementation of Price Increases for Above-Quota Fuel Consumption', *WZGL*, no. 2 (1982): 11–12, in *JPRS*, 254 (1982): 4–6.
110 'No. 4 Instruction for Energy Conservation, 24 July 1982', *CIYB 1983*, pp. 84–5.

5 Central planning in the industry giving way to marketization

1 Deng Xiaoping, 'Women de hongwei mubiao he genben zhengce' (Our Magnificent Targets and Basic Policies), in Deng Xiaoping, *Deng Xiaoping wenxuan* (Selections of Deng Xiaoping), Beijing: Renmin chubanshe, 1993, pp. 77–80.
2 *TJNJ 2000*, p. 55.
3 These changes were duly made in March 1998 at the Eighth NPC.
4 Y. Y. Kueh, 'Prospects for a Transition to a Market Economy without Runaway Inflation', in Y. Y. Kueh, Joseph C. H. Chai, and Gang Fan (eds), *Industrial Reform and Macroeconomic Instability in China*, Studies in Contemporary China, Oxford: Oxford University Press, 1998, p. 284, quoting *Da Gong Bao*, 28 Feb. 1996.
5 Ibid., quoting *Da Gong Bao*, 24 Aug. 1995 and 26 Feb. 96; Wanda Tseng *et al., Economic Reform in China: A New Phase*, Washington, DC: IMF, 1994, p. 38.
6 Susan V. Lawrence, 'Unfinished Business: President Jiang Puts His Own Stamp on Reforms', *FEER*, 17 Dec. 1998, p. 22, citing a table devised by Chang Xiuze and Gao Minghua in *Economics Research Journal*, Nov. 1998.
7 'China's Economic Sectors Re-Classified', *BR*, 4–10 Jan. 1999, pp. 18–19.
8 Jiang Zemin, 'Hold High the Great Banner of Deng Xiaoping Theory for an All-Round Advancement of the Cause of Building Socialism with Chinese Characteristics into the 21st Century', Report delivered at the 15th NPC, 12 Sept. 1997, *BR*, 6–12 Oct. 1997, p. 17.
9 Ibid., pp. 14–16.
10 Tai Ming Cheung, 'Military Business Machines Here to Stay', *SCMP*, 26 July 1998, p. 3.
11 Li Peng, 'Report on the Work of the Government', delivered at the First Session of the 9th NPC, 5 Mar. 1998, *BR*, 6–12 Apr. 1998, p. 16.
12 Willy Wo-Lap Lam, 'Zhu's Battle Beginning', *SCMP*, 11 Mar. 1998, p. 19; 'State Firms to Be Out of Trouble', *BR*, 23–9 Mar. 1998, p. 5.
13 Zeng Peiyan (Minister of the State Development Planning Commission), 'Report on the Implementation of the 1998 Plan for National Economic and Social Development and on the Draft 1999 Plan for National Economic and Social Development' (Excerpts), delivered at the Second Session of the 9th NPC, 6 Mar. 1999, *BR*, 12–18 Apr. 1999, pp. 21–31; China, State Economic and Trade Commission, 'Reform and Development of China's State-Owned Enterprises', *BR*, 11–17 Jan. 1999, pp. 16–19;

Zhu Rongji, 'Report on the Work of the Government', delivered at the Second Session of the 9th National People's Congress, 5 Mar. 1999, *BR*, 5–11 Apr. 1999, p. 17.

14 Jiang Wandi, 'Further Reform Expected in 1999', *BR*, 29 Mar.–4 Apr. 1999, pp. 18–19.

15 'Zhu Targets Financial Reform Amid Slow-Growth Warning', *SCMP*, 6 Mar. 1999; 'Zhu Gambles on Massive Deficit to Save Economy', *SCMP*, 6 Mar. 1999.

16 Zhang Yi, 'Burdensome Population a Headache for China', *BR*, 21 June 1999, p. 20.

17 First discussed at 15th CPC in Sept. 1997. Mark O'Neill, 'Private Business Lobbies for Same Status and Legal Protection as State Sector', *SCMP* Business Section, 9 Mar. 1998, p. 3; 'Smash the Iron Rice Bowl, Says Business', *SCMP*, 14 Mar. 1998, p. 8; 'Amendments Wait for Approval', *BR*, 22–8 Feb. 1999, pp. 4–5; 'NPC Adopts Amendments', *BR*, 29 Mar.–4 Apr., 1999, p. 5; Lin Ji, 'Hot Topics of NPC and CPPCC Sessions', *BR*, 8–14 Mar. 1999, p. 4.

18 'Zhonghua renmin gongheguo guomin jingji he shehui fazhan di qige wu nian jihua (zhaiyao)' (Summary of the People's Republic of China Seventh Five Year Plan for Economic and Social Development), *Zhonghua renmin gongheguo guowuyuan gongbao* (Bulletin of the State Council of the People's Republic of China), 11 (10 May 1986): 319–20.

19 'Wei wancheng "qi wu" shiqi de renwu er nuli fendou' (Struggle Hard Towards Completing Tasks for Seventh Five Year Plan), *MTB*, 11 Dec. 1985, p. 1.

20 Shang Daijiang, 'Mustering All Forces to Develop Shanxi's Coal', *LW Overseas Edition*, 50 (15 Dec. 1986):13, in *JPRS*, 25 (1987): 50–2.

21 Huang Xiang, 'Blueprint of Energy Development', *CD Business Weekly Supplement*, 25 Sept. 1989, p. 4.

22 'Meitan yao shixing yi xiao dingchan de fangzhen' (The Coal Industry Must Follow Production Plans Based on Consumption Levels Guideline), *SXRB*, 4 Mar. 1986, p. 1.

23 'Supply and Demand', *CD*, 15 Sept. 1987.

24 'Guonei meitan zou qiao meijia reng zai shang zhang' (Coal in Great Demand, High Prices in Shanghai), *SXRB*, 29 Mar. 1988, p. 1; Zhou Ruyan and Feng Yongping, 'Jianli shuang xiang yueshu jizhi huan jie que dian maodun' (Establish Mechanism for Controlling the Power Shortage Contradiction), *JJGL*, no. 7 (1988): 48.

25 Liang Weilie, 'Nengyuan duanque fenxi ji sikao' (Analysis and Consideration of Energy Shortages), *NY*, no. 2 (1989): 6; Jiang Xianrong, 'Wo guo jieneng gongzuo de huigu yu zhanwang' (Retrospects and Prospects in China's Energy Conservation Work), *NY*, no. 5 (1989): 40.

26 *CD*, 21 Sept. 1988; Gong Gui, 'Meitan shichang huan zhong qu jin' (Coal Market Postponed Amidst Tight Supplies), *SXRB*, 6 Nov. 1988, p. 2.

27 'Wo guo nengyuan quekou bu xiao, jieneng gongzuo dai zhuajin', (The Energy Shortage Problem Is Not Small, Must Implement Energy Conservation), *RMRB*, 17 June 1989, p. 4.

28 'Wo sheng wai yun meitan hunluan shengfu yanjiu zhunbei zhengdun' (Research into Rectifying Shanxi Province's Coal Transport Chaos', *SXRB*, 19 Mar. 1988, p. 1; Zhao Keming, 'Zhi zhi mei dao' (Tackle Coal Losses), *SXRB*, 8 Nov. 1988; 'Meitan caigou renyuan de kunao' (Anxious Coal Purchasing Personnel), *SXRB*, 4 Nov. 1988.

29 Li Zhongcheng, 'Jiaqiang meitan yunxiao guanli, quebao guojia jihua wancheng' (Strengthen Coal Transport and Sales Management, Guarantee Fulfilment of State Plans), *SXRB*, 15 Apr. 1988.

30 'Meitan yao shixing yi xiao', p. 1.

31 'Quanmin suoyou zhikuang shan qiye caikuang dengji guanli zanxing banfa' (Provisional Regulations on the Registration of Coal Mines), *SXRB*, 1 May 1987, p. 2.

32 'Shanxi lan wa meitan zhi feng jiben shazhu' (Shanxi Must Halt Excessive Digging of Coal), *JJRB*, 4 Mar. 1987, p. 2.

33 Guo Zhongshi, 'State Acts to Reclaim Small Pits This Year', *CD*, 19 Mar. 1988, p. 1.

34 Hu Fuguo replaced Yu Hongen in 1989.

35 *MTGJNJ 1987*, p. 74.
36 The webpage address for this is: www.edu.cn/undp/ccp/cp3/coalprog.htm.
37 For example, Merle Goldman, *Sowing the Seeds of Democracy in China: Political Reform in the Deng Xiaoping Era*, Cambridge: Harvard University Press, 1994; James A. R. Miles, *The Legacy of Tiananmen: China in Disarray*, Ann Arbor: University of Michigan Press, 1996; Andrew G. Walder, 'Workers, Managers and the State: The Reform Era and the Political Crisis of 1989', *CQ*, 127, 1991, pp. 467–92; Yan Sun, 'The Chinese Protests of 1989: The Issue of Corruption', *Asian Survey*, Aug. 1991, pp. 762–82.
38 Yu Yuanchao, 'Production of Energy Up Slightly', *CD*, 28 June 1989, p. 2.
39 FBIS Sept. 1995, p. 33 quoting *Xinhua*, 1411 GMT 9 Sept. 1995.
40 'Shang bannian meitan chukou jiang wancheng wu bai wan dun lizheng "qi wu" qijian leiji dadao yi yi dun' (During the First Six Months Coal Exports Will Reach 5 Million Tons, Struggle Hard During the 7th Five Year Plan Period to Reach One Billion Tons), *RMRB*, 29 June 1986, p. 1.
41 The information here pertaining to the Antaibao project came from the following articles: 'Hammer Offers Chinese Coal to Taiwan', *The Financial Times Ltd.* (Newswire), 17 June 1988; 'Taiwan Firms Cool to PRC Coal', *CWI*, 9, 32, 10 Aug. 1988, p. 7; 'Taiwan Market Looks Good to Oxy for An Tai Bao Coal Output', *CWI*, 9, 24, 15 June 1988, p. 1; 'Taiwanese Coal Users Take Cautious Attitude About Chinese Trade', *CWI*, 9, 27, 6 July 1988, p. 1.
42 Huang Xiang, 'Taiwan's Help Called for by Coal Industry', *CD*, 12 July 1991, p. 2.
43 'Hong Kong Paper on China's Opening Trade Relations with South Korea', translation of excerpts from 'South Korea Has Become China's New Trade Partner', *Chiushih nientai*, 1 Jan. 1988, in BBC SWB, Part 3, The Far East, FE/0042/A3/1; 'Korean Trading Companies Deal with Chinese Delays', *CWI*, 9, 28, 13 July 1988, p. 2; 'South Korea May Experience Trouble in Obtaining Chinese Coal', *CWI*, 10, 5, 31 Jan. 1989, p. 8.
44 Huang Xiang, 'Measures to Strictly Control Coal Market', *CD*, 10 Apr. 1990, p. 2; 'Price Ceiling for "Coal Outside Plans" Issued', broadcast on *Xinhua*, 1000 GMT, 5 Mar. 1990, in *FBIS*, 16 Mar. 1990, p. 26.
45 Michael Kuby, Shi Qingqi and Thawat Watanatada, 'Planning China's Coal and Electricity Delivery System', *Interfaces*, 25, 1 (1 Jan. 1995): 41–68.
46 Chen Dong and He Yuanyuan, 'Zhuazhu ji yu – jinyibu fazhan tielu duo zhong jingying' (Seizing the Opportunity to Make Further Steps in Developing the Diversified Economic Activities of the Railways), *TDYSJJ*, 3 (1998): 28–9.
47 Miao Tianjie, 'Guanyu wo guo di qige wu nian jihua jieneng lu de kanfa' (Analysis of Energy Conservation Rates in China's Seventh Five Year Plan), *NY*, no. 1 (1985): 7.
48 Tian Hongbin and Yan Tao, *Nengyuan zhishi 500 ti* (Energy Knowledge 500 Questions), Lanzhou: Gansu renmin chubanshe, 1985, p. 51; Xu Shoubo and Wu Jiapei, 'Jiaqiang nengyuan kaifa, da li jieyue neng hao' (Strengthen Energy Development, Vigorously Conserve Energy), *RMRB*, 21 Jan. 1983, p. 5; Gong Guangyu, 'Wo guo nengyuan de xianzhuang he weilai' (China's Energy Present and Future), in 2000 nian yanjiu xiao zuzhu (Year 2000 Research Group), *Gongyuan 2000 nian de Zhongguo* (China in the Year 2000), Beijing: Kexue jishu wenxian chubanshe, 1984, p. 40; Miao Tianjie, 'Guanyu wo guo di qige wu nian jihua jieneng lu de kanfa', p. 8. Direct conservation refers to the introduction or expanded use of technically superior energy materials or consuming equipment, and indirect refers to changes in the sectoral or product structure of consumption and the more efficient use of discarded or waste materials.
49 Qu Shiyuan, 'Wo guo 2000 nian nengyuan gongxu yuce he jiegou fenxi' (Forecast and Structural Analysis of China's Demand and Supply of Energy in the Year 2000), *NY*, no. 3 (1987): 2.
50 Gong Guangyu, 'Wo guo nengyuan de xianzhuang he weilai' (China's Energy Present and Future), p. 41.

51 'Jieyue nengyuan guanli zanxing tiaoli' (Provisional Regulations on Energy Conservation), issued by the State Council on 12 Jan. 1986, *RMRB*, 24 Jan. 1986, p. 2.

52 Cogeneration systems in electricity generation capture waste heat which would otherwise be lost in cooling water and from the fuel stack. Depending on the size and type of plant, cogeneration improves overall efficiency by about 50 per cent, makes more power available, and greatly reduces costs and pollution.

53 China State Economic and State Planning Commissions, 'Regulations for Intensifying Conservation of Electricity', broadcast on *Xinhua*, 0134 GMT, 5 April 1987, in *JPRS*, 41 (1987): 84–6. No definition of 'special occasions' is given.

54 For comparison, western households consume 8–10 times this amount per month.

55 Chen Yun and Sun Jie, 'Major Target Readjustment for Eighth Five Year Plan', *BR*, 10–16 May 1993, p. 20.

56 'Planners Set Targets in Coal, Power', *Zhongguo xinwen she* (China News Service) 0447 GMT, 31 Dec. 1994, in *FBIS*, 3 Jan. 1995, p. 62.

57 'Opinions on Stopping the Indiscriminate Extraction of Small Coal Mines and on Ensuring Safety in Coal Mining', broadcast on *Xinhua*, 0205 GMT, 17 Jan. 1993, in *FBIS*, 27 Jan. 1993, pp. 21–2.

58 Introduction to *MTGYNJ 1995*.

59 'Administrative Procedures for Permits on Coal Production', broadcast on *Xinhua*, 0049 GMT, 27 Dec. 1994, in *FBIS*, 13 Jan. 1995, pp. 34–7; 'Ministry Drafts Plan to Boost Coal Industry', broadcast on *Xinhua*, 1244 GMT, 10 Jan. 1994, in *FBIS*, 12 Jan. 1994, p. 48.

60 'Xiangcun meikuang guanli tiaoli shishi banfa' (Implementing LNS Mine Management Statutes), *MTGYNJ 1995*, p. 130.

61 'New Coal Production Figure for 1994 Published', broadcast on *Xinhua*, 0747 GMT, 12 May 1995, in *FBIS*, 12 May 1995, p. 28.

62 Zhang Chaowen, 'Meitan gongye bu xuanbu chengli' (Announcement of Establishment of Coal Ministry), *RMRB*, 4 June 1993, p. 2.

63 Wang Senhao, 'Meitan gongye zouxiang shichang jingji de gaige silu' (Thoughts on Reforming Coal Industry Towards Market Economy), *RMRB*, 3 Jan. 1994, p. 5.

64 *MTGYNJ 1994*, p. 154.

65 'PLA Gets $1.16 Billion Payment for Mines', *SCMP*, 18 Aug. 1994, p. 11. (At the time the article was written 1.3 billion yuan was equivalent to HK$1.16 billion.)

66 The original fourteen 'bureaux' were: The China National Coal Distribution and Transport Corporation, China National Coal Materials Corporation, China Coal Comprehensive Utilization, Energy-Saving Development Corporation, China Coal Mine Equipment (Group) Corporation, China National Local Coal Mine Corporation, China Coal Complete Equipment Corporation, Sida Mining Corporation, Beijing Huamei Industry and Trade Corporation, China International Technology Consulting Development Corporation, China National Coal Construction and Development Corporation, China Coal Diversified Economy Industry and Trade Corporation, China Coal Production Technology Development and Service Corporation, and China Coal Mine Safety and Insurance Corporation.

67 Ching Chi, 'Miners in Heilongjiang Protest Wage Payment Default, One Commits Suicide', *Ming Bao*, 21 Feb. 1993, p. 25, in *FBIS*, 23 Feb. 1993, pp. 80–1.

68 Heath's administration had been brought down by coal miners' strikes, while Thatcher's and Gorbachev's had been seriously threatened.

69 'Coal Consumer Debt Taxes Industry', broadcast on *Zhongguo xinwen she*, 1240 GMT, 14 Jan. 1994, in *FBIS*, 14 Jan. 1994, p. 38; Willy Wo-Lap Lam, 'Fear of Unrest as Money Runs Out', *SCMP*, 17 May 1994; Han Ying (Deputy Minister of Coal), 'Meitan gongye bu fu bu zhang Han Ying tongzhi zai quan guo meitan gongye gongzuo huiyi shang de zongjie' (Summary of Speech at the Coal Work Meeting Held 10 Jan. 1995), *MTGYNJ 1995*, p. 111.

70 'Guanyu niu kui zeng ying de jige wenti' (Some Problems in Turning around Losses), *MTGYNJ 1995*, p. 118.

71 'Mei, zhen yao daoru hai ma?' (Should Coal be Poured into the Sea?), *WZB*, 11 Mar. 1994, p. 1.

72 'Jingjian shi wan ren jian kui jin shi yi wo guo zhongdian meikuang shang bannian' (Cut 100,000 Workers and Reduce Losses at the Main Coal Mines), *RMRB*, 3 Oct. 1993, p. 2.

73 Chang Weimin, 'Massive Layoffs in the Coal Sector', *CD*, 28 Dec. 1992, p. 2.

74 'Meitan qiye duo zhong jingying xingcheng guimo' (The Scale of Coal Industry Diversified Activities), *RMRB*, 12 June 1996, p. 2; Chang Weimin, 'Deficit in Coal Mining Nears End', *CD*, 9 Jan. 1995, p. 1.

75 Introduction to *MTGYNJ 1995*.

76 Ibid.

77 'China "Greatly" Cuts Coal Supply to DPRK', *Tong-a ilbo*, 1 June 1993, p. 2, in *FBIS*, 2 June 1993, p. 9.

78 'Meitan shichang jinqi you zhuan wang jixiang' (Recent Indication of Coal Market Becoming Prosperous), *MTB*, 25 Aug. 1994, p. 4; 'Nanfang mei shihua jin, Shanghai gang cun mei bu zu' (Southern Coal Market Tight, Stocks at Shanghai Port Storage Yard Insufficient), *ZMB*, 29 June 1995.

79 China, State Planning Commission and Ministry of Foreign Trade and Economic Cooperation, 'Compilation of Major Technology Projects Scheduled for The People's Republic of China between 1993 and 2000', *BR*, 28 Nov.–4 Dec. 1994, p. 16.

80 'Coal Ministry Offers 300 Projects to Foreign Investors', *Chingchi taobao*, 23 Oct. 1995, p. 44, in *FBIS*, 6 Dec. 1995, p. 86; 'Coal Industry Woos Foreign Investments', *BR*, 2–8 Oct. 1995, p. 7.

81 'China Publishes Interim Provisions on Guidance for Foreign Investment Along with an Industrial Catalog Guiding Foreign Investment', *BR*, 18–24 Sept. 1995, pp. 17–26.

82 'Oxy Gets Out of Antaibao', *ICR*, 14 June 1991.

83 Lana Wong, 'Reform Vital to Put Country Back on Track', *SCMP China Business Review Section*, 11 July 1996, p. 3.

84 'Shanxi ti chufa mei zhan bu dian yaoqiu' (Shanxi Announces Requirements of Coal Transport Stations), *MTB*, 21 Apr. 1994.

85 Geoffrey Murray, 'Funding Hampers Rail Push', *SCMP Money Section*, 14 Mar. 1996, p. 2; Lana Wong, 'Reform Vital to Put Country Back on Track', p. 3.

86 The China Pipeline Holdings Ltd joint venture was to be operated by China Coal Construction and Development Corporation (49 per cent), and a consortium (51 per cent) led by Customs Coal Corporation and Medical Research International (a unit of China Strategic Holdings Group). The consortium included Goldman Sachs, Williams Technologies Services, and Willbros Butler Engineers Inc. 'Group Builds Pipeline for Coal', *SCMP*, 19 Aug. 1994; 'China to Build Major Coal Slurry Project with Aid of Consortium', *CWI*, 15, 35 (30 Aug. 1994): 2; 'China in $900 Million BOT Coal Slurry Pipeline Deal', *Power Asia*, 5 Sept. 1994.

87 Zhu Daqiang, 'Tapping the Yangzi River's Transportation Capacity', *CM*, 3 (1990): 27–8; Li Guoguang, 'Construction of Water Transport Facilities To Be Expanded', *CM*, 6 (1990): 37, 39.

88 'Waterway Network Under State Plan', *BR*, 30 Oct.–5 Nov. 1995, p. 7.

89 'Energy Ministry Works to Combat Power Shortages', broadcast on *Zhongguo tongxun she*, 0637 GMT, 20 Jan. 1992, in *FBIS*, 29 Jan. 1992, p. 48.

90 Chen Yun and Sun Jie, 'Major Target Readjustment for Eighth Five Year Plan', p. 20.

91 'Zhonghua renmin gongheguo jieyue nengyuan fa (qicao gongzuo zu)' (Energy Conservation Laws of the People's Republic of China), *ZGNY*, no. 9 (1993): 3–8; See also 'Qiye jieyue nengyuan guanli shengji (dingji) guaiding (Regulations for Upgrading Enterprise Energy Management), *NY*, no. 5 (1991): 30–1, 41.

92 Jessica Hamburger, 'Lighting the Way for Energy Savings', *CBR*, Nov.–Dec. 1993, pp. 42–4. Webpage of the Beijing Energy Efficiency Centre: www.gcinfo.com/becon. (accessed Sept. 1999). Another website is www.pnl.gov/china/aboutcen.htm.

93 See the 'Quality' section of the next chapter for an explanation of clean coal technology.
94 Chen Qiuping, 'Environmental Protection in Action', *BR*, 7–13 Sept. 1998, p. 9.
95 'Zhongguo 21 shiji yicheng' (China's Agenda for the 21st Century), *RMRB*, 19 Sept. 1994, p. 3.
96 Christopher Flavin, 'The Legacy of Rio', in Linda Stark (ed.), *State of the World 1997*: A Worldwatch Institute Report on Progress Toward a Sustainable Society, New York: W. W. Norton, 1997, p. 9.
97 Li Xuesheng, 'Zhengque yunyong jihua shouduan he chanye zhengce duo cengci tuijin meitan gongye jingji zengzhang fangshi zhuanbian' (Rightly Apply Plans, Means and Production Policies to Push Forward the Economic Expansion and Transformation of the Coal Industry), *MTJJYJ*, 12 (1996): 5–9; 'Yikao keji jinbu, tuijin jiyue hua shengchan wei shixian meitan jingji zengzhang fangshi zhuanbian er nuli fendou' (Rely on Science and Technological Advances, Push Forward and Persevere in Transforming Economic Growth in the Coal Industry), *MTGYNJ 1997*, pp. 61–8; Qu Yinghua, Zhang Xiaojun, and Wang Lijie, 'Tan meitan gongye jianshe "jiu wu" fazhan silu' (Thoughts on Coal Industry Construction and Development During the Ninth Five Year Plan), Part I, in *MTJJYJ*, no. 4 (1996): 12–15 and Part II, in *MTJJYJ*, no. 5 (1996): 5–7; Introduction to *MTGYNJ 1996*; Li Xuesheng, 'Guanyu meitan gongye "jiu wu" jihua he 2010 nian yuanjing mubiao de zongti silu he kuangjia' (Thoughts and Framework for Coal Industry Targets over the Ninth Five Year Plan and Longterm Plan to 2010), *MTGYNJ 1997*, p. 145.
98 'PRC: Coal-Producing "Giants" Map Out Production Strategy', broadcast on *Xinhua*, 1541 GMT, 10 Mar. 1996, in *FBIS*, 11 Mar. 1996, p. 69.
99 'Coal Mines Get $4 Billion', *SCMP*, 26 May 1997, p. 5.
100 Wu Shiyu, 'Developing Coal Preparation Processing, Increasing Coal Utilization Efficiency and Reducing Pollution', in IEA, *Energy Efficiency Improvements in China: Policy, Measures, Innovative Finance and Technology Deployment*, Proceedings of a Conference held in Beijing, 3–4 Dec. 1996, Paris: OECD/IEA, 1997, pp. 77–80.
101 'Meitan zonghe liyong, duo zhong jingying he di san chanye "jiu wu" jihua he 2010 nian yuanjing guihua' (Comprehensive Use of Coal and the Development of Diversified Activities and Service Trades During the Ninth Five Year Plan and Longterm Plan to 2010), *MTJJYJ*, no. 8 (1996): 13; 'Meitan qiye duo zhong jingying xingcheng guimo' (Scale of Coal Industry's Diversified Activities), *RMRB*, 12 June 1996, p. 2; Pan Weier, 'Yi 50 wan meikuang zhigong zai jiu ye' (Discussion on the Re-employment of 500,000 Coal Mine Staff and Workers) *MTJJYJ*, no. 4 (1996): 15–18.
102 Introduction to *MTGYNJ 1996*.
103 'Meitan bu bushu meikuang anquan shengchan cuoshi' (Coal Ministry Arranges Coal Mine Safety Measures), *RMRB*, 13 May 1997, p. 2; Ivan Tang, 'Drive to Increase Mining Safety Levels after Deaths', *SCMP*, 3 Apr. 1997, p. 10.
104 '4,000 Mining Deaths "Alright"', *SCMP*, 20 Oct. 1999, citing *Legal Daily*.
105 Agatha Ngai, 'Mines Told to Insure Workers', *SCMP*, 2 Apr. 1997, p. 8.
106 Helen Luk and Agencies, 'Officials Face the Music Over Coal Mine Deaths', *SCMP*, 13 June 1998, p. 8.
107 Chang Weimin, 'Xinji Mine Digs Way to Market Transition', *CD*, 16 Apr. 1996, p. 5; 'Wang Senhao zai meitan gongye niu kui zeng ying gongzuo huiyi' (Wang Senhao Discusses Turning Around the Losses in the Coal Industry at Work Meeting), *JJRB*, 20 May 1996.
108 Renee Lai, 'Mining Firm Sets Hopes on Hong Kong Float', *SCMP Business Section*, 15 Jan. 1997, p. 11.
109 'Zhengzhou meitan gongye jituan you xianzeren gongsi chengli' (Zhengzhou Coal Industry Limited Liability Company Established), *MTB*, 20 Jan. 1996.
110 'Zhonghua renmin gongheguo meitan fa' (People's Republic of China Coal Law), *RMRB*, 16 Sept. 1996, p. 2.

111 The CCIEG was established by the merger of four coal companies: China National Local Coal Mines Company, China National Coal Marketing and Transportation Company, China National Coal Production Technology Development Company, and Pingshuo Coal Industry Company. With the creation of the CCIEG there were three bodies licensed to import and export coal, the others being the Shanxi Coal Import and Export Company and the Shenhua Group.

112 Shao Qin, 'Industry Group Plans to Increase Coal Exports', *CD Business Weekly*, 7–13 Sept. 1997; 'Hospitality Sector Sees Its Future in Pudong', *CD*, 22 Mar. 1998, p. 6.

113 Daniel Kwan, 'Cutback Tackles Red-Tape Malaise', *SCMP*, 7 Mar. 1998, p. 8; Jasper Becker, 'Fear Invades the Halls of Bureaucracy', *SCMP*, 8 Mar. 1998, p. 11. The other ministries and commissions were: power; chemicals; internal trade; posts and telecommunications; labour; radio, film and TV; geology and mineral resources; forestry; metallurgy; machine-building; electronics; The Commission of Science, Technology and Industry for National Defence (COSTIND); The State Commission for Restructuring the Economy; and The State Commission for Physical Culture and Sports.

114 The SCIB created a new department called 'Enterprise Down-Level Transfer' to organize and implement the transfer of SPMAs, enterprises and institutions to provincial or other lower administrative levels. The other six departments were: Foreign Affairs, Planning and Development, Industrial Management, Enterprise and Institution Reforms, Personnel, and Local Township Mines Rectification. IEA, Coal Industry Advisory Board, *Coal in the Energy Supply of China*, Report of the CIAB Asia Committee, Paris: OECD/IEA, 1999, p. 28 citing China Coal, English Supplement, vol. 3.

115 'Xiafang guanjing jianguan' (Orderly Decentralization and Closure), *MTJJYJ*, no. 8 (1998): 4; 'Zhongyang meitan qiye xiafang difang fangan queding' (Programme to Decentralize Central Coal Enterprises), *JJCKB*, 16 July 1998, p. 1.

116 'Zhongguo meitan gongye xiehui zu jian guapai' (China Coal Industrial Association Established), *MTB*, 20 Mar. 1999; 'Profile – China's Coal Industry (Mar. 1999), *Asia Pulse*, 31 Mar. 1999; World Coal Institute, *China Update*, 9 July 1999, www.wci-coal.com/gen_chinaupdateJuly99.htm, citing China–Britain Trade Review, May 1999, accessed Sept. 1999.

117 Willy Wo-Lap Lam, 'Five-Year Plan Will Spur Market Forces', *SCMP*, 8 May 1999.

118 Li Xuesheng, 'Zhengyue yunyong jihua', p. 9.

119 Lana Wong, 'Rule Eased to Attract Funding for Coal Mines', *SCMP* Business Section, 7 June 1996, p. 4.

120 'Mine Development Cash Sought', *SCMP*, 17 Mar. 1999.

121 P. T. Bangsberg, 'ASEAN Members Form Coal Forum', *Journal of Commerce*, 26 Apr. 1999.

122 IEA webpage: www.iea.org/iea/goals/files/nmchina.htm, accessed Oct. 1999.

123 China, SPC, 'Industrial Catalogue Guiding Foreign Investment', jointly promulgated by the State Planning Commission and the Ministry of Foreign Trade and Economic Cooperation, *BR*, 2–8 Mar. 1998, p. 16.

124 Tsukasa Furukawa, 'Trading Firms Buy Mine Stake', *Information Access Company*, 28 May 1998.

125 IEA, Coal Industry Advisory Board, *Coal in the Energy Supply of China*, p. 29.

126 *TJNJ*, 1997–2000.

127 'Coal Mines Face Safety Measures Blitz', *SCMP*, 12 Feb. 1997, p. 7.

128 Li Peng, 'Zhongguo de nengyuan zhengce' (China's Energy Policy), *RMRB*, 29 May 1997.

129 'Zhongguo meitan gongye ju de san xiang zhongdian gongzuo' (Three Important Tasks of China's Coal Industry), *Zhongguo meitan*, 22 June 1998.

130 Ministry of Coal official interviewed by the author in August 1998.

131 *TJNJ 1999*, p. 33.

132 'Coal Output to be Cut', *SCMP Business Section*, 6 July 1998, p. 3; Zhao Shaoqin, 'Mine Closure Scheme to Overcome Coal Glut', *CD Business Weekly*, 23–9 Aug. 1998.

133 'Small Mines Close in Overhaul', *SCMP Business Section*, 13 Nov. 1998, p. 4; [China], SETC, 'Reform and Development of China's State-Owned Enterprises', *BR*, 11–17 Jan. 1999, p.18; Zhao Shaoqin, 'China's Coal Sector Seeks to Get Back into Black', *CD*, 8 Jan. 1999; 'Illegal Coal Mines to Close in Shanxi', broadcast on *Xinhua*, 0752 GMT, 22 Dec. 1998.

134 'China Cracks Down on Oil Smuggling', *China Business Information Network*, 28 July 1997; 'China Moves to Offset Declining Oil Prices', Xinhua, 4 May 1998.

135 Cui Cheng, Hu Xiulian, and Liu Jingru, 'Wo guo nengyuan jiegou tiaozheng yu jin chukou zhanbei de jingjixue fenxi' (Economic Analysis of Strategy for Adjusting China's Energy Structure and Imports and Exports), *ZGNY*, no. 4 (1998): pp. 8–12; Yang Qing, 'Yau cong zhanlue gaodu zhongshi LNG jinkou' (Adopt Strategy of Raising LNG Imports', *ZGNY*, no. 5 (1998): 5–8; Yang Zhaohong, 'Liyong guoji youqi ziyuan gaibian zhongguo nengyuan jiegou' (Use International Oil and Gas, Change China's Energy Structure), *ZGNY*, no. 11 (1998): 12–16. See also, 'Energy Use Must Shift from Coal Towards Oil', *CD*, 23 Jan. 1998.

136 Mark O'Neill, 'Output of Coal Falls with Low Demand', *SCMP*, 22 Aug. 1998, p. 3.

137 'Offshore Natural Gas Development Accelerated', broadcast on *Xinhua*, 0927 GMT, 29 Apr. 1999, in SWB Part 3: Asia-Pacific, FEW0587, 5 May 1999.

138 'China Encourages Use of Clean Energy', *Xinhua*, 2 May 1999.

139 'Coal Production Falls 14.5 pc', *SCMP*, 24 Apr. 2000, citing *Xinhua*.

140 'Bufen sheng qu fa dian liang 20 nian lai shouci fu zengzhang' (For the First Time in 20 Years Electricity Production Has Not Increased in Some Provinces and Districts), *JJCKB*, 25 Dec. 1998, p. 1; Zhao Shaoqin 'China's Coal Sector Seeks to Get Back into Black', *CD*, 8 Jan. 1999; 'Oversupply Results in Cuts in Coal Production', broadcast on *Xinhua*, 0103 GMT, 21 Jan. 1999, in SWB Part 3: Asia-Pacific, FEW0573, 27 Jan. 1999.

141 'Deficits and Surpluses in Coal Production Targeted', *CD*, 29 June 1999.

142 'Coal Sector Makes Plans to Suppress Production', *Zhongguo xinwen she*, 19 Apr. 1999, in SWB Part 3: Asia-Pacific, FEW0586, 28 Apr. 1999; 'Zhongguo meitan gong da yu qiu shangwei huan jie' (China Coal Supply and Demand Still Not Solved), *RMRB*, 25 June 1999, p. 2.

143 Zhang Baoming, '1999 nian meitan gongye de zhongdian gongzuo' (Main Work for the Coal Industry in 1999), *MTJJYJ*, no. 2 (1999): 5.

144 'Premier: Friendship with North Korea Will Continue to Develop', broadcast on *Xinhua*, 0802 GMT, 4 June 1999, on SWB, Part 3, Asia Pacific, North Korea, FE/D3554/G; 'Korean Peace Backed', *CD*, 5 June 1999.

145 Shen Bin, 'Closure of Small Coal Mines Progressing', *CD*, 13 May 1999.

146 Mark O'Neill, 'Liuzhi Miners Feel Market Blast', *SCMP*, 3 Dec. 1999.

147 'Zhongguo 99 nian meitan jingji yunxing reng shifen kunnan' (China's Coal Industry in 1999 Still in Economic Difficulties), *MTB*, 24 June 1999; 'Zhongguo meitan chan-liang xiajiang zongliang yiran guosheng' (China's Coal Output Decreases but Still Have Surplus), *MTB*, 19 June 1999.

148 'China Succeeds in Reducing Coal Output', *MTB*, 17 Jan. 2000, on www.china-online.com, accessed 28 June 2000.

149 'Coal, Steel, Sugar Face Output Curbs', *SCMP*, 2 Dec. 1999.

150 'The Coal Shoulder: China Seeks to Mine its Own Business – On a Smaller Scale', on www.chinaonline.com, accessed 22 May 2000, citing *ZGJJSB*, 11 May 2000.

151 Jasper Becker, 'Unpaid Miners Blocking Rail Lines', *SCMP*, 10 Mar. 1998, p. 8.

152 Oliver Chou, 'Coal Miners Protest over Unpaid Wages', *SCMP*, 18 Mar. 1999.

153 '20,000 Miners Riot Over Low Severance Payments', *SCMP*, 4 Apr. 2000.
154 'Pension Protest Blocks Rail', *SCMP*, 29 Apr. 1999.
155 'Miners Block Railway over Bankruptcy', *SCMP*, 16 Sept. 1999.
156 'Zhongguo 99 nian meitan hangye guanbi 40 ge kuangjing pochan 14 ge qiye' (Coal Industry to Shut 40 Mines and Force 14 into Bankruptcy in 1999), *RMRB*, 28 June 1999, p. 2; 'Guojia meitan ju "1440" gongcheng quanmian qidong' (State Coal Bureau Starts '1440' Project), *MTB*, 10 Apr. 1999, p. 1; 'Zhongguo meitan hangye 99 nian nei jiang qiangxing po chan yi pi qiye lizheng quannian chukou 4000 wan dun' (China Coal Industry in 1999 Forces Bankruptcies, Struggles Hard to Export 40 Million Tons), *GRRB*, 28 June 1999, p. 2; 'Deficits and Surpluses in Coal Production Targeted', *CD*, 29 June 1999.
157 Zhang Baoming, '1999 nian meitan gongye de zhongdian gongzuo', p. 5.
158 Introduction to *MTGYNJ 1996*.
159 Peng Jialing, 'Xiang kui sun xuanzhan: meitan zhanxian de zhuanji shi zenme lai de?' (Declare War on Harmful Losses: How Can the Situation on the Coal Front Be Improved?), *RMRB*, 13 Jan. 1997, p. 1.
160 'Coal Officials Pledge Their Firms Will Show Profits', *CD*, 7 Aug. 1996.
161 'Guo you zhongdian meikuang zongti niu kui wei ying' (State Coal Mines Turn Around Losses to Make Profits), *RMRB*, 26 Dec. 1997, p. 4; Zhao Shaoqin, 'Mine Closure Scheme to Overcome Coal Glut', *CD Business Weekly*, 23–9 Aug. 1998; 'Coal in the Black', *SCMP Business*, 27 Dec. 1997, p. 3.
162 Zhao Shaoqin 'China's Coal Sector Seeks to Get Back into Black', *CD*, 8 Jan. 1999.
163 Zhang Baoming, '1999 nian meitan gongye de zhongdian gongzuo', p. 5.
164 Ibid.
165 Ibid.
166 'Infrastructure Offers Market for Long-Term Investors', *BR*, 7 June 1999, p. 27.
167 'China Fixes Focus of Energy Development', *Xinhua*, 20 Oct. 1999. This is not a direct translation of announcement.
168 'China Produces 1.023 Billion Tons of Coal', *Xinhua*, 24 Jan. 2000.
169 Wang Zhaocheng, 'Gao hao "jiu wu" tielu jianshe' (Railway Construction During the Ninth Five Year Plan), *TDYSJJ*, no. 1 (1996):15–18; Hua Maokun, '"Jiu wu" tielu fazhan zhanwang' (Prospects for Railway Construction During the Ninth Five Year Plan), *TDYSJJ*, no. 1 (1996): 6–9; Cao Jing, '1998 nian tielu jianshe zhanwang' (Prospects for Railway Construction in 1998), *TDYSJJ*, 1 (1998): 9–10; 'Electric Railways Speed Up to Eighth Place', *BR*, 12–18 June 1998, p. 26.
170 Geoffrey Crothall, 'Rail Link Heralds New Era', *SCMP* Money Section, 21 Jan. 1996, p. 1; Chen Dong and He Yuanyuan, 'Zhuazhu ji yu – jinyibu fazhan tielu duo zhong jingying' (Seizing the Opportunity to Make Further Steps in Developing the Diversified Business Activities of the Railways), *TDYSJJ*, no. 3 (1998): 28–9.
171 Huang Xibiao, 'Tielu zouxiang shichang yao yi caiwu guanli wei zhongxin' (Financial Management Is the Core Consideration as the Railways Enter the Market), *TDYSJJ*, no. 5 (1997): 14–15; Zhou Wangjun, ' "Jiu wu" tielu yunjia shuiping yu yunjia gaige' (Ninth Five Year Plan Railway Rate Levels and Reform), *TDYSJJ*, no. 8 (1996): 1–3.
172 'Railway Market to Open Further', *BR*, 28 Dec.–3 Jan. 1999, p. 28.
173 Mark O'Neill, 'Massive System Bastion of Planned Economy', *SCMP China Review Pullout*, 12 Feb. 1998, p. 3.
174 Ibid.
175 Wang Kuizhong, 'Shenhua gaige qianghua guanli shixian yunshu qiye san nian niu kui wei ying' (Deepening Reform and Strengthening Management to Make Transportation Enterprises Turn Their Losses into Profits within Three Years), *TDYSJJ*, no. 1 (1998): 7.
176 Mark O'Neill, 'Reform Plan Sets Break-Even Date', *SCMP*, 3 June 1998, p. 4; 'China to Increase Railway Funds', *BR*, 13–19 Apr. 1998, p. 5.

177 Cao Zhengyan, 'Tigao nengyuan xiaolu, Zhongguo qude de chengjiu he zhengfu de fangzhen cuoshi' (Raise China's Energy Efficiency, Achieve Results from Government Policies and Measures), *ZGNY*, no. 3 (1997):10. Mr Cao was Director General, Dept. of Communications and Energy, SPC; Guojia jiwei jiaotong nengyuan si (State Planning Commission, Dept. of Communications and Energy), '"Jiu wu" jieneng jihua ji 2010 nian yuanjing mubiao shexiang' (Plans to Save Energy During the Ninth Five Year Plan Period and Over the Longterm to 2010), *ZGNY*, no. 9 (1996): 1–2.

178 Li Peng, 'Zhongguo de nengyuan zhengce', *RMRB*, 29 May 1997, p. 2; China Unified Distribution Coal Mine Corporation, Energy Conservation Office, 'Deal with Concrete Matters Relating to Work and Innovate, Open Up and Advance, Push Energy Conservation Work to a New Level During the Eighth Five Year Plan', *ZGNY*, 9 (1991): 3–7, in *JPRS*, 11, 22 Nov. 1991, p. 27.

179 'Zhonghua renmin gongheguo jieyue nengyuan fa' (Energy Conservation Law), *RMRB*, 3 Nov. 1997, p. 2.

180 'Highly Efficient Energy Use Advocated', *CD*, 28 July 1999.

181 World Bank, *Clear Water, Blue Skies: China's Environment in the New Century*, China 2020 Series, Washington, DC: IBRD/WB, 1997, p. 30, Box 3.1, citing NEPA, SPC, and SETC 1996.

182 Liu Yinglang, 'City Trying to Cut Coal Pollution' *CD*, 16 Nov. 1998; Huang Wei, 'Air Quality Surveillance Tightened Up', *BR*, 14–20 Sept. 1998, p. 18; Han Guojian, 'We Don't Want a Foggy London', *BR*, 8–14 Feb. 1999, pp. 10–13.

183 'Beijing's Biggest Polluter to Move', *FEER*, 10 August 2000, p. 10.

184 Ye Lou, 'A Probe into the Sandstorm', *BR*, 1 May 2000, pp. 14–15.

6 Economic assessment of the industry after twenty years of reform

1 *MTGYNJ 1997*, p. 321.

2 Calculated from IEA, *Coal Information 1998*, Paris: OECD/IEA, 1999, pp. I.182–3. Hard coal production is calculated as the sum of coking coal and steam coal.

3 IEA, Coal Advisory Board, *Coal in the Energy Supply of China*, Report of the CIAB Asia Committee, Paris: OECD/IEA, 1999, p. 41; and Raoul Oreskovic, 'The Emergence of China as a Major Coke Supply Source', paper presented at a conference, 'Coping with the Tightening Coke Supply: Is a Crisis Looming?', held in Charlotte, NC, 5–7 Mar. 1997, (see www.chinaenergyresources.com/article.html); *MTGYNJ*, various years.

4 'Cong duo zhong kan shao' (Plenty Admist Shortages), *RMRB*, 23 Apr. 1991, p. 2; Li Junfeng and Rong Tao, 'Zanshi weiji, hai shi zhangqi duanque: tantan wo guo nengyuan weihe quanmian jinzhang' (Temporary Crisis or Longterm Shortage: A Discussion About China's Energy Deficits), *ZGNY*, no. 3 (1989): 1–4, 29.

5 Hong Shijie, 'Dianli – Zhongguo jingji you yi ge "pingjing"' (Electricity: Another Economic Bottleneck), *RMRB*, 11 Oct. 1993, pp. 1–2; 'Coal Demand Expected to Exceed Supply', broadcast on *Zhongguo xinwen she*, 0455 GMT, 23 Sept. 1995, in *FBIS*, 25 Sept. 1995, p. 37; Qu Weiying *et al.* 'Mei jia bao zhang: lang zhi feng weixi' (Sudden Rise in Coal Prices: Waves Stop but Wind Still Blows), *JJCKB*, 2 Apr. 1996, p. 1.

6 'Shanghai mei jia si ge yue shang zhang wu ci, zhang jia you yin' (Factors Explaining Why the Price of Coal in Shanghai Increased Five Times in Four Months), *MTB*, 10 June 1993.

7 'Gongxu reng jiang qu jin nian shiji qin meitan' (Coal Supply and Demand Will Be Tight Until the End of the Century), *MTB*, 26 May 1994.

8 'Mei, zhen yao daoru hai ma?' (Coal, Do We Really Want to Pour It into the Sea?), *WZB*, 11 Mar. 1994, p. 1; 'Report on Concern Over Energy Production', *Zhongguo xinwen she*, 0505 GMT, 26 June 1993, in *FBIS*, 28 June 1993, pp. 38–9.

9 'Meitan jiegou xing gongqiu maodun fu chu bingshan yi yong zhu qudao jing ying weisuo fazhan qian jing kan yu' (Structural Supply of Coal Is Contradictory, Only Tip of Iceburg), *WZB*, 8 June 1994; Wu Wenyue and Li Wei, 'Dangqian wo guo meitan gongqiu guanxi zhi qianxi' (Analysis of the Present Coal Supply and Demand Situation in China), *MTJJYJ*, no. 5 (1998): 38–41.

10 Gong Weicai and Zhang Xiaojian, 'Shichang jingji tiaojian xia: meikuang qiye jihua zhibiao tixi sheji' (Part I) (Economic Market Conditions: Designing Coal Mine Enterprise System Target Plans) *MTQYGL*, no. 3, 1998, pp. 30–3; Guo Tingjie, 'Yi tan guo you zhongdian meikuang zhengye jinru shichang de nandian yu duice' (Discussion of the Difficulties Faced by State Mines Trying to Enter the Market), *ZGNY*, no. 8, 1996, pp. 11–13, 23; Meitan bu zhengce fagui si keti zu (Coal Ministry Policy and Regulation Research Group), 'Jiushi nian dai zhonghou qi wo guo meitan shichang jiben zou shi ji duice' (Trends in China's Coal Market in the Mid and Late 1990s), *MTJJYJ*, Part I in no. 1, 1995, pp. 4–8; Part II in no. 5, 1995, pp. 4–7; Wang Senhao, ' "Jiu wu" hou san nian meitan gongye zongti gongzuo shexiang he 1998 nian gongzuo zhongdian' (Discussion of the Work to be Done in the Coal Industry During the Final Three Years of the Ninth Five Year Plan and for 1998), *MTQYGL*, no. 1, 1998, pp. 5–9.

11 The Coal Industry Yearbooks do not use the terminology TVEs. Some Western writers have applied this generic term to all or some of the LNS mines.

12 Some of the decline in production at the provincial mines can be explained by the upgrading of their classification to CMAs.

13 Tai Ming Cheung, 'Military-Business Machine Here to Stay', *SCMP*, 26 July 1998, p. 3.

14 Author's interview with an official at the Ministry of Coal in Sept. 1998.

15 Zhang Lin and Li Yongping, 'Meitan gongye ke chixu fazhan wenti de tantao' (Investigation into Continuing Problems in the Development of the Coal Industry), *ZGNY*, no. 10, 1998, p. 16; Zhao Shaoqin, 'Mine Closure Scheme to Overcome Coal Glut', *CD Business Weekly*, 23–9 Aug. 1998 gives 2,500 LS and 75,000 LNS; According to 'Illegal Mining Targeted', *SCMP*, 3 July 1996, p. 9, there were 150,000 collective mines and 75,000 individually-run mines; According to 'Coal Industry to Further Open to Outside', broadcast on *Xinhua*, 1411 GMT, 9 Sept. 1995, in *FBIS*, 11 Sept. 1995, p. 33, in 1995 there were 104 CMAs, 1,600 LS, and over 80,000 township mines in 1995.

16 *XZQYNJ*, various years.

17 Wu Yin, 'Meitan chanye jingzheng jiegou fenxi' (Analysis of the Competition Structure in Coal Production), *ZGNY*, no. 1, 1999, p. 21.

18 'China to Control Coal Output within 870 Million Tons in 2000', *Xinhua*, 21 Jan. 2000; 'China Produces 1.023 Billion Tons of Coal', *Xinhua*, 24 Jan. 2000.

19 The *MTGYNJ* provide considerable data for the CMAs, but very little apart from the production figures for the LS and LNS mines. They divide the data into two main categories: (a) *guo you zhongdian meikuang* in the Chinese versions, translating this as 'state-controlled coal mines' in the English versions, and (b) *difang meikuang*, translating this as 'local coal mines'.

20 William S. Peirce, *Economics of the Energy Industries*, 2nd edn, Westport, CN: Praeger, 1996, quoting Keystone Coal Industry Manual, 1995, p. S-4.

21 IEA, *Coal Information 1997*, Paris: OECD/IEA, 1998, p. I.173, 176.

22 'Zhongguo meitan gan nian de licheng: cong "duanque" dao "guosheng"' (China's Coal Industry Progresses from Shortages to Surpluses), *MTB*, 1 Oct. 1998, p. 1.

23 *MTGYNJ*, various years.

24 Author's interview with an official at the Ministry of Coal in Sept. 1998.

25 Wang Senhao, 'Renzhen guanche luoshi dang de shiwu da jingshen ba chi xu jiankang fazhan de meitan gongye quanmian tui xiang ershiyi shiji' (Earnestly Ensure Policies of Fifteenth Congress, Vigorously Control the Healthy Development of the Coal Industry as It Pushes Towards the 21st Century), *ZGNY*, no. 3, 1998, p. 7.

26 Yang Zhan and Xi Rubao, 'Local Coalmines – An Important Part of China's Coal Industry', *CIYB 1982*, pp. 133–4.

27 It was pointed out earlier that the production figure for 1994 was revised upwards in May 1995 by some 20 million tons after a more thorough survey of the dispersed LNS mines' output had been carried out. 'Meitan yao shixing yi xiao dingchan de fangzhen (The Coal Industry Must Follow Production Plans Based on Consumption Levels Guideline), *SXRB*, 4 Mar. 1986, p. 1; Wang Senhao, 'Jianchi, fuchi, gaizao, zhengdun, lianhe, tigao de fazhen zujin xiangzhen meikuang jiankang fazhan' (Persist in the Supporting, Transforming, Rectifying, Uniting and the Healthy Development of the Village and Township Coal Mines), *ZGNY*, no. 5, 1994: 8–14; Zou Jihua, 'Guo wuyuan zhaokai quan guo xiangzhen meikuang gongzuo huiyi guanche luoshi dui xiangzhen meikuang "fuchi, gaizao, zhengdun, lianhe, tigao" de shizi fangzhen' (State Council Calls for the Implementation of 'Support, Reform, Rectify, Unite, and Raise' the Management of Village and Town Coal Mines), *ZGNY*, no. 5 1994: 1–7; 'Speculation in Coal Has Made Coal Shortage Even More Serious', *Zhongguo tongxun she*, 1147 GMT, 1 Sept. 1989, in *FBIS*, 14 Sept. 1989, p. 31; Fu Lihong, 'Xiangzhen meikuang cunzai de wenti yu fazhan qianxi' (Analysis of the Problems in Developing Village and Township Mines), *MTJJYJ*, no. 7 (1995): 60–1, 63.

28 Christine P. W. Wong, 'Ownership and Control in Chinese Industry: The Maoist Legacy and Prospects for the 1980s', in *China's Economy Looks Toward the Year 2000, Vol. I, The Four Modernizations*, JEC Report, 1986, p. 590. A good quality washed coal has about 7,000 kilocalories per kg.

29 The total washed in 1997 was the highest ever in one year. In 1996, 212.90 million tons were washed, accounting for 15.5 per cent and 216.40 million tons were washed in 1998, accounting for 17.6 per cent.

30 Zhang Baoming, 'Clean Coal Production and Rational Use of Coal in China', in IEA, *The Clean and Efficient Use of Coal and Lignite: Its Role in Energy, Environment and Life*, Proceedings of a Conference held in Hong Kong, 30 Nov.–3 Dec. 1993, Paris: OECD/IEA, 1994, p. 75.

31 Zhang Zhijian, Wang Jiacheng, and Xin Dingguo, 'Wo guo de nengyuan xingshi he renwu' (Our Country's Energy Situation and Tasks), *NY*, no. 5 (1984): 3.

32 Canadian Energy Research Institute and TATA Energy Research Institute (CERI/TERI), *Planning for the Indian Power Sector: Environmental and Development Considerations*, Study No. 62, New Delhi: The Institutes, 1995, p. 9, citing [India] Central Electricity Authority, *Performance Review of Thermal Power Stations, 1989/90*, New Delhi: 1995.

33 Wu Shiyu, 'Developing Coal Preparation Processing, Increasing Coal Utilization Efficiency and Reducing Pollution', in IEA, *Energy Efficiency Improvements in China: Policy, Measures, Innovative Finance and Technology Deployment*, Proceedings of a Conference held in Beijing, 3–4 Dec. 1996, Paris: OECD/IEA, 1997, pp. 77–80. Mr Wu is Director, Division of Coal Preparation, Dept of Production and Coordination, MIC.

34 Martin Daniel, *Chinese Coal Prospects to 2010*, IEAPER/11, London: IEA, 1994, p. 29, citing a private communication with S. Hu of the MCI in Dec. 1993.

35 *MTGYNJ 1998*, p. 22.

36 Ibid., p. 25; World Bank, *Clear Water, Blue Skies: China's Environment in the New Century*, China 2020 Series, Washington, DC: IBRD/WB, 1997, p. 46; IEA, *Coal Information 1998*, Paris: OECD/IEA, pp. II.19–21. A ton of coal equivalent (tce) has an ash content of about 7–8 per cent. *Coal Information 1998*, p. I.5.

37 IEA, Coal Advisory Board, *Coal in the Energy Supply of China*, p. 56.

38 United States, Energy Information Administration, *The United States Coal Industry, 1970–1990: Two Decades of Change*, Washington, DC: 1992, p. 35; India, Planning Commission, *Eighth Five Year Plan, 1992–97*, Vol. II, Sectoral Programmes of Development, New Delhi: The Commission, 1992, p. 181.

39 World Bank, *Clear Water, Blue Skies*, p. 46.

40 Zhang Baoming, 'Clean Coal Production and Rational Use of Coal in China', in *The Clean and Efficient Use of Coal and Lignite: Its Role in Energy, Environment and Life*, Proceedings of a Conference, Hong Kong, 30 Nov.–3 Dec. 1993, Paris: OECD/IEA, 1994, p. 75.

41 Zhang Mingchuan, 'Control of SO2 and NOx Emissions from Fossil Power Plants: Research and Practice of TPRI', in *The Clean and Efficient Use of Coal and Lignite: Its Role in Energy, Environment and Life*, Proceedings of a Conference, Hong Kong, 30 Nov.–3 Dec. 1993, Paris: OECD/IEA, 1994, p. 906.

42 Wu Shiyu, 'Developing Coal Preparation Processing, Increasing Coal Utilization Efficiency and Reducing Pollution', pp. 77–80.

43 Yu Ertie, 'Role and Prospects of the Coal-Dressing Industry under a Market Economy', *Meitan kexue jishu* (Coal Science and Technology) 22, 1 (Jan. 1994): 53–6, in *FBIS*, 3 May 1994, p. 33.

44 'Overview of Our Nation's Energy Resources', *LW*, 10 (1981): 9–11, in *JPRS China Report*, 228 1982: 16.

45 Masaki Takahashi, Stratos Tavoulareas, and Joseph Gilling, 'The World Bank Clean Coal Initiative', in IEA, *Energy Efficiency Improvements in China: Policy Measures, Innovative Finance and Technology Deployment*, Proceedings of a Conference held in Beijing, 3–4 Dec. 1996, Paris: OECD/IEA, 1997, p. 156, quoting World Bank Report No. 12687-CHA, *China: Investment Strategies for China's Coal and Electricity Delivery System*, 8 Mar. 1995.

46 Peng Fangchun and Zhi Luchuan, 'Jin, Shaan, Mo jiequ de nengyuan gongye: zhanbei dili, wenti ji duice' (Energy Industry in Contiguous Areas of Shanxi, Shaanxi, and Inner Mongolia: Key Strategical Position, Problems, and Countermeasures), *ZGNY*, no. 2 (1994): 19–21.

47 Susan V. Lawrence, 'Parched Province: Shanxi Pins Hopes on Water from the Yellow River', *FEER*, 22 Oct. 1998, pp. 24, 28.

48 Ibid.

49 Liu Changming, 'Environmental Issues and the South-North Water Transfer Scheme', *CQ*, 156 (1998): 900; Zhang Zhiping, 'Debates on South-to-North Water Diversion Project', *BR*, 28 Aug. 2000, pp. 26–8.

50 Richard Louis Edmonds, 'The Environment in the People's Republic of China 50 Years On', *CQ*, 159 (1999): 642.

51 IEA, *International Coal Trade: The Evolution of a Global Market*, Paris: OECD/IEA, 1997, p. 113.

52 Michel B. McElroy, 'Industrial Growth, Air Pollution, and Environmental Damage: Complex Challenges for China', in Michael B. McElroy, Chris R. Nielson and Peter Lydon (eds), *Energizing China: Reconciling Environmental Protection and Economic Growth*, Cambridge, MA: Harvard University Committee on the Environment, 1998, p. 243.

53 World Bank, *Clear Water, Blue Skies*, p. 46. (Mid-1990s figures.)

54 World Resources Institute at www.wri.org, accessed Oct. 1998 IEA, Coal Advisory Board, *Coal in the Energy Supply of China*, p. 53 says it is 5 out of 10.

55 Alex Vanderpol, 'China: Clean Coal Technology', *Market Report: Industry Sector Analysis*, 19 Apr. 1995, p. 129.

56 Ibid.

57 *NYFZBG 1997*, p. 232; World Bank, *Clear Water, Blue Skies*, pp. 8–9.

58 The success or lack thereof of the government's efforts to reduce pollution is examined in detail in articles published in an issue of *The China Quarterly* specially devoted to China's environment. *CQ*, no. 156, 1998.

59 Yu Yongnian, 'Overview of Coal Use and Efficient Future Technology', in IEA, *The Clean and Efficient Use of Coal and Lignite: Its Role in Energy, Environment and Life*, Proceedings of a Conference held in Hong Kong, 30 Nov.–3 Dec. 1993, Paris: OECD/IEA, Table IV, p. 727. See also p. 720.

60 Zhang Baoming, 'Clean Coal Production and Rational Use of Coal in China', p. 77.
61 Of the large-scale power plants, 60 per cent had electrostatic precipitators at the end of 1996. IEA, Coal Advisory Board, *Coal in the Energy Supply of China*, p. 56, citing, *Electric Power Industry Development and Environmental Protection in China*, Ministry of Electric Power Industry, PRC, 1997, section 3.1.1.
62 Ni Weidou and Nien Dak Sze, 'Energy Supply and Development in China', in Michael B. McElroy, Chris R. Nielson and Peter Lydon (eds), *Energizing China: Reconciling Environmental Protection and Economic Growth*, pp. 108 and 110. Scrubbers remove nitrogen and sulphur oxides.
63 World Bank, *Clear Water, Blue Skies*, p. 35.
64 'Acid Rain Takes Growing Toll', *SCMP*, 8 Dec. 1998.
65 *MTGYNJ 1998*, p. 354.
66 IEA, Coal Advisory Board, *Coal in the Energy Supply of China*, p. 56.
67 Zhang Mingchuan, 'Control of SO_2 and NOx Emissions', p. 906.
68 China would rank third after the US and the former Soviet Union.
69 Yu Yongnian, 'Coal Consumption Status and Strategy of Energy Development in China', in IEA, *Coal, the Environment and Development: Technologies to Reduce Greenhouse Gas Emissions*, Proceedings of a Conference held in Sydney 18–21 Nov. 1991, Paris: OECD/IEA, 1992, p. 235.
70 'Mainland Fires Fuelling the Greenhouse Effect', *SCMP*, 22 Sept. 1997, p. 24; Zhang Baoming, 'Clean Coal Production and Rational Use of Coal in China', p. 85.
71 World Bank, *Clear Water, Blue Skies*, pp. 8 and 17, quoting World Bank, *China: Chongqing Industrial Pollution Control and Reform Project*, Staff Appraisal Report, Washington, DC: 1996.
72 World Bank, *Clear Water, Blue Skies*, p. 8.
73 Ibid., p. 47.
74 Eduard Vermeer, 'Industrial Pollution in China and Remedial Policies', *CQ*, 156 Dec. 1998: 983, quoting Guojia huanbaoju ziran baohusi, *Zhongguo xiangzhen gongye huanjing wuran jiqi fangzhi duice* (Environmental Pollution by Township and Village Industries and Remedial Treatment Policies), Beijing: Zhongguo huanjing kexue chubanshe, 1995, pp. 18–19.
75 Eduard Vermeer, 'Industrial Pollution in China', p. 983.
76 Calculated from Jonathan E. Sinton et al. (eds), *China Energy Databook*, Berkeley, CA: Ernest Orlando Lawrence Berkeley National Laboratory Report LBL-32822 Rev. 4, Sept. 1996, Table II.9, p. II.38.
77 World Bank, *Clear Water, Blue Skies*, p. 70.
78 An IGCC plant is one in which 'coal is converted to fuel gas which is then burned in a gas turbine, the hot exhaust from which is used to generate steam for a steam cycle'. An PFBC/CC plant is one in which 'coal is burned in a fluidized bed at pressure, with the flue gas being passed through a gas turbine, and the waste heat being used to raise steam'. IEA, Coal Industry Advisory Board, *Industry Attitudes to Steam Cycle Clean Coal Technologies: Survey of Current Status*, Paris: OECD/IEA, 1994, p. 31.
79 A sub-critical PF-fired plant is one in which 'pulverized coal is burned in a boiler at atmospheric pressure with the steam produced being at all points below a pressure of 221.2 bar, the steam being used to drive a steam turbine'. A supercritical PF-fired plant is one in which 'pulverized coal is burned in a boiler at atmospheric pressure, the steam produced being at some point above a pressure of 221.2 bar (but which is not classed as ultra super-critical)'. An ultrasuper-critical PF-fired plant is one which 'has a maximum steam temperature above 566°C or a maximum steam pressure above 248 bar'. An AFBC plant is one in which 'coal is burned in a fluidized bed (either bubbling or circulating) at atmospheric pressure, with heat being recovered to power a steam cycle'. IEA, Coal Industry Advisory Board, *Industry Attitudes to Steam Cycle Clean Coal Technologies*, p. 31.

80 IEA, Coal Industry Advisory Board, *Factors Affecting the Take-Up of Clean Coal Technologies: Overview Report*, Paris: OECD/IEA, 1994, p. 9; IEA, Coal Industry Advisory Board, *Industry Attitudes to Combined Cycle Clean Coal Technologies*, Paris: OECD/IEA, 1994, p. 11; The Clean Coal Demonstration Programme in the US had spent over US$5.6 billion between 1986 and 1998. IEA, *Coal Information 1998*, p. I.92. Listed website: www.fetc.doe.gov. Efficiency is measured in net electricity generated, net calorific basis.

81 IEA, Coal Industry Advisory Board, *Industry Attitudes to Steam Cycle Clean Coal Technologies*, p. 8.

82 IEA, Coal Industry Advisory Board, *Industry Attitudes to Combined Cycle Clean Coal Technologies*, p. 12.

83 CERI/TERI, *Planning for the Indian Power Sector*, p. 72.

84 Ibid.; According to the World Bank, the fluidized bed combustion boilers would remove 50–80 per cent of the sulphur. World Bank, *Clear Water, Blue Skies*, p. 50.

85 Fiona E. Murray and Peter P. Rogers, 'Living with Coal: Coal-Based Technology Options for China's Electric Power Generating Sector', in Michael B. McElroy, Chris R. Nielson and Peter Lydon (eds), *Energizing China: Reconciling Environmental Protection and Economic Growth*, p. 176 quoting E. S. Tavroulareas and J. P. Charpentier, *Clean Coal Technologies for Developing Countries*, World Bank Technical Paper, Energy Series, Washington, DC: World Bank, 1995. For a description of the main CCT projects in China see: IEA, Coal Advisory Board, *Coal in the Energy Supply of China*, pp. 75–8.

86 See www.edu.cn/undp/ccp/cp3/coalprog.htm, accessed May 1999.

87 Chris P. Nielson and Michael B. McElroy, 'Introduction and Overview' in Michael B. McElroy, Chris R. Nielson and Peter Lydon (eds), *Energizing China: Reconciling Environmental Protection and Economic Growth*, p. 45 quoting Song Jian, 'China: A Land of Hope', Speech given at Harvard University, Cambridge, MA, 14 Apr. 1994.

88 Cao Zhengyan, 'Tigao nengyuan xiaolu Zhongguo qude de chengjiu he zhengfu de fangzhen cuoshi' (Raise China's Energy Efficiency, Achieve Results from Government Policies and Measures), *ZGNY*, no. 3 (1997): 10; 'Power Industry Uses Foreign Funds', *BR*, 2–8 Nov. 1998, p. 27.

89 See Fiona Elizabeth Murray, *Environment and Technology in Investment Decision-Making: Power Sector Planning in China*, PhD diss., Division of Applied Sciences, Harvard University, 1996; Vaclav Smil, 'China's Energy and Resource Uses: Continuity and Change', *CQ*, 156 (1998): 935–51; Zhou Yigong, 'Liuhua chuang ranshao jishu zai mei ranshao lingyu de diwei' (The Status of FBC Burning Technology in China), *ZGNY*, no. 4, 1998, pp. 13–17.

90 IEA, Coal Advisory Board, *Global Methane and the Coal Industry: A Two-Part Report on Methane Emissions from the Coal Industry and Coalbed Methane Recovery and Use*, Paris: OECD/IEA, 1994, p.12, citing R. Cecerone and R. Oremland, 'Biochemical Aspects of Atmospheric Methane', *Global Biogeochemistry Cycles*, vol. 2, 1998, pp. 299–327.

91 Ibid., p.17.

92 Ibid., p. 42

93 'China's Energy Development Enters New Stage', *Asia Pulse*, 9 Jan. 1998.

94 United Nations Development Programme, 'Development of Coalbed Methane Resources in China', Project Document, June 1992; Xu Dashan, 'Gas Tapped as Major Source of New Energy', *CD*, 10 Nov. 1998.

95 'China to Boost Coal Gas Energy with Preferential Policy Initiative', broadcast on *Xinhua*, 0251 GMT, 20 Nov. 1998.

96 'China Steps Up Exploration of Coal-Bed Methane Resources', *Asia Pulse*, 9 July 1998.

97 For example, Vaclav Smil, Lester Ross, and Zhang Zhongxiang. See entire *CQ*, 156 (Dec. 1998).

98 See webpage: www.worldbank.org.
99 World Bank, *Clear Water, Blue Skies*, pp. 19, 23–8.
100 Ibid., pp. 31–40.
101 Masaki Takahashi, Stratos Tavoulareas, and Joseph Gilling, 'The World Bank Clean Coal Initiative', pp. 153–7.
102 There are six greenhouse gases. The main ones are carbon dioxide, methane, and chlorofluorocarbons.
103 Elisabeth Tacey, 'Rich Shift Onus on Global Warming', *SCMP*, 8 Dec. 1997, p. 25.
104 Frank Ching, 'Kyoto Pact Is a Good First Step', *FEER*, 25 Dec. 1997–1 Jan. 1998, p. 98.
105 Elisabeth Tacey, 'Emission Treaty "Hinges on China"', *SCMP*, 3 Dec. 1997, p. 13.
106 'America Accepts Global-Warming Challenge', *SCMP*, 14 Nov. 1998, p. 12.
107 'IEA to Cooperate with China', *BR*, 1–7 Mar. 1999, p. 27.
108 'Sino-British Energy-Saving Project Reduces Pollution', *Xinhua*, 1428 GMT, 12 June 1999.
109 In the past China had been accused of sharing dual-use nuclear technology with Iran and Pakistan. By this time, however, the US was satisfied that China was sincere in promising not to proliferate such technology and agreeing to permit international inspection of all aspects of nuclear technology.
110 'China Gets Loan to Curb Air Pollution', *BR*, 3 Apr. 2000.
111 Michel B. McElroy, preface, in Michael B. McElroy, Chris R. Nielson and Peter Lydon (eds), *Energizing China: Reconciling Environmental Protection and Economic Growth*, p. i.
112 'Sino-German Energy, Environment Research Institute Set Up', *Xinhua*, 0830 GMT, 13 June 1999.
113 Liu Song, 'Cleaner Air in Guiyang', *BR*, 23 Aug. 1999, p. 28; 'Japan Investing Over US$330 Million on Environmental Projects in China', on World Coal Institute Webpage: www.wci-coal.com/gennews-JapanChina.htm, accessed Nov. 1999.
114 See webpage: www.mhenergy.com/demos/electric/index.html, accessed Oct. 1999.
115 Minanmi Ichikawa, 'Coal Technology Options: International Cooperation and Partnership', in IEA, *The Clean and Efficient Use of Coal and Lignite: Its Role in Energy, Environment and Life*, p. 127.
116 *JTNJ 1997*, p. 580.
117 Calculated from *CIYB 1982*, p. 40, and *MTGYNJ 1998*, p. 23.
118 Kang Shouyong and Cui Gang, 'Meitan gongye xiwang zai keji' (Coal Industry's Hopes Depend on Technology), *GYJJ*, no. 2 (1995): 127.
119 'Coal Industry Attracts Foreign Capital', broadcast on *Xinhua*, 1132 GMT, 8 June 1994, in *FBIS*, 9 June 1994, p. 45.
120 *MTGYNJ 1997*, p. 327.
121 *CIYB 1983*, p. 141, and *MTGYNJ 1998*, p. 32. The figures were slightly lower in 1998.
122 Fei Weiwei, 'Cong dian gao dao zonghe caimei' (The Use of Electric Picks in Coal Mines), *RMRB*, 1 Feb. 1994, p. 2.
123 Yang Baochuan and Zhang Qinwen, 'Xiangzhen meikuang de hongguan guanli wenti tantao' (Investigation into Village and Township Coal Mine Macro-Management Problems), *JDYJ*, no. 4 (1985): 4.
124 'Meitan jishu chukou shixian ling de tupo' (Breakthrough in Exports of Coal Expertise), *SXRB*, 4 Sept. 1987, p. 1.
125 'Coal Mining Equipment Now Being Sold Abroad', *CD*, 21 Nov. 1987, p. 2; 'Vice Minister Addresses Coal Technology Meeting', broadcast on *Xinhua*, 1307 GMT, 25 Sept. 1995, in *FBIS*, 28 Sept. 1995, pp. 55–6.
126 Guy Doyle, *China's Potential in International Coal Trade*, London: IEA International Coal Research, 1987, p. 79.

127 *MTGYNJ 1995*, p. 144; The Hunan International Economic and Technical Corporation bought 62 per cent of the Kiwira mine's shares. The project includes a thermal plant capable of generating 6 MW per year. *BBC Report*, 26 April, 1999.

128 'Energy Industries Top 1990 Production Targets', broadcast on *Xinhua*, 1751 GMT, 6 Jan. 1991, in *FBIS*, 7 Jan 1991, pp. 36–7.

129 Li Kaiming, 'Shanxi meitan nengyuan jidi jianshe zhong yizhi bu ke queshao de liliang' (Shanxi Coal Energy Base Construction: No Shortage of Power), *SXJDYJ*, p. 467; Liu Zaixing, 'Shanxi meitan nengyuan jidi jianshe ruogan wenti tantao' (Investigation into Certain Problems About Shanxi Coal Energy Base Construction), *JJWT*, no. 3 (1980): 5; and Wang Jianrong, 'Tongpei meitan guanli tizhi ying gaige' (Should Urgently Reform Distributed Coal Management System), *SXRB*, 1 Apr. 1988, p. 1.

130 Liu Zaixing, '1990 nian san yi dun de chanliang shi nan dadao de' (The 1990 Target of 300 Million Tons Is Difficult to Reach), *SXJDYJ*, p. 223; 'Meitan bu bu zhang Wang Senhao tongzhi zai quan guo xiangzhen meikuang gongzuo huiyi shang de jianghua', (Speech by Coal Minister Wang Senhao at Local Coal Mine Work Conference), *MTGYNJ 1994*, pp. 122–4.

131 *CIYB 1983*, pp. 220–1; *CIYB 1984*, p. 226; *CIYB 1985*, p. 171.

132 Daniel Kwan, 'Blast Blamed on Mine Rivalry', *SCMP*, 17 Apr. 1998, p. 11.

133 IEA, Coal Industry Advisory Board, *Global Methane and the Coal Industry: A Two-Part Report on Methane Emissions from the Coal Industry and Coalbed Methane Recovery and Use*, Paris: OECD/IEA, 1994, pp. 12, 23.

134 *MTGYNJ 1997*, p. 31; *MTGYNJ 1998*, p. 26.

135 Wang Zhuokun, 'Heli youxiao liyung meitan ziyuan' (Rational and Effective Use of Coal Resources), *NY*, 4, 1992, p. 28; Cui Jizhe and Yu Zhenhai, 'Meitan gongye zouxiang shichang de tiao cha baogao' (Report on the Marketization of the Coal Industry), *MTB*, 8 Oct. 1994, p. 2; 'Meitan bu bu zhang Wang Senhao tongzhi zai quan guo xiangzhen meikuang gongzuo huiyi shang de jianghua' (Speech by Coal Minister Wang Senhao at Local Coal Mine Work Conference), *MTGYNJ 1994*, p. 123.

136 'Shenhua gaige gao hao zhengdun cujin xiangzhen meikuang jiankang fazhan' (Deepen Reform, Rectify, and Promote Sound Development of Township and County Coal Mines), *MTGYNJ 1994*, p. 114.

137 'Heisanyong diqu shengtai riqu ehua' (Daily Deterioration of Heisanyong's Ecology), *RMRB*, 13 Oct. 1993, p. 3.

138 *CIYB 1982*, p. 37; *MTGYNJ 1999*, p. 27.

139 Pan Weier, 'Meikuang shuailao baofei yu kuang gong zai jiuye' (Report of Coal Mine Decommissioning and Re-Employment of Miners), *ZGNY*, no. 2 (1996): 20–3.

140 Wang Senhao, 'Renzhen guanche luoshi dang de shiwu da jingshen ba chi xu jiankang fazhan de meitan gongye quanmian tui xiang ershiyi shiji' (Earnestly Ensure Policies of Fifteenth Congress, Vigorously Control the Healthy Development of the Coal Industry as It Pushes Towards the 21st Century), *ZGNY*, no. 3 (1998): p. 6; Mark O'Neill, 'Reforms Derailed by Small Miners', *SCMP Business Section*, 9 Apr. 1998, p. 4.

141 Jasper Becker, 'Unpaid Miners Blocking Rail Lines', *SCMP*, 10 Mar. 1998, p. 8.

142 Interview with an official from the Ministry of Coal in September 1998.

143 Dong Jiben, 'Guanyu Shanxi nengyuan shengchan de jiazhi buchang wenti' (Problems Concerning the Compensation Value of Shanxi's Energy), *JJWT*, no. 1 (1985): 10.

144 *CIYB 1982*, p. 37 and *MTGYNJ 1998*, p. 25. At various CMAs, such as in Shanxi and Hebei, the OMS was over 6 tons. (The *MTGYNJ* give productivity data for the largest mines in the country.)

145 Pan Weier, 'Meikuang shuailao baofei yu kuang gong zai jiuye' (Report of Coal Mine Decommissioning and Re-Employment of Miners), *ZGNY*, no. 2 (1996): 21;

Pan Weier, 'Yi 50 wan meikuang zhigong zai jiu ye' (Discussion on the Re-Employment of 500,000 Coal Mine Staff and Workers), *MTJJYJ*, no. 4 (1996): 16; India, Planning Commission, *Eighth Five Year Plan 1992–97: Vol. II, Sectoral Programmes of Development*, New Delhi: The Commission, 1992, p. 175.

146 Xiao Chang, 'State Sends Coal Mines to Market', *CD*, 30 Dec. 1996.

147 'Women Tunnel On Despite Ban', *SCMP*, 28 Aug. 1994; United Press International, 'China Acts to End Coal Mine Chaos', *SCMP*, 30 Dec. 1994.

148 *XZQYNJ*, various years.

149 *GYJJTJNJ 1998*, p. 289

150 Liu Jiyang, 'Meitan gongye houjin zai nali?' (Where Has the Coal Industry's Reserve Strength Gone?), *RMRB*, 5 Mar. 1989, p. 5.

151 The average wage in 1996 was 7,000 yuan. The average increase between 1992 and 1997 was 800 yuan. 'Zhongguo meitan hangye niu kui wei ying shuping' (Commentary on Turning Around the Losses in China's Coal Industry), *JJDB*, 9 Jan. 1998; Peng Jialing, 'Xiang kui sun xuanzhan' (Declare War on Losses), *RMRB*, 13 Jan. 1997, p. 1; Introduction to *MTGYNJ 1996*.

152 At the Muchongou state mine in Guizhou in late 2000, face workers were earning 1,000 yuan per month while the surface workers were earning 200–300 yuan. 'Managers Blamed for Fatal Blast', *SCMP*, 14 Nov. 2000.

153 'Quebao meitan zhigong guo hao chunjie' (Ensure Coal Staff and Workers Have a Good Spring Festival), *RMRB*, 14 Feb. 1996, p. 2.

154 Jackie Sheehan, *Chinese Workers: A New History*, London: Routledge, 1998, p. 155.

155 Janet Matthews Information Services 1998, Quest Economics Database, Cambridge International Forecasts Country Report, Dec. 1998, 'India: Mining'.

156 'Meitan bu bu zhang Wang Senhao tongzhi zai quan guo xiangzhen meikuang gongzuo huiyi shang de jianghua' (Speech by Coal Minister Wang Senhao at Local Coal Mine Work Conference), *MTGYNJ 1994*, p. 123.

157 The 1990 figure is from 'Kuangshan shangwang shigu pin fa jin ren youlu wanshan lifa zeng he zhili shi zai bi xing' (It is Imperative to Perfect Legislation and Administration to Stop the Worrying Frequency of Mining Casualties), *GRRB*, 20 Feb. 1992, p. 3; The 1991 figure is from United Press International, 'Dangerous Coal Mines Come Under Scrutiny', *SCMP*, 1 Apr. 1994. See also Chan Waifong, 'Mining Deaths Soar to 10,000', *SCMP*, 15 Oct. 1994, p. 10; The 1993 figure is from '1994 nian quan guo zhuangshan qiye zhizhong shangwang shiqu qingkuang fenxi' (Analysis of 1994 Coal Mine Workers' Accidents), *Laudong nei can*, 21 (Mar. 1995): 12–13; The 1994 figure was calculated from *MTGYNJ 1997*, p. 110; The 1995 figure is from the Introduction to *MTGYNJ 1996*, though the calculated total from *MTGYNJ 1997*, p. 115 is 6,915; The 1996 figure is calculated from *MTGYNJ 1997*, p. 117; The 1997 figure is calculated from *MTGYNJ 1998*, p. 80; The 1998 figure is calculated from *MTGYNJ 1999*, p. 73.

158 Fan Weitang, 'Shijie meitan gongye xianzhuang ji fazhan qushi' (Present Trends in the Development of the World Coal Industry), *ZGNY*, no. 9 (1998): 4.

159 Ni Weidou and Nien Dak Sze, 'Energy Supply and Development in China', in Michael B. McElroy, Chris R. Nielson, and Peter Lydon (eds), *Energizing China: Reconciling Environmental Protection and Economic Growth*, Table 6, p. 81.

160 Introduction to *MTGYNJ 1997*. The figures given by Fan Weitang, 'Shijie meitan gongye xianzhuang ji fazhan qushi', p. 4, are different: 4.02 at the LS mines and 7.70 at the LNS.

161 *MTGYNJ 1997*, pp. 336–7.

162 'Dozens of Pupils Died in Blast', *SCMP*, 4 Feb. 1996, p. 1; 'Missing Blast Man Mystery', *SCMP*, 4 Feb. 1996, p. 5.

163 International Labour Organization, *Safety and Health in Coal Mines, An ILO Code of Practice*, Geneva: ILO, 1986.

164 '100 Chinese Miners Murdered for Money', *ST*, 20 June 2000, p. 17.

165 See Mo Liqi, 'Jishu chuang xin yu tigao zhigong suzhi xiang jiehe shixian jingji zengchang fangshi de zhuanbian' (Raise Quality of Staff and Increase Economic Growth Through Technological Improvements), *MTJJYJ*, no. 1 (1998): 52–3, 57.

166 Webpage www.ccri.ac.cn/mkzy/english/english.html, accessed Sept. 1999.

167 Listed in NIRA's World Directory of Think Tanks 1996 at webpage www.nira.go.jp/ice/tt-info/nwdtt96/1311.htm.

168 Harry Campbell (ed.), *Cinfolink Annual Review of Information Services and the Internet in China*, Toronto: Cinfolink Services, 1996, pp. 64–5.

169 For example the Pingdingshan Coal (Group) Co. Ltd webpage: www.chinavista.com/products/pds/pds.html, accessed Jan. 1999. Data about Dazhangjia Coal Mine (Anhui) can be found at: www.webunion.com/project/coa, accessed Jan. 1999.

170 Peter James, *The Future of Coal*, 2nd edn, London: Macmillan, 1984, pp. 200 and 222; Robert P. Greene and J. Michael Gallagher (eds), *Future Coal Prospects: Country and Regional Assessments*, World Coal Study, Cambridge, MA: Ballinger, 1980.

171 IEA, *Coal Information 1998*, Table 6.11, p. I.250.

172 Zeng Peiyan, 'Report on the Implementation of the 1999 Plan for National Economic and Social Development' (Excerpts), delivered at the Third Session of the 9th NPC, 6 Mar. 2000, *BR*, 10 Apr. 2000, p. 17; Li Rongxia, 'Major Turning Point in SOE Reform and Difficulty Relief', *BR*, 21 Aug. 2000, p. 13.

173 Ye Ruixiang, 'Guanyu meitan, shiyou jiage wenti de yanjiu' (Research into the Problems with Coal and Oil Pricing), *CMJJ*, no. 10 (1983): 39; Yang Jinsheng and Li Zhangzhong, 'Meitan qiye jian kui zeng ying de taolun' (Investigation into Reducing Losses and Increasing Profits in the Coal Industry), *MTJJYJ*, no. 10 (1984): 23.

174 'Guanyu niu kui zeng ying de jige wenti' (Some Problems in Turning Around Losses), *MTGYNJ 1995*, p. 117.

175 *TJNJ 1988*, p. 389.

176 1987 data from 'Meitan gongye hou jin zai nali?', (Where is Coal Industry Strength?), *RMRB*, 5 Mar. 1989, p. 5; 1988 data from Yi Chu, 'Meitan gongye shinian gaige yu fazhan huigu' (Looking Back Over Ten Years of Reform and Development in the Coal Industry), *NY*, no. 5 (1989): 22.

177 'Guo you zhongdian meikuang zongti niu kui wei ying' (State Coal Mines Turn Around Losses to Make Profits), *RMRB*, 26 Dec. 1997, p. 4; 'Coal Back in the Black', *SCMP*, 27 Dec. 1997, p. 68; Zhao Shaoqin, 'Mine Closure Scheme to Overcome Coal Glut', *CD Business Weekly*, 23–9 Aug. 1998.

178 Jasper Becker, 'Unpaid Miners Blocking Rail Lines', *SCMP*, 10 Mar. 1998, p. 8.

179 Mark O'Neill, 'Output of Coal Falls with Low Demand', *SCMP*, 22 Aug. 1998, p. 3; Zhao Shaoqin, 'Mine Closure Scheme to Overcome Coal Glut', *CD Business Weekly*, 23–9 Aug. 1998; 'Coal Market in China Set to Remain Sluggish in 1998', *Asia Pulse*, 1 Sept. 1998.

180 'Zhongguo 99 nian meitan gongye jingji zhuangkuang bu rong leguan' (Conditions in China's Coal Industry in 1999 Not Hopeful), *Xinhua*, 8 Jan. 1999; 'Zhongguo meitan 99 nian rengran di wei yunxing' (China's Coal Industry in 1999 is Still Operating in a Low Position), *MTB*, 15 Apr. 1999, p. 3; 'Overall Cut of Loss by Coal Industry Demanded in China', www.peopledaily.com.cn/english, accessed 19 May 2000.

181 'Zhongguo meitan jingji yunxing xingshi yanjun' (Economic Situation of China Coal Industry Is Severe), *JJSB*, 19 May 1999, p. 7; Mark O'Neill, 'Output Cut for Ailing Producers', *SCMP*, 4 Dec. 1999.

182 'Coal Plan Offers Aid to Miners', *CD*, 11 July 1999.

183 'Overall Cut of Loss by Coal Industry Demanded in China', www.peopledaily.com.cn/english, accessed 19 May 2000.

184 'Meitan bu tuan zhongyang gongtong shishi yangguang gongcheng' (Implementation of Sunshine Project by Central Coal Department), *RMRB*, 11 Feb. 1996, p. 2; Mark

O'Neill, 'Reserves Underpin Sector Dominance', *SCMP China Business Review*, 30 Oct. 1997, p. 6.

185 *MTGYNJ 1998*, p. 31. No definition of these categories is given.

186 Shen Liao, 'Team Supervises Mine Closures', *CD*, 18 Apr. 1999.

187 Han Zongshun, 'Guanyu wo guo difang meikuang chixu fazhan de duice' (Countermeasures for the Continuing Development of Local Coal Mines in China), *NY*, no. 5 (1989): 51–2; 'Datong kuangqu meiji gongsi yu pinkun xiang lian ban tiechang' (Datong Mining Area Coal Machinery Company Unites with Poor Village to Run Iron Factory), *SR*, 10 June 1988, p. 2; Zou Jihua, 'Guo wuyuan zhaokai quan guo xiangzhen meikuang gongzuo huiyi guanche luoshi dui xiangzhen meikuang "fuchi, gaizao, zhengdun, lianhe, tigao" de shizi fangzhen' (State Council Calls for the Implementation of 'Support, Reform, Rectify, Unite, and Raise' the Management of Village and Town Coal Mines), *ZGNY*, no. 5 (1994): 1–2; Yang Zhan and Xi Rubao, 'Local Coalmines – An Important Component Part of China's Coal Industry,' *CIYB 1982*, pp. 133–4.

188 China Local Mines Service Inc., 'Vigorous Development of Village and Town-Run Coal Mines,' *1985 CIYB*, pp. 109–11.

189 Dong Jiben, 'Guanyu Shanxi nengyuan shengchan de jiazhi buchang wenti' (Problems Concerning the Compensation Value of Shanxi's Energy), *JJWT*, no. 1 (1985): 12; Shen Bin, 'Closure of Small Coal Mines Progressing', *CD*, 13 May 1999.

190 The totals were calculated from the year 1985, not 1979 because an adjustment was made to the official electricity investment data after 1984, complicating analysis of the total investment figures.

191 Xiao Wei, 'China Plagued by Energy Shortage in Next Decade', *CM*, no. 3 (1991): 10–11; 'Energy Industries Top 1990 Production Targets', broadcast on *Xinhua*, 1751 GMT, 6 Jan. 1991, in FBIS, 7 Jan. 1991, pp. 36–7.

192 Rong Geng, 'Promoting Coal Industry in China through Innovative Financing', in IEA, *Energy Efficiency Improvements in China: Policy Measures, Innovative Finance and Technology Deployment*, Proceedings of Conference held in Beijing, 3–4 Dec. 1996, Paris: OECD/IEA: 1997, pp. 159–62, Mr Rong is in the State Development Bank of China.

193 Wang Senhao, 'Jin yi bu ba meitan gongye de shiqing wei hao' (Towards Improving the Coal Industry), *LW*, no. 38 (18 Sept. 1994): 15.

194 IEA, Coal Industry Advisory Board, *Coal in the Energy Supply of China*, p. 31, citing Coal Licensing and Production Tax Regimes. IEA Coal Research, London 1998; Organization for Economic Cooperation and Development, *Environmental Taxes: Recent Developments in China and OECD Countries*, Paris: 1999, p. 148.

195 Lu Qikang, 'Guanyu meitan shengchan chengben yu chu chang jiage wenti de tantao' (Investigation Into Coal Production Costs and Producer Prices), *JGLLSJ*, no. 1 (1982): 28–9; Wang Jiachun and Li Minxin, 'Guanyu wo guo nengyuan zhengce yu jiage de maodun' (China's Energy Policy and Price Contradictions), *JJWT*, no. 4 (1984): 17.

196 *FT*, 16 Nov. 1987; *FT*, 1 Feb. 88, p. 2.

197 The author interviewed Ye Ruixiang, Deputy Head of the Department of Costing and Pricing Institute of the Finance and Trade Economics Department of the Chinese Academy of Social Sciences in Beijing and several members of the Coal Price Study Group at the Shanxi Academy of Social Sciences in Taiyuan in September 1987. Professor Ye and his colleagues wrote innumerable theoretical articles on how best to reform the prices of coal and other key commodities. Much of what Ye Ruixiang told me can be found in his article, 'Guanyu meitan, shiyou jiage wenti de yanjiu' (Research into Problems with Coal and Oil Pricing)', *CMJJ*, no. 10 (1983): 39–42.

198 'Fuzhong fenjin wanqiang panbo tujin liangge genben xing zhuanbian shixian niu kui zeng ying mubiao' (Carry Heavy Burden, Fight Hard, Push Forward Two Basic Transformations, Achieve Elimination of Losses Target), *MTGYNJ 1997*, p. 69.

199 Li Hongtang, Meng Xianyu, and Peng Ningzhi, 'Guanyu lishun meitan chanpin bi jia wenti de sikao' (Thoughts on Rationalizing Price Parities of Coal Products), *MTJJYJ*, no. 11 (1994): 20.

200 Chang Weimin, 'Coal Shedding 140,000 Workers in '93', *CD*, 16 Aug. 1993, 2.

201 Fei Weiwei and Yang Jiangyou, 'Meitan jiage mingnian qi quanbu fangkai' (Next Year Coal Prices Will Be Totally Free), *RMRB*, 23 Dec. 1993, p. 2.

202 'Guanyu niu kui zeng ying de jige wenti' (Some Problems in Turning Around Losses), *MTGYNJ 1995*, p. 117.

203 By comparison, the Indian government planned to deregulate all coal prices by January 2000. Janet Matthews Information Services 1998, Quest Economics Database, Cambridge International Forecasts Country Report, Dec. 1998, 'India: Mining'.

204 'Guojia jiwei dui dian mei shixing guojia zhidao jiage' (State Planning Commission Implements National Guidance Prices for Coal Sold for Electricity), *ZGWJ*, no. 3 (1996): 37.

205 'Guojia jiwei fachu guanyu 1997 nian dian mei guojia zhidao jiage you guan wenti de tongzhi' (State Planning Commission Issues Notice Concerning National Guidance Prices for Sold Coal for Electricity), *ZGWJ*, no. 1 (1997): 35; Listing Prospectus of Yanzhou Coal Mining Co. Ltd, 24 Mar. 1998; Renee Lai, 'Mining Firm Sets Hopes on HK Float', *SCMP Business Section*, 15 Jan. 1997, p. 11.

206 'Shanghai meijia si ge yue shang zhang wu zi, zhang jia you yin zhong jian huanjie bu duan jia ma hai jiang kan zhang' (Factors Causing the Price of Coal to Be Raised Five Times in Four Months), *MTB*, 10 June 1993.

207 *TJNJ 2000*, p. 289.

208 'Chinese Army Involved in Illegal Reselling of Coal', broadcast on *Zhongguo tongxun she*, 0634 GMT, 4 Apr. 1989, in *FBIS*, 7 Apr. 1989, p. 34, quoting *WZB*, 4 Apr. 1989; 'Shanxi tichu meijia tiao kong xin silu' (Shanxi Considers A New Way to Control Coal Prices), *WZB*, 15 Apr. 1994, p. 1; 'Speculation in Coal Has Made Coal Shortage Even More Serious', broadcast on *Zhongguo tongxun she*, 1147 GMT, 1 Sept. 1989, in FBIS, 14 Sept 1989, p. 31.

209 'Mei, zhen yao dao ru hai ma?' (Do We Really Want to Pour Coal into the Sea?), *WZB*, 11 Mar. 1994, p. 1.

210 'Meitan gongye bu fubu Zhang Yuying tongzhi zai quanguo meitan gongye gongzuo huiyi shang de zongjie' (Summary of Coal Industry Deputy Minister Zhang Yingyu's Speech at Coal Industry Work Conference), 10 Jan. 1995, *MTGYNJ 1995*, p. 111; 'Guanyu niu kui zeng ying de jige wenti' (Some Problems in Turning Around Losses), *MTGYNJ 1995*, pp.118–19; 'Weak Coal Market Puts Pressure on Producers in Third Quarter', broadcast on *Zhongguo xinwen she* (China News Service), 1239 GMT, 4 Dec. 1997.

211 'Coal Consumer Debt Taxes Industry', broadcast on *Zhongguo xinwen she*, 1240 GMT, 14 Jan. 1994, in *FBIS*, 14 Jan. 1994, p. 38.

212 'Guanyu niu kui zeng ying de jige wenti' (Some Problems in Turning Around Losses), *MTGYNJ 1995*, p. 118.

213 'Coal Ministry Hails Payment Policy, Debt Reduction', *Zhongguo tongxun she*, 0927 GMT, 5 Feb. 1995, in *FBIS*, 16 Feb. 1995, p. 49.

214 'Zhongguo mei kuan tuo qian he qi duo' (State Coal Industry Owed So Much), *RMRB*, 5 Aug. 1999.

215 Wang Xiaozhong, 'Delivery-on-Payment Is Not as Easy as It Seems' (part 1 of 2) *CD*, 7 Apr. 1995, p. 4.

216 'Zhongguo 99 nian meitan gongye jingji zhuangkuang bu rong leguan' (Conditions in China's Coal Industry in 1999 Not Hopeful), *Xinhua*, 8 Jan. 1998; 'Coal Plan Offers Aid to Miners', *CD*, 11 July 1999.

217 Introduction to *MTGYNJ 1995*.

218 Mark O'Neill, 'Output of Coal Falls with Low Demand', *SCMP*, 22 Aug. 1998, p. 3; 'Weak Coal Market Puts Pressure on Producers in Third Quarter', broadcast on *Zhongguo xinwen she*, 1239 GMT, 4 Dec. 1997.

219 'India: Coal Industry to Recover Dues from States via Annual Plan Outlay', broadcast on *PTI News Agency*, 1317 GMT, 6 May 1998.

220 'Beijing's First Coal Wholesale Market Opens', broadcast on *Xinhua*, 0246 GMT, 15 Apr. 1992, in *FBIS*, 15 Apr. 1992, p. 61.

221 'First Coal Commodities Market to Open 28 October', broadcast on *Xinhua*, 0720 GMT, 27 Oct. 1992, in FBIS 27 Oct. 1992, p. 35; Li Xuegang, 'Meitan qiye yao shanyu liyong shichang' (Coal Enterprises Should Fully Utilize the Market), *ZMB*, 12 May 1994.

222 Chang Weimin, 'Price Hikes, Layoffs, Planned in Coal Sector', *CD*, 8 Jan. 1993, p. 2; 'Quanguo chuxian meitan jiaoyi shichang re' (Emerging National Coal Market Transactions Are Hot), *WZB*, 2 Sept. 1992; 'Northeast Coal Trade Centre Opens in Liaoning', broadcast on Shenyang Liaoning People's Radio Network, 2300 GMT, 28 Feb. 1993, in *FBIS* 3 Mar. 1993, p. 44; 'Dongbei meitan jiaoyi suo 1993 nian chengjiao meitan 715.7 wan dun chengjiao e da 14.2 yi yuan' (Attainment of 7,157,000 Tons of Coal Traded and 14.2 Billion Yuan Revenue on Northeast Coal Market in 1993), *MTB*, 13 Jan. 1994.

223 Wei Lin and Tian Huiming, 'Official on Plans to Promote Coal Industry', broadcast on *Zhongguo xinwen she*, 1025 GMT, 12 Oct. 1992, in *FBIS*, 30 Oct. 1992, pp. 31–2; Chen Gongyong, 'Wuhan diqu meitan shichang kan hao' (Wuhan District Coal Market Looks Good), *MTB*, 16 Mar. 1995.

224 Zhang Tongquan, 'Mian dui yunshu zhiyue: meitan zou xiang shichang ying caiqu de duice tantao' (Inquiring How to Solve the Problem of Coal Enterprises Being Unable to Participate in the Market Due to Transport Constraints), *MTJJYJ*, no. 9 (1995): 27–8.

225 'Coal Policy to be Readjusted', *ER*, no. 2 (1994): 32.

226 Qu Weiying *et al.*, 'Mei jia bao zhang: lang zheng feng weixiang' (Sudden Rise in Coal Prices: Waves Stop But Wind Still Blows), *JJCKB*, 2 Apr. 1996, p. 1.

227 '3 yue, meitan shichang gongqiu you bian' (Coal Market Supply and Demand Changed in March), *WZB*, 9 Apr. 1993.

228 Most bureaus always had grown whatever food they could.

229 'Beijing Seeks Foreign Funds for Coal Projects', broadcast on *Xinhua* 1023 GMT, 8 Sept. 1995, in *FBIS*, 12 Sept. 1995, pp. 22–3.

230 *MTGYJMSC*, pp. 451–4.

231 Zhongguo shehui kexue yuan, gongye jingji yanjiu suo (Chinese Academy of Social Sciences, Industrial Economics Research Institute), *GYFZBG 1998*, Beijing: Jingji guanli chubanshe, 1998, p. 158; 'Zhongguo meitan hangye niu kui wei ying shuping' (Commentary on Turning Around the Losses in China's Coal Industry), *JJDB*, 9 Jan, 1998; Mark O'Neill, 'Reserves Underpin Sector Dominance', *SCMP China Business Review*, 30 Oct. 1997, p. 6.

232 For example, the Pingdingshan Coal (Group) Co. Ltd. webpage: www.chinavista.com/products/pds/pds.html, accessed Jan. 1999.

233 Introduction to *MTGYNJ 1995*.

234 Liu Fei, 'China Develops World's Largest Coalfield', *New China Quarterly*, 12 (May 1989): 37–40.

235 The name was changed in 1999 to Yankuang.

236 'A' shares are listed on the Shanghai and Shenzhen stock exchanges and can be purchased only by Chinese citizens. 'B' shares are also listed in yuan on the Shanghai and Shenzen stock exchanges but can be purchased only by foreigners with payment in foreign exchange. 'H' shares are listed on the Hong Kong stock exchange.
    One ADR represents 50 H shares. 'The Bank of New York Selected by Yanzhou Coal Mining Co. Ltd for Its ADR Program', *PR Newswire*, 9 Apr. 1998; 'Yanzhou Coal Mining Co. Ltd to be Listed on Shanghai Stock Exchange', *Business Wire*, 30 June 1998.

237 Renee Lai, 'Yanzhou Coal Shines as Others Struggle', *SCMP*, 25 Aug. 1998, p. 2; Yanzhou webpage: www.yanzhoucoal.com.cn. For particulars of companies listed on the Hong Kong Stock Exchange, see www.irasia.com/listco/hk.

238 Mark O'Neill, 'Slump Takes Shine Off First Black-Power Stock', *SCMP Business Section*, 24 Aug 98, p. 5.

239 Jia Jianfeng, 'Meitan xiaoxing qiye gu fen hezuo zhi de tantao' (Investigation into Small Coal Enterprises Creating Joint Stock Systems), *MTJJYJ*, no. 1 (1998): 21–2.

240 See *MTGYJMSC*, pp. 489–94 and *MTGYNJ*, various years.

241 The author visited the mine in September 1987.

242 'China Suspends Shipments', *ICR*, 23 Sept. 1988.

243 Ibid.

244 Robert Thomson, 'Chinese Coal Mining Hits a Troublesome Seam', *FT*, 5 May 1988, p. 34; 'Antaibao Partners in Marketing Dispute', *CWI*, 13 July 1988; 'Oxy Gets Out of Antaibao', *ICR*, 14 June 1991.

245 IMF, *International Financial Statistics Yearbook*, Washington, DC: June 1995, p. 72.

246 China, SPC, 'Industrial Catalogue Guiding Foreign Investment' jointly promulgated by the State Planning Commission, the State Economic and Trade Commission and the Ministry of Foreign Trade and Economic Cooperation, *BR*, 2–8 March, 1998, pp. 15–23.

247 *The Far East and Australasia 1998*, 29th edn, London: Europa, p. 233.

248 'What's the Reason for the Low Perfection Rate, Great Loss and Waste of the Imported Equipment at the Huolinhe Mine Area', *Economic Daily*, 8 Aug. 1983, and 'An Investigation Group Appointed by the State Council Went to the Huolinhe Mine Area to Investigate and Deal with the Serious Extravagance', *Economic Daily*, 9 Aug. 1983, both translated in *CIYB 1984*, pp. 110–12.

249 Private contact with a retired representative of a UK coalmine machinery company who wished to remain unnamed.

250 'Foreign Investment to Improve Coal Sector', *CD*, 18 Mar. 1999; See also 'Coal Industry to Further Open to Outside', *FBIS*, Sept. 1995, p. 33 quoting *Xinhua*, 1411 GMT 9 Sept. 1995.

251 *MTGYJMSC*, pp. 95–6.

252 Liu Song, 'Japanese Loans Pooled to Build Railways', *BR*, 13 Sept. 1999, p. 27. The main projects were the Yanzhou (Shandong)-Shijiazhuang(Hebei), Hengyang(Hunan)-Guangzhou(Guangdong), Zhengzhou(Henan)-Baoji(Shaanxi), Dalian(Liaoning)-Qinghuangdao(Hebei), Hengyang(Hunan)-Shangqui(Henan), Xian(Shaanxi)-Ankang(Shaanxi), and Guiyang(Guizhou)-Loudi(Hunan) lines.

253 See webpage: www.worldbank.org.

254 The main relevant World Bank publications are listed in the bibliography.

255 IEA, Coal Industry Advisory Board, *Coal in the Energy Supply of China*, p. 73 citing *Asian Energy News*, Nov. 1998, which in turn cited 'Foreign Funds Fuel Power Industry', *CD*, 25 Oct. 1998.

256 Li Rongxia, 'Growing Role of Foreign Capital in Power Industry', *BR*, 19–25 May, 1997, p. 22.

257 Wang Zhenshuan, Zhang Wenshan and Ji Yangrui, 'Wo guo meitan gongye liyong waizi cunzai wenti ji duice' (Problems and Countermeasures in the Utilization of Foreign Capital in China's Coal Industry), *MTJJYJ*, Part I in no. 9 (1997): 18–21, and Part II in no. 10, pp. 17–20, 32; 'Guanyu "Meitan gongye jiu wu jihua he 2010 nian yuanjing mubiao" de zongti silu he kuangjia' (Thoughts on the Framework of the Coal Industry over the 'Ninth Five Year Plan and for the 2010 Longterm Targets'), *MTGYNJ 1997*, pp. 145–6; Lin Jian, 'Promoting International Cooperation of Coal Patent Technology Development', in IEA, *The Clean and Efficient Use of Coal and Lignite: Its Role in Energy, Environment and Life*, pp. 811–14.

258 Under the BOT system the project is funded and operated completely by the foreign company for an agreed time period, and then given over to the local company to operate. It relieves the host country of paying large amounts of money in its early stages of modernization. Before BOT, foreign developers were permitted a maximum 50 per cent stake in power plants.

259 IEA, *Coal Information 1998*, Paris: IEA/OECD, 1999, pp. I.132, I.134, I.136.
260 Ibid., p. I.134.
261 Yanzhou Coal Mining Company Financial Statement, 31 Dec. 1999 www.yanzhoucoal.com.cn. The price for 2000 is the average for the six months ended 30 June.
262 US Government, Department of Energy website: www.eia.doe.gov/cneaf/coal/quarterly/html/t17p01p1.html, accessed Nov. 1999.
263 Raoul Oreskovic, 'The Emergence of China as a Major Coke Supply Source', paper presented at conference 'Coping with the Tightening Coke Supply: Is a Crisis Looming?', held in Charlotte, NC, 5–7 Mar. 1997. Available at www. chinaenergyresources.com/article.html.
264 IEA/Organization for Economic Cooperation and Development, *International Coal Trade: The Evolution of a Global Market*, Paris: 1997, p. 113.
265 'India: Coal Industry to Recover Dues From States via Annual Plan Outlay', *PTI News Agency*, New Delhi, 1317 GMT, 6 May 1998.
266 Chang Weimin, 'Reform in Store for Coal Industry', *CD*, 5 Jan. 1997; Jasper Becker, 'Crises Deal Blow to Zhu's Targets', *SCMP*, 30 Aug. 1998, p. 7; IEA, *Coal Information 1997*, p. I.183; 'Coal Industry to Further Open to Outside', broadcast on *Xinhua*, 1411 GMT, 9 Sept. 1995, printed in *FBIS*, 11 Sept. 1995, p. 33.
267 Lana Wong, 'Yanzhou Coal Confident of Target', 26 Aug. 1998, *SCMP Business Section*, p. 3.
268 The website address of the CNCIEC is www.chinacoal.com.
269 CNCIIEC, *Annual Report 1997*, Beijing: 1997, p. 7; 'Top 500 in Foreign Trade', *BR*, 14 Aug. 2000, p. 29.
270 *TJNJ*, various years.
271 Angus Maddison, *Chinese Economic Performance in the Long Run*, Paris: OECD Development Centre, 1998, p. 51.
272 'Electric Railways Speed Up to Eighth Place', *BR*, 12–18 Jan. 1998, p. 26.
273 'Railway Equipment Exports to Iran', *BR*, 12–18 Oct. 1998, p. 25.
274 Xiao Rui and He Yiwen, 'Railway Construction to Enter a New Stage – An Interview of Sun Yongfu, Vice-Minister of Railways', *ER*, 7 (1993): 30; '5 yue, dongnan yanhai meishi jiao jin' (Southeast Coastal Cities Faced with Very Tight Coal Supply Situation in May), *WZB*, 18 June 1993.
275 India, Planning Commission, *Eighth Five Year Plan 1992–1997*, p. 174.
276 Reported on BBC TV World Service in late 1998.
277 Hou Bangan, 'Under-Funded Transport Network Fails to Keep up with Economic Growth', *Keji daobao* (Science and Technology Review), 10 Dec. 1996, pp. 46–50, published in SWB, Part 3: Asia–Pacific, FEW0482, 16 Apr. 1997, p. 7.
278 Geoffrey Crothall, 'Rail Link Heralds New Era', *SCMP Money Section*, 21 Jan. 1996, 1.
279 Weng Zhensong, 'Forecast and Analysis of China's Railway Transport', in Liu Guoguang *et al.* (eds), *PRC 1998 Economic Blue Book*, Hong Kong: Centre for Asian Studies, University of Hong Kong, 1998. In the late 1990s there were thirteen railway bureaux or group corporations under the direct jurisdiction of the Ministry of Railways – Harbin (Heilongjiang), Shenyang (Liaoning), Beijing, Hohhot (Inner Mongolia), Zhengzhou (Henan), Jinan (Shandong), Shanghai, Nanchang (Jiangxi), Guangzhou (Guangdong), Liuzhou (Guanxi), Chengdu (Sichuan), Lanzhou (Gansu), and Urumqi (Xinjiang). The five trunk railways were: Beijing-Shanghai, Beijing-Guangzhou, Harbin-Dalian, Beijing-Shenyang, and Lianyungang-Lanzhou.
280 Hou Bangan, 'Under-Funded Transport Network', p. 2.
281 Ibid., p. 3.
282 Xiao Rui and He Yiwen, 'Railway Construction to Enter a New Stage', p. 30.
283 Hou Bangan, 'Under-Funded Transport Network', p. 4.
284 Jing Wei, 'China's Biggest Energy-Producing Centre', *BR*, 3 Dec. 1984, pp. 16–17, citing *SXRB*.

285 'Meitan yao shixing yi xiao dingchan de fangzhen' (The Coal Industry Must Follow Production Plans Based on Consumption Levels Guideline), *SXRB*, 4 Mar. 1986, p. 1.
286 Huang Xiang, 'Coal Mines See Red over Sales', *CD*, 6 Aug. 1991, p. 2.
287 Shanxi shehui kexue yuan, meitan jiage yanjiu keti zu (Shanxi Academy of Social Sciences, Coal Pricing Research Group), *Meitan jiage gaige yu Shanxi meitan jiage*, p. 20.
288 Hu Benzhe, 'Guanyu Shanxi nengyuan jidi meitan shusong fangshi de tantao' (Investigation into Methods of Coal Transportation in Shanxi Energy Base), *NY*, no. 4 (1983): 1.
289 'Meitan yao shixing yi xiao ding chan de fangzhen' (The Coal Industry Must Follow Production Plans Based on Consumption Levels Guideline), *SXRB*, 4 Mar. 1986, p. 1; Li Kaiming, 'Shanxi meitan nengyuan jidi jianshe zhong yizhi bu ke queshao de liliang' (Shanxi Coal Energy Base Construction: No Shortage of Power), *SXJDYJ*, p. 467.
290 Hou Bangan, 'Under-Funded Transport Network', p. 4.
291 Ibid., p. 4.
292 Ralph Huenemann, 'Modernizing China's Transport System', in US Congress, Joint Economic Committee, *China's Economic Dilemmas in the 1990s*, Armonk, NY: M. E. Sharpe, 1993, pp. 458–9; 'Railway Equipment Exports to Iran', p. 25; 'Locomotive Industry Heading on Right Track', *CD*, 1 Mar. 1998.
293 'Railway Market to Open Further', *BR*, 28 Dec.–3 Jan. 1999, p. 28; 'Railways to Use Overseas Investment for Construction', *Xinhua*, 0627 GMT, 27 Dec. 1998.
294 *TJNJ 2000* pp. 515–6.
295 *TJNJ*, various years.
296 Author's interview with an official of the Ministry of Coal in Sept. 1998.
297 China Local Coal Mines Service Inc., 'Vigorous Development of Village- and Town-Run Coal Mines', *CIYB 1985*, p. 110; Yan Dayu and Yu Xiaomiao, 'Meitan liutong huanjie wenti yuanyin ji duice qianlun' (Discussion of the Causes of the Coal Circulation Problem and Solutions), *MTJJYJ*, no. 4 (1996): 19.
298 Xu Jingyi, 'Cong meitan ziyuan zengjia jiage xiajiang kan jiazhi guilu de zuoyong' (The Effect of Raising Coal Resource Prices on Law of Value) *WZGL*, no. 2 (1985): 2; *CIYB 1984*, p. 25.
299 'Coal Policy To Be Readjusted', *ER*, 2 (1994): 32.
300 Hu Benzhe, 'Guanyu Shanxi nengyuan jidi meitan shusong fangshi de tantao', p. 2.
301 Lian Yitong, 'The Development and Distribution of Transport and Communications', in Sun Jingzhi (ed.), *The Economic Geography of China*, Hong Kong: Oxford University Press, 1988, pp. 355–6.
302 Hou Bangan, 'Under-Funded Transport Network', p. 4.
303 *TJNJ 1999*, p. 519.
304 'Wei jing huan zhao de meitan yunshu qiye bude jixu jingying chusheng meitan' (Illegal Transport of Coal Cannot Continue), *SXRB*, 13 Oct. 1988; Zhao Keming, 'Zhizhi meidao' (Wipe Out Coal Loss), *SXRB*, 8 Nov. 1988; 'State Coal Moves Hit by Side Deals', *CD*, 8 Sept. 1988, p. 3; 'Meitan caigou renyuan de kunao' (The Vexation of Coal Buyers), *SXRB*, 4 Nov. 1988; 'Shanxi zhengdun chu sheng meitan jingying qiye' (Shanxi Re-Organizes Coal Transport Management), *RMRB*, 5 Sept. 1988, p. 1.
305 Zhu Changbo and Me Peiji, 'Yunjia gaige shi tielu zouxiang shichang de guanjian' (Pricing Reform Is the Key for Railways Entering the Market), *TDYSJJ*, 2 (1995): 15–17.
306 Li Rongxia, 'Railway Industry: Turning from Deficit to Profit', *BR*, 13 Mar. 2000, p. 14.
307 *TJNJ 1999*, p. 504.
308 Hou Bangan, 'Under-Funded Transport Network', p. 5.
309 *NYFZBG 1997*, p. 204.

310 The trunk of the Changjiang was to handle 1,000-ton ships up to Shuifu (Yunnan), 3,000-ton to Yichang (Hubei), 5,000-ton to Wuhan (Hubei), and 25,000-ton to Nanjing (Jiangsu). The Minjiang, Jialingjiang, Xiangjiang, Hanjiang, Ganjiang, Xinjiang, and Nanfei tributaries were to handle 300- to 500-ton ships. On the Pearl River (Zhujiang) 5,000-ton ships were to be able to travel as far as Nanning (Guangxi). The Xijiang, a tributary of the Pearl River, was also being dredged to handle larger coal ships, as was the Liangshan (Shandong)-Qiantangjiang (Zhejiang) section of the Grand Canal, the Huai River which connects with the Grand Canal, and the Songhua and Mudan Rivers (both in Heilongjiang).

311 Hou Bangan, 'Under-Funded Transport Network', p. 4.

312 IEA, *Coal Information 1998*, Paris: OECD/IEA, 1999, pp. 1.88, I.90, I.98.

313 *TJNJ 1988*, p. 423 and *TJNJ 2000*, p. 239. The series of figures presented in the latter have undergone considerable adjustment from those in previous yearbooks.

314 Robert Thomson, 'Import Growth Likely to Decrease', *FT*, 29 Sept. 1986, p. VI.

315 Carolyn Dowling, 'Producing Aluminum against the Odds', *CBR*, May–June 1986, p. 62; Stefan Wagstyl, 'Price Fall Cuts Tin Trading', *FT*, 29 Sept. 1986, p. IV.

316 *TJNJ 1988*, p. 808; *TJNJ 2000*, p. 327.

317 The author has examined the availability of electricity in different areas of China on an annual basis since 1994.

318 Calculated from *TJNJ 1992*, pp. 77, 469 and *TJNJ 1999*, pp. 111, 247.

319 IEA, Coal Industry Advisory Board, *Coal in the Energy Supply of China*, Report of the CIAB Asia Committee, Paris: OECD/IEA, 1999, p. 23.

320 Sun Shangqing, 'Nengyuan jiegou' (Energy Structure), in Ma Hong and Sun Shangqing (eds), *Zhongguo jingji jiegou wenti yanjiu* (Problems in China's Economic Structure), Beijing, 1980, p. 277; Yu Quili, 'Arrangements for the 1979 National Economic Plan', *BR*, 20 July 1979, pp. 10–11; Wu Zhexin, 'Hubei meitan shichang yuan he zhouran jinzhang' (Causes of Sudden Angst in the Hubei Coal Market) *MTB*, 25 May 1995, p. 4; Han Jie, 'Meitan gongxu reng you yiding quekou, shui lai zuobi wenzhang' (Coal Shortages Are Certain: Who Can Provide the Solution) *MTB*, 11 May 1995, p. 4.

321 Derivation of these ratios is explained in Chapter 3.

322 IEA, *Global Energy: Changing the Outlook*, Paris: OECD/IEA, 1992, pp. 110–111.

323 *World Development Report 1996*, pp. 202–3.

324 Calculated from information in Jonathan E. Sinton *et al.* (eds), *China Energy Databook*, Table IV.21, p. IV–76.

325 *NYFZBG 1997*, p. 226.

326 Final consumption means direct use of coal, while intermediate means consumption in transformation into electricity, heat, coke, or gas.

327 No definition of 'other' is provided in the *TJNJ*.

328 *TJNJ 1988*, p. 808; *TJNJ 2000*, pp. 316, 247.

329 *TJNJ 2000*, p. 352.

330 Preeti Soni and Vishal Narain, 'India: Energy and Environment Situation', in Preeti Soni (ed.), *Energy and Environmental Challenges in Central Asia and the Caucasus: Windows for Cooperation*, New Delhi: Tata Energy Research Institute, 1996, pp. 167–8, quoting *Current Energy Scene in India*, Bombay: Centre for Monitoring Indian Economy, 1994.

331 IEA, Coal Industry Advisory Board, *Coal in the Energy Supply of China*, p. 71.

332 Author's interview of an official from the China Power Corporation, Sept. 1998.

333 *NYFZBG 1997*, p. 128, quoting State Statistical Bureau, Ministries of Agriculture and Electric Power.

334 CERI/TERI, *Planning for the Indian Power Sector: Environmental and Development Considerations*, p. 13.

335 IEA, *Energy Balances of Non-OECD Countries 1996–1997*, pp. II490–95.

336 IEA, *Coal Information 1998*, p. I.116. The definition of 'thermal' for the OECD data includes coal, oil, natural gas and combustible renewable and wastes. Another category not included here is geothermal, solar, wind and tide and wave.

337 CERI/TERI, *Planning for the India Power Sector*, p. 4.

338 The Qinshan Nuclear Power Station, with an installed capacity of 300,000 kW, began operation in 1992, and the Daya Bay Nuclear Power station, with an installed capacity of 1.8 million kW, in 1994. Wu Naitao, 'Energy Sector Seeks More Foreign Cooperation', *BR*, 24–30 April, 1995, p. 7.

339 Lin Senmu, Zhou Shulian, and Qi Mingchen, 'Industry and Transport', in Yu Guangyuan (ed.), *China's Socialist Modernization*, Beijing: Foreign Languages Press, 1984, p. 311.

340 Calculated from Jonathan E. Sinton *et al.*, (eds), *China Energy Databook*, Table IV.12, p. IV–59. The other fuels used included gasoline, diesel oil, LPG, refinery gas, natural gas, coal gas, and other coking products.

341 CERI/TERI, *Planning for the India Power Sector*, p. 71.

342 Jonathan E. Sinton *et al.* (eds) *China Energy Databook*, Table II.21, p. II–70.

343 Ibid., Table IV.13, p. IV–62; *GYJJTJNJ 1998*, p. 59.

344 Ye Qing, *ZGNY*, 5 (1989): 40, quoted by David Fridley, 'China's Energy Outlook', in US Congress, JEC, *China's Economic Dilemmas in the 1990s: The Problems of Reforms, Modernization and Interdependence*, Armonk, NY: M. E. Sharpe, 1993, p. 518.

345 *NYFZBG 1997*, p. 102.

346 *GYJJTJNJ 1998*, p. 59.

347 Ibid., and *DLNJ 1998*, p. 672. (The two sources do not agree on the 1997 figure.)

348 *DLNJ 1998*, p. 758.

349 Vincent Chan, *Asian Power Sector: Make Way For Change*, Hong Kong: Peregrine Brokerage Ltd., 1995, p. 29.

350 Ibid., p. 31; IEA, Coal Industry Advisory Board, *Coal in the Energy Supply of China*, citing IEA, *World Energy Outlook 1998*, OECD, Paris, 1998, p. 275.

351 Author's interview of an official from the China Power Corporation. For perspective, the average gross efficiency of Indian thermal power stations was about 28 per cent, the average net efficiency was only 25 per cent, and about 75 per cent of the country's thermal capacity operated at efficiencies below 30 per cent. CERI/TERI, *Planning for the India Power Sector*, p. xlvi. (For reference, the maximum attainable thermal efficiency anywhere is about 55 per cent.)

352 Vincent Chan, *Asian Power Sector: Make Way For Change*, p. 30.

353 Organization for Economic Cooperation and Development/IEA, Ad Hoc Group on International Energy Relations, China [draft paper], Paris: 1987, p. 48.

354 Huang Zhiming, 'The Development of High Voltage Overhead Transmission Line Technology in China', *Electricity* (Chinese Society for Electrical Engineers), vol. 10, no. 1 (1999): 39.

355 *DLNJ 1998*, p. 762.

356 CERI/TERI, *Planning for the India Power Sector*, p. 134.

357 *GYJJTJNJ 1998*, p. 59; Vincent Chan, *Asian Power Sector: Make Way For Change*, pp. 30–1. Chan quotes an unspecifed World Bank report published in 1993 which states that, for various reasons, the true figure is more likely to be in the 16–20 per cent range. A spokesman for the Northeastern Electricity Group is reported to have said that about half of the companies in the service industry including restaurants, hotels, and ballrooms, were stealing electricity, some simply by plugging into transformers. 'Chinese Factory Official Detained for Electricity Theft', *Xinhua*, 0249 GMT, 2 Nov. 1998. According to CERI/TERI, *Power Planning for the India Power Sector*, p. 135, of the 23 per cent transmission and distribution losses in India, 4 per

cent are transmission and 19 per cent distribution. Of the latter, non-technical commercial losses amounted to 5 per cent.

358 Hao Fengyin, 'Meitan gongye de jieneng yu cuoshi' (Coal Industry Conservation and Measures), *NY*, no. 1 (1984): 7–9; Liu Zhiping, 'Wo guo jin shi nian jie neng xiangmu de touzi, xiaoyi pingjia' (Appraisal of China's Investment in Energy Conservation Over the Past Ten Years), *ZGNY*, no. 6 (1993): 35–8, 43; Wang Xianglin, 'Jieyue nengyuan sho tigao jingji xiaoyi de zhongyau tujing' (Saving Energy is an Important Way of Enhancing Economic Performance), *NY*, no. 3 (1984): 6–8; Zhou Peinian, 'Luelun jieneng guanli gaige' (Brief Discussion on Reforming Energy Conservation Management), *NY*, no. 1 (1985): 1–4.

359 Cao Zhengyan, 'Tigao nengyuan xiaolu Zhongguo qude de chengjiu he zhengfu de fangzhen cuoshi' (Raise China's Energy Efficiency: Achieve Results from Government Policies and Measures), *ZGNY*, no. 3 (1997): 9.

360 'An Effective Means of Easing the Strain on Energy Resources Is to Save Energy', *JJCK*, 29 June 1989, p. 1, in *FBIS*, 14 July 1989, pp. 33–4.

361 Ye Qing, 'Xiang guowuyuan jieneng bangong huiyi di wuzi huiyi de huibao tigong' (Report on the Fifth Official Energy Conservation Meeting of the State Council), *ZGNY*, no. 5 (1989): 5.

362 Tatsu Kambara, 'The Energy Situation in China', *CQ*, 131 (1992): 616, citing information provided by the Institute.

363 Jonathan E. Sinton and Mark D. Levine, 'Changing Energy Intensity in Chinese Industry: the Relative Importance of Structural Shift and Intensity Change', *Energy Policy* (Mar. 1994): 252–3.

364 World Bank, *Clear Water, Blue Skies*, p. 47, citing World Bank, *China: Energy Conservation Study*, Report 10813-CHA, Washington, DC: 1993.

365 Chan Hing-lin and Lee Shu-kam, 'Energy Consumption in Reforming China', Working Paper Series no. CP94015, Hong Kong: Hong Kong Baptist College, School of Business, Business Research Centre, Nov. 1994.

366 Lin Xiannuan. *China's Energy Strategy: Economic Structure, Technological Choices and Energy Consumption*, Westport, CT: Praeger, 1996, p. 167.

367 IEA, *World Energy Outlook 1998*, pp. 276–80 citing Angus Maddison, *Measuring Chinese Economic Growth and Levels of Performance*, Paris: OECD, 1997 and Ren Ruoen, *China's Economic Performance in an International Perspective*, Paris: OECD, 1997.

368 Richard F. Garbaccio, Mun S. Ho, and Dale W. Jorgenson, 'Why Has the Energy-Output Ratio Fallen in China?', *The Energy Journal*, vol. 20, no. 3 (1999): 63–91.

369 Ye Qing, 'Muqian jieneng xing shi he jin yibu jiaqiang jieneng gongzuode ruogan cuoshi' (Present Situation and Measures for Strengthening Energy Conservation Work) (Report Excerpts of the State Council's Sixth Official Energy Conservation Meeting), *ZGNY*, no. 1 (1991): 3–5.

370 Song Xuefeng, 'The Role of Clean Coal and Efficient Use of Coal in the Future Energy and Environment of China', in IEA, *The Clean and Efficient Use of Coal and Lignite: Its Role in Energy, Environment and Life*, Proceedings of a Conference held in Hong Kong, 30 Nov.–3 Dec. 1993, Paris: OECD/IEA, 1994, p. 1084; John M. Topper, 'Improving Coal Use in Developing Countries through Technology Transfer', in IEA, *The Clean and Efficient Use of Coal and Lignite*, p. 779; Wolfgang Michalski, Riel Miller, and Barrie Stevens, 'China in the Twenty-First Century: An Overview of the Long-Term Issues', in Organization for Economic Cooperation and Development, *China in the 21st Century: Long-Term Global Implications*, Paris: 1996, p. 15; World Bank, *Clear Water, Blue Skies*, p. 49, quoting Wang Qingyi, Jonathan Sinton, and Mark Levine, *China's Energy Conservation Policies and Their Implementation, 1980 to the Present, and Beyond*, Lawrence Berkeley National Laboratory, Berkeley, CA, 1995.

371 World Bank, *Clear Water, Blue Skies*, p. 48–9 quoting Wang Qingyi, Jonathan Sinton, and Mark Levine, *China's Energy Conservation Policies and Their Implementation, 1980 to the Present, and Beyond*, Berkeley, CA: Ernest Orlando Lawrence Berkeley National Laboratory, 1995.

7 Retrospect and prospect

1 Calculated from *TJNJ 1999*, p. 57. There is disagreement among economists as to the relative size of the economy. According to economists at the United Nations, after pricing parity adjustments, China was the world's second largest economy in 1993. (*The Economist*, 15 May 1993). Ren Ruoen also believes China's economy is actually much bigger than usual indicators suggest. Ren Ruoen, *China's Economic Performance in an International Perspective*, Paris: OECD Development Centre, 1997.
2 Tong Ji and Bei Qing, 'Changes in 50 Years', *BR*, 27 Sept. 1999, p. 14.
3 'China's Top 500 Import and Export Enterprises in 1998', *BR*, 6 Sept. 1999, p. 24, citing statistics from the World Trade Organization Secretariat.
4 Embassy of the People's Republic of China, Washington, DC, www.china-embassy.org/eng/index.html, accessed June 1997.
5 Bruce Gilley, 'A Mine Full of Possibilities', *FEER*, 13 July 2000, pp. 62–3.
6 Ray Bashford and Mark O'Neill, 'Lack of Investment Poses Major Challenge', *SCMP China Business Review*, 30 Oct. 1997, p. 2.
7 Ibid.
8 China exported the technology to build a 300 MW plant in Pakistan. Li Qi, 'Nuclear Power in China's Five Year Plan', *Zhongguo jingji daobao*, 30 Dec. 1998, p. B3, translated in SWB Part 3: Asia-Pacific, FEW0575, 10 Feb. 1999.
9 Li Qi, 'Nuclear Power in China's Five Year Plan', 'Official Says Country Not to Start Nuclear Power Projects in Next Three Years', broadcast on *Xinhua*, 0826 GMT, 29 Apr. 1999, in SWB Part 3: Asia-Pacific, FED3522, 30 Apr. 1999.
10 'Official Says Country Not to Start Nuclear Power Projects in Next Three Years', broadcast on *Xinhua*, 0826 GMT, 29 Apr. 1999, in SWB Part 3: Asia-Pacific, FED3522, 30 Apr. 1999.
11 'Lack of Anti-Nuclear Lobby Gives China Chance for Cleaner Power', broadcast on *Kyodo News Service*, 0104 GMT, 30 May 1999, in SWB Part 3: Asia-Pacific, FEW0591, 2 June 1999. It should be noted that Hong Kong residents have periodically protested the Daya Bay facility.
12 World Bank, *Clear Water Blue Skies: China's Environment in the New Century*, China 2020 Series, Washington: IBRD/WB, 1997, Table 4.3, p. 49.
13 'Opportunities for Foreign Investors in China's Offshore Oil Sector', *Xinhua*, 10 Mar. 1999.
14 'Natural Gas Seminar Looks Ahead', *Asiainfo Daily China News*, 16 Apr. 1999; 'China Set to Markedly Boost Gas Output by Year 2000', *Asia Pulse*, 18 Nov. 1998.
15 'More Renewable Energy to be Used', *BR*, 24 July 2000, p. 6.
16 'Use of Solar Energy to be Widened', *BR*, 13 Mar. 2000, pp. 31–2.
17 Zhang Zhongxiang, *The Economics of Energy Policy in China: Implications for Global Change*, Cheltenham, UK: Edward Elgar, 1998, p. 53, citing the [China] State Economic and Trade Commission, 1996; IEA, *World Energy Outlook 1998*, Paris: OECD/IEA, 1998, p. 290 citing Ministry of Power; Ray Bashford and Mark O'Neill, 'Lack of Investment Poses Major Challenge', *SCMP China Business Review*, 30 Oct. 1997, p. 2.
18 Zhang Zhongxiang, *The Economics of Energy Policy in China*, p. 53, citing [China] State Economic and Trade Commission 1996.

19 This is as much a cultural question as it is an economics one. Different countries, for a variety of reasons, have different aesthetics when it comes to the use of electricity.

20 'China's Challenge', *SCMP*, 10 Feb. 1998, p. 16; Jiang Zemin, 'Hold High the Great Banner of Deng Xiaoping Theory for an All-Round Advancement of the Cause of Building Socialism with Chinese Characteristics into the 21st Century', Report Delivered at the 15th National Congress of the Communist Party of China, 12 Sept. 1997, Printed in *BR*, 6–12 Oct. 1997, p. 12.

21 *TJNJ 2000*, p. 239.

22 *CD Business Weekly*, 2 Mar. 1998.

23 Mark Buczek, 'Fueling China's Growth', *CBR*, Sept.–Oct. 1996, pp. 8–15.

24 'Acid Rain Takes High Toll', *SCMP*, 8 May 1999.

25 IEA, *Coal Information 1997*, Paris: OECD/IEA, 1998, p. I–202.

26 The Chinese oil industry was greatly affected by the dramatic fall in international oil prices in 1998. The drop to about half of the level of Chinese prices, caused smuggling, especially of diesel fuel, to become rampant. Planned imports of crude oil were cut, imports of diesel and gasoline were banned, and smugglers who were caught were treated severely in order to restore balance in the nationwide production and stocks.

27 IEA, *World Energy Outlook 1998*, Paris: IEA/OECD, 1998, Table 7.2, p. 85.

28 John V. Mitchell and Christiaan Vrolijk, 'Closing Asia's Energy Gaps', Energy and Environmental Programme, Briefing No. 41, March 1998, p. 5, The Royal Institute of International Affairs, on webpage: www.riia.org/briefingpapers/bp41.htm, accessed Oct. 1999.

29 Officials from the Ministry of Coal interviewed by the author in August 1998.

30 'Coal Production Falls 14.5pc', *SCMP*, 24 Apr. 2000.

31 IEA, *World Energy Outlook 1998*, pp. 280 and 437.

Epilogue

1 Raymond Li, 'Guangdong Looks West for Solution to Power Problem', *SCMP*, 12 Sept. 2000.

2 *CMS*, no. 3 (2001).

3 *ZGWJ*, 2000 and 2001 issues.

4 *CMS*, no. 3 (2001).

5 For more information on the status of CCTs in China see: www.cct.org.cn/eng/index.htm and tradeport.org/ts/countries/china/isa/isar0034.html.

6 Tim Mitchell, 'Mainland Plans to Turn Coal into Oil', *SCMP*, 22 Sept. 2000; 'China Boosts "Coal-for-Oil" to Curb Pollution, Oil Shortage', www.peopledaily.com.cn, accessed Sept. 2000.

7 *Zhongguo xinwen she*, 23 Mar. 2001, reported on www.chinaonline.com.

8 The Kyoto Protocol resulted from the Kyoto Conference on Climate Change held November 1997. According to the Protocol the EU would cut greenhouse gas emissions before 2012 by 8 per cent of 1990 levels, the US by 7, and Japan by 6.

9 Some scientists contend it does not exist at all, and some believe it is a natural, unavoidable phenomenon. Others are certain that man-made emissions of greenhouse gases are causing global warming to accelerate, while others believe that they are not the problem.

10 IEA, *Electricity Information 2000*, Paris: OECD/IEA, 2000, p. II.683.

11 In July 2000 the Chinese government began to adjust domestic oil prices on a monthly basis in an attempt to keep them on par with international oil prices. Oil prices were low in 1998 due to a mild winter in North America and the fall in demand in Asia due

to the Asian crisis. In February 1999, the international price was only US$10. The following month OPEC cut production by about 4.3 million barrels per day, amounting to about 6 per cent of world output. By then Asia was recovering from the crisis, and early winter struck North America. By the end of 1999, the world was consuming 1.4 million barrels per day more than it was producing and prices reached US$38.

12 *CMS*, no. 3 (2001).

13 Jasper Becker, 'Coast Pushes for Nuclear Power', *SCMP*, 15 Mar. 2001.

Bibliography

Primary sources

In September 1987 the author interviewed some twenty-five officials from the Chinese Academy of Social Sciences (Beijing), the Shanxi Academy of Social Sciences (Taiyuan), the Shanxi Coal Administration Bureau of the Ministry of Coal Industry, the Shanxi Provincial Economic Commission, the Shanxi Provincial Science and Technology Commission, the Shanxi Mining College, and the China Pingshuo Surface Coal Corporation. In August 1998 she interviewed two officials from the Coal Industry Administration, one from China Power Corporation, and a senior faculty member at Beijing Mining University.

Books and articles in English

'Administrative Procedures for Permits on Coal Production', broadcast on *Xinhua*, 0049 GMT, 27 Dec. 1994, in *FBIS*, 13 Jan. 1995, pp. 34–7.

Albouy, Yves, *Coal Pricing in China: Issues and Reform Strategy*, World Bank Discussion Papers, China and Mongolia Department, Series no. 138, Washington, DC: International Bank for Reconstruction and Development/World Bank, 1991.

Ambler, John *et al.*, 'Soviet Railways: Lethargy or Crisis?', in John Ambler, Denis J. B. Shaw and Leslie Symons (eds), *Soviet and East European Transport Problems*, London: Croom Helm, 1985, pp. 24–89.

'Antaibao Mine Partners Work to Overcome Development Obstacles', *CWI*, 24 Feb. 1988.

'Antaibao Partners in Marketing Dispute', *CWI*, 13 July 1988.

Astakhov, A. and A. Grubler, *Resource Requirements and Economics of the Coal-Mining Process: A Comparative Analysis of Mines in Selected Countries*, Laxenburg, Austria: International Institute for Applied Systems Analysis, 1984.

Baum, Richard, 'The Fifteenth National Party Congress: Jiang Takes Command?', *CQ*, 153 (Mar. 1998), pp. 141–56.

'Beijing Seeks Foreign Funds for Coal Projects', broadcast on *Xinhua*, 1023 GMT, 8 Sept. 1995, in *FBIS*, 12 Sept. 1995, pp. 22–3.

'Beijing's First Coal Wholesale Market Opens', broadcast on *Xinhua*, 0246 GMT, 15 Apr. 1992, in *FBIS*, 15 Apr. 1992, p. 61.

Berliner, Joseph, *Soviet Industry from Stalin to Gorbachev: Essays on Management and Innovation*, Ithaca, NY: Cornell University Press, 1988.

Bornstein, Morris, 'Soviet Price Policy in the 1970s', in *Soviet Economy in a New Perspective*, JEC Report, 1976, pp. 17–66.

—— 'Soviet Price Theory and Policy', in *New Directions in the Soviet Economy*, JEC Report, 1966, pp. 65–98.

British Petroleum Amoco, *British Petroleum Statistical Review of World Energy*, London, various years.

Buczek, Mark, 'Fuelling China's Growth', *CBR*, Sept.–Oct. 1996, pp. 8–15.

Buxton, Neil K., *The Economic Development of the British Coal Industry*, London: B. T. Batsford, 1978.

Campbell, Harry (ed.), *Cinfolink Annual Review of Information Services and the Internet in China*, Toronto: Cinfolink Services, 1996.

Campbell, Robert W., *Soviet Economic Power: Its Organization, Growth and Challenge*, London: Stevens, 1960.

—— *Soviet Energy Technologies: Planning, Policy Research and Development*, Bloomington: Indiana University Press, 1980.

—— *Trends in the Soviet Oil and Gas Industry*, Baltimore: John Hopkins Press, 1976.

Canadian Energy Research Institute (Calgary, Alberta, Canada) and TATA Energy Research Institute (New Delhi), *Planning for the Indian Power Sector: Environmental and Development Considerations*, Study No. 62, New Delhi: The Institutes, 1995.

Carin, Robert, *Power in Communist China, Communist China Problem Research Series EC44*, Hong Kong: Union Research Institute, 1969.

Carlsen, Eric Hills, *Soviet Aid and the Development of Communist China's Coal Industry, 1949–1960*, Ph.D. diss., Cornell University, 1969.

Chai, C. H., *China: Transition to a Market Economy*, Oxford: Clarendon, 1997.

'The Challenges and Opportunties Faced by China's Coal Industry', *SJMTJS*, 2 (Feb. 1985): 3–6, in *JPRS*, 2 (Feb. 1985): 51–60.

Chan Hing-lin and Lee Shu-kam, 'Energy Consumption in Reforming China', Working Paper Series no. CP94015, Hong Kong: Hong Kong Baptist College, School of Business, Business Research Centre, Nov. 1994.

Chan, Vincent, *Asian Power Sector: Make Way for Change*, Hong Kong: Peregrine Brokerage Ltd., 1995.

Chang, John K., *Industrial Development in Pre-Communist China: A Quantitative Analysis*, Edinburgh: University of Edinburgh Press, 1969.

Chao Yu-shen, 'The Thirty-Point Decision on Industry: An Analysis', *Issues and Studies*, 15, 1 (Jan. 1979): 48–63.

Chen Yun, 'Talk at the Forum on Coal Work (Oct. 1961)', in Nicholas R. Lardy and Kenneth Lieberthal (eds), *Chen Yun's Strategy for China's Development: A Non-Maoist Alternative*, Armonk, NY: M. E. Sharpe, 1983.

Cheng Chu-yuan, *China's Economic Development: Growth and Structural Change*, Boulder, CO: Westview, 1982.

—— *The Demand and Supply of Primary Energy in Mainland China, Mainland China Economic Series no. 3*, Seattle: University of Washington Press, 1985.

'China Encourages Use of Clean Energy', *Xinhua*, 2 May 1999.

China Facts and Figures Annual, Gulf Breeze, FL: Academic International Press, 1978-. Various years.

'China "Greatly" Cuts Coal Supply to DPRK', *Tong-a ilbo*, 1 June 1993, p. 2, in *FBIS*, 2 June 1993, p. 9.

China Handbook Editorial Committee, *Geography, China Handbook Series*, Beijing: Foreign Languages Press, p. 108.

'China: Illegal Coal Mines to Close in Shanxi', broadcast on *Xinhua*, 0752 GMT, 22 Dec. 1998.

'China in $900 Million BOT Coal Slurry Pipeline Deal', *Power Asia*, 5 Sept. 1994.

'China Steps up Exploration of Coal-Bed Methane Resources', *Asia Pulse*, 9 July 1998.

'China Suspends Shipments', *ICR*, 23 Sept. 1988.

'China to Boost Coal Gas Energy with Preferential Policy Initiative', broadcast on *Xinhua*, 0251 GMT, 20 Nov. 1998.

'China to Build Major Coal Slurry Project with Aid of Consortium', *CWI*, 15, 35 (30 Aug. 1994): 2.

China Unified Distribution Coal Mine Corporation, Energy Conservation Office, 'Deal with Concrete Matters Relating to Work and Innovate, Open Up and Advance, Push Energy Conservation Work to a New Level During the Eighth Five Year Plan', *ZGNY*, no. 9 (1991): 3–7, in *JPRS*, 11, 22 Nov. 1991, p. 27.

'Chinese Army Involved in Illegal Reselling of Coal', broadcast on *Zhongguo tongxun she*, 0634 GMT, 4 Apr. 1989, in *FBIS*, 7 Apr. 1989, p. 34.

Ching, Frank, 'Kyoto Pact Is a Good First Step', *FEER*, 25 Dec. 1997–1 Jan. 1998, p. 98.

Ching Chi, 'Miners in Heilongjiang Protest Wage Payment Default, One Commits Suicide', *Ming Bao*, 21 Feb. 1993, p. 25, in *FBIS*, 23 Feb. 1993, pp. 80–1.

Choe Boum-Jong, *A Model of World Energy Markets and OPEC Pricing*, World Bank Staff Working Paper, no. 633, Washington, DC: International Bank for Reconstruction and Development/World Bank, 1984.

Clarke, William, 'China's Electric Power Industry', in *Chinese Economy Post-Mao*, JEC Report, 1978, pp. 427–33.

'Coal Consumer Debt Taxes Industry', broadcast on *Zhongguo xinwen she*, 1240 GMT, 14 Jan. 1994, in *FBIS*, 14 Jan. 1994, p. 38.

'Coal Demand Expected to Exceed Supply', broadcast on *Zhongguo xinwen she*, 0455 GMT, 23 Sept. 1995, in *FBIS*, 25 Sept. 1995, p. 37.

'Coal Industry Attracts Foreign Capital', broadcast on *Xinhua*, 1132 GMT, 8 June 1994, in *FBIS*, 9 June 1994, p. 45.

'Coal Industry To Further Open to Outside', broadcast on *Xinhua*, 1411 GMT, 9 Sept. 1995, in *FBIS*, 11 Sept. 1995, p. 33.

'Coal Ministry Hails Payment Policy, Debt. Reduction', *Zhongguo tongxun she*, 0927 GMT, 5 Feb. 1995, in *FBIS*, 16 Feb. 1995, p. 49.

'Coal Ministry Offers 300 Projects to Foreign Investors', *Chingchi taobao*, 23 Oct. 1995, p. 44, in *FBIS*, 6 Dec. 1995, p. 86.

'Coal Policy To Be Readjusted', *ER*, 2 (1994): 32.

'Coal Sector Makes Plans to Suppress Production', *Zhongguo xinwen she*, 19 Apr. 1999, in SWB Part 3: Asia-Pacific, FEW0586, 28 Apr. 1999.

Cohn, Stanley H., 'Sources of Low Productivity in Soviet Capital Investment', in *Soviet Economy in the 1980s: Problems and Prospects, Part 1*, JEC Report, 1982, pp. 169–94.

Crosfield (Joseph G.) and Sons Ltd. See Joseph G. Crosfield and Sons Ltd.

Daniel, Martin, *Chinese Coal Prospects to 2010*, IEAPER/11, London: International Energy Agency, 1994.

Dienes, Leslie and Theodore Shabad, *The Soviet Energy System: Resource Use and Policies*, Washington, DC: V. H. Winston, 1979.

Dorian, James P. and David. G. Fridley (eds), *China's Energy and Mineral Industries: Current Perspectives*, Westview Special Studies in Natural Resources and Energy Management, Boulder: Westview, 1988.

Dowling, Carolyn, 'Producing Aluminium Against the Odds', *CBR*, May–June 1986, p. 62.

Doyle, Guy, *China's Potential in International Coal Trade*, London: International Energy Agency Coal Research, 1987.

'Draft Decision Concerning Some Problems in Speeding Up the Development of Industry' (The '30-Point Decision on Industry'), Part II, *Issues and Studies*, xv, 1 (1979): 69–98.

Edmonds, Richard Louis, 'The Environment in the People's Republic of China 50 Years On', *CQ*, 159 (1999): 640–9.

'An Effective Means of Easing the Strain on Energy Resources Is to Save Energy', *JJCK*, 29 June 1989, p. 1, in *FBIS*, 14 July 1989, 33–4.

'Eighth Five-Year Plan To Be Readjusted', *ER*, no. 11 (1993).

Elliot, Iain F., *The Soviet Energy Balance: Natural Gas, Other Fossil Fuels and Alternative Power Sources*, New York: Praeger, 1974.

'Energy Industries Top 1990 Production Targets', broadcast on *Xinhua*, 1751 GMT, 6 Jan. 1991, in *FBIS*, 7 Jan. 1991, pp. 36–7.

'Energy Ministry Works To Combat Power Shortages', broadcast on *Zhongguo tongxun she*, 0637 GMT, 20 Jan. 1992, in *FBIS*, 29 Jan. 1992, p. 48.

Fan Qimiao and Peter Nolan, *China's Economic Reforms: The Costs and Benefits of Incrementalism*, New York: St. Martin's, 1994.

The Far East and Australasia, 29th ed, *1998*, London: Europa.

Field, Robert Michael and J. A. Flynn, 'China: An Energy-Constrained Model of Industrial Performance through 1985', in *China Under the Four Modernizations*, JEC Report, 1982, pp. 334–64.

Field, Robert Michael, Nicholas R. Lardy, and John P. Emerson, *A Reconstruction of the Gross Value of Industrial Output by Province in the People's Republic of China: 1949–1973*, Foreign Economic Report 7 (July 1976), Washington, DC: US. Dept. of Commerce, Bureau of Economic Analysis, 1976, pp. 2–8.

'First Coal Commodities Market To Open 28 October', broadcast on *Xinhua*, 0720 GMT, 27 Oct. 1992, in *FBIS*, 27 Oct. 1992, p. 35.

Flavin, Christopher, 'The Legacy of Rio', in Linda Stark (ed.), *State of the World 1997: A Worldwatch Institute Report on Progress Toward a Sustainable Society*, New York: W. W. Norton, 1997, pp. 9–22.

Forney, Matt, 'Jiang's Big Bet', *FEER*, 25 Sept. 1997, pp. 14–15.

Fridley, David, 'China's Energy Outlook', in US Congress, JEC, *China's Economic Dilemmas in the 1990s: The Problem of Reforms, Modernization and Interdependence*, Armonk, NY: ME Sharpe, 1993, pp. 495–526.

Gao Yangwen, 'On the Issue of China's Energy Policy', *MTKXJS*, no. 4 (Apr. 1981): 4–8, in *JPRS*, 166 (1981): 13–22.

—— 'Minister Unveils Grand Strategy to Revamp Coal Industry', *MTKXJS*, 10 (1982): 2–4, in *JPRS*, 336 (1983): 82–7.

Garbaccio, Richard F., Mun S. Ho, and Dale W. Jorgenson, 'Why Has the Energy–Output Ratio Fallen in China?', *The Energy Journal*, vol. 20, no. 3 (1999): 63–91.

Gerschenkron, Alexander, *Economic Backwardness in Historical Perspective*, New York: Frederick A. Praeger, 1962.

Goldman, Merle, *Sowing the Seeds of Democracy in China: Political Reform in the Deng Xiaoping Era*, Cambridge: Harvard University Press, 1994.

Goldman, Merle and Roderick MacFarquhar (eds), *The Paradox of China's Post-Mao Reforms*, Cambridge, MA: Harvard University Press, 1999.

Granick, David, *Management of the Industrial Firm in the USSR: A Study of Soviet Economic Planning*, New York: Columbia University Press, 1954.

Greene, Robert P. and J. Michael Gallagher (eds), *Future Coal Prospects: Country and Regional Assessments, World Coal Study*, Cambridge, MA: Ballinger, 1980.

Gu Dalin, 'Does China Really Have Too Much Coal? – A Survey of Coal Markets, Shipping and Sales', *MTB*, 21 May 1986, pp. 1–2, in *JPRS*, 8 (1987): 61.

Gustafson, Thane, *Crisis Amid Plenty: The Politics of Soviet Energy under Brezhnev and Gorbachev*, Princeton, NJ: Princeton University Press, 1989.

Haberstroh, John R., 'Eastern Europe: Growing Energy Problems', in *East European Economies Post-Helsinki*, JEC Report, 1977, pp. 379–95.

Hamburger, Jessica, 'Lighting the Way for Energy Savings', *CBR*, Nov–Dec. 1993, pp. 42–4.

Hammer, Armand with Neil Lyndon, *Hammer Witness to History*, London: Simon and Schuster, 1987.

Hanan, Kate, *Industrial Change in China: Economic Restructuring and Conflicting Interests*, London/New York: Routledge, 1998.

Hattori, Kenji, 'Trends in China's Coal Development and Japan-China Coal Trade', *China Newsletter*, no. 43 (Mar.–Apr. 1983): 14–21.

Hay, Donald, Derek Morris, Guy Liu and Shujie Yao, *Economic Reform and State-Owned Enterprises in China, 1979–1987*, Oxford: Clarendon, 1994.

Henderson, P. D. *India: The Energy Sector*, Delhi: Oxford University Press, 1975.

Hewett, Ed A., *Energy, Economics and Foreign Policy in the Soviet Union*, Washington, DC: Brookings Institute, 1984.

Hollingworth, B., *Railways of the World*, New York: Gallery, 1979.

Hou Bangan, 'Under-Funded Transport Network Fails to Keep up with Economic Growth', *Keji daobao* (Science and Technology Review), 10 Dec. 1996, pp. 46–50, published in SWB, Part 3: Asia-Pacific, FEW0482, 16 Apr. 1997.

Hou Chi-ming, 'Manpower, Employment and Unemployment', in Alexander Eckstein, Walter Galenson, and Ta-Chung Liu (eds), *Economic Trends in Communist China*, Edinburgh: Edinburgh University Press, 1968, pp. 329–96.

Howe, Christopher, *China's Economy: A Basic Guide*, London: Granada, 1978.

Howe, Christopher and Kenneth R. Walker, *The Foundations of the Chinese Planned Economy: A Documentary Survey, 1953–65*, London: Macmillan, 1989.

—— 'Mao the Economist', in Dick Wilson (ed.), *Mao Tse-Tung in the Scales of History*, Cambridge: Cambridge University Press, 1977, pp. 174–222.

Hsieh, C. Y. and M. C. Chu, 'Foreign Interest in the Mining Industry', in China Institute of Pacific Relations, *Economic Trends and Problems in the Early Republican Period*, Preliminary Paper Prepared for the 4th Biennial Conference of the Institute of Pacific Relations, Hangzhou, 21 Oct. – 4 Nov. 1931, Shanghai: The Institute, July 1931.

Huang Zhijie, 'Analysis of Our Nation's Energy Shortage', *Ziran bianzhengfa tongxun* (Journal of Dialectics of Nature), 4, 1 (1982): 2–4, in *JPRS*, 261 (1982): 3–8.

Huang Zhiming, 'The Development of High Voltage Overhead Transmission Line Technology in China', *Electricity* (Chinese Society for Electrical Engineering), 10, 1 (1999): 38–41.

Huenemann, Ralph W., *The Dragon and the Iron Horse: The Economics of Railroads in China 1876–1937*, Cambridge, MA: Council on East Asian Studies, Harvard University, 1984.

—— 'Modernizing China's Transport System', in US Congress, JEC, *China's Economic Dilemmas in the 1990s*, Armonk, NY: ME Sharpe, 1993, pp. 455–68.

Huenemann, Ralph W. and Nicholas H. Ludlow, 'China's Railroads', *CBR*, 4, 2 (1977): 27–43.

Hughes, T. J. and D. E. T. Luard, *The Economic Development of Communist China*, London: Oxford Univerisity Press, 1961.

Hunter, Holland, 'Transport in Soviet and Chinese Development', *Economic Development and Cultural Change*, 14 (Oct. 1965): 71–84.

—— 'Transportation in Soviet Development', in Gary Fromm (ed.), *Transport Investment and Economic Development*, Washington, DC: Brookings Institution, 1965, pp. 123–43.

Ichikawa, Minanmi, 'Coal Technology Options: International Cooperation and Partnership', in International Energy Agency, *The Clean and Efficient Use of Coal and Lignite: Its Role in Energy, Environment and Life*, Proceedings of a Conference held in Hong Kong, 30 Nov. –3 Dec. 1993, Paris: OECD/IEA, 1994, pp. 123–9.

Ikonnikov, Alexander B., *The Coal Industry of China*, Canberra: Research School of Pacific Studies, Australian National University, 1977.

'India: Coal Industry to Recover Dues from States via Annual Plan Outlay', broadcast on *PTI News Agency*, 1317 GMT, 6 May 1998.

'Is it Good to Have Excess Stored Materials?', *Da gongbao*, 19 Mar. 1966, p. 3, in *JPRS*, 262 (29 Apr. 1966): 1–5.

James, Peter, *The Future of Coal*, 2nd edn, London: Macmillan, 1984.

Janet Matthews Information Services 1998, Quest Economics Database, *Cambridge International Forecasts Country Report*, Dec. 1998: 'India: Mining'.

Janossy, Ferenc, 'The Origins of Contradictions in Our Economy and the Path to Their Solution' (translation of *Közgazdasagri Szemle*, 6, 1969), in *Eastern European Economics*, 8, 4 (Summer 1970): 357–90.

Jiang Zhenping, 'Energy Efficiency in China', a paper presented at the *Industrial Ecology and Global Change* conference sponsored by the Global Change Institute, held in Snowmass, CO, July 1992. The complete proceedings, including this paper, were published in Hong Kong by the Chinese University Press in 1994.

Jones, Dori, 'The Dawning of Coal's "Second Age"', *CBR*, May–June 1980, pp. 38–46.

Jordan, Constantin, *The Romanian Oil Industry*, New York: New York University Press, 1955.

Joseph G. Crosfield and Sons Ltd., *The Chinese Coal Industry*, Warrington, UK: 1961–62.

Kambara, Tatsu, 'The Energy Situation in China', *CQ*, 131 (1992): 608–36.

Keidel, Albert III, 'China's Coal Industry', in *China's Economy Looks Toward the Year 2000, Vol. II, Economic Openness in Modernizing China*, JEC Report, 1986, pp. 60–86.

'Korean Trading Companies Deal with Chinese Delays', *CWI*, 9, 28, 13 July 1988, p. 2.

Kornai, Janos, *Economics of Shortage*, vol. A, Amsterdam: North-Holland, 1980.

Kuby, Michael, Shi Qingqi and Thawat Watanatada, 'Planning China's Coal and Electricity Delivery System', *Interfaces*, 25, 1 (1 Jan. 1995): 41–68.

Kueh Y. Y., ' Prospects for a Transition to a Market Economy without Runaway Inflation', in Y. Y. Kueh, Joseph C. H. Chai, and Gang Fan (eds), *Industrial Reform and Macroeconomic Instability in China, Studies in Contemporary China*, Oxford: Oxford Univerisity Press, 1998.

Kuo Tzu-cheng, 'Transport Planning', *JHJJ*, no. 8 (1955): 30–3, in Nicholas R. Lardy (ed.), *Chinese Economic Planning: Translations from Chi-hua Ching-chi*, New York: M. E. Sharpe, 1978, pp. 37–48.

Kwong, Julia, *The Political Economy of Corruption in China*, Studies on Contemporary China, Armonk, NY: M. E. Sharpe, 1997.

'Lack of Anti-Nuclear Lobby Gives China Chance for Cleaner Power', broadcast on *Kyodo New Service*, 0104 GMT, 30 May 1999, in SWB Part 3: Asia-Pacific, FEW0591, 2 June 1999.

Lamb, Malcolm, *Directory of Officials and Organizations in China: A Quarter Century Guide*, Australian National University, Contemporary China Papers, Armonk, NY: M. E. Sharpe, 1994.

Lands, Thomas, 'New Future for Coal: China's Coal Industry', *Petroleum Economist*, 59, 12 (1992): 8.

Lardy, Nicholas R., *China in the World Economy*, Washington, DC: Institute for International Economics, 1994.

—— *China's Unfinished Economic Revolution*, Washington, DC: Brookings Institute, 1998.

Lawrence, Susan V., 'Parched Province: Shanxi Pins Hopes on Water from the Yellow River', *FEER*, 22 Oct. 1998, pp. 24, 28.

—— 'Unfinished Business: President Jiang Puts His Own Stamp on Reforms', *FEER*, 17 Dec. 1998, p. 22.

Leach, Gerald *et al.*, *Energy and Growth: A Comparison of Thirteen Industrial and Developing Countries*, London: Butterworths, 1986.

Lee Chae-Jin, *China and Japan: New Economic Diplomacy*, Stanford, CA: Hoover Institution Press, 1984.

Leggett, Robert, 'Soviet Investment Policy in the Eleventh Five Year Plan', in *Soviet Economy in the 1980s: Problems and Prospects, Part I*, JEC Report, 1982.

Leung Chi-keung, *China: Railway Patterns and National Goals*, University of Chicago, Dept. of Geography Research Paper no. 195 (1980).

Levine, Herbert S., 'Possible Causes of the Deterioration of Soviet Productivity Growth in the Period 1976–1980', in *Soviet Economy in the 1980s: Problems and Prospects, Part 1*, JEC Report, 1982, pp. 153–68.

—— 'Pressure and Planning in the Soviet Economy', in Morris Bornstein and Daniel Fusfield (eds), *The Soviet Economy: A Book of Readings*, Homewood, IL: Richard D. Irwin, 1974, pp. 43–61.

Li Choh-Ming, *The Statistical System of Communist China*, Berkeley, CA: University of California Press, 1962.

Li Guoguang, 'Construction of Water Transport Facilities to be Expanded', *CM*, 6 (1990): 37, 39.

Li Jingwen, 'Energy Economics in Building a Modern China', in Michael B. McElroy, Chris R. Nielson and Peter Lydon (eds), *Energizing China: Reconciling Environmental Protection and Economic Growth*, Cambridge, MA: Harvard University Committee on the Environment, 1998, pp. 344–68.

Li Qi, 'Nuclear Power in China's Five Year Plan', *Zhongguo jingji daobao*, 30 Dec. 1998, p. B3, in SWB Part 3: Asia-Pacific, FEW0575, 10 Feb. 1999.

Li Ta M. and K. P. Wang, 'China's Emerging Mining Industry: A Long March Toward Modernization', *Mining Engineering*, 32, 3 (Mar. 1980): 273–7.

Lian Yitong, 'The Development and Distribution of Transport and Communications', in Sun Jingzhi (ed.), *The Economic Geography of China*, Hong Kong: Oxford University Press, 1988, pp. 311–68.

Libbon, J. D. and M. D. Boehji, 'International Structure of the US Coal Industry', *American Journal of Agricultural Economics*, 59 (1977): 456–66.

'Liberalization of Policies for Accelerated Small Coal Mine Development, State Council Approves and Forwards Ministry of Coal Industry's "Report on Eight Measures for Accelerating the Development of Small Coal Mines", Requiring All Jurisdictions to Implement It', *MTB*, 11 May 1983, p. 1, in *JPRS*, 389 (1983): 34–7.

Lieberthal, Kenneth and Michel Oksenberg, *Policy Making in China: Leaders, Structures, and Processes*, Princeton: Princeton University Press, 1988.

Lin, Cyril Chiren, 'The Reinstatement of Economics in China Today', *CQ*, 85 (March 1981): 1–48.

Lin Jian, 'Promoting International Cooperation of Coal Patent Technology Development', in International Energy Agency, *The Clean and Efficient Use of Coal and Lignite: Its Role in Energy, Environment and Life*, Proceedings of a Conference held in Hong Kong, 30 Nov.–3 Dec. 1993, Paris: OECD/IEA, 1994, pp. 811–14.

Lin Senmu, Zhou Shulian, and Qi Mingchen, 'Industry and Transport', in Yu Guangyuan (ed.), *China's Socialist Modernization*, Beijing: Foreign Languages Press, 1984, pp. 271–349.

Lin Xiannuan, *China's Energy Strategy: Economic Structure, Technological Choices and Energy Consumption*, Westport, CT: Praeger, 1996.

Liu Changming, 'Environmental Issues and the South-North Water Transfer Scheme', *CQ*, 156 (Dec. 1998): 899–910.

Liu Fei, 'China Develops World's Largest Coalfield', *New China Quarterly*, 12 (May 1989): 37–40.

Liu Guoguang and Wang Riusan, 'Restructuring of the Economy', in Yu Guangyuan (ed.), *China's Socialist Modernization*, Beijing: Foreign Languages Press, 1984.

Liu Suinian and Wu Qungan (eds), *China's Socialist Economy: An Outline History 1949–1984*, Beijing: Beijing Review, 1986.

Lu Yingzhong, *Fuelling One Billion: An Insider's Story of Chinese Energy Policy Development*, Washington Institute Press, 1993.

Lyons, Thomas P., *Economic Integration and Planning in Maoist China*, New York: Columbia University Press, 1987.

Maddison, Angus, *Chinese Economic Performance in the Long Run*, Paris: OECD Development Centre, 1998.

McElroy, Michael B., Chris R. Nielson and Peter Lydon (eds), *Energizing China: Reconciling Environmental Protection and Economic Growth*, Cambridge, MA: Harvard University Committee on the Environment, 1998.

'Measures for Implementation of Price Increases for Above-Quota Fuel Consumption', *WZGL*, no. 2 (1982): 11–12, in *JPRS*, 254 (1982): 4–6.

'Measures to Improve Energy and Transport', *Peking Home Service*, 1200 GMT, 24 Mar. 1983, *SWB Part 3, The Far East, C*, FE/7305/C/1.

Michalski, Wolfgang, Riel Miller, and Barrie Stevens, 'China in the Twenty-First Century: An Overview of the Long-Term Issues', in *Organisation for Economic Co-operation and Development, China in the 21st Century: Long-Term Global Implications*, Paris: 1996.

Mieczkowski, Bogdan, *Transportation in Eastern Europe: Empirical Findings*, New York: East European Quarterly, 1978.

Miles, James A. R., *The Legacy of Tiananmen: China in Disarray*, Ann Arbor: University of Michigan Press, 1996.

'Ministry Drafts Plan to Boost Coal Industry', broadcast on *Xinhua*, 1244 GMT, 10 Jan. 1994, in *FBIS*, 12 Jan. 1994, p. 48.

Munasinghe, Mohan and Gunter Schramm, *Energy Economics, Demand Management and Conservation Policy*, New York: Van Nostrand Reinhold, 1983.

Munasinghe, Mohan with Peter Meier, *Integrated National Energy Planning and Management: Methodology and Application to Sri Lanka*, World Bank Technical Paper, no. 86, Industry and Energy Series, Washington, DC: World Bank, 1988.

Murray, Fiona Elizabeth, *Environment and Technology in Investment Decision-Making: Power Sector Planning in China*, Ph.D. diss., Division of Applied Sciences, Harvard University, 1996.

Murray, Fiona E. and Peter P. Rogers, 'Living with Coal: Coal-Based Technology Options for China's Electric Power Generating Sector', in Michael B. McElroy, Chris R. Nielson and Peter Lydon (eds), *Energizing China: Reconciling Environmental Protection and Economic Growth*, Cambridge, MA: Harvard University Committee on the Environment, 1998, pp. 167–2000.

Nakajima, Seiichi, 'China's Energy Problems: Present and Future', *Developing Economies*, 20, 4, (1982): 472–98.

Naughton, Barry, *Growing Out of the Plan: Chinese Economic Reform, 1978–1993*, Cambridge (UK); New York: Cambridge University Press, 1996.

—— 'The Third Front: Defence Industrialization in the Chinese Interior', *CQ*, 115 (1988): 351–86.

'New Coal Production Figure for 1994 Published', broadcast on *Xinhua*, 0747 GMT, 12 May 1995, in *FBIS*, 12 May 1995, p. 28.

Ni Weidou and Nien Dak Sze, 'Energy Supply and Development in China', in Michael B. McElroy, Chris R. Nielson and Peter Lydon (eds), *Energizing China: Reconciling Environmental Protection and Economic Growth*, Cambridge, MA: Harvard University Committee on the Environment, 1998, pp. 67–117.

Nielson, Chris P. and Michael B. McElroy, 'Introduction and Overview' in Michael B. McElroy, Chris R. Nielson, and Peter Lydon (eds), *Energizing China: Reconciling Environmental Protection and Economic Growth*, Cambridge, MA: Harvard University Committee on the Environment, 1998.

'Northeast Coal Trade Centre Opens in Liaoning', broadcast on Shenyang Liaoning People's Radio Network, 2300 GMT, 28 Feb. 1993, in *FBIS*, 3 Mar. 1993, p. 44.

'Official Says Country Not to Start Nuclear Power Projects in Next Three Years', broadcast on *Xinhua*, 0826 GMT, 29 Apr. 1999, in SWB Part 3: Asia-Pacific, FED3522, 30 Apr. 1999.

'Offshore Natural Gas Development Accelerated', broadcast on *Xinhua* 0927 GMT, 29 Apr. 1999.

Ono, Shuichi, *Sino-Japanese Economic Relationships: Trade, Direct Investment, and Future Strategy*, World Bank Discussion Paper, No. 146, China and Mongolia Dept. Series, Washington, DC: IBRD/WB, 1992.

'Opinions on Stopping the Indiscriminate Extraction of Small Coal Mines and on Ensuring Safety in Coal Mining', broadcast on *Xinhua*, 0205 GMT, 17 Jan. 1993, in *FBIS*, 27 Jan. 1993, pp. 21–2.

'Oversupply Results in Cuts in Coal Production', broadcast on *Xinhua*, 0103 GMT, 21 Jan. 1999, in SWB Part 3: Asia-Pacific, FEW0573, 27 Jan. 1999.

'Overview of Our Nation's Energy Resources', *LW*, 10 (1981): 9–11, in *JPRS*, 228 (1982): 10–18.

'Oxy Gets Out of Antaibao', *ICR*, 14 June 1991.

Peirce, William S., *Economics of the Energy Industries*, 2nd edn., Westport, CN: Praeger, 1996.

Perkins, Dwight H., 'Industrial Planning and Management', in A. Eckstein, W. Galenson, and Ta-Chung Liu (eds), *Economic Trends in Communist China*, Edinburgh: Edinburgh University Press, 1968, pp. 597–636.

—— *Market Control and Planning in Communist China*, Cambridge, MA: Harvard University Press, 1966.

Peterson, Albert S., 'China: Transportation Developments, 1971–80', in *China Under the Four Modernizations*, JEC Report, 1982, pp. 138–70.

Philip, George, *The Political Economy of International Oil*, Edinburgh: Edinburgh University Press, 1994.

Platte, Erika, 'The Role of China in Japan's Quest for Energy Security', *Hitotsubashi Journal of Economics*, 37 (1996): 69–86.

'PRC: Coal-Producing "Giants" Map Out Production Strategy', broadcast on *Xinhua*, 1541 GMT, 10 Mar. 1996, in *FBIS*, 11 Mar. 1996, p. 69.

'Premier: Friendship with North Korea Will Continue to Develop', broadcast on *Xinhua*, 0802 GMT, 4 June 1999, on SWB, Part 3, Asia Pacific, North Korea, FE/D3554/G.

'Price Ceiling for "Coal Outside Plans" Issued', broadcast on *Xinhua*, 1000 GMT, 5 Mar. 1990, in *FBIS*, 16 Mar. 1990, p. 26.

'Proposals of the Eighth National Party Congress for the Second Five Year Plan for Development of the National Economy, 1958–1962', (adopted 27 Sept. 1956), in Stuart

Kirby (ed.), *Contemporary China, Vol. II, 1956–1957*, Oxford: Oxford University Press, 1958, pp. 142–61.

Rath, Steven, *The Hungarian Coal Industry*, National Committee for a Free Europe, Mid-European Studies Centre, Research Document no. 280, Washington: Photoduplication Service, Library of Congress, 1955.

Rawski, Thomas G., *Economic Growth in Prewar China*, Berkeley: University of California Press, 1989.

Ren Ruoen, *China's Economic Performance in an International Perspective*, Paris: OECD Development Centre, 1997.

'Report on Concern Over Energy Production', *Zhongguo xinwen she*, 0505 GMT, 26 June 1993, in FBIS, 28 June 1993.

'Report on the Organization of the Coal Market for 1957, 18 March 1957', *Xinhua banyue kan* (New China Semi-Monthly), 9 (10 May 1957): 110. Extracts published in Chao Kuo-chun, *Economic Planning and Organization in Mainland China: A Documentary Study, 1949–1957*, Cambridge, MA: East Asian Research Centre, Harvard University, 1963, vol. 2, pp. 41–3.

Rong Geng, 'Promoting Coal Industry in China through Innovative Financing', in International Energy Agency, *Energy Efficiency Improvements in China: Policy Measures, Innovative Finance and Technology Deployment*, Proceedings of a Conference held in Beijing, 3–4 Dec. 1996, sponsored by IEA, Chinese State Planning Commission, and the Environmental and Resources Protection Committee of the Chinese NPC, Paris: OECD/IEA, 1997, pp. 159–62.

Sarkar, Sunit Kumar and Subhankar Sarkar, *State of Environment and Development in Indian Coalfields*, Calcutta: Oxford and IBH, 1996, p. 8.

Seth, Ram P., *Pricing and Related Policies of Publicly Owned Electrical Utilities*, Halifax: Institute of Public Affairs, 1984.

Shang Daijiang, 'Mustering All Forces to Develop Shanxi's Coal', *LW Overseas Edition*, 50 (15 Dec. 1986): 13, in *JPRS*, 25 (1987): 50–2.

Shao Shiwei *et al.* (eds), *China: Power Sector Regulation in a Socialist Market Economy*, World Bank Discussion Paper, no. 361, Washington: International Bank for Reconstruction and Development/World Bank, 1997.

Sheehan, Jackie, *Chinese Workers: A New History*, London: Routledge, 1998.

Shi Zulin and Liu Deshun, 'Socio-Economy, Energy and the Environment', in Preeti Soni (ed.), *Energy and Environmental Challenges in Central Asia and the Caucasus, Windows for Co-operation'*, New Delhi: Tata Energy Research Institute, 1996.

Sinton, Jonathan E. *et al.* (eds), *China Energy Databook*, Berkeley, CA: Ernest Orlando Lawrence Berkeley National Laboratory Report LBL-32822 Rev. 4, Sept. 1996.

Sinton, Jonathan E. and Mark. D. Levine, 'Changing Energy Intensity in Chinese Industry: the Relative Importance of Structural Shift and Intensity Change', *Energy Policy*, Mar. 1994, 239–55.

Smil, Vaclav, *China's Energy: Achievements, Problems, Prospects*, New York: Praeger, 1976.

—— 'China's Energy and Resource Uses: Continuity and Change', *CQ*, 156 (Dec. 1998): 935–51.

—— 'Energy' in Leo A. Orleans and Caroline Davidson (eds), *Science in Contemporary China*, Stanford: Stanford University Press, 1980.

—— *Energy in China's Modernization: Advances and Limitations*, Armonk, New York: M. E. Sharpe, 1988.

Song Xuefeng, 'The Role of Clean Coal and Efficient Use of Coal in the Future Energy and Environment of China', in International Energy Agency, *The Clean and Efficient*

Use of Coal and Lignite: Its Role in Energy, Environment and Life, Proceedings of a Conference held in Hong Kong, 30 Nov. – 3 Dec. 1993, Paris: OECD/IEA, 1994, pp. 1081–88.

Soni Preeti and Vishal Narain, 'India: Energy and Environment Situation', in Preeti Soni (ed.), *Energy and Environmental Challenges in Central Asia and the Caucasus: Windows for Co-operation*, New Delhi: Tata Energy Research Institute, 1996, pp. 161–79.

'South Korea May Experience Trouble in Obtaining China Coal', *CWI*, 10, 5, 31 Jan. 1989, p. 8.

'Speculation in Coal Has Made Coal Shortage Even More Serious', broadcast on *Zhongguo tongxun she*, 1147 GMT, 1 Sept. 1989, in *FBIS*, 14 Sept. 1989, p. 31.

'State Owned Mines Boost Output', *Xinhua*, 21 Sept. 1985, in *JPRS*, 96 (1985): 44.

'Statistics of Energy Utilization in China', *ER*, Feb. 1981, p. 9.

Steinfeld, Edward, *Forging Reform in China: The Fate of State-Owned Industry*, Cambridge: Cambridge University Press, 1998.

Sun Jingzhi (ed.), *The Economic Geography of China*, Hong Kong: Oxford University Press, 1988.

Sun Shu and Sun Yiyin, 'The Study and Development of Mineral and Energy Resources in China', in James P. Dorian and David G. Fridley (eds), *China's Energy and Mineral Industries: Current Perspectives*, Boulder, CO: Westview, 1988, pp. 99–104.

Szuprowicz, B. O., 'Coal: China's Economic Backbone', *Coal Age*, 82, 6 (June 1977): 114–22.

'Taiwan Firms Cool to PRC Coal', *CWI*, 9, 32, 10 Aug. 1988, p. 7.

'Taiwan Market Looks Good to Oxy for Antaibao Coal Output', *CWI*, 9, 24, 15 June 1988, p. 1.

'Taiwanese Coal Users Take Cautious Attitude About Chinese Trade', *CWI*, 9, 27, 6 July 1988, p. 1.

Takahashi, Masaki, Stratos Tavoulareas, and Joseph Gilling, 'The World Bank Clean Coal Initiative', in International Energy Agency, *Energy Efficiency Improvements in China: Policy Measures, Innovative Finance and Technology Deployment*, Proceedings of a Conference held in Beijing 3–4 Dec. 1996, sponsored by the International Energy Agency, Chinese State Planning Commission, and the Environmental and Resources Protection Committee of the Chinese NPC, Paris: OECD/IEA, 1997, pp. 153–7.

Tang Jie, 'Port Construction in China', *New China Quarterly*, 13 Aug. 1989, pp. 30–2.

Taylor, Robert P., *Rural Energy Development in China*, Washington, DC: Resources for the Future, 1981.

Thomson, Elspeth, *The Development of China's Coal Industry, 1949–1978: Towards An Analytical Model*, Ph.D. diss., School of Oriental and African Studies, University of London, 1993.

—— 'Reforming China's Coal Industry', *CQ*, 147, (1996): 726–50.

Toohey, Alex, 'Country Experiences and Lessons Learned', in International Energy Agency, *Energy Efficiency Improvements in China: Policy Measures, Innovative Finance and Technology Deployment*, Proceedings of a Conference held in Beijing 3–4 Dec. 1996, Paris: OECD/IEA, 1997, pp. 71–5.

Topper, John M., 'Improving Coal Use in Developing Countries through Technology Transfer', in International Energy Agency, *The Clean and Efficient Use of Coal and Lignite: Its Role in Energy, Environment and Life*, Proceedings of a Conference in held in Hong Kong, 30 Nov.–3 Dec. 1993, Paris: OECD/IEA, 1994. pp. 775–92.

Tseng, Wanda *et al.*, *Economic Reform in China: A New Phase,* Washington, DC: International Monetary Fund, 1994.

Tyner, Wallace E., *Energy Resources and Economic Development in India*, Leiden: Martinus Nijhoff, 1978.

'US Invests \$4 B in Coal Mining, Transport', broadcast on *Xinhua*, 1236 GMT, 10 Aug. 1995, in *FBIS*, 11 Aug. 1995, p. 41.

Vanderpol. Alex, 'China: Clean Coal Technology', *Market Report: Industry Sector Analysis*, 19 Apr. 1995, pp. 129–45.

Vermeer, Eduard, 'Industrial Pollution in China and Remedial Policies', *CQ*, 156 (Dec.1998): 952–85.

'Vice Minister Addresses Coal Technology Meeting', broadcast on *Xinhua*, 1307 GMT, 25 Sept. 1995, in *FBIS*, 28 Sept. 1995, p. 55–6.

Walder, Andrew G., 'Workers, Managers and the State: The Reform Era and the Political Crisis of 1989', *CQ*, 127 (1991): 467–92.

Wang Kung-ping, *Controlling Factors in the Future Development of the Chinese Coal Industry*, Morningside Heights, NY: King's Crown Press, 1947.

Wang Senhao, 'Some Preliminary Conceptions of Technical Progress in China's Coal Industry During the Next Eighteen Years', *SJMTJS*, no. 5 (1983): 2–6, in *JPRS* 377 (1983): 52–62.

Wang Zhenzhi and Wang Yongshi, 'Epilogue: Prices in China', in Wei Lin and Arnold Chao (eds), *China's Economic Reforms*, Philadelphia: University of Pennsylvania Press, 1982, pp. 220–34.

'Weak Coal Market Puts Pressure on Producers in Third Quarter', broadcast on *Zhongguo xinwen she*, 1239 GMT, 4 Dec. 1997.

Wei Lin and Tian Huiming, 'Official on Plans to Promote Coal Industry', broadcast on *Zhongguo xinwen she*, 1025 GMT, 12 Oct. 1992, in *FBIS*, 30 Oct. 1992, pp. 31–2.

Weil, Martin, 'China's Troubled Coal Sector', *CBR*, Mar.–Apr., 1982, pp. 23–34.

—— 'Coal's Promises and Problems', *CBR*, Mar.–Apr. 1984, p. 40.

Weng Zhensong, 'Forecast and Analysis of China's Railway Transport', in Liu Guoguang *et al.* (eds), *PRC 1998 Economic Blue Book*, Hong Kong: Centre for Asian Studies, University of Hong Kong, 1998.

Wirtshafter, Robert M. and Ed Shih, 'Decentralization of China's Electricity Sector: Is Small Beautiful?', *World Development*, 18, 4 (April 1990): 505–12.

Wong, Christine P. W., 'Ownership and Control in Chinese Industry: The Maoist Legacy and Prospects for the 1980s', in *China's Economy Looks Toward the Year 2000, Vol. I, The Four Modernizations*, JEC Report, 1986, pp. 571–603.

Woodard, Kim, *The International Energy Relations of China*, Stanford, CA: Stanford University Press, 1980.

Wright, Tim, *Coal Mining in China's Economy and Society 1895–1937*, Cambridge Studies in Chinese History, Literature and Institutions, Cambridge: Cambridge University Press, 1984.

—— 'Growth of the Modern Chinese Coal Industry: An Analysis of Supply and Demand, 1896–1936', *Modern China*, 7, 3 (July 1981): 317–50.

Wu Shiyu, 'Developing Coal Preparation Processing, Increasing Coal Utilization Efficiency and Reducing Pollution', in International Energy Agency, *Energy Efficiency Improvements in China: Policy Measures, Innovative Finance and Technology Deployment*, Proceedings of a Conference held in Beijing, 3–4 Dec. 1996, Paris: OECD/IEA, 1997.

Wu Yuan-li, *An Economic Survey of Communist China*, New York: Twayne, 1956.

Wu Yuan-li and H. C. Ling, *Economic Development and the Use of Energy Resources in Communist China*, New York: Praeger, 1963.

Xia Guocai, 'Accelerating the Development of an Energy-Heavy-Chemical Industrial Base in Shanxi Is a Major Strategic Decision in China's Modernization Programme', *JJRB*, 3 (15 Sept. 1983): 28, in *JPRS*, 14 (1984): 3.

Xiao Rui and He Yiwen, 'Railway Construction to Enter a New Stage – An Interview of Sun Yongfu, Vice-Minister of Railways', *ER*, no. 7 (1993): 30.

Xiao Wei, 'China Plagued by Energy Shortage in Next Decade', *CM*, no. 3 (1991): 10–11.

Xue Muqiao, *Current Economic Problems in China*, (edited and translated by K. K. Fung), Boulder, CO: Westview, 1982.

Yan Sun, 'The Chinese Protests of 1989: The Issue of Corruption', *Asian Survey*, Aug, 1991, pp. 762–82.

Yangzhou Coal Mining Co. Ltd., Listing Prospectus, 24 Mar. 1998.

Yenny, Jacques and Lily V. Uy, *Transport in China: A Comparison of Basic Indicators with Those of Other Countries*, World Bank Staff Working Paper, no. 723, Washington, DC: International Bank for Reconstruction and Development/World Bank, 1985.

Yu Ertie, 'Role and Prospects of the Coal-Dressing Industry under a Market Economy', *MTKXJS*, 22, 1 (Jan. 1994): 53–6, in *FBIS*, 3 May 1994, p. 33.

Yu Guangyuan (ed.), *China's Socialist Modernization*, Beijing: Foreign Languages Press.

Yu Yongnian, 'Coal Consumption Status and Strategy of Energy Development in China', in International Energy Agency, *Coal, the Environment and Development: Technologies to Reduce Greenhouse Gas Emissions*, Proceedings of a Conference held in Sydney 18–21 Nov. 1991, Paris: OECD/IEA, 1992, pp. 235–42.

—— 'Overview of Coal Use and Efficient Future Technology', in International Energy Agency, *The Clean and Efficient Use of Coal and Lignite: Its Role in Energy, Environment and Life*, Proceedings of a Conference held in Hong Kong, 30 Nov.–3 Dec. 1993, Paris: OECD/IEA, 1994, pp. 719–28.

Zhang Baoming, 'Clean Coal Production and Rational Use of Coal in China', in International Energy Agency, *The Clean and Efficient Use of Coal and Lignite: Its Role in Energy, Environment and Life*, Proceedings of a Conference held in Hong Kong, 30 Nov.–3 Dec. 1993, sponsored by the IEA, Australian Dept. of Primary Industries and Energy, Ministry of International Trade and Industry of Japan US Dept. of Energy, US Agency for International Development; US-Asia Environmental Partnership; and United Nations, Paris: OECD/IEA, 1994, pp. 71–86.

Zhang Hongshun, 'Building Shanxi Energy Base Hinges on Upgrading Existing Technology', *Jishu jingji yu guanli yanjiu* (Technology Economics and Management Research), 4 (31 Dec. 1983): 44–7, in *JPRS*, 32 (1984): 40.

Zhang Mingchuan, 'Control of SO2 and NOx Emissions from Fossil Power Plants: Research and Practice of TPRI', in International Energy Agency, *The Clean and Efficient Use of Coal and Lignite: Its Role in Energy, Environment and Life*, Proceedings of a Conference held in Hong Kong, 30 Nov.–3 Dec. 1993, Paris: OECD/IEA, 1994, pp. 905–14.

Zhang Qinwen and Ren Boping, 'The Choice of Strategy for the Construction of the Shanxi Energy Base', *Shehui jingji daobao* (Social Economic Herald), 4 June 1985, p. 5, in *JPRS*, 69 (1985): 1–4.

Zhang Zhongxiang, *The Economics of Energy Policy in China: Implications for Global Climate Change*, Cheltenham, UK: Edward Elgar, 1998.

Zhu Daqiang, 'Tapping the Yangzi River's Transportation Capacity', *CM*, no. 3 (1990): 27–8.

Official publications in English

Beijing Review Press, *The Development of China 1949–1989, China Issues and Ideas*, Beijing: 1989.

Chen Yun and Sun Jie, 'Major Target Readjustment for Eighth Five Year Plan', *BR*, 10–16 May 1993, pp. 18–21.

China, State Economic and State Planning Commissions, 'Regulations for Intensifying Conservation of Electricity', broadcast on *Xinhua*, 0134 GMT, 5 Apr. 1987, in *JPRS*, 41 (1987): 84–6.

China, State Economic and Trade Commission, 'Reform and Development of China's State-Owned Enterprises', *BR*, 11–17 Jan. 1999, pp. 16–19.

China, State Planning Commission, *First Five Year Plan for the Development of the National Economy of the People's Republic of China in 1953–1957*, Peking: Foreign Languages Press, 1956.

—— 'Industrial Catalogue Guiding Foreign Investment', jointly promulgated by the State Planning Commission, the State Economic and Trade Commission and the Ministry of Foreign Trade and Economic Cooperation, *BR*, 2–8 Mar., 1998, pp. 15–23.

China, State Planning Commission and Ministry of Foreign Trade and Economic Cooperation, 'Compilation of Major Technology Projects Scheduled for The People's Republic of China Between 1993–2000', *BR*, 28 Nov.–4 Dec. 1994, pp. 14–22.

China, State Statistical Bureau, *Ten Great Years: Statistics of the Economic and Cultural Achievements of the People's Republic of China*, Peking: Foreign Languages Press, 1960. (Reprinted as Occasional Paper no. 5 of the Programme in East Asian Studies, West Washington State College, 1974).

China Coal Industry Yearbook, (Ministry of Coal Industry Compiling Committee), Hong Kong: Economic Information and Agency, 1982–1999. See *Zhongguo meitan gongye nianjian* for Chinese version.

China National Coal Industry Import and Export (Group) Corporation, *Annual Report 1997*, Beijing: 1997.

China Unified Distribution Coal Mine Corporation Energy Conservation Office, 'Deal with Concrete Matters Relating to Work and Innovate, Open Up and Advance, Push Energy Conservation Work to a New Level during the Eighth Five Year Plan', *ZGNY*, 9 (1991): 3–7, in *JPRS*, 11 (1991): 26–30.

'China Publishes Interim Provisions on Guidance for Foreign Investment Along with an Industrial Catalog Guiding Foreign Investment', *BR*, 18–24 Sept. 1995, pp. 17–26.

'China's Economic Sectors Re-Classified', *BR*, 4–10 Jan. 1999, pp. 18–19.

'Communique of the Third Plenary Session of the 11th Central Committee of the Communist Party of China', *PR*, 52, 29 Dec. 1978, p. 11.

Gt. Brit., House of Commons, Energy Committee, Session 1986–87, First Report, *The Coal Industry*, London: HMSO, 28 January 1987.

Gt. Brit., National Coal Board, *The Coal Industry of the USSR: A Report by the Technical Mission of the NCB*, London: HMSO, 1957.

—— [Report on the 1976 Delegation's Visit to China] [London: 1976], paragraph 2.6.

Hua Guofeng, 'Report on the Work of the Government', delivered 26 Februry 1978, *PR*, 10 Mar. 1978, pp. 7–40.

India, Planning Commission, *Eighth Five Year Plan 1992–97, Vol. II, Sectoral Programmes of Development*, New Delhi: The Commission, 1992.

International Energy Agency, *Coal Information* [annual], Paris: OECD/IEA. Various years.

—— *Coal Prospects and Policies in IEA Countries, 1987 Review*, Paris: OECD/IEA, 1988.

—— *Electricity Information 2000*, Paris: OECD/IEA, 2000.

—— *Energy Balances of Non-OECD Countries, 1996–1997*, Paris: OECD/IEA, 1999.

—— *Energy Policies of IEA Countries, 1993 Review*, Paris: OECD/IEA, 1994.

—— *Global Energy: Changing the Outlook*, Paris: OECD/IEA, 1992.

—— *International Coal Trade: The Evolution of a Global Market*, Paris: OECD/IEA, 1997, p. 113.

—— *World Energy Outlook*, Paris: OECD/IEA. Various years.

International Energy Agency, Coal Industry Advisory Board. *Factors Affecting the Take-Up of Clean Coal Technologies: Overview Report*, Paris: OECD/IEA, 1994.

—— Coal Industry Advisory Board, *Global Methane and the Coal Industry: A Two-Part Report on Methane Emissions from the Coal Industry and Coalbed Methane Recovery and Use*, Paris: OECD/IEA, 1994.

—— Coal Industry Advisory Board. *Industry Attitudes to Combined Cycle Clean Coal Technologies*, Paris: OECD/IEA, 1994.

—— Coal Industry Advisory Board. *Industry Attitudes to Steam Cycle Clean Coal Technologies: Survey of Current Status*, Paris: OECD/IEA, 1994.

—— Coal Industry Advisory Board, *Coal in the Energy Supply of China*, Report of the CIAB Asia Committee, Paris: OECD/IEA, 1999.

International Energy Agency/Organization for Economic Cooperation and Development, *International Coal Trade: The Evolution of a Global Market*, Paris: 1997.

International Labour Organization, *Safety and Health in Coal Mines, An ILO Code of Practice*, Geneva: ILO, 1986.

—— Coal Mines Committee, [*Report on the 1st Session*], *3rd Sitting*, Geneva: ILO, 1947.

—— Coal Mines Committee, [*Report on the 4th Session*], *Productivity in Coal Mines*, Item 3, Geneva: ILO, 1951.

—— Eleventh Session. *Employment and Training with Reference to Health and Safety at Coal Mines*. Programme of Industrial Activities, Geneva: 1982.

International Monetary Fund, *International Financial Statistics Yearbook*, Washington, DC: various years.

Jiang Zemin, 'Hold High the Great Banner of Deng Xiaoping Theory for an All-Round Advancement of the Cause of Building Socialism with Chinese Characteristics into the 21st Century', Report delivered at the 15th NPC, 12 Sept. 1997, *BR*, 6–12 Oct. 1997, pp. 10–33.

Li Peng, 'Report on the Work of the Government', delivered at the First Session of the 9th NPC, 5 March 1998, *BR*, 6–12 Apr. 1998, pp. 9–24.

Liu Shaoqi, 'Report on the Work of the Central Committee of the Communist Party of China' delivered at the Second Session of the 8th NPC, 5 May 1958', *PR*, no. 14, 1958, pp. 6–22.

'NPC Adopts Amendments', *BR*, 29 Mar.–4 Apr., 1999.

Organization for Economic Cooperation and Development, *Energy Balances of OECD Countries*, Paris: 1980.

—— *Applying Market-Based Instruments to Environmental Policies in China and OECD Countries*, Paris: 1997.

—— *Environmental Taxes: Recent Developments in China and OECD Countries*, Paris: 1999.

Organization for Economic Cooperation and Development/International Energy Agency, *Ad Hoc* Group on International Energy Relations, *China* [draft paper], Paris: 1987.

—— *Energy Balances for Developing Countries, 1971–1982*, Paris: OECD/IEA; Washington, DC: OECD Publications and Information Centre, 1984.

United Nations, *Industrial Statistics Yearbook*, New York: Dept. of Economic and Social Affairs, State Statistical Office of the United Nations: various years.

—— Dept. of International Economic and Social Affairs, Statistical Office, *World Energy Supplies*, New York: various years.

United Nations, Dept. of International Economic and Social Affairs, Statistical Office, *World Energy Supplies, 1950–1976*, Statistical Papers, Series J, 19, New York: 1976.

United Nations Development Programme, 'Development of Coalbed Methane Resources in China', Project Document, June 1992.

United Nations, Economic Commission for Europe, *European Economy in 1979*, New York: 1980.

United States, Central Intelligence Agency, *Chinese Coal Industry: Prospects Over the Next Decade*, ER 79-10092, Washington, DC: National Foreign Assessment Centre, 1979.

United States, Congress, House, Committee on Interstate and Foreign Commerce, Subcommittee on Energy and Power, *The Energy Factbook: Data on Energy Resources, Reserves, Production, Consumption, Prices, Processing and Industry Structure*, Washington, DC: 1980.

United States, Energy Information Administration, *Documentation of the Project Independence Energy Evaluation System*, vols I–V, Washington, DC: 1979.

—— *The United States Coal Industry, 1970–1990: Two Decades of Change*, Washington, DC: 1992.

World Bank. *China: Efficiency and Environmental Impact of Coal Use*, Washington, DC: IBRD/WB, 1991.

—— *China Engaged: Integration with the Global Economy*, China 2020 Series, Washington, DC: IBRD/WB, 1997.

—— *China: The Energy Sector. Annex 3 to China: Long-Term Development Issues and Options*, Washington, DC: IBRD/WB, 1985.

—— *China: Investment Strategies for China's Coal and Electricity Delivery System*, Washington, DC: IBRD/WB, 1995.

—— *China: Long-Term Development Issues and Options*, Baltimore: Johns Hopkins University Press, 1985.

—— *China: Power Sector Reform: Toward Competition and Improved Performance*, Washington, DC: IBRD/WB, 1994.

—— *China: Railway Investment Study*, Washington, DC: IBRD/WB, 1992.

—— *China: Renewable Energy for Electric Power*, Washington, DC: IBRD/WB, 1996.

—— *China: The Transport Sector. Annex 6 to China: Long-Term Development Issues and Options*, Washington, DC: IBRD/WB, 1985.

—— *China's Railway Strategy*, Washington, DC: IBRD/WB, 1993.

—— *Clear Water, Blue Skies: China's Environment in the New Century*, China 2020 Series, Washington, DC: IBRD/WB, 1997.

—— *Energy Options and Policy Issues in Developing Countries*, Staff Working Paper 350, Washington, DC: IBRD/WB, August 1979.

—— *Transport in China: an Evaluation of World Bank Assistance*, Washington, DC: IBRD/WB, 1999.

—— *World Development Report* [annual], Washington, DC: various years.

Zeng Peiyan, 'Report on the Implementation of the 1998 Plan for National Economic and Social Development and on the Draft 1999 Plan for National Economic and Social Development' (Excerpts), delivered at the Second Session of the 9th NPC, 6 Mar. 1999, *BR*, 12–18 Apr. 1999, pp. 21–31.

—— 'Report on the Implementation of the 1999 Plan for National Economic and Social Development and on the Draft 2000 Plan for National Economic and Social Development' (Excerpts), delivered at the Third Session of the 9th NPC, 6 Mar. 2000, *BR*, 10 Apr. 2000, pp. 16–23.

Zhu Rongji, 'Report on the Work of the Government', delivered at the Second Session of the 9th NPC, 5 March 1999, *BR*, 5–11 Apr. 1999, pp. 10–25.

Books and articles in Chinese

'1992 nian jieneng jiang hao he ziyuan zong heli yong chengxiao xianzhu' (Remarkable Progress in Resource Conservation and Use in 1992), *ZGNY*, no. 9 (1993): 24.

'1994 nian quan guo zhuangshan qiye zhizhong shangwang shiqu qingkuang fenxi' (Analysis of 1994 Coal Mine Workers' Accidents), *Laudong nei can*, 21 (Mar. 1995): 12–13.

'Bixu jiuzheng hushi chanpin zhiliang de pianxiang' (Must Correct the Tendency to Neglect Product Quality), *MTGY*, no. 10 (1956): 10–11.

Cao Jing, '1998 nian tielu jianshe zhanwang' (Prospects for Railway Construction in 1998), *TDYSJJ*, no. 1 (1998): 9–10.

Cao Zhengyan, 'Tigao nengyuan xiaolu Zhongguo qude de chengjiu he zhengfu de fangzhen cuoshi' (Raise China's Energy Efficiency: Achieve Results from Government Policies and Measures), *ZGNY*, no. 3 (1997): 9–10, 15.

Chen Bingqiang, 'Zhanwong 2000 nian de wo guo meitan gongye' (Prospects of China's Coal Industry towards the Year 2000), in Peng Shiji (ed.), *Zhongguo meitan gongye sishi nian* (40 Years of China's Coal Industry), Beijing: Meitan gongye chubanshe, 1990, pp. 11–20.

Chen Dianmo, 'Fahui Shanxi jingji youshi de genben tujing' (Bring into Play Shanxi's Superiority), *SXJDYJ*, pp. 248–55.

Chen Dong and He Yuanyuan, 'Zhuazhu ji yu – jinyibu fazhan tielu duo zhong jingying' (Seizing the Opportunity to Make Further Steps in Developing the Diversified Economic Activities of the Railways), *TDYSJJ*, no. 3 (1998): 28–9.

Chen Dongsheng, 'Tiaozheng jingji jiegou, jiasu meitan nengyuan jidi jianshe' (Transform Economic Structure, Increase Speed of Coal Base Construction), *SXJDYJ*, pp. 256–75.

Chen Hang, 'Fujian meitan jie yungang de fazhan jishen heli buju' (The Development and Rational Location of a Coal Import-Transfer Port in Fujian), *JJDL*, no. 3 (1984): 226–31.

Chen Heping, 'Guanyu bianzhi 'shi wu' jieneng jihua de sikao' (Thoughts on Devising a Plan for Saving Energy During the Tenth Five Year Plan Period), *ZGNY*, no. 11 (1998): 6–9, 22.

—— 'Jieneng gongzuo gaige chutan' (An Initial Look at Reforming Energy Conservation Work), *NY*, no. 2 (1985): 7–9.

Chen Zhicheng, 'Dui dangqian qiye nengyuan guanli gongzuo de jidian kanfa' (Some Points About Current Enterprise Energy Management Work), *NY*, no. 5 (1982): 12–14.

Chen Zhu, 'Da li fazhan meitan xixuan jieyue nengyuan' (Increase Coal Processing to Save Energy), *NY*, no. 4 (1984): 27–8.

Cheng Zhiping (ed.), *Zhongguo wujia wushi nian, 1949–1998* (Fifty Years of Prices in China, 1949–1998), Beijing: Zhongguo wujia chubanshe, 1998.

Cui Cheng, Hu Xiulian, and Liu Jingru, 'Wo guo nengyuan jiegou tiaozheng yu jin chukou zhanbei de jingjixue fenxi' (Economic Analysis of Strategy for Adjusting China's Energy Structure and Imports and Exports), *ZGNY*, no. 4 (1998): 8–12.

Dangdai Zhongguo congshu bianzhi bu (Contemporary China Series Editorial Dept.), *Dangdai Zhongguo de meitan gongye* (Contemporary China Coal Industry), Beijing: Zhongguo shehui kexue chubanshe, 1988.

Deng Xiaoping, 'Women de hongwei mubiao he genben zhengce' (Our Magnificent Targets and Basic Policies), in *Deng Xiaoping wenxuan* (Selections of Deng Xiaoping), Beijing: Renmin chubanshe, 1993.

Dong Jiben, 'Guanyu Shanxi nengyuan shengchan de jiazhi buchang wenti' (Problems Concerning the Compensation Value of Shanxi's Energy), *JJWT*, no. 1 (1985): 9–12.

Dong Jiben and Zhai Ligong (eds), *Meitan gongye qiye wuzi guanli* (Coal Industry Enterprise Materials Management), Taiyuan: Shanxi renmin chubanshe, 1985.

Dong mei lianhe gongsi, meitan jingji yanjiu hui chanpin xiaoshou zu (Coal Economics Research Meeting of the Product Sales Group, East Coal Company), 'Gaige meitan yunxiao gongzuo de jidian yijian' (Some Views on Reforming Coal Marketing Work), *MTJJYJ*, no. 9 (1984): 7–8.

Fan Jie, 'Meikuang de diqu guotu zongti guihua de chubu yanjiu' (A Preliminary Approach to the Comprehensive Territorial Planning in the Coal Mine Area), *DLYJ*, no. 7 (1988): 46–51.

Fan Shiqing, 'Wo guo meitan shichang cunzai de wenti ji duice' (Problems and Countermeasures in Our Country's Coal Market), *MTJJYJ*, no. 4 (1999): 38–9.

Fan Weitang, 'Shijie meitan gongye xianzhuang ji fazhan qushi' (Present Trends in the Development of the World Coal Industry), *ZGNY*, no. 9 (1998): 4.

Fan Weitang (ed.), *Zhongguo meitan gongye jianming shouce* (Concise Handbook of China's Coal Industry), Beijing: Meitan gongye chubanshe, 1995.

Fang Rukang, 'Wo guo nengyuan ziyuan de xianzhuang ji heli liyong wenti' (Rationalizing China's Current Use of Energy Resources), *DLYJ*, no. 3 (Dec. 1984): 25–38.

Fang Weizhong (ed.), *Zhonghua renmin gongheguo jingji da shiji 1949–1980* (Major Economic Events in Communist China 1949–1980), Beijing: Zhongguo shehui kexue chubanshe, 1984.

Feng Baogen, 'Daxing meitan zhengye: jianli xiandai zhengye zhiding zhong de yi ge tuchu nandian ji duice' (Difficulties in Establishing Modern Enterprise System in Large Coal Mines), *MTJJYJ*, no. 9 (1995): 29–30.

Fu He, 'Tan Guangdong sheng diaoyun meitan de yunshu fangshi he lujing de xuanze' (Discussion on Direct Means of Transporting Coal to Guangdong Province), *NY*, no. 1 (1988): 8–11, 16.

Fu Lihong, 'Xiangzhen meikuang cunzai de wenti yu fazhan qianxi' (Analysis of the Problems in Developing Village and Township Mines), *MTJJYJ*, no. 7 (1995): 60–1, 63.

'Fuzhong fenjin wanqiang tujin liange genben xing zhuanbian shixian niu kui zeng ying mubiao' (Carry Heavy Burden, Fight Hard, Push Forward Two Basic Transformations, Achieve Elimination of Losses Target), *MTGYNJ 1997*, pp. 68–74.

'Gaige meitan yunxiao gongzuo de jidian yijian' (Some Views on Reforming Coal Marketing Work), *MTJJYJ*, no. 9 (1984): 7–8.

Gao Huimin, 'Jianshe Shanxi nengyuan jidi touzi guimo yuce' (The Scale of Investment in the Shanxi Energy Base), *SXJDYJ*, pp. 393–406.

Gao Yangwen, 'Quanli kaichuang meitan gongye xiandai hua jianshe de xin jumian' (Considerable Effort Must Be Devoted to Modernizing the Coal Industry), *NY*, no. 2 (1983): 1–7, 13.

—— 'Yikao kexue jishu jinbu shixian meitan chanliang fan yi fan' (Depend on Improvement of Science and Technology to Realize Twofold Increase of Annual Gross Output Value of Coal), *NY*, no. 6 (1983): 1–8.

—— 'Chuang li you Zhongguo tese de chongman shengji he huoli de meitan gongye jingji tizhi' (Creating a Distinctly Chinese-Style and Vigourous Coal Industry Economic System), *NY*, no. 2 (1985): 1–6.

Gao Zongwen (ed.), *Jiben jianshe tongji: meikuang qiye guanli jichu duwu* (Basic Construction Statistics: Coal Mine Enterprise Management Basic Reading Material), Beijing: Meitan gongye chubanshe, 1984.

Ge Fu, 'Mei jia gaige xuyao you yi ge zongti anpai' (Need Thorough Arrangements for Coal Pricing Reform), *JJGL*, no. 8 (1989): 16–17.

Geng Sunsan, 'Luetan meitan de xianxing jiage' (The Strategy of Current Coal Prices), *JJYJ*, no. 3 (1959): 28–30.

Gong Guangyu, 'Wo guo nengyuan de xianzhuang he weilai' (China's Energy Present and Future), in 2000 nian yanjiu xiao zuzhu (Year 2000 Research Group), *Gongyuan 2000 nian de Zhongguo* (China in the Year 2000), Beijing: Kexue jishu wenxian chubanshe, 1984, pp. 32–47.

Gong Weicai and Zhang Xiaojian, 'Shichang jingji tiaojian xia meikuang qiye jihua zhi-biao tixi sheji' (Part I) (Economic Market Conditions: Designing Coal Mine Enterprise System Target Plans) *MTQYGL*, no. 3, 1998, pp. 30–3.

'Gu qi geming ganjin shixian shengchan jianshe da yaojin' (Revolutionary Energy to Achieve Great Leap Forward in Production and Construction), *MTGY*, no. 4 (1958): 4–6.

'Guanyu bianzhi diqu neng pinghang de jiuge wenti' (Nine Problems Concerning Drawing Up Regional Energy Balances), *GYJJ*, no. 4 (1985): 153–5.

'Guanyu meitan gongye "jiu wu" jihua he 2010 nian yuanjing mubiao de zongti silu he kuangjia' (Train of Thought and Framework for the Coal Industry's Ninth Five Year Plan and 2010 Long Term Plan), *MTGYNJ 1997*, pp. 143–9.

Guo Tingjie, 'Yi tan guo you zhongdian meikuang zhengye jinru shichang de nandian yu duice' (Discussion of the Difficulties Faced by State Mines Trying to Enter the Market), *ZGNY*, no. 8 (1996): 11–13.

Han Yongchun, 'Jianshe Shanxi meitan nengyuan jidi yiding yao changfen fahui difang meikuang de zuoyong' (In Building the Shanxi Energy Base, a Major Role Should Be Given to the Local Mines), *SXJDYJ*, pp. 452–65.

Han Zongshun, 'Guanyu wo guo difang meikuang chixu fazhan de duice' (Countermeasures for the Continuing Development of Local Coal Mines in China), *NY*, no. 5 (1989): 51–6.

Hao Fengyin, 'Meitan gongye de jieneng yu cuoshi' (Coal Industry Conservation and Measures), *NY*, no. 1 (1984): 7–9.

Hao Shengde, 'Shehui dui meikuang yuan mei chengben bianhua tan jiage, shuishou tiaozheng de biyao xing' (Discussion on the Adjustment of Production Costs, Prices and Taxes for Commune-Run Coal Mines), *JJWT*, no. 5 (1982): 41–2.

He Baisha, 'Fahui difang jijixing, da li fazhan difang meikuang' (Bring into Play the Positive Aspects of the Development of Local Coal Mines), *JHJJ*, no. 2 (1958): 12–14.

He Shui, 'Cong meikuang jianshe gongzuo de chengjiu kanzhong suliang guo renmin de weida youyi' (Coal Mine Construction Achievements Resulting from Friendship with the Soviet Union), *MTGY*, no. 21 (1957): 12–14.

Hu Benzhe, 'Guanyu Shanxi nengyuan jidi meitan shusong fangshi de tantao' (Investigation into Methods of Coal Transportation in Shanxi Energy Base), *NY*, no. 4 (1983): 1–3, 17.

Hu Changnuan, 'Dui wo guo meitan jiage de kaocha' (An Investigation into China's Coal Prices), *SHKX*, no. 6 (1991): 145–58.

Hu Guangrong, 'Shanxi meitan jidi jianshe zhong de yunshu wenti' (Transport Problems in the Construction of Shanxi Coal Base), *NY*, no. 3 (1982): 11–12.

Hua Maokun, '"Jiu wu" tielu fazhan zhanwang' (Prospects for Railway Construction During the Ninth Five Year Plan), *TDYSJJ*, no. 1 (1996): 6–9.

Huang Xibiao, 'Tielu zouxiang shichang yao yi caiwu guanli wei zhongxin' (Financial Management Is the Core Consideration as the Railways Enter the Market), *TDYSJJ*, no. 5 (1997): 14–15.

Huang Zaiyao, 'Dui Shanxi sheng fazhan shedui meikuang jige wenti de shangtao' (Discussion about Problems with Commune Coal Mine Production in Shanxi Province), *JJWT*, no. 5 (1984): 20–2.

—— 'Dui wo guo jinnian lai nengyuan taolun zhong jige wenti de shangtao' (Recent Discussions about Energy Problems), *GYJJGLCK*, no. 1 (1981): 39–43.

Huang Zhipei, 'Tantan nei di gongye jianshe zhong de wenti' (Discussing Some Problems in the Construction of Industry in Inland Areas), *JJGL*, no. 5 (1979): 14–15, 24.

Jia Jianfeng, 'Meitan xiaoxing qiye gu fen hezuo zhi de tantao' (Investigation into Small Coal Enterprises Creating Joint Stock Systems), *MTJJYJ*, no. 1 (1998): 21–2.

Jiang Hao, 'Ji dian sikao: guanyu wo guo shenghua yong nengzhi mei hua de' (Points to Ponder Regarding the Use of Coal in China), *ZGNG*, no. 10 (1998): 28–30, 42.

Jiang Xianrong, 'Wo guo jieneng gongzuo de huigu yu zhanwang' (Retrospects and Prospects in China's Energy Conservation Work), *NY*, no. 5 (1989): 37–42.

Jiang Ying, 'Da hao dian gongye de heli buju wenti' (Rational Distribution of Electricity Consumption), *DL*, no. 4 (1963): 160–4.

Jianshe yinhang Shanxi sheng fen hang zonghe chu (Comprehensive Section of the Shanxi Branch of the Construction Bank), 'Guanyu Shanxi jijian touzi jiegou de tantao' (Investigation into the Structure of Construction Investment in the Shanxi Base), *SXJDYJ*, pp. 354–66.

Kang Shouyong and Cui Gang, 'Meitan gongye xiwang zai keji' (Coal Industry's Hopes Depend on Technology), *GYJJ*, no. 2 (1995): 127–8.

Kou Zongji and Li Sufang, 'Xietong nuli, jinkuai jiejue Huoxian kuang wuji meitan chaochu jiya wenti' (Strive Towards Solving As Soon As Possible the Problem of Overstocked Coal at Huoxian Mine), *JJWT*, no. 7 (1985): 44.

Lei Ding and Liang Kuangbai, 'Jiaotong yunshuye zai wo guo guomin jingji zhong de zuoyong' (The Role of Communications and Transportation in the National Economy), *JJYJ*, no. 2 (1965): 39–43.

Lei Jihua (ed.), *Shengchan jihua guanli: meikuang qiye guanli jichu duwu* (Production Planning Management: Coal Mine Enterprise Management Basic Reading Material), Beijing: Meitan gongye chubanshe, 1986.

Li Beiling, 'Qianghua guoyou dazhong xing meikuang zhanlue guanli de sikao' (Thoughts on Strengthening Strategy for Management of State Coal Mines), *ZGNY*, no. 3 (1998): 22–4.

Li Chengrui, 'Shi nian nei luan qijian wo guo jingji qingkuang fenxi' (Analysis of Economic Conditions in the Ten Years of Chaos), *JJYJ*, no. 1 (1984): 23–31.

Li Guangchun, 'Gonggu he fazhan dangqian meitan gongqiu de da hao xingshi' (Strengthen and Develop the Current Good Demand and Supply Situation in the Coal Industry), *WZGL*, no. 2 (20 Feb. 1986): 8–10.

Li Hongtang, Meng Xianyu, and Peng Ningzhi, 'Guanyu lishun meitan chanpin bi jia wenti de sikao' (Thoughts on Rationalizing Price Parities of Coal Products), *MTJJYJ*, no. 11 (1994): 20–3.

Li Junfeng and Rong Tao, 'Zanshi weiji, hai shi zhangqi duanque: tantan wo guo nengyuan weihe quanmian jinzhang' (Temporary Crisis or Longterm Shortage: A Discussion about China's Energy Deficits), *ZGNY*, no. 3 (1989): 1–4.

Li Kaiming, 'Shanxi meitan nengyuan jidi jianshe zhong yizhi bu ke queshao de liliang' (Shanxi Coal Energy Base Construction: No Shortage of Power), *SXJDYJ*, pp. 466–71.

Li Qin, 'Jianshe Shanxi huodian jidi de tantao' (Investigation into the Construction of the Shanxi Thermal Electric Power Base), *SXJDYJ*, pp. 472–85.

Li Shiyi, 'Meitan jiage gaige de chubu shexiang' (Preliminary Ideas on Coal Price Reform), *JDYJ*, no. 22 (1986): 1–8.

Li Wenyan, 'Shilun wo guo de nengyuan fazhan zhanbei' (Development Strategy of China's Energy), *JJDL*, no. 6 (1986): 86–91.

Li Xuesheng, 'Zhengque yunyong jihua shouduan he chanye zhengce duo cengci tuijin meitan gongye jingji zengzhang fangshi zhuanbian' (Rightly Apply Plans, Means and Policies to Push Forward the Economic Expansion and Transformation of the Coal Industry), *MTJJYJ*, no. 12 (1996): 5–9.

Li Yinglu, 'Wuzi guanli tizhi bixu gaige' (Materials Management System Needs Reform), *JJGL*, no. 10 (1979): 23–5.

Liang Rencai, 'Dianli yu gongye buju' (Electricity and Industrial Distribution), *DL*, no. 1 (1963): 3–38.

Liang Weilie, 'Nengyuan duanque fenxi ji sikao' (Analysis and Consideration of Energy Shortages), *NY*, no. 2 (1989): 6–9.

Liu Kaimo, 'Wuzi xiaohao dinge zhong cunzai de jige wenti' (Problems Concerning Materials Consumption Quotas), *JHJJ*, no. 1 (1956): 33–4.

Liu Zaixing, '1990 nian san yi dun de chanliang shi nan dadao de' (The 1990 Target of 300 Million Tons Is Difficult to Reach), *SXJDYJ*, pp. 219–23.

—— 'Shanxi meitan nengyuan jidi jianshe ruogan wenti tantao' (Investigation into Certain Problems about Shanxi Coal Energy Base Construction), *JJWT*, no. 3 (1980): 1–6.

Liu Zhiping, 'Wo guo jin shi nian ji eneng xiangmu de touzi, xiaoyi pingjia' (Appraisal of China's Investment in Energy Conservation Over the Past Ten Years), *ZGNY*, no. 6 (1993): 35–8, 43.

Lu Dadao, 'Erlinglingling nian wo guo gongye shengchanli buju zongtu di kexue jichu' (The Scientific Foundation for an Overall Plan of Industrial Location in China by A.D. 2000), *DLKX*, no. 6 (1986): 110–18.

Lu Qikang, 'Guanyu meitan shengchan chengben yu chu chang jiage wenti de tantao' (Investigation into Coal Production Costs and Producer Prices), *JGLLSJ*, no. 1 (1982): 28–31.

Lu Ying, 'Qiantan meitan chan xiao xingshi he jiage gaige de silu' (Discussions on Coal Marketing and Thoughts on Price Reform), *NY*, no. 2 (1989): 20–2.

Luo Genji and Fan Lianfen, 'Chongfen fahui jingji ganggan de zuoyong cujin jin jieneng' (Abundantly Bring into Play Economic Levers to Promote the Conservation of Energy), *NY*, no. 6 (1985): 9–10, 13.

Luo Guibo, 'Gaodu zhongshi shuiziyuan de heli kaifa liyong' (Paying Attention to the Rational Opening and Use of Water Resources), *SXJDYJ*, pp. 219–23.

Ma Hong, 'Guanyu Shanxi jingji jiegou de yanjiu' (Research into the Economic Structure of Shanxi), *SXJDYJ*, pp. 7–28.

Mei Yan, 'Fazhan meitan xixuan jiagong shi jieneng de you xiao tujing' (Development of Coal Washing and Dressing Is an Effective Method of Energy Conservation), *JJGL*, no. 12 (1981): 28–30.

—— 'Tiaozheng meitan jiage shi zai bi xing' (Transforming Coal Prices Is Imperative), *GYJJGLCK*, no. 11 (1983): 55–6, 45.

Meitan bu zhengce fagui si keti zu (Coal Ministry Policy and Regulation Research Group), 'Jiushi nian dai zhonghou qi wo guo meitan shichang jiben zou shi ji duice' (Trends in

China's Coal Market in the Mid and Late 1990s), *MTJJYJ*, Part I in no. 1 (1995): 4–8; Part II in no. 5 (1995): 4–7.

Meitan gongye bu ban gong ting (Ministry of Coal Industry Administration), 'Meitan gongye di yige wu nian jihua de juda chengjiu' (Enormous Achievements of the Coal Industry in the First Five Year Plan), *MTGY*, no. 19 (1957): 2–3.

Meitan jiage gaige xueshu taolun hui (Coal Price Reform Study Group), 'Shenru tantao meitan jiage gaige de lilun he shijian wenti', (Thorough Investigation into the Problems and Practice of Coal Pricing Reform), *JDYJ*, no. 22 (1986): 1–8.

'Meitan jishu chukou shixian ling de tupo' (Breakthrough in Exports of Coal Expertise), *SXRB*, 4 Sept. 1987, p. 1.

'Meitan zonghe liyong, duo zhong jingying he di san chanye "jiu wu" jihua he 2010 nian yuanjing guihua' (Comprehensive Use of Coal and the Development of Diversified Activities and Service Trades During the Ninth Five Year Plan and Longterm Plan to 2010), *MTJJYJ*, no. 8 (1996): 11–15.

Meng Renlun and Shen Changzhi, 'Jinhou gangtie gongye de jieneng fangxiang' (Future Directions in Iron and Steel Industry Energy Conservation), *NY*, no. 5 (1983): 1–2, 18.

Miao Tianjie, 'Guanyu wo guo di qige wu nian jihua jieneng lu de kanfa' (Analysis of Energy Conservation Rates in China's Seventh Five Year Plan), *NY*, no. 1 (1985): 7–9.

Mo Liqi, 'Jishu chuang xin yu tigao zhigong suzhi xiang jiehe shixian jingji zengchang fangshi de zhuanbian' (Raise Quality of Staff and Increase Economic Growth Through Technological Improvements), *MTJJYJ*, no. 1 (1998): 52–3.

Mu Qing and Yang Luwu, 'Zhongshi mei cengqi jichu yanjiu jiasu mei cengqi chanye fazhan' (Paying Attention to Basic Research on Coalbed Methane and Speeding Up Development of the Industry), *ZGNY*, no. 4 (1999): 7–8.

'Nengyuan fenpei lun' (Discussion on Energy Distribution), *GYJJ*, no. 10 (1985): 159–64.

Ning Mingdong, 'Meitan de qingjie shuyong ji fazhan jiejiang mei jishu zhanbei' (Application of Clean Coal Technologies and Developing a Strategy), *ZGNY*, no. 2 (1999): 45–7.

Pan Weier, 'Meikuang shuailao baofei yu kuang gong zai jiuye' (Report of Coal Mine Decommissioning and Re-Employment of Miners), *ZGNY*, no. 2 (1996): 20–3.

—— 'Yi 50 wan meikuang zhigong zai jiu ye' (Discussion on the Re-Employment of 500,000 Coal Mine Staff and Workers), *MTJJYJ*, no. 4 (1996): 15–18.

Peng Fangchun and Zhi Luchuan, 'Jin, Shaan, Mo jiequ de nengyuan gongye: zhanbei dili, wenti ji duice' (Energy Industry in Contiguous Areas of Shanxi, Shaanxi and Inner Mongolia: Key Strategical Position, Problems and Countermeasures), *ZGNY*, no. 2 (1994): 19–21.

Peng Shiji, *Zhongguo meitan gongye sishi nian* (40 Years of China's Coal Industry), Beijing: Meitan gongye chubanshe, 1990.

Qi Guang, 'Luelun gongye qiye jian de guding xiezuo yu dingdian gongying' (Analysis of Cooperation and Supply among Designated Industries), *JJYJ*, no. 9 (Sept. 1964): 22–6.

Qin Meng, 'Meitan jiage gaige yu Shanxi meitan jiage guandian zhaiyao' (A Point of View on Coal Price Reform and Coal Prices in Shanxi), *JDYJ*, no. 19 (1986): 3–5.

Qin Yuzheng, 'Jianshe xin mei jing yingdang zhuyi diqu chan xu pingheng' (In Building New Coal Mines, Attention Must Be Paid to the Regional Balance of Production), *JHJJ*, no. 6 (1958): 10–11.

'Qiye jieyue nengyuan guanli shengji (dingji) guaiding' (Regulations for Upgrading Enterprise Energy Management), *NY*, no. 5 (1991): 30–1, 41.

Qu Shiyuan, 'Wo guo 2000 nian nengyuan gongxu yuce he jiegou fenxi' (Forecast and Structural Analysis of China's Demand and Supply of Energy in the Year 2000), *NY*, no. 3 (1987): 1–4.

Qu Yinghua, Zhang Xiaojun, and Wang Lijie, 'Tan meitan gongye jianshe "jiu wu" fazhan silu' (Thoughts on Coal Industry Construction and Development During the Ninth Five Year Plan), Part I in *MTJJYJ*, no. 4 (1996): 12–15; Part II in no. 5 (1996): 5–7.

Shanxi shehui kexue yuan, meitan jiage yanjiu keti zu (Shanxi Academy of Social Sciences, Coal Pricing Research Group), *Meitan jiage gaige yu Shanxi meitan jiage* (Coal Pricing Reform and Shanxi Coal Pricing), an unpublished research report prepared by the Academy, Taiyuan: Aug. 1986.

Shanxi sheng shehui kexue yuan nengyuan jingji yanjiu suo (Shanxi Academy of Social Sciences, Energy Economics Research Institute), 'Meitan jiage gaige de chubu shexiang' (Preliminary Considerations about Coal Price Reform), *JDYJ*, no. 22 (1986): 1–8.

Shanxi sheng shehui kexue yuan, Taiyuan shi jishu jingji yanjiu (Shanxi Academy of Social Sciences, Taiyuan City Technological Economics Research Centre), *Shanxi nengyuan zhonghua gong jidi zonghe kaifa yanjiu* (Research on the Development of the Shanxi Energy and Chemicals Base), Taiyuan: Shanxi renmin chubanshe, 1984.

Shanxi sheng sheke yuan nengyuan jingji yanjiu suo (Shanxi Academy of Social Sciences, Energy Economics Research Institute), 'Meitan jiage de lilun yu shijian' (Coal Pricing Theory and Practice), *JJWT*, no. 3 (1987): 33–9.

Shao Tonghan, 'Ti gao meitan zhiliang jieyue nengyuan' (Raise Coal Quality, Save Energy), *NY*, no. 3 (1984): 32–3.

'Shenhua gaige gao hao zhengdun cujin xiangzhen meikuang jiankang fazhan' (Deepen Reform, Rectify, and Promote Sound Development of Township and County Coal Mines), *MTGYNJ 1994*, pp. 113–17.

Shi Zemin (ed.), *Liudong zijin guanli: meikuang qiye guanli jichu duwu* (Working Capital Management: Coal Mine Enterprise Management Basic Reading Material), Beijing: Meitan gongye chubanshe, 1983.

Shi Zhigui, 'Meitan qiye er ci chuangye' (Coal Enterprise Pioneering Work, a Second Time), *ZGNY*, no. 3 (1999): 32–3.

Song Shangda, 'Tantan chengben guanli de liangge wenti' (Discussion of Two Production Cost Management Problems), *MTJJYJ*, no. 7 (1983): 36–40.

Song Yuxiang, 'Dongbei diqu meikuang chengzhen fazhan yu buju de yanjiu' (Research on the Distribution of Northeast Coal Mining Towns), *DLKX*, no. 11 (Feb. 1991): 94–6.

Su Xiaojun and Xue Xinmin, 'Guonei wai tielu yong neng fenxi' (Analysis of Energy Utilization by Railways in China and Abroad), *NY*, no. 4 (1988): 5–7.

Sun Hongzheng, 'Zuzhi jingji yunxing shi jieneng de zhongyao huanjie' (Organizing Economic Systems Is an Important Link in Energy Conservation), *NY*, no. 5 (1983): 38–9.

Sun Maoyun, 'Mei cengqi zai wo guo nengyuan fazhan zhong de zhanbei diwei ji qi fazhan yao su' (The Position of Coalbed Methane in China's Energy Development Strategy and Countermeasures), *ZGNY*, no. 4 (1999): 4–6.

Sun Shangqing, 'Nengyuan jiegou' (Energy Structure), in Ma Hong and Sun Shangqing (eds), *Zhongguo jingji jiegou wenti yanjiu* (Problems in China's Economic Structure), Beijing: s.n., 1980, pp. 261–94.

Sun Shangqing and Zhai Ligong, *Zhongguo nengyuan jiegou yanjiu* (Research into China's Energy Structure), Taiyuan: Shanxi renmin chubanshe/Zhongguo shehui kexue chubanshe, 1987.

Tan Weiping and Cai Weitian, 'Tantao shanxi nengyuan jidi jianshe de jingji xiao guo' (Investigation into the Economic Results of Shanxi Energy Base Construction), *GYJJGLCK*, no. 1 (1981): 44–6.

Tang Xingxia, 'Guanyu Fujian sheng nengyuan jiegou yu buju wenti' (The Energy Structure and Allocation Problem in Fujian Province), *JJDL*, no. 6 (1986): 98–103.

Tian Hongbin and Yan Tao, *Nengyuan zhishi 500 ti* (Energy Knowledge 500 Questions), Lanzhou: Gansu renmin chubanshe, 1985.

Tiao Tianbao, 'Fazhan wo guo meitan gongye zhong de jige wenti' (Some Problems in Developing Our Country's Coal Industry), *GYJJGL*, no. 11 (1983): 17–24, 38.

Tian Yuan and Qiao Gong (eds), *Zhongguo jiage gaige yanjiu, 1984–1990* (Research on Price Reform in China, 1984–1990), Beijing: Dianle zhengye chubanshe, 1991.

Tong Boyin, 'Jige langfei yuan cailiao de qiye' (Enterprises That Waste Materials), *JHJJ*, no. 3 (1956): 27–8.

Wan Jing, 'Wajue qiye qianli, jiasu wuzi zhouzhuan' (Tap Enterprise Potential, Speed up Materials Turnover), *JJGL*, no. 8 (1979): 26–9.

Wang Jiacheng, 'Cong wo guo nengyuan de xiaofei goucheng kan jieneng de qianli he tujing' (Discussion of the Means and Potential for Energy Saving from Observing the Consumption Structure of Energy in Our Country), *GYJJGLCK*, no. 3 (1981): 1–8.

Wang Jiachun and Li Minxin, 'Guanyu wo guo nengyuan zhengce yu jiage de maodun' (China's Energy Policies and Price Contradictions), *JJWT*, no. 4 (1984): 17–21.

Wang Jinlin, 'Meitan gongye zhixing di yige wu nian jihua de zhuyao jingyan jiaoxun' (Principal Learning Experiences of the Coal Industry During the First Five Year Plan), *MTGY*, no. 19 (1957): 4–6.

Wang Kangmao, 'Zhongguo nengyuan jiegou de jingji fenxi' (Economic Analysis of China's Energy Structure), in Nengyuan jingji weiyuanhui de zhongguo nengyuan yanjiu weiyuanhui (Energy Economics Group of Chinese Energy Research Committee), *Zhongguo nengyuan jingji wenti wenji* (Collected Works of Chinese Energy Economics Problems), Beijing: Nengyuan chubanshe, 1984, pp. 37–46.

Wang Kuizhong, 'Shenhua gaige qianghua guanli shixian yunshu qiye san nian niu kui wei ying' (Deepening Reform and Strengthening Management to Make Transportation Enterprises Turn Their Losses into Profits Within Three Years), *TDYSJJ*, no. 1 (1998): 7.

Wang Maolin, 'Jiejue meitan waiyun shi jianshe shanxi nengyuan jidi de guanjian' (The Crux of the Construction of the Shanxi Energy Base Is to Solve the Transport Problem), *SXJDYJ*, pp. 597–607.

——(ed.), *Meitan gongye qiye guanli* (Coal Industry Enterprise Management), Taiyuan: Shanxi renmin chubanshe, 1982.

—— 'Zhuajin xian you qiye jishu gaizao jiakuai Shanxi meitan fazhan' (Must Modernize and Improve Enterprise Technology in Order to Raise Shanxi's Coal Production Quickly), *SXJDYJ*, pp. 486–97.

Wang Senhao, 'Cong shixing liangge genben zhuanbian de gaodu lai renshi he tuiguang xin ji jingyan jaikuai meitan jiben jianshe gaige bufa' (Achieve Two Basic Transformations, Recognize and Popularize Experience, Quickly Reform Coal Industry Construction), *MTGYNJ 1997*, pp. 137–40.

—— 'Jianchi, fuchi, gaizao, zhengdun, lianhe, tigao de fazhen zujin xiangzhen meikuang jiankang fazhan' (Persist in the Supporting, Transforming, Rectifying, Uniting and the Healthy Development of the Village and Township Coal Mines), *ZGNY*, no. 5 (1994): 8–14.

——'Jin yi bu ba meitan gongye de shiqing wei hao' (Towards Improving the Coal Industry), *LW*, no. 38 (18 Sept. 1994): 15.

—— '"Jiu wu" hou san nian meitan gongye zongti gongzuo shexiang he 1998 nian gongzuo zhongdian' (Discussion of the Work to be Done in the Coal Industry During the Final Three Years of the Ninth Five Year Plan and for 1998), *MTQYGL*, no. 1 (1998): 5–9.

—— 'Renzhen guanche luoshi dang de shiwu da jingshen ba chi xu jiankang fazhan de meitan gongye quanmian tui xiang ershiyi shiji' (Earnestly Ensure Policies of Fifteenth Congress, Vigorously Control the Healthy Development of the Coal Industry as It Pushes Towards the 21st Century), *ZGNY*, no. 3 (1998): 4–10, 15.

—— 'Shenhua gaige cujin meitan zhengye jiakuai xiang shichang jingji guodu' (Deepen the Reformation, Improve and Speed Up the Transition of the Coal Industry Towards the Market Economy), *ZGNY*, no. 3 (1994): 3–7.

Wang Senhao and Liu Guanwen, *Xiang zhen meikuang jingji guanli*. (Economic Management of Township Coal Mines), Taiyuan: Shanxi renmin chubanshe, 1986.

Wang Shaoxiong, 'Da pian kuang de mei zhi jiancha wang' (Coal Quality Inspections to Take Place Everywhere), *MTGY*, no. 14 (1956): 27–9.

Wang Xianglin, 'Jieyue nengyuan shi tigao jingji xiaoyi de zhongyao tujing' (Saving Energy Is an Important Way of Enhancing Economic Performance), *NY*, no. 3 (1984): 6–8.

Wang Yujing, 'Wo guo jiaotong yunshu ye xianzhuang yu 2000 nian yuce' (China's Transport System Now and in 2000), in 2000 nian yanjiu xiao zu (Year 2000 Research Group), *Gongyuan 2000 niande zhongguo* (China in the Year 2000), Beijing: Kexue jishu wenxian chubanshe, 1984.

Wang Zhaocheng, 'Gao hao "jiu wu" tielu jianshe' (Railway Construction During the Ninth Five Year Plan), *TDYSJJ*, no. 1 (1996): 15–18.

Wang Zhenshuan, Zhang Wenshan, and Ji Yangrui, 'Wo guo meitan gongye liyong waizi cunzai wenti ji duice' (Problems and Countermeasures in the Utilization of Foreign Capital in China's Coal Industry), *MTJJYJ*, Part I in no. 9 (1997): 18–21; Part II in no. 10 (1997): 17–20, 32.

Wang Zhuokun, 'Heli youxiao liyung meitan ziyuan' (Rational and Effective Use of Coal Resources), *NY*, no. 4 (1992): 28–9.

Wu Dechun, 'Jin mei kaifa sudu he guimo de tantao' (Investigation into the Speed and Scale of Coal Mine Openings), *SXJDYJ*, pp. 232–47.

Wu Dechun and Zhai Ligong, 'Kaichuang Shanxi nengyuan zonghua gong jidi jianshe xin jumian' (The Conditions at Shanxi Energy Base at the Outset), *SXJDYJ*, pp. 311–32.

Wu Runtao, 'Jin mei yunshu yu qita wuzi yunshu de bili guanxi' (Relationship Between Coal Tranport and Other Materials Transport), *SXJDYJ*, pp. 643–53.

Wu Wenyue *et al.*, 'Guoyou meitan zhengye yunying mianlin de xingshi he tuchu wenti' (The Present Situation and Problems Faced by State-Owned Coal Enterprises), *MTJJYJ*, no. 6 (1998): 41–2.

Wu Wenyue and Li Wei, 'Dangqian wo guo meitan gongqiu guanxi zhi qianxi' (Analysis of the Present Coal Supply and Demand Situation in China), *MTJJYJ*, no. 5 (1998): 38–41.

Wu Yin, 'Meitan chanye jingzheng jiegou fenxi' (Analysis of the Competition Structure in Coal Production), *ZGNY*, no. 1 (1999): 17–21.

'Xiafang guanjing jianguan' (Orderly Decentralization and Closure), *MTJJYJ*, no. 8 (1998): 4.

Xu Jingyi, 'Cong meitan ziyuan zengjia jiage xiajiang kan jiazhi guilu de zuoyong' (The Effect of Raising Coal Resource Prices on Law of Value), *WZGL*, no. 2 (1985): 2–3.

Xu Shoubo, *Nengyuan jishu jingji xue* (Energy Technology Economics), Changsha: Hunan renmin chubanshe, 1981.

Xu Sifu, 'Shanxi sheng shuiziyuan de kaifa he liyong wenti' (Problems in the Opening and Use of Shanxi's Water Resources), *SXJDYJ*, pp. 663–9.

Xu Yi, Chen Baosen, and Liang Wuxia, *Shehui zhuyi jiage wenti* (Socialist Pricing Problems), Beijing: Zhongguo caizheng chubanshe, 1982.

Xue Muqiao, 'Shehui zhuyi shehui chanpin jiage zhong de jige shang dai taolun de wenti' (Several Problems in Socialist Economy Commodity Pricing Still Awaiting Discussion), *JJYJ*, no. 5 (1963): 1–5.

Yan Dayu and Yu Xiaomiao, 'Meitan liutong huanjie wenti yuanyin ji duice qianlun' (Discussion of the Causes of the Coal Circulation Problem and Solutions), *MTJJYJ*, no. 4 (1996): 18–21.

Yang Baochuan and Zhang Qinwen, 'Xiangzhen meikuang de hongguan guanli wenti tantao' (Investigation into Village and Township Coal Mine Macro-Management Problems), *JDYJ*, no. 4 (1985): 1–15.

Yang Jinsheng and Li Zhangzhong, 'Meitan qiye jian kui zeng ying de taolun' (Investigation into Reducing Losses and Increasing Profits in the Coal Industry), *MTJJYJ*, no. 10 (1984): 23–5, 28.

Yang Qing, 'Yau cong zhanlue gaodu zhongshi LNG jinkou' (Adopt Strategy of Raising LNG Imports), *ZGNY*, no. 5 (1998): 5–8.

Yang Zhaohong, 'Liyong guoji youqi ziyuan gaibian zhongguo nengyuan jiegou' (Use International Oil and Gas, Change China's Energy Structure), *ZGNY*, no. 11 (1998): 12–16.

Yao Xitang and Jin Xingren, 'Nengyuan xiaohao yu jingji jiegou de guanxi' (The Relationship Between Energy Consumption and Economic Structure), in Nengyuan jingji weiyuanhui de zhongguo nengyuan yanjiu weiyuanhui (Energy Economics Group of Chinese Energy Research Committee), *Zhongguo nengyuan jingji wenti wenji* (Collected Works of Chinese Energy Economics Problems), Beijing: Nengyuan chubanshe, 1984, pp. 46–51.

'Yau qie shi jiaqing zongliang kongzhi' *(Benkan pinglunyuan)* (Control, Cut and Strengthen Total Amount) (Journal's Commentator), *MTJJYJ*, no. 2 (1999): 4.

Ye Qing, 'Muqian jieneng xing shi he jin yibu jiaqiang jieneng gongzuode ruogan cuoshi' (Present Situation and Measures for Strengthening Energy Conservation Work) (Report Excerpts of the State Council's Sixth Official Energy Conservation Meeting), *ZGNY*, no. 1 (1991): 3–6.

—— 'Xiang guowuyuan jieneng bangong huiyi di wuzi huiyi de huibao tigong' (Report on the Fifth Official Energy Conservation Meeting of the State Council), *ZGNY*, no. 5 (1989): 4–8.

Ye Ruixiang, 'Guanyu meitan, shiyou jiage wenti de yanjiu' (Research into the Problems with Coal and Oil Pricing), *CMJJ*, no. 10 (1983): 39–42.

—— 'Wo guo de nengyuan jiage yanbian ji qi gaige' (Our Country's Energy Pricing Development), in Wang Zhenzhi (ed.), *Jiage gaige yu jiage guanli* (Pricing Reform and Pricing Management), Beijing: Zhongguo wuzi chubanshe, 1987, pp. 235–48.

—— 'Wo guo nengyuan jiage wenti de yanjiu' (Research into China's Energy Pricing Problems), in Zhongguo nengyuan jingji yanjiu weiyuanhui (China Energy Economics Research Committee), *Zhongguo nengyuan jingji wenti wenji* (China Energy Economics Problems: Collected Works), Beijing: Nengyuan chubanshe, 1984, pp. 24–9.

Yi Chu, 'Meitan gongye shinian gaige yu fazhan huigu' (Looking Back Over Ten Years of Reform and Development in the Coal Industry), *NY*, no. 5 (1989): 18–23.

Yu Xiaomiao and Wang Liang, 'Gong guo yu qiu xingshi xia meitan fazhan celue sikao' (Thoughts on Developing a Strategy to Handle the Over-Supply of Coal), *MTJJYJ*, no. 1 (1998): 18–20.

Zeng Yingjun and Xia Zhenpeng, 'Hen zhua "bao, guan, gai" jieneng zeng xiaoyi' (Aggressively Promote Energy Conservation Through 'Contact, Management and Reform'), *ZGNY*, no. 6 (1991): 36, 37, 45.

Zhai Ligong, 'Guanyu tiaozheng meitan jiage de tantao' (Investigation into Adjusting Coal Prices), *SXJDYJ*, pp. 333–53.

Zhai Ligong and Dong Jiben (eds), *Meitan gongye qiye laudong gongzi guanli* (Management of Coal Enterprise Labour Wages), Taiyuan: Shanxi renmin chubanshe, 1986.

Zhang Baoming, '1999 nian meitan gongye de zhongdian gongzuo' (Main Work For the Coal Industry in 1999), *MTJJYJ*, no. 2 (1999): 5–7, 11.

Zhang Benlian, 'Yingxiang guoying meitan gongye yingli shuiping tigao de jige wenti' (Some Questions Influencing Raising of the Profit Level in the State Mining Coal Industry), *CZ*, no. 7 (1957): 30–1.

Zhang Hongshun, 'Ye jianshe nengyuan zhong hua gong jidi wei zhongxin lishun Shanxi jingji' (Build Up the Shanxi Energy, Heavy Industry and Chemical Base as the Centre of a Well-Balanced Economy), *JJWT*, no. 7 (1984): 12–16.

Zhang Hongyu, 'Guanyu jiakuai Shanxi meitan gongye fazhan bufa wenti de chubu tantao' (Preliminary Investigation into Quickly Raising Coal Industry Production in Shanxi), *SXJDYJ*, pp. 424–39.

Zhang Lin and Li Yongping, 'Meitan gongye ke chixu fazhan wenti de tantao' (Investigation into Continuing Problems in the Development of the Coal Industry), *ZGNY*, no. 10, 1998, pp. 15–19.

Zhang Liqing, 'Shanxi meitan gongye zai chuanguo de diwei' (The Position of Shanxi's Coal Industry in the Whole Country), *SXJDYJ*, pp. 407–12.

Zhang Mengceng, 'Guoying meitan gongye chanpin jiage wenti' (Problems with the Prices of Products from the State Owned Coal Industry), *CZ*, no. 3 (1958): 14–15.

Zhang Shouyi, 'Gaige guanli tizhi jiakuai Shanxi nengyuan jidi jianshe de bufa' (Measures to Hasten the Reform Management System of Construction for the Shanxi Energy Base), *SXJDYJ*, pp. 283–9.

Zhang Tongquan, 'Mian dui yunshu zhiyue: meitan zou xiang shichang ying caiqu de duice tantao' (Inquiring How to Solve the Problem of Coal Enterprises Being Unable to Participate in the Market Due to Transport Constraints), *MTJJYJ*, no. 9 (1995): 27–8.

Zhang Yangming, 'Wo guo tieluwang buju de bianhua yu jinzhan' (China's Railway Network: Its Change and Evolution), *JJDL*, no. 5 (1985): 11–16.

Zhang Yonglai, 'Qianxi meikuang wasi shigu ji qi fangzhi duice' (Analysis of the Control of Accidents Caused by Gas in Coal Mines), *MTQYGL*, no. 3 (1998): 36–8.

Zhang Zhijian, Wang Jiacheng, and Xin Dingguo, 'Wo guo de nengyuan xingshi he renwu' (Our Country's Energy Situation and Tasks), *NY*, no. 5 (1984): 1–4.

Zhang Zhixiong, *Meitan gongye qiye caiwu guanli* (Coal Industry Enterprise Financial Affairs Management), Taiyuan: Shanxi renmin chubanshe, 1985.

Zhao Shanqing and Wang Guihua, 'Shidu guimo: meitan chukou de fazhan mubiao' (Appropriate Coal Export Targets), *ZGNY*, no. 2 (1996): 24–8.

Zhong Mingan (ed.), *Chengben hesuan yu guanli* (Cost Accounting and Management), Beijing: Meitan gongye chubanshe, 1985.

'Zhongguo 98 nian shengchan yuan mei 12.36 yi dun' (China's 1998 Coal Production Is 1,236 Million Tons), *Xinhua*, 25 Feb. 1999.

'Zhongguo 99 nian meitan gongye jingji zhuangkuang bu rong leguan' (Conditions in China's Coal Industry in 1999 Not Hopeful), *Xinhua*, 8 Jan. 1998.

Zhongguo jindai meikuang shi bianxue zu (The History of Coal Mining in Contemporary China Compiling Study Group), *Zhongguo jindai meikuang shi* (The History of Coal Mining in Contemporary China), Beijing: Meitan gongye chubanshe, 1990.

'Zhongguo nengyuan jingji quhua de chubu' (Preliminary Research into China's Economic Energy Districts), *GYJJ*, no. 1 (1984): 149–58.

Zhou Peinian, 'Luelun jieneng guanli gaige' (Brief Discussion on Reforming Energy Conservation Management), *NY*, no. 1 (1985): 1–4.

Zhou Ruyan and Feng Yongping, 'Jianli shuang xiang yueshu jizhi huan jie que dian maodun' (Establish Mechanism for Controlling the Power Shortage Contradiction), *JJGL*, no. 7 (1988): 48–50.

Zhou Shulian, 'Lun tigao touzi xiaoguo' (On Enhancing Investment Results), *JJYJ*, no. 6 (1980): 26–32.

Zhou Shulian and Pei Shuping, *Zhongguo gongye fazhan zhanlue wenti yanjiu* (Research Concerning Problems in China's Industrial Development Strategy), Tianjin: Tianjin chubanshe, 1981.

Zhou Wangjun, '"Jiu wu" tielu yunjia shuiping yu yunjia gaige' (Ninth Five Year Plan Railway Rate Levels and Reform), *TDYSJJ*, no. 8 (1996): 1–3.

Zhou Yigong, 'Liuhua chuang ranshao jishu zai mei ranshao lingyu de diwei' (The Status of FBC Burning Technology in China), *ZGNY*, no. 4 (1998): 13–17.

Zhou Yimin and Zhang Du, 'Jianchi gaige gaohao meitan gongye de wuzi gongying guanli' (Persist in Reforming Coal Industry's Materials Supply Management), *WZGL*, no. 4 (1986): 15.

Zhu Changbo and Me Peiji, 'Yunjia gaige shi tielu zouxiang shichang de guanjian' (Pricing Reform Is the Key for Railways Entering the Market), *TDYSJJ*, no. 2 (1995): 15–17.

Zou Jihua, 'Guo wuyuan zhaokai quan guo xiangzhen meikuang gongzuo huiyi guanche luoshi dui xiangzhen meikuang "fuchi, gaizao, zhengdun, lianhe, tigao" de shizi fangzhen' (State Council Calls for the Implementation of 'Support, Reform, Rectify, Unite, and Raise' the Management of Village and Town Coal Mines), *ZGNY*, no. 5 (1994): 1–7.

Official publications in Chinese

Ditu chubanshe bianzhi (Map Publishing House Unit), *Zhongguo jiaotong tuce* (China Transportation Handbook), Beijing: Ditu chubanshe, 1984.

Guojia dianli gongsi ([China] State Power Company), *Zhongguo dianli shichang fenxi yu yanjiu 1999* (China Electricity Market Analysis and Research 1999), Beijing: Zhongguo dianli chubanshe, 1999.

'Guojia jiwei dui dian mei shixing guojia zhidao jiage' (State Planning Commission Implements National Guidance Prices for Coal Sold for Electricity), *ZGWJ*, 3, 1996, p. 37.

'Guojia jiwei fachu guanyu 1997 nian dian mei guojia zhidao jiage you guan wenti de tongzhi' (State Planning Commission Issues Notice Concerning National Guidance Prices for Coal Sold for Electricity), *ZGWJ*, 1, 1997, p. 35.

Guojia jiwei jiaotong nengyuan si (State Planning Commission, Dept. of Communications and Energy), '"Jiu wu" jieneng jihua ji 2010 nian yuanjing mubiao shexiang' (Plans to Save Energy During the Ninth Five Year Plan Period and Over the Longterm to 2010), *ZGNY*, no. 9 (1996): 1–3, 12.

Guojia tongji ju, gongjiao wuzi si (China Materials Statistical Department) *Zhongguo nengyuan tongji nianjian*, 1986, (1986 China Energy Statistical Yearbook), Beijing: Nengyuan chubanshe, 1987.

Guojia tongji ju, gongye jiaotong tongji si (State Statistical Bureau, Department of Industrial and Transportation Statistics) (ed.), *Zhongguo gongye jingji tongji nianjian*

1998 (China Industrial Economics Statistics 1998), Beijing: Zhongguo tongji chuban-she, 1998.

Guojia tongji ju, gongye jiaotong tongji sibian (State Statistical Bureau, Department of Industrial and Transportation Statistics) (ed.), *Zhongguo nengyuan tongji nianjian 1991* (China Energy Statistics Yearbook 1991), Beijing: Zhongguo tongji chubanshe, 1992.

Guojia tongji ju, gongye jiaotong tongji sibian (State Statistical Bureau, Department of Industrial and Transportation Statistics) (ed.), *Zhongguo nengyuan tongji nianjian 1991–1996* (China Energy Statistics Yearbook 1991–1996), Beijing: Zhongguo tongji chubanshe, 1998.

Guojia tongji ju, gongye jiaotong wuzi tongji sibian (State Statistical Bureau, Department of Industrial and Transportation Statistics) (ed.), *Gongye jingji tongji ziliao 1949–1984* (Industrial Economics Statistics 1949–1984), Beijing: Zhongguo tongji chubanshe, 1985.

Guojia tongji ju, guding zichan touzi tongji ju (State Statistics Bureau, National Fixed Investments Section), *Zhongguo guding zichan touzi tongji ziliao 1950–1985* (National Fixed Asset Investment Statistical Data 1950–1985), Beijing: Guojia tongji chubanshe, 1987.

Guojia tongji ju, shehui tongji sibian (State Statistical Bureau, Social Statistics Department), *Guojia laodong gongzi tongji ziliao 1949–1985* (National Labour and Wages Statistical Data, 1949–1985), Beijing: Guojia tongji chubanshe, 1987.

'Jieyue nengyuan guanli zanxing tiaoli' (Provisional Regulations on Energy Conservation), issued by the State Council on 12 Jan. 1986, *RMRB*, 24 Jan. 1986, p. 2.

Nongye bu xiangzhen qiye sibian (Agriculture Department Village and Township Enterprise Editorial Committee), *Zhongguo xiangzhen qiye tongji zhaiyao 1993* (China Village and Township Enterprise Statistics Summary 1993), Beijing: 1993.

Shanxi jingji nianjian 1985 (Shanxi Province Economic Yearbook, 1985), Taiyuan: Shanxi renmin chubanshe, 1985.

Yan Changyue, *Zhongguo nengyuan fazhan baogao 1997* (China Energy Development Report 1997), Beijing: Jingji guanli chubanshe, 1997.

Ye Qing, 'Muqian jieneng xing she he jin yibu jiaqiang jieneng gongzuode ruogan cuoshi' (Present Situation of Energy Conservation and Some Measures for Strengthening the Work of Energy Conservation), Report Outline of the State Council's Sixth Official Energy Conservation Meeting, *ZGNY*, no. 1 (1991): 3–6.

Zhongguo dianli nianjian bianji weiyuan hui (China Electricity Yearbook Editorial Committee), *Zhongguo dianli nianjian* (China Electricity Yearbook), Beijing: Zhongguo dianli chubanshe, various volumes.

Zhongguo dui wai jingji maoyi nianjian (China Foreign Economic Trade Yearbook), Beijing: Zhongguo shehuizhuyi chubanshe, 1988.

Zhongguo gongye jingji tongji ziliao 1949–1984 (China Industrial Economics Statistical Data, 1949–1984), Beijing: Zhongguo tongji chubanshe, 1985.

Zhongguo gongye jingji tongji ziliao 1986 (China Industrial Economics Statistical Data, 1986) Beijing: Zhongguo tongji chubanshe, 1986.

Zhongguo huanjing nianjian bianji weiyuan hui (China Environment Editorial Committee), *Zhongguo huanjing nianjian 1998* (China Environment Yearbook 1998), Beijing: Zhongguo huanjing nianjian chubanshe, 1998.

Zhongguo jiaotong nianjian bianji weiyuan hui (China Transport and Communications Yearbook Editorial Committee), *Zhongguo jiaotong nianjian 1997* (China Transport and Communications Yearbook 1997), Beijing: Wenwu chubanshe, 1997.

Zhongguo jingji nianjian (Almanac of China's Economy), Beijing: Jingji guanli chuban-she, various volumes.

Zhongguo meitan gongye nianjian (China Coal Industry Yearbook), Beijing: Ministry of Coal Industry, 1982–1999.

Zhongguo shehui kexue yuan, gongye jingji yanjiu suo (China Social Sciences Academy, Industrial Economics Research Institute), *Zhongguo gongye fazhan baogao 1998* (China Industrial Development Report 1998), Beijing: Jingji guanli chubanshe, 1998.

Zhongguo tongji jubian (State Statistical Bureau), *Zhongguo tongji nianjian* (China Statistical Yearbook), Beijing: Zhongguo tongji chubanshe, 1981–2000. All volumes.

Zhongguo xiangzhen qiye nianjian bianji weiyuanhui (China Village and Township Enterprise Yearbook Editorial Committee), *Zhongguo xiangzhen qiye nianjian* (China Village and Township Enterprise Yearbook), Beijing: Zhongguo nongye chubanshe, various volumes.

Zhonghua renmin gongheguo guojia jihua weiyuanhui jiaotong nengyuan si (State Planning Bureau, Department of Communications and Energy) [China Energy White Paper] Beijing: Zhongguo wujia chubanshe, 1997.

Zhonghua renmin gongheguo guojia tongji ju gongye tongji sibian ([China] State Statistical Bureau, Industrial Statistical Department), *Wo guo gangtie, dianli, meitan, jixie, fangzhi, zaozhi gongye de jinxi* (The Past and Present State of China's Steel, Electric Power, Coal, Machinery, Textile, and Paper Industries), Beijing: Tongji chubanshe, 1958.

'Zhonghua renmin gongheguo jieyue nengyuan fa (qicao gongzuo zu)' (Energy Conservation Laws of the People's Republic of China), *ZGNY*, no. 9 (1993): 3–8.

'Zhonghua renmin gongheguo guomin jingji he shehui fazhan di qige wu nian jihua (zhaiyao)' (Summary of the People's Republic of China Seventh Five Year Plan for Economic and Social Development), *Zhonghua renmin gongheguo guowuyuan gongbao* (Bulletin of the State Council of the People's Republic of China), 11 (10 May 1986): 319–20.

'Zhonghua renmin gongheguo guomin jingji he shehui fazhan di liuge wu nian jihua' (The People's Republic of China Economic and Socialist Development Sixth Five Year Plan), *Zhonghua renmin gongheguo guowuyuan gongbao* (Bulletin of the State Council of the People's Republic of China), 9 (10 June 1983): 340–1.

'Zhonghua renmin gongheguo dianli fa' (People's Republic of China Electric Power Law), *Zhonghua renmin gongheguo guowuyuan gongbao* (Bulletin of the State Council of the People's Republic of China), 32 (1995): 1301.

'Zhonghua renmin gongheguo guomin jingji he shehui fazhan shi nian guihua he di bage wu nian jihua gangyao' (Outlines of the People's Republic of China Economic and Socialist Development Ten-Year Long-Term Plan and Eighth Five Year Plan), *Zhonghua renmin gongheguo guowuyuan gongbao* (Bulletin of the State Council of the People's Republic of China), 12 (10 May 1991): 389.

'Zhonghua renmin gongheguo meitan fa' (People's Republic of China Coal Law), *RMRB*, 16 Sept. 1996, p. 2.

Index